COMPREHENSIVE CHEMICAL KINETICS

COMPREHENSIVE

Section 1. THE PRACTICE AND THEORY OF KINETICS

Volume 1 The Practice of Kinetics

Volume 2 The Theory of Kinetics

Volume 3 The Formation and Decay of Excited Species

Section 2. HOMOGENEOUS DECOMPOSITION AND ISOMERISATION REACTIONS

Volume 4 Decomposition of Inorganic and Organometallic Compounds

Volume 5 Decomposition and Isomerisation of Organic Compounds

Section 3. INORGANIC REACTIONS

Volume 6 Reactions of Non-metallic Inorganic Compounds

Volume 7 Reactions of Metallic Salts and Complexes, and Organometallic Compounds

Section 4. ORGANIC REACTIONS (6 volumes)

Volume 8 Proton Transfer

Volume 9 Addition and Elimination Reactions of Aliphatic Compounds

Volume 10 Ester Formation and Hydrolysis and Related Reactions

Volume 12 Electrophilic Substitution at a Saturated Carbon Atom

Volume 13 Reactions of Aromatic Compounds

Section 5. POLYMERISATION REACTIONS (3 volumes)

Volume 14 Degradation of Polymers

Volume 14A Free-radical Polymerisation

Volume 15 Non-radical Polymerisation

Section 6. OXIDATION AND COMBUSTION REACTIONS (2 volumes)

Volume 17 Gas-phase Combustion

Section 7. SELECTED ELEMENTARY REACTIONS (1 volume)

Volume 18 Selected Elementary Reactions

Additional Sections

HETEROGENEOUS REACTIONS

KINETICS AND TECHNOLOGICAL PROCESSES

CHEMICAL KINETICS

EDITED BY

C. H. BAMFORD

M.A., Ph.D., Sc.D. (Cantab.), F.R.I.C., F.R.S.
Campbell-Brown Professor of Industrial Chemistry,
University of Liverpool

AND

C. F. H. TIPPER

Ph.D. (Bristol), D.Sc. (Edinburgh)
Senior Lecturer in Physical Chemistry,
University of Liverpool

VOLUME 14A

FREE-RADICAL POLYMERISATION

ELSEVIER SCIENTIFIC PUBLISHING COMPANY
AMSTERDAM – OXFORD – NEW YORK
1976

ELSEVIER SCIENTIFIC PUBLISHING COMPANY
335 JAN VAN GALENSTRAAT
P.O. BOX 1270, AMSTERDAM, THE NETHERLANDS

AMERICAN ELSEVIER PUBLISHING COMPANY, INC.
52 VANDERBILT AVENUE
NEW YORK, NEW YORK 10017

Library of Congress Cataloging in Publication Data

Bamford, C H
Free-radical polymerisation.

(Their Comprehensive chemical kinetics ; v. 14A)
Includes bibliographies and index.
1. Polymers and polymerization. 2. Radicals (Chemistry) I. Tipper, Charles Frank Howlett, joint author. II. Title.
QD501.B242 vol. 14A [QD381] 541'.39'08 [547'.28]
ISBN 0-444-41486-X 76-28370

ISBN 0-444-41486-X

WITH 56 ILLUSTRATIONS AND 80 TABLES

Transferred to digital printing 2005

COMPREHENSIVE CHEMICAL KINETICS

Contributors to Volume 14A

N. C. BILLINGHAM	The Chemical Laboratory, The University of Sussex, Brighton, England
G. C. EASTMOND	The Donnan Laboratories, The University, Liverpool, England
A. D. JENKINS	The Chemical Laboratory, The University of Sussex, Brighton, England
E. G. SMITH	Unilever Research, Port Sunlight, Wirral, England
H. TOMPA	11 Boulevard Leopold III, 1030 Brussels, Belgium

Preface

Section 5 is concerned with polymerization reactions in the liquid and vapour phases, including Zeigler–Natta polymerizations, emulsion, suspension and precipitation polymerizations, but not solid phase polymerizations which are dealt with in a later section. It seemed reasonable to include polymer degradation reactions of all types, despite the fact that most of these involve solid polymers. Calculation of molecular weight distributions and averages is included, but there is no discussion of the experimental techniques for determining molecular weights. The microstructure of polymer chains is considered when relevant to the kinetics. Rate parameters of individual elementary steps, as well as of overall processes, are given when available.

Volume 14A deals with polymerization processes involving free radicals as intermediates. In Chapter 1, the kinetics of the straightforward homogeneous polymerization of vinyl monomers is discussed, in particular the the determination of the rate coefficients of the individual steps in the chain process and the modes of initiation, propagation and termination and the relation to polymer structure. Chapter 2 covers chain transfer, retardation and inhibition, and in Chapter 3, the reliable data on rate coefficients of chain propagation, termination and transfer for various monomers are summarized. Copolymerization is reviewed in Chapter 4 and reactivity ratios listed from a critical examination of the literature. The patterns of reactivity of monomers and radicals is considered in Chapter 5 and polymerization in heterogeneous systems, in which the propagating radicals pass from one phase to another during growth, in Chapter 6. Finally, the calculation of mole-weight distributions from kinetic schemes is discussed in Chapter 7.

Liverpool
June, 1976

C. H. Bamford
C. F. H. Tipper

Contents

Chapter 2 (G. C. Eastmond)

Chapter 3 (G. C. Eastmond)

Chapter 4 (G. C. Eastmond)

Chapter 5 (N. C. Billingham and A. D. Jenkins)

Chapter 6 (N. C. Billingham and A. D. Jenkins)

Chapter 7 (H. Tompa)

Chapter 1

The Kinetics of Free-Radical Polymerization of Vinyl Monomers in Homogeneous Solution

G. C. EASTMOND

1. Introduction

Of all branches of polymerization free-radical processes are commercially the most important and scientifically the most thoroughly investigated. Among the reasons for this is the fact that useful high-molecular-weight polymers and copolymers can be prepared from a wide variety of monomers. Also, these processes are generally the easiest to carry out and control. Thus, free-radical polymerization provides a convenient route to polymers with a wide range of properties.

In general, free-radical polymerizations are the most amenable to scientific investigation, requiring less sophisticated techniques than their ionic counterparts, being free from the effects of counterions and not requiring traces of co-catalyst, essential to many cationic polymerizations. The intensive, systematic study of these reactions dates from the mid-1930s when the chain reaction nature of some polymerizations was identified, and it became apparent that free radicals could be the active intermediates [1]. Evidence for this arose from the use of known free-radical sources as initiators and from the general agreement between early kinetic experiments and theories of radical-chain reactions.

The kinetics of a straightforward free-radical polymerization are those of a simple radical-chain reaction in which the chain propagates by the addition of a monomer to the active chain carrier, regenerating the active radical intermediate. In the free-radical polymerizations of vinyl monomers of the types $CH_2{=}CHX$ and $CH_2{=}CXY$, with which we are mainly concerned, propagation involves the addition of the radical to one end of the double bond of the monomer, as in

$$\sim CH_2{-}\dot{C}XY + CH_2{=}CXY \longrightarrow \sim CH_2{-}CXY{-}CH_2{-}\dot{C}XY$$

Thus, at each step in the chain the free-radical intermediate increases in size by one monomer unit, and the average number of monomer molecules incorporated into each polymer molecule, or the degree of polymerization, is directly related to the length of the kinetic chain. As with most chain reactions, possible side-reactions exist. Over the years a number of complicating features have been identified and their influence on the rates of polymerization, and on the molecular weights of the final polymers are understood. Indeed, with some monomers the nature of the component reactions are sufficiently well understood to allow polymerization kinetics to be used as a means of determining rates of radical formation and hence, as a tool for investigating mechanisms of radical formation in new initiating systems. With many other monomers, however, the particular combinations of complicating features are not so well identified and the fundamental kinetic parameters are not known with any degree of certainty.

In this chapter we shall be mainly concerned with the processes which are inherent in the homogeneous polymerization of a single vinyl monomer, namely initiation, propagation and termination of the kinetic chain. Transfer to monomer, which may occur during the polymerization, will be considered along with other transfer reactions in a subsequent chapter dealing with the reactions of the propagating radicals with added substrates. We shall first consider the overall simple kinetics of free-radical polymerization and then consider the individual aspects in turn.

2. The simple kinetic equations

In this section we present a simple derivation of the basic kinetic relations, following the standard pattern adopted in various accounts of free-radical polymerizations. Neglecting the possible presence of such processes as primary radical termination, inhibition and transfer, three fundamental processes must be considered. These are *initiation* of the chain reaction by the "slow" formation of initiating radicals, *propagation*, involving growth of the chain carriers, and *bimolecular termination* of the chain reaction, forming unreactive polymer. The component reactions, in a "catalysed" polymerization, may be written, along with their rate coefficients, as

Initiation

$$C \longrightarrow 2R_c\cdot$$

$$R_c\cdot + M \xrightarrow{k_i} R_1\cdot$$

Propagation

$$R_1\cdot + M \xrightarrow{k_{p1}} R_2\cdot$$

$$R_2\cdot + M \xrightarrow{k_{p2}} R_3\cdot \qquad (1)$$

$$R_n\cdot + M \xrightarrow{k_{pn}} R^{\cdot}_{(n+1)}$$

Termination (a) by combination

$$R_n\cdot + R_m\cdot \xrightarrow{k_{tc}} P_{(n+m)}$$

(b) by disproportionation

$$R_n\cdot + R_m\cdot \xrightarrow{k_{td}} P_n + P_m$$

where C and M represent molecules of initiator and monomer, respectively, $R_c\cdot$ is a primary (initiating) radical derived from C, and $R_n\cdot$ and P_n represent growing (propagating) radicals and "dead" polymer molecules containing n monomer units, respectively.

In deriving the simple kinetic relations appertaining to reaction scheme (1), four simplifying assumptions are normally made.

(i) The length of the propagating chain is large, so that the total rate of monomer consumption ($-$d[M]/dt) may be equated to the rate of consumption in the propagation reaction alone, when

$$-\frac{d[M]}{dt} = \sum_n k_{pn}[R_n\cdot][M] \tag{2}$$

(ii) The rate coefficients for both propagation and termination are independent of chain length, i.e. independent of the size of the propagating polymer molecule, and of conversion, so that $k_{p1} = k_{p2} = k_{pn} = k_p$, and similarly for individual values of k_{tc} and k_{td}. This assumption is reasonable, especially for long chains, since the reactivity of a radical is determined by the molecular structure in the vicinity of the unpaired electron and in a homopolymerization all propagating radicals have the same structure, differing only in the lengths of the polymer chains of which they form the terminal unit (see Rates of Termination, Section 4.2).

(iii) A stationary state in radical concentration is established, allowing the rate of change of concentration of any radical intermediate to be equated to zero. Such a situation is closely approximated to in many systems although it is never strictly obeyed since, as the radical concentration builds up (rapidly in the initial stages) at a constant rate of radical formation, an increase in termination follows the increase in radical concentration and the rates of the two processes are equal at infinite time. In practice, using normal polymerization conditions, the stationary state is effectively established in about one minute. A more serious limitation on this assumption is that as reaction proceeds the reactants are consumed and the rate of initiation varies with time. The stationary-state approximation therefore only applies under conditions where such changes are negligible.

(iv) Termination involving primary radicals is negligible, i.e. bimolecular termination between $R_c\cdot$ and $R_n\cdot$. This assumption is valid for most systems under conditions of long chain length, since the concentration of $R_c\cdot$ is small compared with the total radical concentration.

In deriving the simple kinetic expressions, with the aid of the above assumptions, we may take account of different mechanisms of initiation and the fact that in many cases the primary radicals are not 100 % efficient in initiation by referring to the total rate of formation of radicals $R_c\cdot$ which ultimately give rise to propagating radicals as the rate of initiation, which we shall denote $\mathscr{I}$. Thus, a stationary-state equation for each type of radical may be written, viz.

$$\frac{d[R_c\cdot]}{dt} = \mathscr{I} - k_i[R_c\cdot][M] = 0$$

$$\frac{d[R_1\cdot]}{dt} = k_i[R_c\cdot][M] - k_p[R_1\cdot][M] - (k_{tc} + k_{td})[R_1\cdot]\sum_n [R_n\cdot] = 0$$

$$\frac{d[R_2\cdot]}{dt} = k_p[R_1\cdot][M] - k_p[R_2\cdot][M] - (k_{tc} + k_{td})[R_2\cdot]\sum_n [R_n\cdot] = 0$$

$$\vdots \qquad \vdots \qquad \vdots \qquad \vdots$$

$$\frac{d[R_n\cdot]}{dt} = k_p[R^{\cdot}_{(n-1)}][M] - k_p[R_n\cdot][M] - (k_{tc} + k_{td})[R_n\cdot]\sum_n [R_n\cdot] = 0$$

Addition of these equations gives

$$\mathscr{I} = (k_{tc} + k_{td})[R\cdot]^2 + k_p[R_n\cdot][M] \tag{3}$$

where $[R\cdot] = \sum_n [R_n\cdot]$. In writing the stationary-state equations we have written the rate of termination according to the convention

$$-\frac{d[R\cdot]}{dt} = k_t[R\cdot]^2$$

rather than as $2k_t[R\cdot]^2$. While the latter convention is often used in many fields, the former is the most widely used in free-radical polymerization kinetics, and is therefore adopted here. This convention is not universally used by polymer chemists, however, and equations in the literature may differ by this factor of two.

Chain termination can occur at all stages of growth and the population of radical species decreases as their size increases. That is, as n increases $[R_n\cdot]$ decreases so that for large n $[R_n\cdot]$ is negligible and the right-hand term in eqn. (3) may be neglected. Then, substitution of $[R\cdot]$ from eqn. (3) into eqn. (2) gives the overall rate of polymerization as

$$-\frac{d[M]}{dt} = k_p[M]\left(\frac{\mathscr{I}}{k_{tc} + k_{td}}\right)^{1/2} \tag{4}$$

The mean kinetic chain length, ν, defined as the average number of monomer molecules consumed per initiation, or the average number of monomer units which add to one radical before termination, is given by

$$\nu = -\frac{d[M]}{dt}\frac{1}{\mathscr{I}} = k_p[M]\{\mathscr{I}(k_{tc} + k_{td})\}^{-1/2} \tag{5}$$

The average degree of polymerization, DP, or the average number of monomer molecules consumed per inactive polymer molecule formed is

$$\begin{aligned} DP &= \left(-\frac{d[M]}{dt}\right) \frac{1}{\left(\frac{k_{tc}}{2} + k_{td}\right)[R\cdot]^2} \\ &= \frac{k_p[M]}{\left(\frac{k_{tc}}{2} + k_{td}\right)} \left(\frac{k_{tc} + k_{td}}{\mathscr{I}}\right)^{1/2} \end{aligned} \tag{6}$$

The factor of two appears in the term for combination in eqn. (6) since only one polymer molecule is formed from two propagating chains if termination is by combination. If termination is wholly by disproportionation each propagating radical yields one polymer molecule and DP = ν, while if termination is wholly by combination DP = 2ν.

The average lifetime of a kinetic chain, denoted τ, i.e. the time for each primary radical to add an average of ν monomer units, is

$$\tau = \frac{[R\cdot]}{\mathscr{I}} = \{\mathscr{I}(k_{tc} + k_{td})\}^{-1/2} \tag{7}$$

Values of τ cannot be determined directly by experiment.

Since termination can occur at all stages of growth of a polymer chain, individual radicals continue to grow for different periods of time and the resulting polymer contains a distribution of molecular sizes. The quantities defined in eqns. (5)—(7) are therefore average quantities. The distribution of molecular sizes is controlled by the mechanism of termination, and other factors, and is considered in detail in a subsequent chapter.

The parameters defined by eqns. (4)—(6) are those which may normally be determined experimentally. A further important kinetic parameter can be calculated from the products of the rate and degree of polymerization, when we obtain

$$\left(-\frac{d[M]}{dt} \cdot \frac{DP}{[M]^2}\right)^{1/2} = \frac{k_p}{\left(\frac{k_{tc}}{2} + k_{td}\right)^{1/2}} \tag{8}$$

Experimental values of DP often require correction for transfer (Chap. 2) before being used in eqn. (8). The parameter on the right-hand side of eqn. (8) is usually considered to be a characteristic constant for the free-radical polymerization of a given monomer at a given temperature, and is often denoted simply as $k_p/k_t^{1/2}$. In the absence of information on the termination reaction this parameter is often incorporated into eqns. (4) and (6) and is tantamount to equating k_t to both $(k_{tc} + k_{td})$ and $(k_{tc}/2 + k_{td})$, simultaneously. This can of course lead to errors in, for example, rates of initiation calculated from eqn. (4). We shall sometimes use k_t in the text as an abbreviation for $(k_{tc} + k_{td})$ and $(k_{tc}/2 + k_{td})$. Care should also be exercised in assuming that $k_p/k_t^{1/2}$ is constant under all conditions. As we shall see later the value of this parameter can vary under conditions of diffusion-controlled termination and, in some cases, in the presence of additives.

Many polymerizations show deviations from these simple relations due to various complicating factors. It is unrealistic to carry out a completely general derivation of the kinetic relations and in practice the complicating features are often identified by the nature of the deviations from the simple relations produced under various conditions and their effect on the parameter $k_p/k_t^{1/2}$.

Since the kinetic relations appertaining to stationary-state studies contain only the parameter $k_p/k_t^{1/2}$ a separation of the rate coefficients for propagation and termination from these studies alone is impossible.

2.1 REMOVAL OF THE SIMPLIFYING ASSUMPTIONS

We may now examine the consequences of relaxing the simplifying assumptions used in the derivation of eqns. (4)—(7) in turn.

2.1.1 *Small kinetic chain lengths*

If the kinetic chain length is small, the rate of consumption of monomer in initiation may be a significant fraction of the total rate of monomer consumption. Under such conditions eqn. (2) is not obeyed and, since each act of initiation consumes one monomer molecule, the rate of monomer consumption in initiation is equal to the rate of initiation and the overall rate of consumption becomes

$$-\frac{\mathrm{d}[\mathrm{M}]}{\mathrm{d}t} = \mathscr{I} + \frac{k_p[\mathrm{M}]\mathscr{I}^{1/2}}{(k_{tc} + k_{td})^{1/2}}$$

The derivation of expressions for ν and DP involve the rate of consumption of monomer in initiation and must be modified accordingly, thus the expression for ν becomes

$$\nu = 1 + k_p[\mathrm{M}]\{\mathscr{I}(k_{tc} + k_{td})\}^{-1/2}$$

and that for the degree of polymerization

$$\mathrm{DP} = \frac{k_{tc} + k_{td}}{\frac{k_{tc}}{2} + k_{td}} + \frac{k_p \mathrm{M}}{\left(\frac{k_{tc}}{2} + k_{td}\right)} \left(\frac{k_{tc} + k_{td}}{\mathscr{I}}\right)^{1/2}$$

The equation for τ remains unaltered.

2.1.2 Rate coefficients for propagation

The validity of the assumption that the rate coefficients are independent of the length of the propagating radical has been questioned from time to time, but the assumption is usually considered valid. A complete removal of this assumption is not practicable. Gee and Melville [2] considered the effect of replacing this assumption by one which states that, "although the velocity coefficients may be dependent on chain length, the ratio of the propagation coefficient to some quantity related to the termination coefficient is independent of chain length", and have shown that the form of the kinetic relations remains unaltered. The equations remain the same when the treatment is extended to include transfer to monomer [3].

2.1.3 Stationary-state approximation

The stationary-state approximation, which assumes constancy of both [M] and [R•], may fail for various reasons. This assumption is not applicable to polymerizations which take place over prolonged periods of time during which appreciable amounts of reactants are consumed, decreasing both [M] and [R•]. The time variation in [M] is obtained by integration of eqn. (4). In the cases where initiator consumption is important, the rate of initiation, and hence the rate of polymerization, will decrease with time, while the average lifetime of a radical, the average degree of polymerization and kinetic chain length will tend to increase. The mean values of these parameters $\bar{\tau}$, $\overline{\mathrm{DP}}$, $\bar{\nu}$ are obtained by integration of the appropriate equations. Unless the initiator lifetime is very short the stationary-state situation will be approximated to at any instant in time, and the instantaneous degree of polymerization will be given by eqn. (6). The shape of the distribution curve for instantaneous degrees of polymerization will also remain approximately the same (Chap. 7), but the mean will vary with time, so that the overall distribution of molecular weights obtained by reaction for a finite time will be a superposition of distribution curves with different means.

The kinetics of the situation in which initiator is consumed rapidly has been discussed by various workers. The kinetics predict that, in general and in the absence of chain transfer and "gel-effects", such polymerizations are characterized by limiting conversions, and are termed "dead-end polymerizations". This situation is discussed later (p. 29).

a. The pre-effect

Stationary-state conditions do not normally hold at the beginning or end of the reactions. Indeed, immediately following any abrupt change in rate of initiation, as at the beginning or end of a photo- or radiolytically induced polymerization, non-stationary-state kinetics apply. We may consider the situation in which there is a steady thermal rate of initiation of rate $\mathscr{I}_1$, and photolytically induced initiation of rate $\mathscr{I}_2$. At the start of the experiment, in the absence of $\mathscr{I}_2$, an initial stationary state will be established with radical concentration $[\mathrm{R}\cdot]_0$, given by

$$[\mathrm{R}\cdot]_0 = \{\mathscr{I}_1/(k_{tc} + k_{td})\}^{1/2}$$

If we increase the rate of initiation by $\mathscr{I}_2$ (i.e. to a total rate of $\mathscr{I}_1 + \mathscr{I}_2$) at $t = 0$, eventually a new stationary state will be established with radical concentration $[\mathrm{R}\cdot]_s$, given by

$$[\mathrm{R}\cdot]_s = \{(\mathscr{I}_1 + \mathscr{I}_2)/(k_{tc} + k_{td})\}^{1/2} \tag{9}$$

In the intervening non-stationary-state period the radical concentration will increase with time according to

$$\frac{\mathrm{d}[\mathrm{R}\cdot]}{\mathrm{d}t} = \mathscr{I}_1 + \mathscr{I}_2 - (k_{tc} + k_{td})[\mathrm{R}\cdot]^2 \tag{10}$$

At any time t, the radical concentration is obtained by integration of eqn. (10) when

$$[\mathrm{R}\cdot] = \theta\left\{\frac{\mathscr{I}_1}{(k_{tc} + k_{td})}\right\}^{1/2}\left[\frac{(\theta + 1)\exp\{2\theta\mathscr{I}_1^{1/2}(k_{tc} + k_{td})^{1/2}t\} - \theta + 1}{(\theta + 1)\exp\{2\theta\mathscr{I}_1^{1/2}(k_{tc} + k_{td})^{1/2}t\} + \theta - 1}\right]$$

where

$$\theta = \left(\frac{\mathscr{I}_1 + \mathscr{I}_2}{\mathscr{I}_1}\right)^{1/2}$$

In many systems the rate of thermal initiation is zero or negligible, and under such conditions

$$[\mathrm{R}\cdot] = \left\{\frac{\mathscr{I}_2}{(k_{tc} + k_{td})}\right\}^{1/2} \tanh\{\mathscr{I}_2(k_{td} + k_{td})\}^{1/2}t$$

and the ratio of the radical concentration at time t to that in the stationary state is given by

$$\frac{[\mathrm{R}\cdot]}{[\mathrm{R}\cdot]_s} = \tanh\{\mathscr{I}_2(k_{tc} + k_{td})\}^{1/2}t$$

Stationary-state conditions apply when this ratio is unity, i.e. when

$$\{\mathscr{I}_2(k_{tc} + k_{td})\}^{1/2} \gg 1 \tag{11}$$

Although relation (11) provides an estimate of the time interval over which the non-stationary conditions apply, a quantity of more practical use is the magnitude of the pre-effect. This quantity, denoted ΔM_{pre}, is defined as the difference between the amount of reaction which has occurred at the time the stationary state is established and that which would have occurred had the stationary state been established instantaneously at $t = 0$.

Since at any instant the relation

$$-\frac{d[M]}{dt} = k_p[M][R\cdot]$$

holds (i.e. eqn. (2) holds) where $[R\cdot]$ is the instantaneous radical concentration, the magnitude of the pre-effect is given by

$$\Delta M_{pre} = k_p[M] \int_0^{\infty} ([R\cdot]_s - [R\cdot])\, dt \tag{12}$$

Combining eqns. (12) and (10) we obtain

$$\Delta M_{pre} = k_p[M] \int_{[R\cdot]_0}^{[R\cdot]_s} \frac{[R\cdot]_s - [R\cdot]_0}{\mathscr{I}_1 + \mathscr{I}_2 - k_t[R\cdot]^2}\, d[R\cdot]$$

and by use of relation (9)

$$\Delta M_{pre} = \frac{k_p[M]}{(k_{tc} + k_{td})} \ln \frac{2\theta}{\theta + 1} \tag{13}$$

Under conditions where $\mathscr{I}_1$ is zero or negligible eqn. (13) reduces to

$$\Delta M_{pre} = \frac{k_p[M]}{(k_{tc} + k_{td})} \ln 2 \tag{14}$$

From eqn. (14) it can be seen that the magnitude of the pre-effect is independent of the rate of initiation when $\mathscr{I}_1$ is zero, and when $\mathscr{I}_1$ is not zero eqn. (13) shows that ΔM_{pre} is determined by the ratio, and not the absolute values, of the two rates of initiation.

As seen from eqn. (12) the magnitude of the pre-effect is directly proportional to the integrated difference of radical concentration. To illustrate this we may consider the case where $\mathscr{I}_1 = 10^{-9}$, $\mathscr{I}_2 = 8 \times 10^{-8}$ mole l^{-1} sec^{-1} and $(k_{tc} + t_{td}) = 10^7$ mole^{-1} l sec^{-1}. In Fig. 1, the hori-

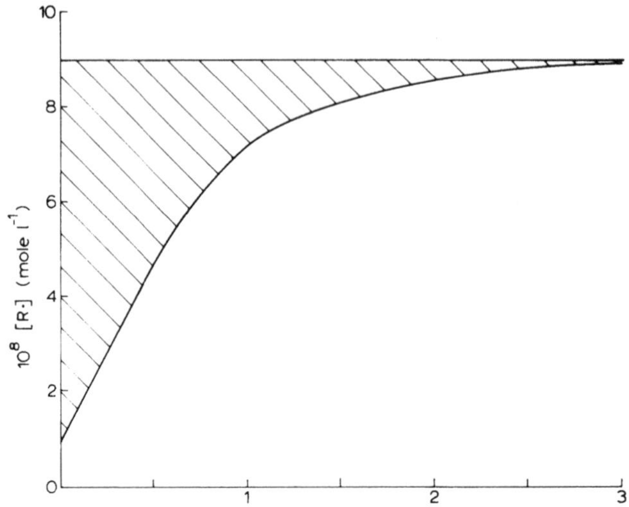

Fig. 1. Variation of radical concentration with time during the pre-effect. Curve calculated assuming a thermal rate of initiation $\mathscr{I}_1 = 10^{-9}$ mole l^{-1} sec^{-1}, photo-initiation of rate $\mathscr{I}_2 = 8 \times 10^{-8}$ mole l^{-1} sec^{-1} and $(k_{tc} + k_{td}) = 10^7$ mole^{-1} l sec^{-1}. The magnitude of the pre-effect is proportional to the shaded area.

zontal line represents the situation where the stationary state with rate of initiation ($\mathscr{I}_1 + \mathscr{I}_2$) is established instantaneously, while the curve shows the variation in [R·] in the hypothetical experiment. The magnitude of the pre-effect is proportional to the area between the horizontal line, the curve and the ordinate.

b. The after-effect

Directly analogous to the pre-effect is the after-effect which occurs as the radical concentration decreases from the initial stationary state corresponding to rate of initiation ($\mathscr{I}_1 + \mathscr{I}_2$) to the final stationary state with rate of initiation $\mathscr{I}_1$. If $\mathscr{I}_2$ is removed at time $t = 0$, the initial and final stationary-state radical concentrations, $[R\cdot]_0$ and $[R\cdot]_s$ are given by

$$[R\cdot]_0 = \left(\frac{\mathscr{I}_1 + \mathscr{I}_2}{k_{tc} + k_{td}}\right)^{1/2}$$

$$[R\cdot]_s = \left(\frac{\mathscr{I}_1}{k_{tc} + k_{td}}\right)^{1/2}$$

respectively, the change in radical concentration with time is given by

$$\frac{d[R\cdot]}{dt} = \mathscr{I}_1 - (k_{tc} + k_{td})[R\cdot]^2$$

and the radical concentration at time t is

$$[\mathrm{R}\cdot] = \left(\frac{\mathscr{I}_1}{k_{\mathrm{tc}} + k_{\mathrm{td}}}\right)^{1/2}\left[\frac{(\theta + 1)\exp\{2\mathscr{I}_1^{1/2}(k_{\mathrm{tc}} + k_{\mathrm{td}})^{1/2}t\} + \theta - 1}{(\theta + 1)\exp\{2\mathscr{I}_1^{1/2}(k_{\mathrm{tc}} + k_{\mathrm{td}})^{1/2}t\} - \theta + 1}\right]$$

The after-effect, ΔM_{aft} is defined as the excess reaction which occurs by the time the new stationary state is established over that which would have occurred had the new stationary state been established instantaneously at $t = 0$, and is given by the corresponding relation to eqn. (12), viz.

$$\Delta M_{\mathrm{aft}} = k_{\mathrm{p}}[\mathrm{M}] \int_0^{\infty} ([\mathrm{R}\cdot] - [\mathrm{R}\cdot]_{\mathrm{s}})\, \mathrm{d}t$$

$$= \frac{k_{\mathrm{p}}[\mathrm{M}]}{(k_{\mathrm{tc}} + k_{\mathrm{td}})} \ln \frac{\theta + 1}{2} \qquad (15)$$

As in the case of the pre-effect, the magnitude of the after-effect is determined by the ratio of the two rates of initiation, but, unlike the pre-effect, if $\mathscr{I}_1$ is zero ΔM_{aft} does not take on a finite value but becomes infinite, as a result of the asymptotic decay of radical concentration.

The curve in Fig. 2 shows the variation in radical concentration during the after-effect, while the horizontal straight line at $10^8\,[\mathrm{R}\cdot] = 1$ shows the radical concentration assuming instantaneous adjustment to the new stationary state. The same parameters are used as for Fig. 1, and the magnitude of the after-effect is proportional to the area between the

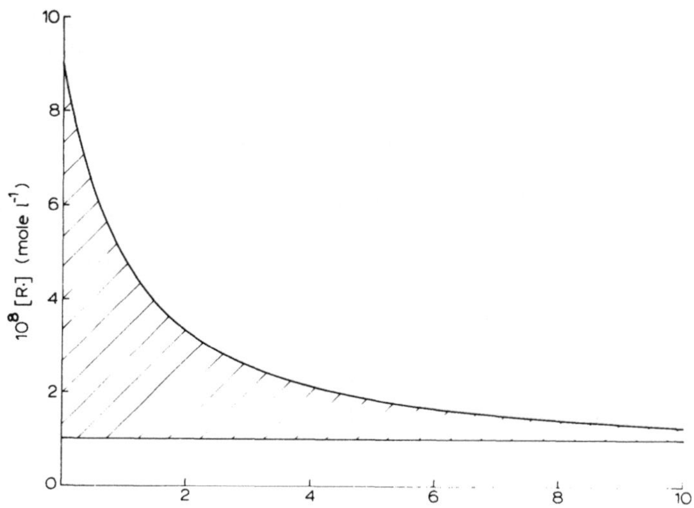

Fig. 2. Variation of radical concentration with time during the after-effect, curve calculated using the same parameters as used for Fig. 1. The magnitude of the after-effect is proportional to the shaded area.

horizontal line, the curve and the ordinate. As is obvious from Figs. 1 and 2 both the duration and the magnitude of the after-effect are greater than those of the pre-effect. This is always the case since the difference in the magnitude of the effects is given by the difference between eqns. (14) and (15). The condition for $\Delta M_{aft} > \Delta M_{pre}$ is $\theta > 1$, which it is by definition.

It should be noted that, while relationships derived from stationary-state conditions contain the parameter $k_p/k_t^{1/2}$, a study of non-stationary-state kinetics allows evaluation of the parameter k_p/k_t. Hence, by a simultaneous investigation of stationary- and non-stationary-state kinetics a separation of the individual rate coefficients for propagation and termination can be achieved.

2.1.4 Primary-radical termination

Primary-radical termination occurs when the rate of radical formation is greater than that of initiation and a fraction of the primary radicals take part in terminating growing chains. Under these conditions, as the rate of radical formation increases the rate of initiation does not increase proportionally and both rates and degrees of polymerization are lower than anticipated on the basis of simple kinetics. A detailed discussion of primary-radical termination is presented later.

2.1.5 Other assumptions

In all the preceding discussion it has been implicitly assumed that polymerization takes place under isothermal conditions. However, polymerization is an exothermic process, and if the heat of polymerization is not effectively removed the temperature of the reaction mixture will increase with the possibility of a thermal explosion. Semenov's [4] treatment of thermal explosions has been applied to the radiation induced explosive polymerization of formaldehyde by Chachaty et al. [5]. The internal temperature varies with time according to the equation

$$dT_i/dt = -\{KS(T_i - T_e) + QV[M][M^*]A_p \exp(E_p/RT_i)\}/\gamma V[M]$$

where γ is the specific heat of monomer M, [M*] is the concentration of growing chains; A_p and E_p are the pre-exponential factor and activation energy for propagation; Q is the heat of polymerization; V and S are the volume and surface area of the reaction mixture; T_i and T_e are the internal and external temperatures and K is the coefficient for total thermal exchange. The critical condition for explosion is

$$T_{i,crit} - T_e = RT_e^2/E_p$$

and

$$\phi = (VQA_p/KS)[M][M^*](E_p/RT_e^2)\exp(E_p/RT_i) = 1/e$$

explosion occurring when $\phi > 1/e$ ($e = 2.718$).

Explosive polymerizations of acrylic acid have been described for polymerizations catalysed by azoisobutyronitrile [6], radiation induced polymerizations during the formation of laminated films [6] and on repeated freezing and thawing [7].

Experimental studies of free-radical polymerization are normally carried out as batch processes. Denbigh [8] has considered the problem of molecular-weight distributions under conditions of continuous flow and Jenkins [9] has considered the overall kinetics of such processes, including the pre-stationary-state period.

2.2 DETERMINATION OF RATE COEFFICIENTS

As we have seen in the preceding section, determination of absolute rate coefficients from either stationary- or non-stationary-state kinetics is not possible, since the former only yields the parameter $k_p/k_t^{1/2}$ and the latter k_p/k_t. Separation of k_p and k_t from stationary-state measurements alone requires prior knowledge of τ or [R·]. Direct determination of τ is not possible, but requires knowledge of [R·] or k_t. Unfortunately electron spin resonance techniques are not yet capable of accurate determination of [R·] under normal experimental conditions where typical values of [R·] are of the order of 10^{-7} mole l^{-1}. Consequently absolute rate coefficients are normally obtained from a comparison of stationary- and non-stationary-state data.

2.2.1 The stationary-state measurements

The three quantities which are amenable to direct experimental determination are the rates and degrees of polymerization and the kinetic chain lengths. Determination of the kinetic chain length requires knowledge of the rates of polymerization and of initiation, and there is often uncertainty as to the value of the latter quantity. It is normal practice to determine rates and degrees of polymerization, and hence $k_p/k_t^{1/2}$, or, more accurately, $k_p/(k_{tc}/2 + k_{td})^{1/2}$ from eqn. (8).

a. Rate of polymerization

Measurement of the stationary rate of polymerization requires measurements of initial rates of reaction, i.e. determination of the rate of polymer formation at very low conversions. Two methods are normally employed, dilatometry and gravimetric measurements.

Polymers are normally significantly denser than the corresponding monomers and reaction is associated with a contraction in volume. This contraction can be conveniently and accurately measured by carrying out the polymerization in a dilatometer fitted with a narrow capillary, and observing the level of the meniscus in the capillary. Knowledge of the densities of monomer and polymer solutions allows the contraction to be converted to conversion. Various dilatometer designs have been described elsewhere [10]. It is common practice to assume ideal mixing of both monomer and polymer in the reaction mixture and that conversion is proportional to contraction. Treloar [11] has pointed out that, while the densities of solutions of styrene or polystyrene in carbon tetrachloride vary linearly with concentration, deviations from linearity occur in the three-component system styrene—polystyrene—carbon tetrachloride, and that conversions determined from dilatometer contractions may be overestimated. Although carbon tetrachloride would not normally be used as a solvent in free-radical polymerizations of styrene, in view of the high transfer constant, the same effects may occur in other systems and care should be taken when accurate rate determinations are required.

Under certain circumstances it may not be possible to carry out the polymerization in a dilatometer, or it may not be possible to produce sufficient polymer for molecular weight measurements. Gravimetric techniques may then be employed to determine the total conversion at finite time. The polymers are isolated by pouring the reaction mixture into a large volume of suitable precipitant for the polymer, which is isolated and weighed. It is usually possible to plan gravimetric experiments so that a suitable amount of polymer is prepared under steady-state conditions.

b. Degrees of polymerization

The degrees of polymerization are determined from molecular weight measurements on the polymer formed during the rate experiments. We have remarked (p. 5) on the fact that the polymer molecules have a distribution of molecular weights, and the degree of polymerization defined in eqn. (6) is the number-average degree of polymerization (i.e. number-average molecular weight of the polymer divided by the molecular weight of the monomer). Of the two usual absolute techniques for molecular weight determinations of high polymers, osmometry provides the number average of the distribution and light scattering the weight average. The viscosity of a polymer solution of a given concentration (w/v) is strongly dependent on the molecular weight of the polymer and it has been normal practice to determine polymer molecular weights from viscosity measurements, converting the intrinsic viscosity of the polymer to molecular weight using some previously published viscosity—molecular weight relation. It is, of course, important to use a relationship calibrated for the same molecular-weight distribution as that produced during the polymerization experiment in question; the distributions can vary appreciably depending on the detailed nature of the polymerization kinetics. Insufficient attention to these aspects, uncertainties of molecular-weight distributions and lack of suitable viscosity—molecular weight data has led to errors in the determination of kinetic parameters. With the development of osmometric and other techniques, such as gel-permeation chromatography, for rapid and accurate determinations of molecular weights and distributions, there is now less reason to rely on uncertain molecular weight data.

2.2.2 The non-stationary-state measurements

Direct experimental determinations of absolute values of the pre-effect and after-effect defined in eqns. (13) and (15) are difficult, since conversions involved are small and the magnitudes of the effects are very sensitive to trace impurities, such as inhibitors and retarders. The reaction times involved are normally small and photochemical techniques are usually used to effect rapid changes in rates of initiation. In studying after-effects care must be taken to ensure that there is no thermal initiation from by-products of photoinitiation.

Melville and co-workers investigated a number of techniques which were potentially suitable for investigating reactions over short periods. These included temperature rise [12], interferometry [13] and dielectric constant measurements [14]. Application of these techniques is difficult since changes associated with the non-stationary state may be smaller than effects occurring in other ways.

Burnett [15] estimated the magnitude of the pre-effect in styrene by plotting conversion against time at low conversions and under conditions of low rate of initiation and negligible thermal rate. In this way ΔM_{pre} was obtained as a negative intercept on the conversion axis. Since conversions were low, [M] could be considered constant and the parameter $k_p/(k_{tc} + k_{td})$ was obtained from eqn. (14).

a. The viscosity method

The viscosity method, developed by Bamford and Dewar [16], is a technique for investigating the stationary state and the non-stationary after-effect by following changes in viscosity of the reaction mixture. This is made possible by the high sensitivity of the viscosity of the reaction mixture to both concentration and molecular weight of the polymer at low polymer concentrations.

Under stationary-state conditions and at low conversions the degree of polymerization is time-independent and increases in viscosity are the result of increasing conversion. The intrinsic viscosity, $[\eta]$, of the polymer is related to the degree of polymerization by the Mark—Houwink equation

$$[\eta] = K\{M_0(\mathrm{DP})\}^{\alpha}$$

where M_0 is the molecular weight of the monomer and K and α are constants. The dependence of viscosity on polymer concentration is given by the equation

$$\eta = c[\eta]$$

where η is the "ideal specific viscosity" and can be determined from a single measurement of the specific viscosity by use of the Schulz—Sing relationship [17]. With the aid of these expressions it is possible to derive relationships describing the variation in viscosity with time in terms of rate coefficients in appropriate kinetic schemes. To effect separation of rate coefficients it is also necessary to study viscosity changes during the non-stationary state, and this is achieved by studying the viscosity after-effect following photochemical initiation. During the after-effect the radical concentration and hence, the instantaneous degree of polymerization varies with time. The photochemical viscosity after-effect is defined as the difference between the actual value of the ideal specific viscosity at time t and that calculated assuming instantaneous change in rate to the thermal rate when photoinitiation is interrupted. This measurement provides the necessary additional relationship between the kinetic parameters. In addition to the rate coefficients for propagation and termination this technique allows evaluation of the rate coefficients for thermal initiation and transfer.

The equations involved in this method are rather cumbersome and the reader is referred to the original paper [16]. Experimentally the viscometric rates are determined for thermal initiation and various rates of photoinitiation; it is only necessary to know relative rates of initiation. The viscosity after-effect is measured and the molecular weight of one polymer. The viscosity measurements are made during reaction in vacuum viscometers.

In the original derivation of the basic relationships for this technique it was assumed that initiation occurred through the formation of diradicals and that termination was solely by disproportionation. This scheme is incorrect, initiation probably involves monoradicals and termination with some monomers occurs almost exclusively by combination (Sect. 4.1.2). In its original form the viscosity method was applied to styrene [16], methyl methacrylate [18], vinyl acetate [19] and *p*-methoxy styrene [20], and in modified form to 4-vinyl pyridine [21]. Onyon [21] has recalculated the original data obtained with styrene in terms of initiation by monoradicals and termination by combination with little effect on the final values of the rate coefficients.

b. The rotating-sector technique

The sector technique is by far the most widely used method of determining absolute rate coefficients in radical polymerizations. This technique is in no way restricted to polymerization reactions, and was originally developed for gas-phase radical-chain reactions involving bimolecular termination [22, 23]. The principle of the method is to generate a series of non-stationary-state periods by periodic interruption of photochemical initiation. This is normally achieved mechanically by placing a rotating disc with sector-shaped apertures in the light beam. During a short period of illumination the radical concentration and rate of polymerization increase, as in the pre-effect, and during a dark period the radical concentration and rate of polymerization decrease, as in the after-effect.

Assuming the total rate of initiation during illumination is $\mathscr{I}$, under steady-state conditions the radical lifetime τ will be given by eqn. (7). If the light and dark periods are extremely long compared with τ the light and dark periods will be essentially independent; in a light period the rate of polymerization will increase rapidly to its stationary-state value and in a dark period the rate will decay to zero, assuming there is no thermal rate. Hence, if the light periods constitute $1/n$ of the total time the average rate of polymerization will be $1/n$ of the normal stationary-state value under constant illumination.

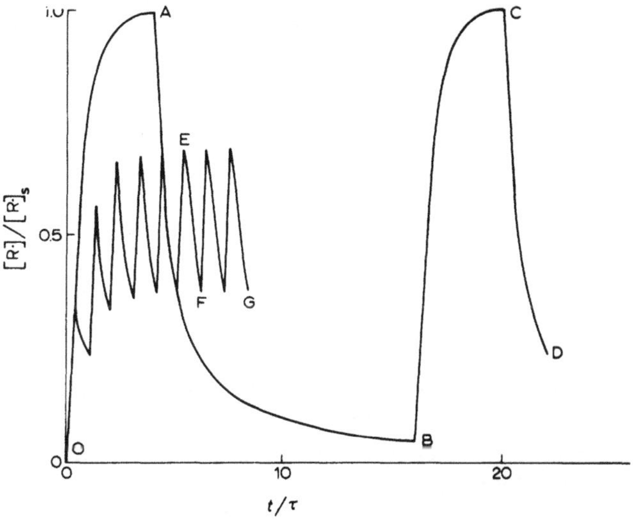

Fig. 3. Variations of radical concentration with time for intermittent illumination with light periods of two different lengths. In both cases the duration of the light period is one third that of the dark period. (Reproduced from W. A. Noyes and P. A. Leighton, The Photochemistry of Gases, ACS Monograph Series No. 28, Reinhold, New York, 1941.)

By decreasing the length of the light and dark periods we first reach the situation shown in curve OABCD of Fig. 3, which shows the variation in the ratio $[R\cdot]/[R\cdot]_s$ (where $[R\cdot]$ is the radical concentration at time t and $[R\cdot]_s$ the stationary-state value under continuous illumination) with t/τ. In the first light period the radical concentration increases along the curve OA until it approaches the stationary-state value, interruption of the illumination at A causes the radical concentration to decay along AB, at the commencement of the next light period at B not all radicals generated during the first light period have decayed. The radical concentration then increases along BC and so on. In this situation the light and dark periods are not completely independent and, since the after-effect is of longer duration than the pre-effect, the average rate of polymerization will be greater than $1/n$ of the stationary-state value.

Further reduction in the duration of the light and dark periods, maintaining the original value of n, causes the radical concentration to follow the serrated curve OFGE, where the average radical concentration in a light and dark period gradually increases until a pseudo-stationary state is reached with the radical concentration varying between the fixed limits F and G. As can be seen from Fig. 3 for rapid flashing of the initiating illumination the average radical concentration, and hence the rate of polymerization, approximates to one half of the value of that under continuous illumination.

The derivation of the final equations has been presented a number of times [3, 23—26]. Here we shall describe the method of the treatment, the final equations and the method of application. The rates of change of radical concentration in the light and dark periods are given by the equations

$$\frac{\mathrm{d}[\mathrm{R}\cdot]}{\mathrm{d}t} = \mathscr{I} - k_t[\mathrm{R}\cdot]^2$$

and

$$\frac{\mathrm{d}[\mathrm{R}\cdot]}{\mathrm{d}t} = -k_t[\mathrm{R}\cdot]^2$$

respectively. These equations are integrated to obtain the variations in radical concentration with time. Re-integration allows the mean radical concentrations to be derived in terms of the lengths of the light and dark periods, the radical concentrations at points F and G (Fig. 3), the stationary radical concentration $[\mathrm{R}\cdot]_s$ and $(k_{tc} + k_{td})$. The mean radical concentration for one (light + dark) cycle $[\overline{\mathrm{R}\cdot}]$ is obtained and suitable manipulation gives the ratio $[\overline{\mathrm{R}\cdot}]/[\mathrm{R}\cdot]_s$ as

$$\frac{[\overline{\mathrm{R}\cdot}]}{[\mathrm{R}\cdot]_s} = \frac{1}{p+1}\left\{1 + \frac{1}{t_1(k_t\mathscr{I})^{1/2}}\log\left(1 + \frac{pt_1(k_t\mathscr{I})^{1/2}}{1 + \dfrac{[\mathrm{R}\cdot]_s}{[\mathrm{R}\cdot]_F}}\right)\right\}$$

where t_1, pt_1, are the lengths of light and dark periods, respectively, $k_t = (k_{tc} + k_{td})$ and $[\mathrm{R}\cdot]_F$ is the radical concentration at point F. The ratio $[\mathrm{R}\cdot]_F/[\mathrm{R}\cdot]_s$ is given by

$$\frac{[\mathrm{R}\cdot]_F}{[\mathrm{R}\cdot]_s} = \frac{pt_1(k_t\mathscr{I})^{1/2}\tanh t_1(k_t\mathscr{I})^{1/2}}{2\{pt_1(k_t\mathscr{I})^{1/2} + \tanh t_1(k_t\mathscr{I})^{1/2}\}} \times$$

$$\left\{1 + \left(1 + \frac{4}{pt_1(k_t\mathscr{I})^{1/2}\tanh t_1(k_t\mathscr{I})^{1/2}} + \frac{4}{p^2 t_1{}^2 k_t\mathscr{I}}\right)^{1/2}\right\}$$

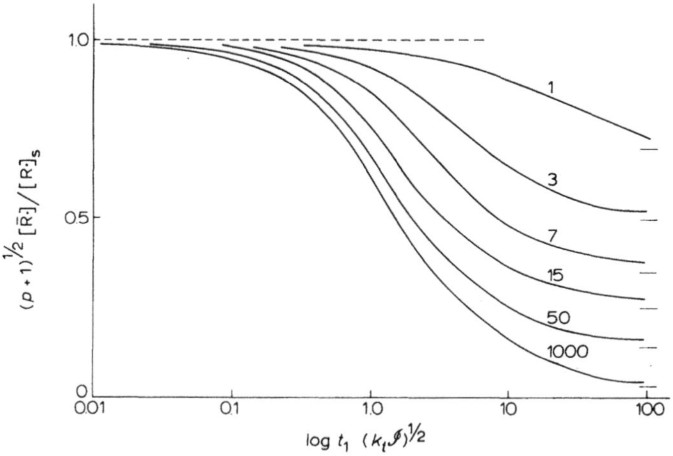

Fig. 4. Ratios of the rates of polymerization under intermittent illumination to those under continuous illumination as a function of the length of the light period for various values of p, the ratio of durations of light and dark periods. Values of p are indicated on the curves. The thermal rate of initiation is assumed to be zero. (Reproduced from W. A. Noyes and P. A. Leighton, The Photochemistry of Gases, ACS Monograph Series No. 28, Reinhold, New York, 1941.)

Thus $[\overline{R\cdot}]/[R\cdot]_s$ can be expressed in terms of $(\mathscr{I}k_t)^{1/2}$ or τ, and since rates of polymerization are proportional to radical concentration, this ratio is equal to the ratio of rates of polymerization under intermittent and continuous illumination. To obtain the parameter τ or $(\mathscr{I}k_t)^{1/2}$ a calculated curve is drawn of $(p+1)^{1/2}[\overline{R\cdot}]/[R\cdot]_s$ against log $t_1(k_t\mathscr{I})^{1/2}$; p is determined by the geometry of the rotating disc. Figure 4 shows a series of such curves for various values of p, the factor $(p+1)^{1/2}$ is a normalizing factor. Rates of polymerization under constant illumination and under intermittent illumination for various t_1 are determined and a plot of $(p+1)^{1/2}[\overline{R\cdot}]/[R\cdot]_s$ against log t_1 is made on transparent graph paper. This graph is placed over theoretical curves and moved horizontally to obtain the best fit between the calculated and experimental curves for the appropriate value of p and thus the value of $(k_t\mathscr{I})^{1/2}$ is obtained directly. Then, if the value of $\mathscr{I}$ is known k_t can be determined and from stationary-state measurements a separation of the rate coefficients can be effected.

In the derivation of the basic equations for the rotating-sector treatment it has been assumed that thermal initiation is zero. Matheson et al. [25] have considered the case where thermal initiation cannot be neglected and have shown that thermal initiation influences the shapes of the curves described above. Figure 5 shows a series of such curves calculated by Matheson et al. [25] for various contributions to the total rates of initiation from thermal initiation for a single value of $p = 3$. The influence of thermal initiation is most noticeable at large t_1, i.e. at slow sector speeds.

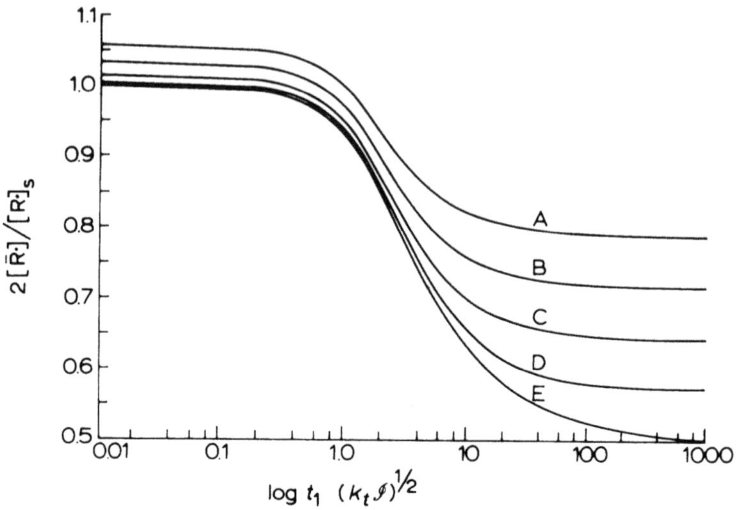

Fig. 5. Ratio of rates of polymerization under intermittent illumination to those under continuous illumination as a function of the length of the light period, for p = 3. Individual curves refer to different rates of thermal initiation. A, $q^{1/2}$ = 0.20; B, $q^{1/2}$ = 0.15; C, $q^{1/2}$ = 0.10; D, $q^{1/2}$ = 0.05; E, $q^{1/2}$ = 0.00; q is the ratio of the rate of thermal initiation to the total rate of initiation under continuous illumination. (After Matheson et al. [25], by courtesy of J. Am. Chem. Soc.)

A further basic assumption in the derivation of the equations is that the transition from light to dark is instantaneous. Although such conditions may be approximated to in practice, Burns and Dainton [27] considered the case of a trapezoidal model as a more accurate description of the shape of the light pulse. These workers concluded that the errors introduced by assuming a step function are negligible except at low sector speeds. Bamford et al. [3] have considered the case of sinusoidal variation of light intensity. Burns and Dainton [27] also calculated and tabulated possible errors arising from non-uniform light-absorption in the reaction mixture, assuming zero diffusion.

c. *Emulsion technique*

It is relevant to consider here the application of emulsion systems to the determination of absolute rate coefficients [28—30]. The method is based on the Smith—Ewart theory of emulsion polymerization [28]. According to this mechanism initiating radicals are generated in the aqueous phase and diffuse into the emulsified polymer particles. In these particles the primary radicals may initiate or terminate polymerization. It is assumed that termination between radicals in a single polymer particle is rapid compared with the rate of entry of the particles. As a consequence of this there can be only one radical in any polymer particle, and at any given time only one half of the polymer particles contain propagating radicals. Therefore, the concentration of propagating radicals is one half the concentration of polymer particles and this latter quantity can be determined by electron microscopy of the polymer latex. The rate of polymerization is simply $k_p[M][R\cdot]$, and since $[R\cdot]$ has been determined k_p can be calculated from the rate of polymerization if [M] is known. [M] is, of course, the concentration of monomer in the polymer particle and is determined in a separate experiment by measuring the uptake of monomer by polymer. Knowing k_p, k_t can then be determined from conventional stationary-state studies.

d. Comparison of techniques

Determinations of absolute rate coefficients by direct study of the pre-effect or by the viscosity method have one advantage over the sector technique; namely, that a knowledge of the absolute values of rates of initiation are not necessary. In using the sector technique, if the absolute value of $\mathscr{I}$ has not been determined by some independent method it is often obtained from stationary-state measurements by calculating $k_p/(k_{tc}/2 + k_{td})^{1/2}$ from eqn. (8) and feeding this value into eqn. (4). This immediately brings into question the nature of the termination reaction, since if this substitution is made directly by equating $(k_{tc}/2 + k_{td})$ and $(k_{tc} + k_{td})$ it is being assumed that termination occurs by disproportionation. If in fact termination is wholly by combination then the value of $\mathscr{I}$ computed from eqn. (4) is in error by a factor of 2; if termination is by both combination and disproportionation the factor lies between 1 and 2. Use of the incorrect value of $\mathscr{I}$ in the sector technique then produces an error in $(k_{tc} + k_{td})$, again by a factor of up to 2, the apparent value being too large. Feeding the incorrect value of $(k_{tc} + k_{td})$ into the parameter $k_p/(k_{tc}/2 + k_{td})^{1/2}$ then produces a value of k_p which may be in error by $\sqrt{2}$. In all determinations of absolute rate coefficients from a comparison of stationary- and non-stationary-state data errors may arise through equating $(k_{tc} + k_{td})$ and $(k_{tc}/2 + k_{td})$. Accurate values of the absolute rate coefficients can only be determined through accurate knowledge of the proportions of combination and disproportionation termination and/or rates of initiation, and it is with these quantities that there is considerable uncertainty in many systems.

The emulsion technique also appears to have an advantage in that it provides a direct determination of k_p without requiring prior knowledge of either the rate of initiation or the mechanism of termination. However, in using the emulsion technique care should be taken to ensure that the polymerization kinetics do obey simple theory and that the correct value of [M] is employed. Furthermore, it is possible that the value of k_p obtained in this way may differ from that in solution polymerization as a result of physical factors under the unusual environmental conditions.

3. Initiation

In discussing kinetics of initiation two factors require consideration, (i) the absolute rates of radical formation and initiation and (ii) the nature of the initiating radicals. The absolute rate of initiation determines the rate of formation of polymer molecules and a knowledge of this value is useful in determining the value of the kinetic parameter $k_p/(k_{tc} + k_{td})^{1/2}$ unambiguously through eqn. (4). Conversely, if the kinetic parameters are known with sufficient degree of certainty a study of rates of polymerization under various conditions can provide a means of investigating mechanisms of radical formation in hitherto unknown systems. In some cases rates of initiation and radical formation are not equal and studies of polymerization kinetics can provide useful additional information on radical reactions. A knowledge of the nature of initiating radicals is important since rates of formation of polymer molecules with known end-groups can be important in various aspects of polymer science. Furthermore, it is not always the primary radical generated from the initiator which initiates polymerization and a knowledge of the primary radicals can provide information on the rates of conversion of primary to secondary radicals.

When comparing rates of initiation in different solvents from polymerization data care should be exercised since it is known that, in at least some systems, values of k_p and k_t are solvent-dependent.

3.1 RATES OF INITIATION

If the rate coefficients of propagation and termination are known, then rates of initiation are readily calculated from eqn. (4). If the absolute values of the rate coefficients are unknown, values of $\mathscr{I}$ may be computed from a comparison of eqns. (4) and (8) if the mechanism of termination is known and if transfer is absent or negligible. We have discussed in Sect. 2.2.2.d the consequences of not knowing the mechanism of bimolecular termination, when values of $\mathscr{I}$ calculated in this way may be in error by a factor of 2. In cases where transfer is not negligible degrees of polymerization should be corrected for transfer, as explained in Chap. 2, before they are used in eqn. (8).

A further simple method of determining values of $\mathscr{I}$ is to compare rates of polymerization with those obtained using an initiator which decomposes at a known rate or for which its rate of initiation of some other monomer is known. This method requires the assumption that the efficiency of initiation (see below) of the standard initiator is unity or that it is the same for the two monomers. This method does not require prior knowledge of the kinetic parameters for propagation and termination.

For initiators with short half-lives, the half-life of the initiator can be determined from the shape of the conversion—time curve without knowledge of $k_p/k_t^{1/2}$. If in fact $k_p/k_t^{1/2}$ is known then the efficiency of the initiator can be calculated from the rates and the half-life [31, 32].

3.1.1 From kinetic chain length

Rates of initiation may be determined from comparisons of rates and kinetic chain lengths of polymerization, by use of eqns. (4) and (5). The average kinetic chain length is readily obtained, in principle, by estimating the number of initiating fragments in a given amount of polymer if initiation occurs only by the incorporation of initiator fragments into the polymer and such fragments are not incorporated in any other way. This method requires the availability of an analytical technique which will accurately measure the low concentration of initiator fragments. Tracer techniques, using isotopically labelled initiator, often provide a convenient means of analysis. This technique has been employed by Bevington and co-workers, who have detailed the precautions necessary to obtain accurate data [33]. These include complete isolation of the polymer, including low-molecular-weight fractions which contain a relatively high proportion of initiator fragments, and efficient removal of excess initiator.

If some primary radicals undergo a transfer process and the transfer radicals initiate polymerization, not all initiator fragments will be incorporated into the polymer and high values of kinetic chain length and low values of rate of initiation will be obtained. Alternatively, anomalously high rates of initiation are obtained under conditions of primary-radical termination, since initiator fragments are incorporated into the polymer during both initiation and termination.

Simple transfer to monomer or solvent has no effect on the results if the kinetic chains are long; however transfer to initiator results in a high level of incorporation of active fragments in the polymer since the transfer radicals will initiate and be incorporated into polymer chains. In this way more than one initiator fragment will be incorporated per kinetic chain. This difficulty may be overcome by the simultaneous use of two initiators, the initiator under investigation being suitably labelled [34]. In the particular case considered by Bevington and Lewis [34], styrene polymerization was initiated by bis(3,5-dibromo-4-methoxybenzoyl)-peroxide, which we may designate as A—A and which reacts with the propagating styrene radical according to the reaction

$$\sim\!\!\sim S\cdot + A\!-\!A \xrightarrow{k_f} \sim\!\!\sim SA + A\cdot$$

the A· radical subsequently initiating polymerization. This process incorporates two initiator fragments per act of transfer (or induced decomposition). The total rate of incorporation of labelled fragments into the polymer, or apparent rate of initiation $\mathscr{I}_{app}$ is

$$\mathscr{I}_{app} = \mathscr{I}_1 + 2k_f[R\cdot][A{-}A]$$

where $\mathscr{I}_1$ is the true rate of initiation by labelled initiator. Since the rate of polymerization is $k_p[R\cdot][M]$ we have

$$\mathscr{I}_{app} = \mathscr{I}_1 + \frac{2k_f}{k_p}\frac{[A{-}A]}{[M]}\left(-\frac{d[M]}{dt}\right)$$

Maintaining a constant rate of initiation by the labelled initiator the rate of polymerization is varied by addition of a second, unlabelled initiator. A plot of $\mathscr{I}_{app}$ against $(-d[M]/dt)$ should be linear with an intercept, at zero rate of polymerization, of $\mathscr{I}_1$. From the slope of the line it is also possible to obtain the ratio k_f/k_p, the transfer constant to initiator. The extrapolation to zero rate of polymerization is equivalent to determining the rate of initiation at zero concentration of propagating radicals, when transfer cannot occur.

3.1.2 Inhibitors and retarders

A further general method of determining rates of initiation is based on the use of inhibitors and retarders. A detailed discussion of inhibition and retardation is reserved for a later chapter. Here we present the basic principles only, which are necessary for an understanding of the technique.

Inhibitors and retarders are substances which interact with reactive radicals to produce unreactive or less reactive species. There is no intrinsic difference between inhibitors and retarders except in degree. Ideally we may define an inhibitor as a species which interacts rapidly with a reactive radical, either a primary or short propagating radical, to produce a species incapable of supporting further polymerization. Simply, an inhibitor is an efficient radical scavenger which completely suppresses polymerization until it is virtually all consumed. Following consumption of the inhibitor the rate of polymerization rises very rapidly to assume its normal value, consistent with the consumption of initiator during the induction period. The length of the induction period, for low consumption of initiator, is proportional to both the initial concentration of the inhibitor and the number of radicals formed during the induction period; it is inversely proportional to the rate of initiation for constant initial initiator concentration. The rate of initiation is then simply obtained from the length of the induction period and the initial inhibitor concentration. Two conditions must be fulfilled to obtain accurate data in this way: (i) the number of radicals interacting with each molecule of inhibitor must be known and (ii) the inhibitor and monomer must not interfere with geminate recombination (see below). One of the main causes of uncertainties in rates of initiation determined in this way stems from lack of knowledge of the number of radicals consumed per molecule of inhibitor in many cases.

The same information can be obtained, without determining the magnitude of the induction period, by observing the rate of consumption of inhibitor or the rate of formation of a product of inhibition if a suitable analytical technique is available. This is feasible in some cases since many inhibitors, e.g. diphenylpicrylhydrazil, are highly coloured and can be estimated absorptiometrically. Ferric chloride is an efficient inhibitor in some systems, such as styrene polymerizations, although it is only a retarder in other systems, e.g. with acrylonitrile [35], and reacts according to the mechanism

$$R_n\cdot + Fe^{3+} \longrightarrow P_n + Fe^{2+}$$

forming ferrous ion which may be estimated volumetrically [35, 36].

Retarders, being simply less reactive than inhibitors, compete with bimolecular termination for propagating radicals, they thus reduce the rate of polymerization rather than suppress it completely, unless present in high concentration. Low concentrations of efficient retarders will produce low initial rates of polymerization, the rate gradually increasing to its normal value as initiator is consumed. It is possible to analyse the data (Chap. 2) to obtain an "equivalent induction period", which is the time for the retarder to be consumed had it acted as an inhibitor, and hence obtain rates of initiation.

3.2 INITIATOR EFFICIENCY

Polymerizations are often initiated by the dissociation, either thermal or photolytic, of initiators such as azo-compounds (R—N=N—R), with elimination of a molecule of nitrogen, and peroxy compounds (R—O—O—R), by scission of an O—O bond, to yield primary radicals $R_c\cdot$ according to the reactions

$$R{-}N{=}N{-}R \xrightarrow{k_d} 2R_c\cdot + N_2 \tag{16a}$$

$$R{-}O{-}O{-}R \longrightarrow 2R_c\cdot \tag{16b}$$

respectively. The rate of primary-radical formation is $2k_d[C]$ where C is initiator. Initiation then occurs by

$$R_c\cdot + M \xrightarrow{k_i} R_cM\cdot$$

and the rate of initiation is given by

$$\mathscr{I} = k_i[R_c\cdot][M]$$

Frequently, rates of initiation are less than those of primary-radical formation and f, the efficiency of the initiator, defined as

$$f = \frac{\text{Rate of initiation of propagating chains}}{2(\text{Rate of initiator decomposition})}$$

is less than unity. This corresponds to a wastage of primary radicals.

One source of radical wastage is through recombination of primary radicals from different molecules of initiator in the body of the solution. In the presence of monomers this is unlikely since most monomers act as efficient radical scavengers. A more probable process is the termination of a growing chain by a primary radical. This process is usually only observed at high rates of initiation or low monomer concentrations, particularly when an inactive radical is used to initiate the polymerization of an unreactive monomer (see Sect. 4.3). A more serious source of radical wastage, at low rates of initiation, is the recombination of primary radicals derived from the same molecule of initiator, or geminate recombination [37].

Initiator decomposition by reactions of the type (16a) or (16b) produces radicals in pairs and their distribution in solution, at the time of formation, is not random and the probability of recombination is enhanced. Geminate recombination may occur in two ways [37]: (i) by primary recombination within about 10^{-11} sec of dissociation, before diffusion of the radicals has taken place, and with which probably no scavenger can compete; (ii) by secondary recombination within about 10^{-9} sec of dissociation when there is still a finite probability of interaction. Both these processes may be considered as cage reactions. The kinetics of geminate recombination have been considered by Noyes [38] who has tabulated extents of geminate recombination, in some cases in the absence of monomers [39].

Initiation of propagating chains normally takes place in the body of the solution and monomers are, in general, not sufficiently powerful scavengers to interfere with secondary recombination. If a monomer could interfere with secondary recombination the observed rate of initiation would vary with the monomer concentration. This behaviour is normally only observed under conditions of primary-radical termination. If a scavenger, such as bromine [40], capable of interfering with secondary recombination is used to assess rates of initiation, by measuring rates of primary-radical formation, anomalously high rates of initiation would be obtained.

The consequences of geminate recombination are dependent on the nature of the initiator. If primary radicals are formed by simple scission of a weak bond and geminate recombination occurs by reformation of that bond, regenerating a molecule of initiator, the efficiency is unity. Many initiators used in practice dissociate into radicals with the simultaneous elimination of a stable small molecule, e.g. azo-initiators as shown in reaction (16a). Recombination of the primary radicals cannot regenerate the original initiator but usually yields stable molecules, when geminate recombination leads to efficiencies less than unity. In the specific case of initiation by azobisisobutyronitrile (AIBN), an extremely common initiator, geminate recombination of the $(CH_3)_2\dot{C}CN$ radicals leads to a significant proportion of the ketenimine $(CH_3)_2C{=}C{=}N{-}C(CH_3)_2CN$ [41]. This ketenimine dissociates at about the same rate as AIBN, but on dissociation the initial radicals are not separated by a small molecule and recombination is more extensive than with AIBN [42].

Peroxides used in initiation are often of the type RCOO—OOCR which yield RCOO· radicals on dissociation. These, and other suitable radicals, undergo rapid decomposition, splitting off a simple stable molecule and leaving a smaller secondary radical, e.g.

$$R{-}\underset{\underset{O}{\|}}{C}{-}O\cdot \longrightarrow R\cdot + CO_2 \qquad (17)$$

If the half-life for dissociation of the primary radical is very short geminate recombination will involve the secondary R· radicals. In the decomposition of diacetyl peroxide in solution both ethane and methyl acetate have been observed as products of geminate recombination [43]. On the other hand, if the half-life of the primary radical is long compared with 10^{-9} sec geminate recombination will reform the original initiator. In this latter case primary-radical dissociation will then take place in the body of the solution and either primary or secondary radicals will initiate the polymerization. The proportions of initiation by the various radicals will then depend on the nature of the monomer and its concentration. The proportions of primary and secondary initiation may be determined by comparing the rate of initiation and the rate of formation of dissociation products or by radio-tracer techniques using suitably labelled initiators. A common example of this type of initiator is benzoyl peroxide, the decomposition of which forms the benzoyloxy radical which dissociates into phenyl radicals and carbon dioxide [44—46].

Another major source of initiating radicals are redox reactions where radical formation accompanies a one-electron transfer process, e.g. the reduction of a hydroperoxide ROOH by ferrous ions. In such processes radicals are formed singly and do not partake in geminate recombination. However, radical wastage can occur by further reduction of the alkoxy radical [47]

$$RO\cdot + Fe^{2+} \longrightarrow RO^- + Fe^{3+}$$

Under suitable circumstances wastage of primary radicals can occur through transfer processes of the type

$$R_c\cdot + AB \longrightarrow R_cA + B\cdot$$

where B is incapable of initiating polymerization. This process is equivalent to degradative chain transfer and the efficiency of the initiating system is dependent on [AB].

3.3 RATE COEFFICIENTS FOR INITIATION

There is little available information on rate coefficients, k_i, for the addition of primary radicals to monomers. In most practical systems these rate coefficients are fairly large so that the monomer acts as an efficient radical scavenger, the stationary concentration of primary radicals in solution is normally low and these species are not involved in termination. However, primary-radical termination can occur (Sect. 4.3) but is usually only significant at high rates of initiation and low [M]. The conditions under which primary-radical termination becomes apparent are determined by the reactivities of the monomer and radical, i.e. the value of k_i. Thus, in the case of a relatively unreactive monomer, such as vinyl chloride, and initiation by a stabilized radical, such as the isobutyronitrile radical, primary-radical termination sets in at lower rates of initiation than if a more reactive radical, such as $Cl_3C\cdot$ is used [48].

Most of the available data on reactions between primary radicals and monomers is in the form of relative reactivities. The most widely studied system in this connection is probably initiation by benzoyl peroxide. The benzoyloxy radical undergoes two reactions in the presence of monomer

$$C_6H_5COO\cdot + M \longrightarrow C_6H_5COOM\cdot \tag{18a}$$

$$C_6H_5COO\cdot \longrightarrow C_6H_5\cdot + CO_2 \tag{18b}$$

The phenyl radicals also initiate polymerization. By studying the formation of decomposition products of the benzoyloxy radical, Bagdasarjan [49] has investigated the reactivity of monomers with the primary radical. Bevington et al. [33, 45, 46, 50—54] have used the tracer technique, with appropriately labelled peroxides, to measure the relative proportions of initiation by the primary and secondary radicals. In this way the relative reactivities of a series of monomers, using styrene as a standard, were measured for reaction with the benzoyloxy radical at 60 °C [33, 46]. The results quoted (relative values of rate coefficients of reaction (18a)) for reaction in benzene solution are: 2,5-dimethyl styrene, 2.0; styrene, 1.0; 2,4,6-trimethyl styrene, 0.67; vinyl acetate, 0.36; methyl methacrylate, 0.12; and acrylonitrile (in DMF solution), <0.05. In the same way the relative reactivities of a number of aroyloxy radicals to the propagating styrene radical have been determined [45, 46, 52—54].

Similar data can be obtained by comparing rates of initiation with rates of reaction of the primary radicals with a suitable scavenger. Pavlinic and Lazar [55] obtained information on reactivities towards the benzoyloxy radical by comparing rates of initiation with rates of transfer of primary radicals and of propagating radicals. These workers found that the reactivity of benzoyloxy radicals with styrene was eight times greater than with methyl methacrylate at 50 °C; almost identical with the results of Bevington quoted above. The corresponding factor for addition of cumyloxy radicals at 130 °C was 25. Data has also been obtained by Orr and Williams [47], for reactivities of substituted cumyloxy radicals by studying the competition for the radicals by ferrous ion and monomer.

By comparing the reactions of methyl radicals from diacetyl peroxide with monomer and transfer agent (iso-octane), Szwarc [56] has obtained relative reactivities of monomers to the methyl radical. At 65 °C the relative reactivities of various monomers were found to be acrylonitrile, 1540; methyl methacrylate, 1420; styrene, 792; vinyl acetate, 37; ethylene, 34.1 and allyl acetate, 8.4. The order of reactivities is quite different to that obtained with the benzoyloxy radicals. The Hammett σ value for the benzoyloxy radical is about unity, indicating that polar factors are very important in its reactions [33], and this possibly accounts for the different order in monomer reactivities.

Such relative values of rate coefficients for initiation are useful in studies of monomer and radical reactivities.

3.4 DEAD-END POLYMERIZATION

In studies of free-radical polymerization it is normal practice to work under conditions where the half-life of the initiator is sufficiently long that initiator consumption is not a serious problem. If the initiator half-life is short, rates of polymerization decrease as initiator is consumed. Tobolsky [57] has considered the kinetics of such reactions and has shown that for first-order decay of initiator, in the absence of thermal initiation and a gel effect (p. 67), such reactions are characterized by a limiting conversion of monomer to polymer. These reactions are termed dead-end polymerizations. In the presence of thermal initiation the rate of polymerization naturally decreases to the thermal rate on consumption of the initiator.

On the basis of a simple kinetic scheme comprising the reactions (with the appropriate rate coefficients), first-order decomposition of initiator into two radicals (k_d), addition of primary radical to monomer (k_i), propagation (k_p), and termination by combination (k_{tc}), the variation of monomer concentration with time is given by the equation [57]

$$\ln \frac{[M]}{[M]_0} = \frac{2k_p f^{1/2}[I]_0^{1/2}}{k_{tc}^{1/2} k_d^{1/2}} \{1 - \exp(-k_d t/2)\}$$

($[M]_0$, $[I]_0$ are the initial concentrations of monomer and initiator, respectively, and f is the initiator efficiency), from which the monomer concentration at infinite time and limiting conversion can be obtained.

The number-average degree of polymerization is given by

$$\overline{\mathrm{DP}}_n = \frac{[\mathrm{M}]_0 - [\mathrm{M}]}{f([\mathrm{I}]_0 - [\mathrm{I}])}$$

This latter expression has been corrected to allow for chain transfer, and expressions for the weight-average degree of polymerization have been determined. It has been demonstrated that the equations give a good description of polymerizations under ideal conditions [31, 32, 58].

O'Driscoll and McArdle [59] have considered dead-end polymerizations involving bimolecular initiation, as in redox processes. If initiation produces one radical from the initiator components A and B, the limiting monomer concentration $[\mathrm{M}]_\infty$ is given by

$$\frac{[\mathrm{M}]_0}{[\mathrm{M}]_\infty} = \left\{\frac{([\mathrm{A}]_0/[\mathrm{B}]_0)^{1/2} + 1}{([\mathrm{A}]_0/[\mathrm{B}]_0)^{1/2} - 1}\right\}^{(fk_\mathrm{p}^2/k_\mathrm{t}k_\mathrm{d})^{1/2}} \tag{19}$$

and the variation of monomer concentration with time by

$$\frac{[\mathrm{M}]_\infty}{[\mathrm{M}]} = \left[\frac{([\mathrm{A}]_0/[\mathrm{B}]_0)^{1/2}\exp\{k_\mathrm{d}t([\mathrm{A}]_0 - [\mathrm{B}]_0)/2\} - 1}{([\mathrm{A}]_0/[\mathrm{B}]_0)^{1/2}\exp\{k_\mathrm{d}t([\mathrm{A}]_0 - [\mathrm{B}]_0)/2\} + 1}\right]^{(fk_\mathrm{p}^2/k_\mathrm{t}k_\mathrm{d})^{1/2}}$$

From suitable application of these equations to experimental data, k_d and f can be computed, and this has been done for initiation by benzoyl peroxide/dimethyl aniline [59, 60]. Equation (19) predicts that for bimolecular initiation dead-end polymerization does not occur when $[\mathrm{A}]_0 = [\mathrm{B}]_0$.

It has been reported [61] that dead-end effects in the persulphate initiated polymerization of acrylic acid in aqueous solution do not conform to Tobolsky's treatment. In an attempt to interpret the experimental data the influence of primary-radical termination of the propagating chains has been discussed [61].

3.5 INITIATORS

As mentioned in the preceding section azo-compounds and peroxides are common initiators of free-radical polymerization. Radical formation, from such compounds, is usually associated with an activation energy of about 30 kcal mole^{-1}, so that the temperature range over which an individual initiator is useful is limited. The useful temperature range can often be extended by photolysis of the initiator or by its incorporation into a redox system where the activation energy is usually lower. The decompositions of such compounds are described in detail in other volumes of this series (e.g. Vols. 5 and 7) and we shall restrict our attention to aspects which are important in polymerization studies.

In addition to the above initiators there must be several hundred other reports of free-radical initiating systems and it is impossible to provide a comprehensive description of these. In any case, many of the kinetic studies have been cursory, limited to a comparison of relative initiating activities or a simple determination of the overall dependence of rate of polymerization on reactant concentrations. Large numbers of reports contain the results of a few gravimetric experiments for polymerizations carried out over long periods and taken to high conversion. Such studies mask effects due to retardation, "gel-effects", etc., so that the validity of the comparative data obtained may be limited. For the purposes of determining rates and mechanisms of initiation, and making practical use of the systems, such studies are virtually useless. Consequently, we shall restrict our attention to a few representative systems for which meaningful kinetic information is available.

3.5.1 Azo-initiators

a. Thermal initiation

In view of its importance in polymerization studies we may discuss the thermal decomposition of azo-compounds in terms of the decomposition of azobisisobutyronitrile (AIBN). The driving force in the thermal decomposition of these initiators is the scission of the weak C—N bonds in the azo-linkages with the elimination of a molecule of nitrogen. It is usually considered that AIBN and its analogues decompose by the simultaneous cleavage of the two C—N bonds, according to the reaction

$$(CH_3)_2(CN)C{-}N{=}N{-}C(CN)(CH_3)_2 \longrightarrow 2(CH_3)_2\dot{C}CN + N_2 \quad (20)$$

Recent ESR evidence has indicated the transient existence of the $(CH_3)_2(CN)CN_2\cdot$ radical [62], suggesting that decomposition occurs in two stages; for the purpose of polymerization studies eqn. (20) is usually sufficient description of the overall process.

The kinetics of AIBN decompositions are relatively simple since decomposition is first-order to high conversion [63—65], and most workers consider that the rate of decomposition is relatively solvent-independent; although some variations in rates of decomposition in various solvents have been observed the differences are small, especially when compared with those observed for peroxide decompositions. The kinetics of AIBN decomposition have been studied in various solvents using different techniques such as nitrogen evolution, radical scavenging and direct spectroscopic determination of [AIBN]. On the basis of early data [63—70] obtained with various solvents in the temperature range 40—100 °C, van Hook and Tobolsky [71] suggested a "best value" of the rate coefficient for decomposition of $k_d = 1.58 \times 10^{15} \exp(-30{,}800/RT)\ \text{sec}^{-1}$. Betts et al. [72] subsequently reported a value of $k_d = 1.7 \times 10^{15} \exp(-30{,}700/RT)\ \text{sec}^{-1}$ for reaction in *N,N*-dimethylformamide solution. More recently [73—76] variations in k_d with various

TABLE 1
Solvent effects in the thermal decomposition of azobisisobutyronitrile

Solvent	Temp. (°C)	$10^5 k_d$ (sec^{-1})	$\Delta H^\ddagger$ (kcal $mole^{-1}$)	$\Delta S^\ddagger$ (cal $mole^{-1}$ deg^{-1})	E_a (kcal $mole^{-1}$)	$10^{-15}A$ (sec^{-1})	Ref.
Propylene carbonate	72.23	5.82	31.75	14.19			73
Diphenyl methane	72.23	6.06	31.21	12.25			73
N-methyl propionamide	72.23	6.26	31.86	14.19			73
N,N-dimethyl aniline	72.23	6.91	30.08	9.23			73
Decalin	72.23	6.5[a]			33.4	69	76
Cyclohexane	72.23	5.8[a]			33.4	62	76
Dibromomethane	72.23	4.8[a]			30.7	0.94	76
Propionic acid	72.23	8.3[a]			30.6	1.4	76
Formamide	72.23	7.4[a]			31.7	6.4	76
Toluene	72.23	7.7[a]			31.1	2.7	76
Diethyl malonate	70	3.35	31.2	11.7	31.88		75
Methyl methacrylate	70	3.40	31.32	12.0	32.00		75
Xylene	70	4.10	30.96	11.4	31.64		75
Benzene	70	4.10	30.96	11.4	31.64		75
Diethyl oxalate	70	3.60	30.23	8.8	30.91		75
Cyclohexanol	70	4.70	29.81	8.3	30.49		75
γ-Butyrolactone	70	3.80	30.95	11.5	31.64		75
1,4-Dioxan	70	3.4	31.27	11.9	31.95		75
1,2-Dichlorobenzene	70	4.48	31.31	12.5	31.99		75

[a] Values calculated from A and E_a. $k_D = Ae^{-E_a/RT}$

solvents have been reported by a number of workers and some of these data are quoted in Table 1. Representative values of kinetic parameters for some azo-initiators are given in Table 2. Data on additional compounds may be found in other compilations [94, 95]; a number of the azo-derivatives not included in Table 2 decompose rapidly and are associated with low activation energies for dissociation, but often produce highly stabilized radicals which may be susceptible to primary-radical termination. Applications of a number of more unusual azo-initiators have been described elsewhere [95].

The data referred to so far relate, of course, to the decomposition of the initiator and not directly to initiation. Bawn and Mellish [69] found, by using diphenylpicrylhydrazil as a scavenger, that the extent of geminate recombination in carbon tetrachloride is higher than in benzene, hence the rate of formation of useful radicals is less. Efficiencies of

TABLE 2
Kinetic parameters for the thermal decomposition of various azo-initiators

Initiator	Solvent	Temp. (°C)	$10^5 k_d$ (sec^{-1})	E_a (kcal $mole^{-1}$)	Ref.
2-Cyano-2-propyl-azo-formamide	Toluene, xylene	100	2.1	34.5	77
Triazobenzene	Nitrobenzene	105	0.035	33.5	78
	Tetralin	105	0.040	32.5	78
Azobisisobutyramide	Dimethyl sulphoxide	70	0.368		79
Azobisisobutyronitrile	Various common solvents[a]	50	0.20	30.8	71
		60	0.83		
		70	3.36		
		80	12.3		
1,1′-Azobis-1-cyclobutane nitrile	Mesitylene	130.4	5.20	32.1	80
	Xylene	120.4	2.05		80
2,2′-Azobis-2-methyl butyronitrile	Toluene	69.8	2.3	31	64
	Xylene	80	9.97	29.4	63
2,2′-Azobis-methyl-2-methyl propionate	Xylene	80	10.9	35.8	63
2,2′-Azobis-1-carbomethoxy-3-methyl propane	Benzene	36	0.03		81
	Carbon tetrachloride	55	0.38		81
4,4′-Azobis-4-cyano-pentanoic acid	Water	80	8.97	34	63
1,1′-Azobis-cyclopentane nitrile	Toluene	80.3	7.45	33.8	80
2,2′-Azobis-2-cyclopropyl-propionitrile					
(M. pt. 64—65)[b]	Toluene	49.5	7.53	28	82
(M. pt. 76—77)[b]	Toluene	49.5	8.17	26	82
2,2′-Azobis-2,3-dimethyl-butyronitrile	Toluene	69.8	2.6	32	64
2,2′-Azobis-2-methyl valeronitrile	Toluene	69.8	4.2	33	64
2,2′-Azobis-2-cyclobutyl propionitrile					
(M. pt. 38—42)[b]	Toluene	80.5	15.1		83
(M. pt. 82 ± 1)[b]	Toluene	80.5	15.1		83
1,1′-Azobis-1-cyclohexane nitrile	Toluene	80.3	0.65	29.0	80

TABLE 2—*continued*

Initiator	Solvent	Temp. (°C)	$10^5 k_d$ (sec^{-1})	E_a (kcal $mole^{-1}$)	Ref.
2,2′-Azobis-2,3,3-trimethyl butyronitrile					
(M. pt. 114—116)[b]	Toluene	79.9	7.42	35	82
(M. pt. 116—118)[b]	Toluene	79.7	10.5	30	82
2,2′-Azobis-2-methyl-hexylonitrile	Toluene	80.2	15.8		64
2,2′-Azobis-2,4-dimethyl valeronitrile	Toluene	69.8	19.8	29	64
(M. pt. 56—57)[b]	Toluene	69.9	28.9	29	82
(M. pt. 74—76)[b]	Toluene	69.8	19.8	29	82
2,2′-Azobis-2-isopropyl butyronitrile	Toluene	80.5	10.1		84
1,1′-Azobis-1-cyclo-heptanenitrile	Acetic acid	69.5	32.8		80
	Toluene	48.9	2.69	27.5	80
2,2′-Azobis-2-methyl-heptoylnitrile	Toluene	80	16.3		84
	Xylene	80	17.8	30.2	63
1,1′-Azobis-1-(2-methylcyclohexane)-nitrile	Toluene	80.2	0.74		80
1,1′-Azobis-1-cyclo-hexanecarbonitrile	Xylene	80	0.47	39.9	63
2,2′-Azobis-2-cyclopentyl-propionitrile					
(M. pt. 72.2—74.5)[b]	Toluene	80.5	13.1		83
(M. pt. 96.3—97.6)[b]	Toluene	80.5	13.0		83
2,2′-Azobis-2,4,4-trimethyl valeronitrile					
(M. pt. 67.5—69)[b]	Toluene	40	11.8	27	84
(M. pt. 94.5—95.5)[b]	Toluene	40	6.95	29	84
2,2′-Azobis-2-isopropyl-3-methyl butyronitrile	Toluene	80.5	13.3		84
1,1′-Azobis-1-phenylethane	Dodecane	97.3	3.18		73
	Ethylbenzene, toluene	100.4	5.45	32.6	85, 86
	N-methyl-*N*-benzyl-aniline	97.3	4.14		73
	N-methyl propion-amide	97.3	3.69		73
	Diphenyl methane	97.3	4.00		73
	Propylene carbonate	97.3	3.29		73

TABLE 2—*continued*

Initiator	Solvent	Temp. (°C)	$10^5 k_d$ (sec^{-1})	E_a (kcal $mole^{-1}$)	Ref.
3,7-Diphenyl-1,2-diazo-1-cycloheptane	Xylene	80	42.6		87
2,2′-Azobis-2-cyclohexyl propionitrile	Toluene	80.5	22.7		83
1,1′-Azobis-1-cyclo-octanenitrile	Toluene	36.6	5.35	25.9	80
1,1′-Azobis-1-(4-methoxyphenyl)ethane	Ethyl benzene	100.4	7.15	35.8	86
1,1′-Azobis-1-phenyl-propane	Ethyl benzene	100.4	2.35	32.3	86
1,1′-Azobis-1-(4-tolyl)-ethane	Ethyl benzene	100.4	5.9	32.9	86
Phenyl-azo-diphenylmethane	Decalin	124.5	3.44	34.0	88
2,2′-Azobis-2-benzyl-propionitrile	Toluene	80	11.6		89
2,2′-Azobis-2-isobutyl-4-methyl valeronitrile	Toluene	60.1	37.8		84
		80.5	528		84
2,2′-Azobis-2-(4-chloro-benzyl) propionitrile	Toluene	80	8.8		89
2,2′-Azobis-2-(4-nitrobenzyl) propionitrile	Toluene	80	10		89
1,1′-Azobis-1-cyclodecane-nitrile	Toluene	50.8	5.4		80
		60.2	17.0		80
		69.5	59.9		80
1,1′-Azobis-1-phenyl-methylbutane	Ethyl benzene	100.4	7.6	33.3	86
Phenyl-azo-triphenyl-methane	Acetic acid	43.3	5.7	28	90
	Anisole	25.0	0.30	28.28	91
	Toluene	43.8	6.3	27.6	92
	Toluene	45.45	8.48	29.3	93
	Toluene	43.30	6.0	27	90
	Toluene	53.3	22.5	27.0	88

[a] Values calculated from $k_d = A \exp(-E_a/RT)$.
[b] Isomers, identified by their melting or decomposition points.

initiation by AIBN have been assessed a number of times [35, 72, 96–98]. According to Betts et al. [72] the maximum efficiency obtainable with AIBN is 0.73 in *N,N*-dimethylformamide solutions. Bamford et al. [98] have suggested that if the monomer concentration is not too low the efficiencies are probably independent of the nature of the monomer.

Although the primary radical from AIBN is usually written as $(CH_3)_2\dot{C}CN$ some of the radicals react as $(CH_3)_2C{:}CN\cdot$ [66].

b. Photoinitiation

Photolysis of azo-compounds constitutes a convenient and fairly efficient source of initiating radicals at ambient temperatures or below, when the rate of initiation is proportional to the light intensity. A detailed discussion of the photolysis of these compounds has been presented elsewhere [99]. As in thermal initiation AIBN is the most commonly used initiator, usually with light of wavelength 3650 A. Under these conditions photodissociation appears to produce the same primary radicals as thermal initiation [100], with a quantum yield for substrate disappearance of 0.47. Radicals produced by UV irradiation of a number of azo-compounds, and identified by ESR [101], indicate that photolysis occurs according to eqn. (16a). Using 3650 A radiation there is evidence to suggest that the rates of photoinitiation are essentially independent of the nature of the monomer; the same relative rates of initiation being obtained with AIBN and with the manganese carbonyl—ethyl trichloracetate system (λ = 4358 A), under standard conditions, for about ten monomers [102].

Prolonged standing of AIBN solutions in air produces photosensitive products which produce kinetic complications [103].

c. Complex formation

Rates of decomposition of azo-compounds are generally uninfluenced by the presence of metal salts. However, the accelerated decomposition of AIBN by organoaluminium compounds [104] and by silver perchlorate [105, 106] have been reported. In the latter case kinetic data suggested the existence of a 1 : 1 $AgClO_4$—AIBN complex [105], which dissociates four times as rapidly (as judged by nitrogen evolution) and initiates polymerization four times as rapidly as AIBN alone; no unusual kinetic features were observed. The postulated complex was subsequently isolated and identified [106]. Complexing appears to involve interaction between the silver ion and the nitrile groups, thus forming a cyclic complex introducing strain into the azo-linkage. This is consistent with the decrease of 2.7 kcal mole^{-1} in the activation energy for decomposition on complex formation [106]. The frequency factor for complex dissociation is lower than that for AIBN by a factor of 12.5, possibly as a result of a variation in the degrees of freedom available in the transition state. Complex formation involving the nitrile groups has also been suggested to account for the enhanced rate of decomposition in the presence of organo-aluminium compounds [104].

3.5.2 Peroxy initiators

Decompositions of peroxy compounds have been reviewed a number of times [107—110]. The first step in the thermal decomposition of peroxy compounds is the simple scission of the weak O—O linkage, according to reaction (16b). Following this primary process a number of secondary processes can occur in some cases. Many of the kinetic complications arise from induced decomposition, i.e. from the interaction of a radical with the peroxide, according to eqn. (21).

$$R'\cdot + RO{-}OR \longrightarrow R'OR + RO\cdot \qquad (21)$$

The $R'\cdot$ radicals may, in principle, be any radical present in the reaction mixture and formed by any process such as transfer or secondary dissociation of primary radicals. In the presence of induced decomposition the rate of disappearance of the original peroxide is greater than the thermal rate of scission of the peroxide linkage and greater than the rate of radical formation. In polymerization systems, however, the peroxides decompose in the presence of a radical scavenger, the monomer. The propagating radicals of many monomers are stabilized to some extent and are not sufficiently reactive to induce decomposition; reaction (21) could still occur, however, if the propagating radicals are very reactive. Thus, measurements of rates of decomposition in the presence of monomers, or other suitable scavengers, give more reliable information on the rates of spontaneous decomposition than results obtained in the absence of scavengers. This is amply demonstrated by the work of Swain et al. [111] who studied the rates of decomposition of benzoyl peroxide in various solvents at 80 °C. In the absence of a scavenger widely different results were obtained, but in the presence of styrene (1 mole l^{-1}) the rates of decomposition tended to converge to a common value. This is supported by the results of various workers who have found agreement in the rates of spontaneous decomposition in various solvents. Kinetic data for the dissociation of some peroxides are given in Table 3, a much more complete compilation is given in ref. 112. Recently Bains et al. [149] have discussed the thermal decomposition of and initiation by peroxy-carbamates and have listed known kinetic parameters.

We have already referred to secondary dissociation of primary radicals and quoted the example of benzoyl peroxide in which the primary $C_6H_5COO\cdot$ radical dissociates into phenyl radicals and carbon dioxide according to reaction (18b); either radical may initiate polymerization. Bevington and Toole [45] estimated the activation energy for this dissociation as ~12 kcal $mole^{-1}$.

TABLE 3
Kinetic data for the decomposition of some peroxides

Peroxide	Solvent	Temp. (°C)	$10^5 k_d$ (sec^{-1})	E_a (kcal $mole^{-1}$)	Ref.
Acetyl peroxide	Benzene	70	2.38		113
		80	8.7		114
	Carbon tetrachloride	80	5.5		114
	Cyclohexene	80	7.0	31.9	114
	Cyclohexene	80	7.0	32.8	114
	Cumene	80	7.6		114
	1-Hexene	80	8.7	31.7	114
	2-Methyl-1-pentene	80	9.0	30.3	114
Ethyl peroxide	Styrene	60	1.2×10^{-4}	35.2	115
Tert-butyl peracetate	Benzene	100	1.54	36.3	116
Propionyl peroxide	*n*-Hexane	65.0	1.50	29.6	90
	Toluene	65.0	1.87	31.1	90
Tert-butyl peroxide	Benzene(+styrene)	130	2.48—3.04	34	117
	Benzene	120	1.39		118
		130	3.00	35.1	116
	Tri-*n*-butylamine	125	1.7 ± 0.3		119
		120	0.77[a]	35.0	120
		130	2.07[a]		120
	Tert-butyl benzene	125	1.5 ± 0.2		119
		120	0.77[a]	35.0	120
		130	2.07[a]		120
	Carbon tetrachloride	120	0.9		118
	Cumene	125	1.6 ± 0.1		119
		120	0.77[a]	35.0	120
		130	2.07[a]		120
	Cyclohexene	120	0.83		118
	Methyl methacrylate	120	0.77[a]	35.0	120
		130	2.07[a]		120
	Styrene	120	0.77[a]	35.0	120
		130	2.07[a]		120
	Toluene	120	1.34		118
1-Hydroxy-*n*-butyl peroxide	α-Methyl styrene	79.4	1.7	24.6	121

TABLE 3—*continued*

Peroxide	Solvent	Temp. (°C)	$10^5 k_d$ (sec^{-1})	E_a (kcal $mole^{-1}$)	Ref.
Tert-butyl perisobutyrate	Bulk	70	4.12		122
	Benzene	70	6.69	33.6	116
		78	3.77		122
Tert-butylperoxy-isopropyl carbonate	Benzene	100	2.21		123
		110	6.87		123
Ethyl-*tert*-butyl peroxalate	Benzene	55	16.3	26.9[b]	124
Di-isopropylperoxy dicarbonate	Benzene(+styrene)	54.0	5.0		125
	Ethylbenzene (+styrene)	54.3	5.2		125
	2,2′-Oxydiethylene-bis(allyl carbonate)	50	2.28	28.1	126
	Toluene	50	3.03		126
Cumene hydroperoxide	Mesitylene			23.6	127
	Styrene			24.2	127
Pinane hydroperoxide	Benzene	145	2.72	29.6	128
1-Phenyl-2-methyl-propyl-1-hydro-peroxide	Benzene	153.9	2.0	29.2[b]	129
1-Phenyl-2-methyl-propyl-2-hydro-peroxide	Benzene	154.5	1.21	30[b]	129
Di(*tert*-butylperoxy) oxalate	Benzene	35.0	6.77	25.5[b]	130
Bis(2-nitro-2-methylpropyl) peroxybicarbonate	Toluene	50	2.22		126
Tert-butyl perbenzoate	Benzene	100	1.07	34.7	116
		110	3.50	34.5	131
	n-Butyl acetate (+styrene)	110	3.61	35.5	131
	Chlorobenzene	110	3.83	33.9	131
	4-Chlorotoluene	110	3.25	34.5	132
	Phenylether	110.1	2.28	37.5	89
Tert-butyl-*N*-(4-bromophenylperoxy) carbamate	Toluene	70.8	3.32	30.5[b]	133
Tert-butyl-4-chloro-perbenzoate	Phenylether	110.1	1.85	39.3	89

References pp. 92—103

TABLE 3—*continued*

Peroxide	Solvent	Temp. (°C)	$10^5 k_d$ (sec^{-1})	E_a (kcal $mole^{-1}$)	Ref.
Tert-butyl-*N*-(2-chlorophenyl peroxy)carbamate	Toluene	87.0	3.5	37.6[b]	133
Tert-butyl-*N*-(3-chlorophenyl peroxy)carbamate	Toluene	78.0	2.6	27.6[b]	133
Tert-butyl-*N*-(4-chlorophenyl peroxy)carbamate	Toluene	73.0	5.57	28.6[b]	133
Tert-butyl-*N*-(2,5-dichlorophenyl peroxy)carbamate	Toluene	88.0	2.75	30.8[b]	133
Tert-butyl-2-iodo-perbenzoate	Chlorobenzene	85.0	4.02 ± 0.04		134
		102.4	2.58 ± 0.02		134
Tert-butyl-4-nitro-perbenzoate	Phenylether	120.2	3.19	41.3	89
Tert-butyl-*N*-(3-nitrophenylperoxy)-carbamate	Toluene	88.7	2.43	31.8[b]	133
Tert-butyl-*N*-(4-nitrophenylperoxy)-carbamate	Toluene	87.0	3.11	27.2[b]	133
Tert-butyl-*N*-(phenylperoxy)-carbamate	Toluene	67.7	2.31	33.4[b]	133
Cyclobutaneacetyl-peroxide	Carbon tetrachloride	65	1.37		135
		75	3.83		135
Cyclopentaneformyl-peroxide	Carbon tetrachloride	40	1.50		135
		50	4.96		135
Tert-butylphenyl-peracetate	Chlorobenzene	60	0.68	28.7[b]	136
		77.0	6.85	28.1[b]	137
Tert-butyl-α-cumyl peroxide	Dodecane	128	4.44	35	138
Benzoyl peroxide	Acetone	70	2.63	26.6	116
	Acetonitrile	70	1.76		139
	Acetophenone	80	4.32	30.2	140
	Benzene	80	4.39	32	141
		75.8	1.19	29.6	69
	Benzene(+styrene)	70	1.03		113
		70	1.18		139
		80	2.5	29.7	45

TABLE 3—*continued*

Peroxide	Solvent	Temp. (°C)	$10^5 k_d$ (sec^{-1})	E_a (kcal $mole^{-1}$)	Ref.
	Benzene(+methyl methacrylate)	75	1.66	29.7	142
		79	2.58		143
		80	3.35		144
	Benzyl alcohol	80	44.4		145
	Butanol	80	60.6		145
	Butanone	80	4.64		145
	Carbon tetrachloride	75	1.07		145
		79	1.69		146
	Chlorobenzene	70	1.35		139
		80	4.64		145
	Cumene	80	3.69		145
		80	1.70	28.8	147
	Dioxan	70	1.30		139
	Ethylbenzene	80	3.33		145
	Methyl cyclohexane	80	5.25		145
	Nitrobenzene	80	4.58		145
	Styrene	74.8	1.83		148
		100.0	4.58		148
	Toluene	70.3	1.10	29.6	69
3-Bromobenzoyl peroxide	Benzene	80	1.22		54
		80	2.60		144
	Dioxan(+3,4-dichlorostyrene)	80	2.57		111
2-Chlorobenzoylperoxide	Acetophenone	80	38.8	29.4	140
	Benzene	80	31.2		144
3-Chlorobenzoylperoxide	Acetophenone	80	2.85	30.7	140
	Dioxan(+3,4-dichlorostyrene)	80	2.63		111
4-Chlorobenzoylperoxide	Acetophenone	80	3.83	30.4	140
	Benzene	80	2.17		144
		85	6.64	30.8	116
	Dioxan(+3,4-dichlorostyrene)	80	3.62		111
	Styrene	74.8	2.22		148
		100	4.17		148
2,4-Dichlorobenzoyl peroxide	Benzene	50	1.08	28.1	116
	Styrene	49.4	2.39		148
		61.0	7.78		148

References pp. 92—103

TABLE 3—*continued*

Peroxide	Solvent	Temp. (°C)	$10^5 k_d$ (sec^{-1})	E_a (kcal mole^{-1})	Ref.
Cumyl peroxide	Benzene	115	1.56	40.7	116
Lauroyl peroxide	Benzene	60	1.51	30.4	116
		60	0.92		81
		70	4.33		113
	Carbon tetrachloride	70	1.87		81
	Ethyl acetate	70	3.99		81
	Styrene	61.0	1.42		148
		74.8	10.0		148

[a] Calculated from $A \exp(-E/RT)$.
[b] $\Delta H^{\ddagger}$.

As another example of dissociation of primary radicals we quote the case of di-*tert*-butyl peroxide, which is a useful initiator at temperatures about and above 100 °C. This peroxide dissociates to yield *tert*-butoxy radicals which dissociate according to the reaction

$$(CH_3)_3CO\cdot \longrightarrow (CH_3)_2CO + \dot{C}H_3$$

for which process an isotope effect has been reported [117]. *Tert*-butoxy radicals are active in transfer reactions, but if the temperature is not too high and the monomer concentration not too low initiation is mainly by the primary radicals [117].

Although not widely used, photolytic decomposition of peroxy compounds can lead to initiating radicals. However, the wavelengths necessary to dissociate many of the common peroxides may also cause photolysis of the monomer. It is possible for photolysis to lead to primary radicals different to those produced by thermal decomposition. Photodissociation of benzoyl peroxide in the near-UV produces some phenyl radicals as the primary species [150] and, in the photodecomposition of *tert*-butyl peroxide, Frey [151] has reported that the nature of the primary radicals is dependent on the wavelength of the incident radiation.

Sulphur analogues of organic peroxides can be used as initiators. These initiators can lead to complex overall polymerization kinetics as a result of simultaneous initiation and retardation [152]; retardation occurs through degradative chain transfer. Ferrington and Tobolsky [152] have separated the initiation and transfer reactions for tetramethylthiouram disulphide and obtained first-order rate coefficients for initiation of methyl methacrylate and styrene polymerizations at 70 °C of 5.45×10^{-6} sec^{-1} and 3.34×10^{-6} sec^{-1}, respectively.

3.5.3 *Redox reactions*

Redox systems of various types are very important as free-radical initiators. While homolytic decomposition of a peroxy- or azo-initiator must produce radicals in pairs redox reactions, which proceed by one-electron transfer processes, can produce radicals singly. Radical formation, in general, may accompany transfer of an electron to or from an ionic species or a stable molecule, in the latter case with the formation of an ionic and a radical species, or a radical-ion.

a. Redox reactions based on peroxy compounds

Activation energies for redox reactions are of the order of 10 kcal $mole^{-1}$. These reactions often allow high rates of radical formation from peroxides at temperatures below those at which thermal decomposition takes place at a convenient rate. Since a large proportion of redox systems involve ionic species they constitute a useful source of radicals in aqueous solution. Consequently, redox initiators are widely used in emulsion polymerizations and the patent literature contains numerous references to such initiators, often in the form of complex recipes containing additives which may affect the course of the polymerization or the properties of the polymer, e.g. colour. As a result of the complexities of the systems detailed mechanisms of radical formation are often unknown. The nature of the solvation shell of the ionic species, which is dependent on the reaction conditions, may influence the mechanism of radical formation. Complications may also arise since many of the reactions involve ions which are capable of terminating polymerization.

The most well known redox initiator is the ferrous ion—hydrogen peroxide system [153—155]. The first step in the reaction may be written [154]

$$Fe^{2+} + H_2O_2 \longrightarrow Fe^{3+} + OH^- + HO\cdot \qquad (22)$$

or, more correctly

$$Fe^{2+} + H_2O_2 \longrightarrow FeOH^{2+} + HO\cdot$$

with a rate coefficient [154] of $4.45 \times 10^8 \exp(-9400/RT)$ sec^{-1}. The initiating species is the hydroxyl radical [155—157], but in the absence of a suitable radical scavenger this radical may be further reduced to hydroxyl ion by ferrous, or it may enter into a chain reaction with hydrogen peroxide [153]. The ferric ions may be reduced by hydrogen peroxide, according to the reaction [154]

$$Fe^{3+} + H_2O_2 \longrightarrow Fe^{2+} + H^+ + HO_2\cdot$$

References pp. 92—103

Other metals [153], e.g. Ti^{3+} [158], may also partake in reactions analogous to reaction (22), and initiation by hydrogen peroxide and thiourea has been reported [159]. The nature of the initiating radical may be varied by carrying out the reaction in a suitable medium, thus, in the presence of methanol, the initiating radical may be $HOCH_2$ [160].

Peroxydisulphate is a common component in redox systems. The peroxydisulphate ion is reduced by ferrous according to the reaction [47, 161—163]

$$Fe^{2+} + S_2O_8^{2-} \longrightarrow Fe^{3+} + SO_4^{2-} + SO_4^{-}\cdot$$

the sulphate radical-ion is the usual initiating species. This radical-ion may be further reduced to sulphate ion by ferrous [161]

$$Fe^{2+} + SO_4^{-}\cdot \longrightarrow Fe^{3+} + SO_4^{2-}$$

Other metal ions will also dissociate peroxydisulphate to generate initiating radicals. For reduction by silver(I) it has been suggested [165, 166] that the reaction is

$$Ag^{+} + S_2O_8^{2-} \longrightarrow Ag^{2+} + SO_4^{2-} + SO_4^{-}\cdot$$

followed by oxidation of water by silver(II), viz.

$$Ag^{2+} + H_2O \longrightarrow Ag^{+} + H^{+} + HO\cdot$$

while other workers [166—169] have suggested that the first step involves the oxidation of silver(I) to silver(III), with the formation of sulphate ions, and subsequent oxidation of water by silver(III) to produce silver(I) and hydroxyl radicals. End-group studies have indicated [170] that both SO_4^{-} and OH end-groups are incorporated into the polymer.

Polymerization can also be initiated by peroxydisulphate in combination with metal chelates [171] and various reducing agents including sulphite [168], thiosulphate [168], bisulphite [168, 172] and metabisulphite [168, 173, 174]. In these systems the rate of radical formation may be enhanced by the addition of traces of metal ions [167, 168, 175]. The details of the reaction mechanisms and rates in these systems are somewhat uncertain. The chemistry of peroxydisulphate reactions has been reviewed [176], and a number of end-group studies to determine the nature of the initiating radicals have been carried out [177]. The spontaneous decomposition of peroxydisulphate in aqueous solution leads to hydroxyl radical initiation in alkaline solution [176], but the mechanism of initiation in acid solution is unclear [178].

Diacyl peroxides, such as benzoyl and acetyl peroxides, are susceptible to reduction by metal ions, e.g. Fe^{2+}, presumably according to the reaction [179]

$$Fe^{2+} + (RCOO)_2 \longrightarrow Fe^{3+} + RCOO^{-} + RCOO\cdot$$

Rates of polymerization are accelerated by the addition of heavy metal salts or reducing agents [180—182]. Alkylaluminium compounds also induce decomposition of these peroxides to yield initiating radicals [183].

The reduction of hydroperoxides by ferrous ion has received some detailed investigation. Initiating radicals are produced according to the reaction

$$ROOH + Fe^{2+} \longrightarrow Fe^{3+} + OH^- + RO\cdot \qquad (23)$$

Among the hydroperoxides investigated are those of cumene [162], *p*-menthane [184] and *p*-isopropyl cumene [185]. Rate coefficients for a number of such reactions have been tabulated [186] and efficiencies of chain starting have been discussed [47]. In practical systems it is common practice to maintain the ferrous ion concentration and rate of polymerization by the addition of agents, such as sugars or sodium formaldehyde sulphoxylate, which reduce the ferric ions formed in the initiation reaction. With alkaline solutions complexing agents such as disodium ethylenediaminetetracetate may be added. The literature on such systems is extensive [187], but little detailed information is available. The decomposition of *tert*-butyl hydroperoxide by Cu^+ and Cu^{2+} has also been described [188].

While other reducing metal ions, such as Co^{2+}, decompose hydroperoxides by reactions analogous to eqn. (23), oxidizing metal ions, e.g. Co^{3+}, react according to [189]

$$Co^{3+} + ROOH \longrightarrow Co^{2+} + H^+ + ROO\cdot$$

Organic peroxides can partake in wholly organic redox processes in non-aqueous media. The most well-known of these are the peroxide—amine systems. With benzoyl peroxide and *N,N*-dimethyl aniline the first step appears to be the formation of a 1:1 complex (I) [190, 191], viz.

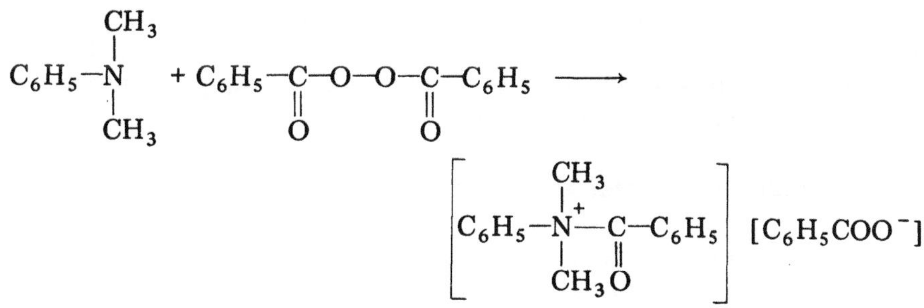

(I)

which dissociates to produce a benzoyloxy radical

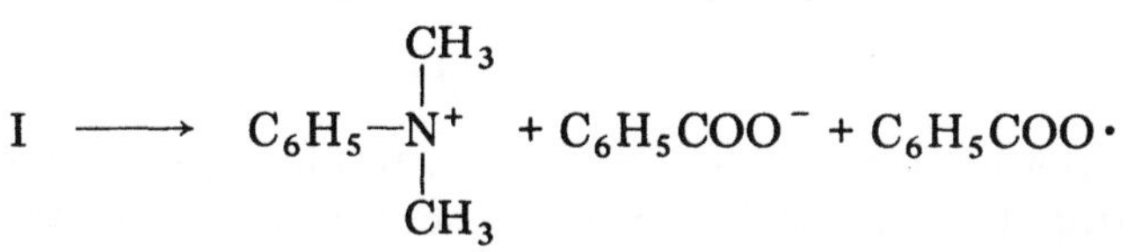

which in turn initiates polymerization. Rates of polymerization are approximately proportional to the square root of the product of the concentrations of the initiator components [59, 60]. Rate coefficients and efficiencies of initiation have been determined by Walling and Indictor [192] and by O'Driscoll et al. [59, 60]; the latter workers used the dead-end polymerization technique. A number of organic reducing agents, including sulphinic acids, α-ketols, formic acid, thiols and hydrazines, have been reported to decompose benzoyl peroxide with the formation of free radicals [191, 193–195]. It has been demonstrated [191, 193, 196–198] that other peroxides (diacyl and hydroperoxides) are susceptible to this kind of attack and that the rates of radical formation are accelerated by the addition of soluble metal derivatives [191, 199, 200].

b. Analogues of peroxy compounds

Nitrogen and sulphur analogues of peroxides undergo similar reactions to some of those described above. Thus, Kolthoff et al. [201] suggested that the mechanism for radical formation in the disulphide—ferrous citrate complex system is

$$Fe^{2+}(\text{complex}) + RSSR \longrightarrow Fe^{3+}(\text{complex}) + RS^- + RS\cdot$$

Strongly reducing metal ions, e.g. Ti^{3+}, have been shown, by Davies et al. [158], to produce $\cdot NH_2$ radicals from hydroxylamine in acid solution by the reaction

$$Ti^{3+} + NH_3OH^+ \longrightarrow Ti^{4+} + H_2O + H_2N\cdot$$

Photolysis of hydrazine, to produce $\cdot NH_2$ radicals, requires wavelengths less than 2600 A [153]. It is possible to achieve initiation from hydrazine in the dark [153], when it appears that initiation, both in the presence and absence of metal ions, proceeds through $H_3N_2\cdot$ radicals [202, 203]. It has also been reported that in the presence of oxygen initiation occurs partially through $HO\cdot$ radicals [204].

3.5.4 *Initiation by metal salts in aqueous solution*

Strong oxidizing metal ions can generate radicals by the thermal decomposition of water [205], e.g.

$$Co^{3+}OH^- \longrightarrow Co^{2+} + HO\cdot$$

or other substrates, e.g.

$$Co^{3+} + HCOOH \longrightarrow Co^{2+} + H^+ + HCOO\cdot$$

Acidic solutions of permanganate have been reported to initiate polymerization by generating $HO\cdot$ radicals through the formation of Mn^{4+} as intermediate [157], viz.

$$Mn^{4+} + H_2O \longrightarrow Mn^{3+} + H^+ + HO\cdot$$

$$Mn^{3+} + H_2O \longrightarrow Mn^{2+} + H^+ + HO\cdot$$

The range of metal ions capable of undergoing such thermal reactions is limited but may be increased by photoactivation. The photolytic decomposition of water in the hydration shells of both cations (Me^{n+}) and anions (A^{n-}) have been studied by Dainton and James [206], who suggested initiation occurred through hydrogen atoms produced in the reactions

$$Me^{n+}\cdot H_2O \xrightarrow{h\nu} Me^{(n+1)+} + OH^- + H\cdot$$

$$A^{n-}\cdot H_2O \xrightarrow{h\nu} A^{(n-1)-} + OH^- + H\cdot$$

$$A^{n-}\cdot H_3O^+ \xrightarrow{h\nu} A^{(n-1)-} + H_2O + H\cdot$$

In the presence of traces of oxygen it is possible that some hydroxyl radicals are produced as primary species [207] when Me^{n+} is cuprous. Mitra et al. [207] have reported that, while cuprous ions will initiate the polymerization of methyl and ethyl methacrylates, methacrylic acid and styrene, they will not initiate the polymerizations of vinyl acetate, acrylonitrile or methyl and ethyl acrylates. Failure to effect initiation in the latter cases was thought to be due to complex formation between the monomer and cuprous.

Numerous other ion pairs have been reported to be photoinitiators [206, 208–212]; typical examples include $Fe^{3+}OH^-$, $Fe^{3+}Cl^-$, $Fe^{3+}F^-$, $Fe^{3+}N_3^-$, $Fe^{3+}C_2O_4^{2-}$, $Fe^{3+}HCitrate^{2-}$, $Pb^{2+}Cl^-$, $Ce^{4+}OH^-$, $[Co(H_2O)_6]$-$(NO_3)_3$. In many of these cases initiation is considered to involve charge transfer to ligand or solvent, e.g.

$$Fe^{3+}X^- \xrightarrow{h\nu} Fe^{2+}X\cdot \longrightarrow Fe^{2+} + X\cdot$$

where $X\cdot$ is the initiating radical. It is probable that in a number of these systems, in aqueous solution, initiation occurs through hydroxyl radicals [213].

3.5.5 *Metal chelates*

A group of initiators, which have been studied recently, are the metal chelates, especially metal acetylacetonates. Arnett and Mendelsohn [214], during a study of chelate decompositions, observed that certain metal chelates initiate the polymerization of styrene. These workers found that the initiating ability of acetylacetonates was in the order Mn(III) > Co(III) ≈ Ce(IV) > Fe(III) while the Cr(III) and Al(III) compounds were inactive and the V(III) derivative was an inhibitor. It was suggested [215] that radical formation involved electron transfer to the metal and loss of a ligand as a radical, in the form of a cyclic peroxide in equilibrium with the more conventional type of radical $(CH_3CO)_2CH$, which could also exist in a resonance form with the unpaired spin on the oxygen atoms.

Kastning et al. [216] demonstrated that the rates of polymerization were proportional to $[\text{chelate}]^{0.5}$, indicating a free-radical mechanism, and claimed that various additives, such as organic halides, acted as activators for radical formation. This latter point has led to some discrepancies in the literature. While some activation has been observed by addition of carbon tetrachloride to systems initiated by Cu(II)acetylacetonate [217], activation by carbon tetrachloride or ammonium trichloracetate in the presence of Mn(III)acetylacetonate was not confirmed [218]. High rates of radical formation in the system Cu(II)acetylacetonate—ammonium trichloracetate appears to be a special case [219] and will be referred to later.

Although the mechanism of radical formation suggested by Arnett and Mendelsohn does not involve participation of any solvent species, it has been shown that the rates of polymerization of methyl methacrylate and styrene initiated by Mn(III)acetylacetonate [218, 220] and of methyl methacrylate initiated by Cu(II)ethylacetoacetate [221], in inert solvents such as benzene, are proportional to $[\text{monomer}]^{1.5}[\text{chelate}]^{0.5}$, indicating that rates of initiation are proportional to [monomer]. Itakura et al. [220] suggested that initiation involved co-ordination of monomer to the chelate while Bamford and Lind [218(a)] suggested that the first step is the formation of an activated species, having a weakened Mn—O bond, followed by reaction with monomer, which facilitates electron transfer and the formation of a free radical and an Mn(II) compound, as in eqn. (24a, b). Bamford and Lind also demonstrated that in the presence of ethyl acetate the order in monomer decreased and suggested that solvents such as ethyl acetate could also be involved in radical formation according to eqn. (24c). Decreases in overall order in monomer (and in activation energy) for polymerization in the presence of donors such as cyclohexanone and caproic aldehyde have been reported by Kaeriyama [222], whose results are also commensurate with participation of monomer and solvent in initiation. With Cu(II)ethylacetoacetate and styrene it has been reported

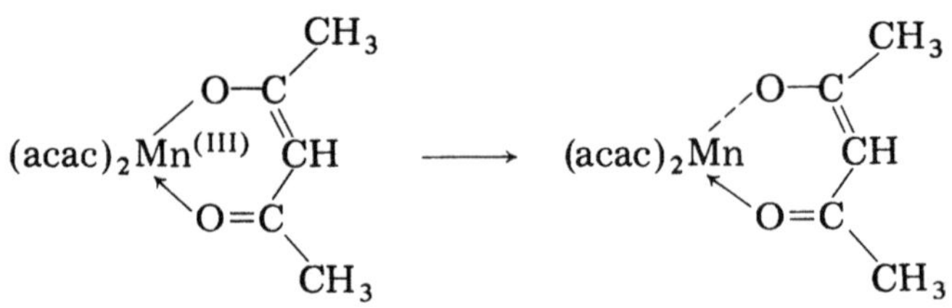

(II) (24a)

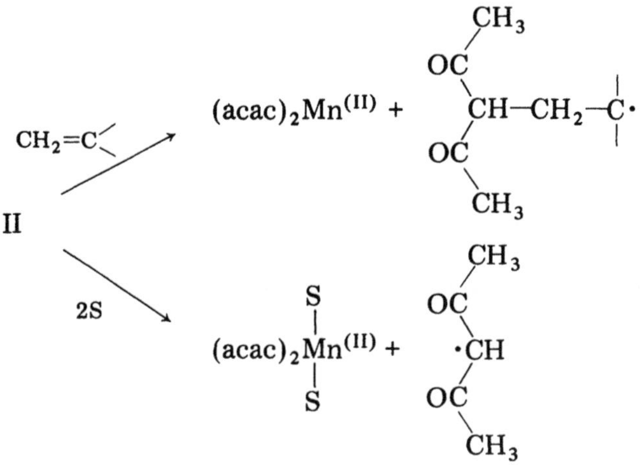

(24b)

(24c)

that the rate of polymerization is proportional to [monomer]^n·[chelate]$^{0.5}$, where n varies from 1.3 to 2.4 depending on the solvent used, and participation of two monomer molecules in initiation has been proposed [223].

The effects of the substituents on the acetylacetonate ligand have been investigated to some extent. Some comparative data on rates of polymerization have been presented for various derivatives of Cu(II)acetylacetonate [223, 224]. Reported data [218(a)] for Mn(III) chelates are collected in Table 4, where k is the rate coefficient for initiation in the expression

$$\mathscr{I} = k[\text{chelate}] \text{ mole l}^{-1} \text{ sec}^{-1}$$

TABLE 4
Kinetic data for initiation by Mn(III) chelates at 80°C [218]

Chelate ligand	Monomer	$10^5 \cdot k$ (sec^{-1})	E_i (kcal mole^{-1})	A_i (sec^{-1})
$(CH_3CO)_2CH$	Methyl methacrylate	5.1	26.2	6.7×10^{11}
	Styrene	5.0		
$(C_6H_5CO)_2CH$	Methyl methacrylate	3.3[a]		
	Styrene	4.3[a]		
$(CH_3CO)_2CC_6H_5$	Methyl methacrylate	15	12.9	4.6×10^4
	Styrene	14		
CF_3COCH–CH_3CO	Methyl methacrylate	38	15.8	2.9×10^6
	Styrene	Low		

[a] At low [chelate].

and E_i, A_i are the activation energies and pre-exponential factors for initiation. The most notable feature of Table 4 is the high value of k for initiation of methyl methacrylate by the trifluorinated chelate and the very low rate of initiation for styrene polymerization. In addition it was shown that the same chelate is a good initiator for acrylonitrile polymerization but not for vinyl acetate [218(a)]. Thus, the chelate is a good initiator for those monomers susceptible to anionic polymerization, although propagation is a free-radical process. It was suggested that initiation occurs through heterolytic dissociation of a Mn—O bond as in

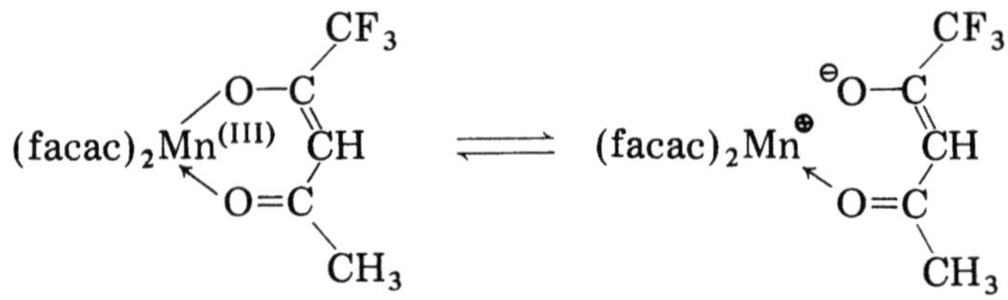

(III)

followed by the addition of monomer to the anion and subsequent electron transfer to produce a radical species, viz.

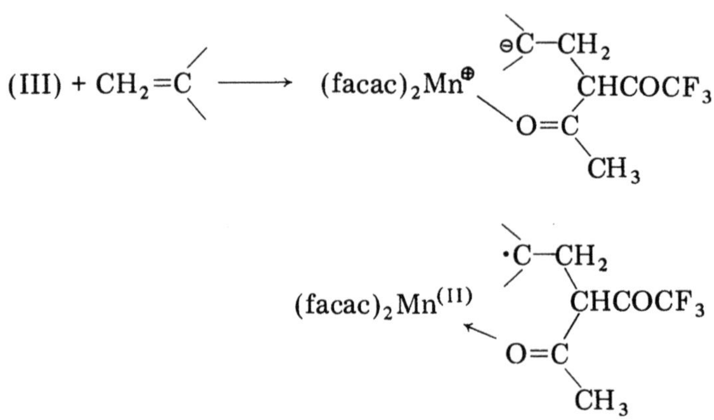

Dimethyl sulphoxide markedly increases rates of initiation of methyl methacrylate by the unsubstituted $Mn(acac)_3$, and the presence of this solvent causes the appearance of selective initiation; a mechanism similar to that proposed for $Mn(facac)_3$, involving participation of solvent, was proposed [218(b)]. Some selectivity has also been reported for polymerizations initiated by copper chelates [225].

A kinetic study of methyl methacrylate polymerization initiated by Co(III) cyclohexanecarboxylate indicates that the mechanism of initiation is similar to that found with metal chelates [226].

For the polymerization of methyl methacrylate at 80 °C initiated by copper acetylacetonate and ammonium trichloracetate it was found [219] that at low initiator concentrations ($<10^{-3}$ mole l^{-1}) the rate of polymerization increased with increasing concentration of initiator components in a manner consistent with formation of a 1 : 1 complex between chelate and ammonium salt, and initiation by decomposition of the complex. From polymerization kinetics an equilibrium constant for complex formation of 2706 $mole^{-1}$ l was calculated (confirmed by optical studies) and a first-order rate coefficient for complex decomposition of 2.18×10^{-5} sec^{-1} determined. At high ammonium trichloracetate concentrations ($>2.5 \times 10^{-2}$ mole l^{-1}) in the presence of the chelate, the rate of polymerization increased commensurate with second-order initiation by the ammonium trichloracetate and retardation by the salt.

At low initiator concentrations it was proposed [219] that initiation at 80 °C involved electron transfer in the complex from the trichloracetate anion to the Cu(II), leading to $\dot{C}Cl_3$ radicals. Ammonium trichloracetate dissociates readily into chloroform, ammonia and carbon dioxide, probably through the intermediate formation of Cl_3C^- ions. It was considered that at high [ammonium trichloracetate] the Cl_3C^- ions could undergo electron transfer to Cu(II), the resulting Cu(I) being reoxidized by abstraction of a Cl atom from the ammonium salt to form another radical, thus generating high rates of initiation at high salt concentrations. Retardation by ammonium trichloracetate was thought to involve abstraction of a CCl_3 group from the salt by the propagating radical.

More recent studies [227], at 65 °C, have indicated that the 1 : 1 complex originally proposed as the initiating species is actually a precursor of the true initiator and that the kinetics at the lower temperature are further complicated by the complex, or a derivative of it, terminating the growing chains. Thus, the details of the initiation mechanism remain uncertain.

3.5.6 The use of metals in their zero oxidation state

Freydlina and Belyavskii [228] reported that metal carbonyls in the presence of high concentrations of carbon tetrachloride (an active transfer agent) induce the telomerization of ethylene. Later, Bamford and Finch [229] demonstrated that in the presence of a vinyl monomer, and low concentrations of certain organic halides, the same metal derivatives yield high-molecular-weight polymers. Subsequently, initiating systems based on a series of metal carbonyls and related derivatives have received detailed kinetic analysis. Preferred organic halides are those containing $-CX_3$ or $-CX_2Y$ groups, where X is Cl or Br, especially if this group is adjacent to a carbonyl group (e.g. ethyl trichloracetate). In general any halide with a relatively labile halogen atom would appear to be suitable.

Metal derivatives which have been studied include the carbonyls of molybdenum [230—233], chromium [229], tungsten [230], cobalt [234], nickel [235], manganese [236, 237] and rhenium [238], hexaphenylisocyanides of molybdenum, chromium and tungsten [239], $Mo(CO)_5$ pyridine [240, 241] and various phosphine and phosphite derivatives of transition metals [242—245]. The individual metal derivatives have different reactivities and by suitable choice of the derivatives a useful series of thermal initiators covering the temperature range 0—100 °C is available. In addition, manganese and rhenium carbonyls can also be used in photoinitiation.

a. Thermal initiation

The essential features of the kinetics and mechanism of initiation using the various metal derivatives are very similar, although the details vary. In all cases the first step is to produce a reactive species from the metal derivative which then reacts with the halide to produce a complex which dissociates into a radical and an oxidized metal derivative.

We shall describe the main features of initiation by molybdenum carbonyl, which has been studied in greatest detail, as an example of a thermal initiator; details for other derivatives may be found in the original literature. The mechanism of initiation is

$$Mo(CO)_6 + M \underset{k_2}{\overset{k_1}{\rightleftharpoons}} \underset{(IV)}{M\text{---}Mo(CO)_5} + CO \tag{25a}$$

$$(IV) \xrightarrow{k_2'} \text{inactive products} \tag{25b}$$

$$(IV) + H \xrightarrow{k_3} \underset{(V)}{\begin{matrix} H \\ M \end{matrix}\!\!\!>\! Mo(CO)_4} + CO \tag{25c}$$

$$(V) \xrightarrow{k_4} R\cdot + Mo(I)\ \text{derivative} \tag{25d}$$

The first and rate determining reaction (the forward reaction in eqn. (25a)) is a ligand exchange process in which a molecule of carbon monoxide is replaced by a molecule of monomer (M), to form the more reactive complex (IV). The second essential step, (25c), is the formation of a second complex (V) from (IV) and a molecule of the halide (H), and finally complex (V) dissociates to form a radical and an oxidized molybdenum derivative in (25d).

With increasing halide concentration rates of initiation (and polymerization) increase rapidly to approach a limiting value at "high" [halide], for many halides (e.g. carbon tetrachloride, chloroform, ethyl trichloroacetate) this limit is independent of the nature of the halide. The form of the dependence is a consequence of the competition between reactions (25b) and (25c); the limiting rate being achieved when the rate of (25b) is negligible, its value is determined by the rate of reaction (25a). The halide concentration at which the limiting rate is attained depends on the nature of the halide and is ~10^{-2} mole l^{-1} for ethyl trichloracetate, ~0.1 mole l^{-1} for CCl_4 and ~1 mole l^{-1} for $CHCl_3$.

Addition of carbon monoxide reduces the rate of initiation [230] through the removal of complex (IV) in the back-reaction in eqn. (25a).

On the basis of the mechanism (25) the rate of initiation is

$$\mathcal{I} = \frac{k_1 k_3 [Mo(CO)_6][M][H]}{k_2 + k_2'[CO] + k_3[H]} \tag{26}$$

At "high" [halide] and in the absence of carbon monoxide eqn. (26) reduces to

$$\mathcal{I} = k_1 [Mo(CO)_6][M][H] \tag{27}$$

Thus, the rate of initiation is dependent on [M] and in reaction mixtures containing an inert diluent, such as benzene, the rate of polymerization is proportional to $[M]^{1.5}$. In the presence of other solvents [231] which are weak donor ligands, the overall order in [M] for polymerization lies between 1 and 1.5. This arises since a weak donor solvent, S, can take the place of M in reaction (25a) and give rise to radical formation by a mechanism identical with reaction (25) at the rate

$$\mathcal{I}_s = k_{1S} [Mo(CO)_6][S][H]$$

where k_{1S} corresponds to k_1 in eqn. (25). In the presence of such solvents rates of initiation are additive. If a strong donor solvent is present, e.g. benzonitrile, rates of initiation are affected by the strong donor removing complex (IV) to form a more stable complex which requires further activation before leading to radical formation [233]. Some kinetic parameters illustrating solvent effects are summarized in Table 5.

On the basis of mechanism (25), rates of initiation are proportional to $[Mo(CO)_6]$ and hence, rates of polymerization are proportional to $[Mo(CO)_6]^{0.5}$. This holds true at low carbonyl concentrations

TABLE 5

Kinetic parameters for molybdenum carbonyl + CCl_4 initiator at 80 °C

Solvent	$10^5 k_{1S}$ (sec^{-1})	k_3/k_2'	Ref.
Methyl methacrylate	1.3	33.3	231
Ethyl acetate	1.4	355	231
Dioxan	0.75	163	233
Acetic anhydride	1.87	230	233
Benzene	0	—	231

($<10^{-4}$ mole l^{-1}), but at higher values of $[Mo(CO)_6]$ the rate of initiation increases less rapidly than anticipated. It has been suggested that the reaction

$$(IV) + (V) \longrightarrow \text{Inactive products}$$

could account for kinetic and chemical observations [232].

Scheme (25) provides a satisfactory description of the kinetics at low $[Mo(CO)_6]$ and short reaction times, when CCl_4 is used as halide, and predicts the formation of one radical per molecule of $Mo(CO)_6$ reacted and presumably the formation of a Mo(I) derivative. Development of an ESR signal accompanies polymerization but at a slower rate, initially, than predicted by the kinetics of initiation [246]. This signal arises from a Mo(V) species produced by subsequent oxidation of the initial Mo(I) derivative [246, 247]. Kinetic studies [246] have demonstrated that the secondary oxidation with halides such as CCl_4 occurs relatively slowly and leads to the formation of two more initiating radicals. The consequence of this is that the rate of initiation does not decay with the half-life of $Mo(CO)_6$ consumption (1.93 h at 80 °C) but maintains an almost constant rate of initiation for approximately three hours. With carbon tetrabromide as the halide the secondary oxidation is rapid [248], and the development of the Mo(V) ESR signal corresponds to the decay of the carbonyl. In the absence of radical scavengers, such as monomer, the $\dot{C}Br_3$ radicals so formed oxidize the parent $Mo(CO)_6$ and lead to a chain reaction for oxidation of the carbonyl.

Schemes analogous to (25) have been used to describe initiation by other metal derivatives. Variations and unusual features encountered with some derivatives (e.g. the low rate of initiation by $Ni(CO)_4$ compared with

TABLE 6
Kinetic parameters for thermal initiation by metal carbonyls and related compounds in association with carbon tetrachloride in methyl methacrylate solution

Metal derivative	Temp. (°C)	$10^5 k_1$ (l mole^{-1} sec^{-1})	$10^2 k_2'/k_3$ (mole l^{-1})	$10^3 k_2/k_3$ [a] (mole l^{-1} torr^{-1})	Ref.
$Mo(CO)_6$	80	1.3	3	11.8	231
$Mo(CO)_5Py$	90	10	10		241
$Ni(CO)_4$	25	0.477	1.8	0.3	235
		0.149 [b]	0.3	0.53	235
$Ni(CO)_2(PPh_3)_2$	25	2.22			243
$Ni\{P(OPh)_3\}_4$	25	2.6			244

Py = pyridine.
[a] For carbon monoxide pressures expressed in torr.
[b] Styrene as monomer.

the rate of CO exchange [235], the low activation energy and pre-exponential factor [245] associated with initiation by $Pt\{P(OPh)_3\}_4$) have been discussed.

Some kinetic data for initiation by such derivatives is presented in Table 6, and relative reactivities of a number of phosphine derivatives have been tabulated [242].

b. Photoinitiation

Manganese and rhenium carbonyls are active photoinitiators (λ = 4358 A) in the presence of organic halides, initiating with a quantum efficiency of unity [249, 250]. The overall features of the polymerization kinetics are similar to those described for thermal initiation. Kinetic analysis suggests that initiation proceeds through photolysis of the carbonyl involving scission of the metal—metal bond to produce the unequal fragments F_1, F_2 (possibly $Mn(CO)_6$ and $Mn(CO)_4$, for example), only one of which is normally involved in initiation [250]. A kinetic scheme proposed for initiation by $Mn_2(CO)_{10}$ and CCl_4 in methyl methacrylate solution is

$$Mn_2(CO)_{10} \xrightarrow{h\nu} Mn(CO)_4 + Mn(CO)_6$$

$$Mn(CO)_6 \rightleftharpoons Mn(CO)_5 + CO$$

$$Mn(CO)_5 + CCl_4 \longrightarrow Mn(CO)_5CCl_4$$

$$Mn(CO)_5CCl_4 + Mn(CO)_5 \longrightarrow Mn_2(CO)_{10} + CCl_4$$

$$2Mn(CO)_5CCl_4 \longrightarrow Mn_2(CO)_{10} + 2CCl_4$$

$$Mn(CO)_5CCl_4 \longrightarrow Mn(CO)_5Cl + \dot{C}Cl_3$$

According to this scheme monomer is not involved in radical formation, and this is in agreement with kinetic observations on benzene solutions of methyl methacrylate [250], where the rate of initiation is independent of [monomer]. Also identical rates of initiation have been observed in the presence of a number of monomers [102]. However, in ethyl acetate and dioxan solutions distinctly lower rates of initiation, under standard conditions, have been observed [251]. It is apparent, therefore, that solvents can influence the rates of initiation and it is likely that methyl methacrylate and other monomers are involved in the radical forming process, albeit in a non-rate-determining manner. The second fragment regenerates manganese carbonyl, so that the overall mechanism for radical formation may be written

$$Mn_2(CO)_{10} + CCl_4 \xrightarrow{h\nu} \dot{C}Cl_3 + \tfrac{1}{2}Mn_2(CO)_{10} + Mn(CO)_5Cl$$

References pp. 92—103

Certain solvents have recently been shown to be capable of capturing the second fragment produced on carbonyl photolysis to produce a complex which slowly reacts with halide to produce radicals [252].

c. Nature of the initiating radicals

In all cases referred to above radical formation results from the decomposition of a complex derived from the metal derivative and a molecule of the halide, and involves oxidation of the metal and reduction of the halide. If we represent the complex as metal^{O} . . . RX, where X is a chlorine or bromine atom, this reaction can occur, in principle, in two ways

$$\text{Metal}^{O} \cdots\cdot \text{R}{-}\text{X} \longrightarrow \text{Metal}^{(I)}\text{X}^{-} + \text{R}\cdot \qquad (28a)$$

$$\text{Metal}^{O} \cdots\cdot \text{R}{-}\text{X} \longrightarrow \text{Metal}^{(I)}\text{R}^{-} + \text{X}\cdot \qquad (28b)$$

For initiation by $Mo(CO)_6$, $Mn_2(CO)_{10}$ (photoinitiation), $Pt\{P(OPh)_3\}_4$, $Ni(CO)_4$, $Ni(CO)_2(PPh_3)_2$ and $Ni\{P(OPh)_3\}_4$, using CCl_4 as halide, tracer studies have established that initiation proceeds exclusively by reaction (28a) [253, 254]; i.e. the initiating radical is $\dot{C}Cl_3$. When ethyl trichloracetate is used as halide reaction (28a) is also the sole route to radical formation for $Mo(CO)_6$, $Mn_2(CO)_{10}$ and $Pt\{P(OPh)_3\}_4$ and the initiating radical is $\dot{C}Cl_2COOEt$. This was demonstrated by tracer studies and also by gelation studies using polyvinyltrichloracetate as the halide component [254]. In all these cases, therefore, initiation is by chlorine atom abstraction from the halide. However, using the nickel derivatives with ethyl and polyvinyltrichloracetates both routes (28a) and (28b) lead to radical formation, presumably initiation occurs partly through halogen atoms [254]. The relative rates of the two processes depends on the ligands of the nickel derivative, the percentages of radicals produced by reaction (28a), the average from gelation and tracer studies, for the various derivatives being $Ni(CO)_4$, 62 %; $Ni(CO)_2(PPh_3)_2$, 79 %; $Ni\{P(OPh)_3\}_4$, 54 %.

The use of metal complexes as described above is effectively a means of using metals in their zero oxidation states in solution. Colloidal metals have also been used to initiate polymerizations in the presence of organic halides [255]. Possibly such systems involve reactions similar to those described above. It has also been reported that metal—organic halide systems will initiate both the radical [256, 257] and non-radical [257, 258] polymerizations under suitable circumstances.

3.5.7 Other initiators involving monomer—metal complexes

Among the monomers capable of forming complexes with metal ions are the vinyl pyridines. The polymerization kinetics of 4-vinylpyridine

(4-VP), initiated by cupric acetate, have been investigated in various solvents, especially methanol [259]. Rates of polymerization vary markedly with the solvent in the order, methanol, acetone > monomer > *N,N*-dimethylformamide > pyridine > *n*-butanol > benzene ~0. In methanol solutions of 4-VP (~1 mole l^{-1}), at low concentrations of cupric (~10^{-2} mole l^{-1}), the cupric is present entirely as the complex $(4\text{-VP})_2Cu(acetate)_2$, and the rate of polymerization is proportional to $[Cu(II)]^{0.5}[4\text{-VP}]$, indicating that the complex is the initiator. In the presence of 4-VP and pyridine three complexes are present, and (at [4-VP] + [pyridine] = constant) the rate of polymerization is proportional to $[Cu(II)]^{0.5}[4\text{-VP}]^2$, consistent with only the $(4\text{-VP})_2Cu(acetate)_2$ complex acting as initiator. Initiation (activation energy, 5.4 kcal $mole^{-1}$) is thought to occur through electron transfer in the complex to produce a monomer radical-cation, propagation being radical in nature. For a homologous series of alcohols as solvents the rates of polymerization show a simple dependence on the dielectric constant of the medium, illustrating the importance of polar factors in initiation. Relative rates of reduction of Cu(II) and rates of polymerization in the presence of other vinylpyridines have been discussed.

N-vinylimidazoles are also strong co-ordinating monomers, and form complexes with many metal ions. The oxidizing metal ions Ce(IV) and UO_2^{2+} produce complexes with strong charge-transfer absorptions and these act as effective photoinitiators [260]. Sodium chloraurate and copper acetate also produce strongly absorbing complexes, but are not effective initiators as a result of termination by the oxidizing metal ions. Non-oxidizing metal ions, such as Zn(II), although forming complexes, do not initiate polymerization.

Acrylamide is a weak co-ordinating monomer, but there is kinetic evidence to indicate its participation in redox initiation [261]. Sodium chloraurate initiates and terminates the thermal and photopolymerization of acrylamide. Although the overall rate expression does not require the involvement of monomer in initiation, photoirradiation in the absence of an amide does not cause reduction of Au(III); the overall results indicate the participation of a weak complex in initiation. Similar kinetic behaviour has been observed for thermal initiation at 30 °C, when a rate coefficient for initiation of 6.8×10^{-7} l $mole^{-1}$ sec^{-1} was found, assuming $\mathscr{I} \propto [Au(III)][acrylamide]$. Uranyl ions also initiate the polymerization of acrylamide [262] and other monomers [263].

3.5.8 Miscellaneous initiators

Oster et al. [264] have demonstrated that photosensitization of dyes in the presence of mild reducing agents and low concentrations of oxygen will induce the free-radical polymerization of vinyl monomers. It has been suggested that the mechanism of initiation involves photoreduction of the dyes to their leuco forms followed by subsequent oxidation to produce free radicals [265]. Alternatively, it has been suggested [266] that the photochemical dye reacts with oxygen to produce hydrogen peroxide which reacts with the reducing agent to produce radicals. It has also been claimed that the polymerization of acrylamide in these systems does not require the presence of oxygen [267]. Other photopolymerizations are given in a review by Oster and Yang [208].

From kinetic and tracer studies it was suggested that initiation of methyl methacrylate polymerization by dimethylbenzeneanilinium chloride occurs by dissociation of the initiator to produce dimethylaniline and benzyl chloride [268], the dimethylaniline then undergoing a redox reaction with another molecule of dimethylbenzeneanilinium chloride to generate initiating phenyl radicals.

Silver ethyl acts as a low temperature thermal initiator characterized by a rate coefficient for decomposition in the temperature range —20 to —60 °C of $1.50 \times 10^{-2} \exp(-6600/RT)$ sec^{-1}. The rate of polymerization of methyl methacrylate is proportional to $[monomer]^{1.5}$; the order in monomer concentration and the initiator efficiency have been discussed [269]. The use of binary mixtures of lead tetraethyl and metal salts as low-temperature initiators has been described [270]. Lead tetraethyl will itself act as a thermal initiator in the temperature range 200—400 °C, and as a photoinitiator at room temperature [271].

Boron alkyls have been reported as initiators of vinyl polymerizations [272]; these reactions require the presence of oxygen [273]. It was shown [274] that the role of the oxygen is to convert a portion of the boron alkyl (R_3B) to the peroxide R_2BOOR, that the rate of initiation of methyl methacrylate polymerization is proportional to $[R_3B]$ and $[R_2BOOR]$, independent of [monomer], and associated with an activation energy of 13 kcal $mole^{-1}$. Neither the boron alkyl nor the peroxide alone initiate polymerization. The proposed mechanism of initiation involves a rapid equilibrium between the alkyl and monomer to form a complex which induces the decomposition of the peroxide to generate radicals [274]. The equilibrium was introduced to allow for different rates of initiation with different monomers. The kinetics of methyl methacrylate polymerization initiated by triethyl boron and oxygen, under conditions of autoxidation of the alkyl, are consistent with the above mechanism [275]. Rates of polymerization of acrylonitrile in homogeneous solution are reported [276] to be proportional to $[monomer]^{1.5}$.

Solid lead dioxide is reported to initiate the polymerizations of methyl methacrylate and vinyl acetate but not those of styrene and acrylonitrile [277]; an overall rate expression has been described for methyl methacrylate polymerization. Solid nickel peroxide has also been reported to effect polymerization of various vinyl monomers [278], and to produce an isotactic polymer from styrene [279].

3.5.9 *Uncatalysed polymerizations*

Monomers are normally stabilized to prevent "spontaneous" polymerization during storage. Often such spontaneous polymerization could arise from the thermal decomposition of peroxides produced during storage in the presence of oxygen, especially if they are stored in the light. Such polymerizations are, of course not truly spontaneous. Another source of initiation during storage can be adventitious impurities introduced during manufacture, e.g. trace metal impurities. Inhibitors are normally added to prevent such polymerizations.

Some highly purified monomers, e.g. styrene, appear to undergo true thermal polymerization while other monomers, notably acrylonitrile, appear to be resistant to such spontaneous initiation. A number of monomers are susceptible to unsensitized photoinitiation. Rates of spontaneous polymerization are usually low and extremely sensitive to the presence of impurities. Consequently, it is difficult to obtain accurate data on such polymerizations and uncertainties as to the existence of a spontaneous rate with some monomers probably arises from the effects of trace impurities. In the presence of an added initiator the rates of catalysed and spontaneous initiation are additive.

Styrene has a relatively high rate of thermal polymerization and can give fairly reproducible data. Early studies gave conflicting results, but, in a subsequent careful study, Mayo [280] found that the rate of initiation was third-order in styrene and that low-molecular-weight products are formed simultaneously; the overall rate of polymerization was proportional to $[\text{styrene}]^{2.5}$. Subsequently, Mayo [281] suggested that thermal initiation involves formation of a dimer which reacts with styrene to produce radicals. Later, in a more detailed publication [282] on the formation of dimers, it was reported that a number of dimerization products can be obtained, depending on the additives present, but that in pure styrene the main products are 1,2-diphenylcyclobutane and 1-phenyltetralin, formed by non-radical processes. Initiation was then considered to proceed by hydrogen atom transfer from a dimer intermediate to styrene, with an activation energy of 29 kcal mole^{-1}. The overall processes leading to radical formation are described in the equations

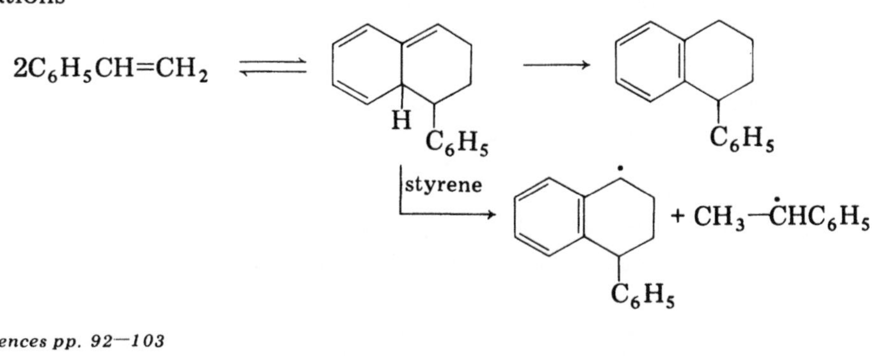

The involvement of the *ortho*-hydrogen is supported by the work of Kirchner [283, 284] who also studied polymerization kinetics in various solvents and found the results to agree with the mechanism of Mayo. It was also suggested that the lack of thermal polymerization with some monomers may reflect their inability to form reactive dimers [281]. Phenyltetralin or the dimer intermediate could act as transfer agents [285] and account for the decrease in molecular weight of the polymer as conversion increases at low conversions [286]. Recently Brown [287] has refuted Müller's claims that phenyltetralin is the major dimer product and claims that the principal dimers are *cis*- and *trans*-1,2-diphenylcyclobutane.

Vinyl furan [288] and vinyl thiophene [289] also undergo thermal polymerization with simultaneous formation of dimers analogous to phenyltetralin. For vinyl thiophene it was found that the rate of initiation is third-order in monomer and associated with an activation energy of 28.2 kcal $mole^{-1}$. In both cases it was considered that initiation occurred by hydrogen atom abstraction from the dimer.

In an attempt to establish the feasibility of initiation by hydrogen abstraction, Aso et al. [290] demonstrated that hydrocarbons such as triphenylmethane, possessing labile hydrogen, will initiate the polymerization of acrylonitrile at temperatures above 130 °C, the rate of polymerization decreasing with time at low conversion as a result of termination by the triphenylmethyl radical.

Rates of thermal polymerization of methyl methacrylate are generally lower than those of styrene. Using carefully purified monomer irreproducible data was obtained for reaction at 100—150 °C by Walling and Briggs [291], and at 25 °C by Bamford and Dewar [18, 292]. The latter workers found that the thermal rate could be enhanced by pre-irradiation with UV light, and decreased by prior thermal treatment. From kinetic studies it was proposed that the former treatment produced a catalyst of relatively short half-life while the latter produced an inhibitor. It was also suggested [18] that both products arose from dimerizations of monomer molecules, the inhibitor being a quinone. Mackey and Melville [293] also obtained evidence for catalyst formation on irradiation.

A number of reports have appeared describing the "spontaneous" polymerization of methyl methacrylate in the presence of cellulose, or other fibres, and carbon tetrachloride and water [294]. However, it is probable that these polymerizations are catalysed by impurities, as in the polymerization of vinyl chloride by methanolamine and carbon tetrachloride where it was found that a necessary constituent of the initiating system was copper extracted from a hypodermic needle used in the experiments [295].

The thermal polymerization of chloroprene has received some attention from Robb and co-workers [296, 297], who could only obtain reproducible rates if the liquid phase was separated from the vapour phase by a mercury plug [296]; heterogeneous polymerization occurs in the vapour phase. The reaction mixture was found to contain a series of cyclic dimers derived from the monomer and having the same density as the polymer. The dimers are formed (in the presence of a polymerization inhibitor) by a non-radical process at a rate equal to that of the thermal (radical) polymerization in the absence of inhibitor. Polymerization of the dimers themselves requires the presence of monomer (~ 5 %) and appears to be initiated specifically by an oxygenated impurity in the monomer [297]. The kinetics of the cyclo-addition process have recently been described [298].

There is relatively little information available on the thermal polymerization of other monomers. It has been reported that in the presence of air and daylight crystalline acrylamide produces a polymeric peroxide which initiates the polymerization when the monomer is dissolved in water [299], and it is known that acrylic and methacrylic acids are liable to undergo spontaneous polymerization on repeated freezing and thawing during degassing procedures [7], but nothing is known about the mechanisms of the reactions.

4. Bimolecular termination

A propagating radical may be destroyed by a variety of processes including termination by added substrates, such as stable radicals or oxidizing metal ions. The latter species may be a constituent of the initiating system. Such reactions will be considered in a later discussion of inhibition and retardation processes (Chap. 2). In this section we are concerned with the mutual destruction of reactive radicals inherent in the system. We must therefore consider the destruction of propagating radicals by identical propagating radicals and by primary radicals. Under normal polymerization conditions primary-radical termination may be neglected since the concentration of primary radicals is very low. Therefore, we shall first consider the mutual termination of propagating radicals.

4.1 THE MECHANISM OF TERMINATION

The mutual destruction of propagating radicals may occur by combination

$$2 \sim CH_2{-}\dot{C}XY \longrightarrow \sim CH_2{-}CXY{-}CXY{-}CH_2 \sim$$

or by disproportionation, which may be represented in two ways, viz.

$$2 \sim\!\!\sim CH_2{-}\dot{C}XY \longrightarrow \sim\!\!\sim CH_2{-}CHXY + \sim\!\!\sim CH{=}CXY \qquad (29a)$$

and

$$2 \sim\!\!\sim CH_2{-}\underset{\underset{\displaystyle CH_3}{|}}{\dot{C}}X \longrightarrow \sim\!\!\sim CH_2{-}\underset{\underset{\displaystyle CH_3}{|}}{C}HX + \sim\!\!\sim CH_2{-}\underset{\underset{\displaystyle CH_2}{\|}}{C}X \qquad (29b)$$

yielding one or two polymer molecules from two propagating chains.

Obviously, the mode of termination controls the molecular weight of the polymer. For disproportionation termination the number-average degree of polymerization of the polymer and its kinetic chain length are equal, but for combination termination the number-average degree of polymerization is twice as large (Chap. 7). Molecular-weight distributions are also affected, the ratio of the weight to number-average degrees of polymerization being 1.5 for combination and 2 for disproportionation. Termination may not occur exclusively by one mechanism or the other, both mechanisms may operate simultaneously and the degrees of polymerization and the distributions are affected accordingly.

The mode of termination can also have a marked effect on other aspects of polymer science. We have already referred (Section 2.2.2.d) to the fact that uncertainty in the relative values of k_{tc} and k_{td}, the rate coefficients of the two processes, can produce errors in the derivations of kinetic parameters. Apart from such scientific considerations the mechanism of termination can have commercial implications, since the ultimate object of making polymers is to fabricate and use the products as materials. The termination reaction introduces into the polymer a low concentration of unusual groupings. Combination termination introduces a head-to-head grouping while disproportionation produces one chain containing the saturated analogue of the head grouping and one chain containing the unsaturated analogue. For monomers having bulky substituent groups the head-to-head grouping is severely congested and both this and the unsaturated group may influence the ultimate properties and applications as a result of the reactivity of these groups. It has been demonstrated that the unsaturated end-group in polymethyl methacrylate is a preferred site for initiation of thermal degradation of the polymer [300]. It therefore appears surprising that over the years relatively few attempts have been made to investigate these termination processes.

4.1.1 Techniques

It is possible, in principle, to determine the ratio k_{tc}/k_{td} from the simple kinetic relations derived in Sect. 2, if sufficient supplementary information is available. Baysal and Tobolsky [301] have pointed out that this can be achieved by a comparison of kinetic parameters, such as kinetic chain length and rate of initiation, determined by different methods. One method, e.g. the initiator fragment technique or methods based on known rates of initiation, should not involve any assumption as to the termination mechanism. The second method should involve the termination mechanism, e.g. determination of kinetic chain length from the average degree of polymerization of the polymer.

Assuming that initiation occurs through addition of the primary radical to monomer and that transfer is absent or negligible, combination termination results in the incorporation of two initiator end-groups in each polymer molecule while disproportionation results in the incorporation of only one initiator end-group. Thus, end-group determinations can provide data on the mechanism of termination. Transfer processes can have a marked influence on the results. Transfer to polymer has no net influence on the number of end-groups incorporated per polymer molecule, while transfer to monomer or solvent decreases and transfer to initiator increases the degree of end-group incorporation. In the latter case the number of fragments incorporated per polymer molecule tends, in the limit, to one or two, irrespective of the mode of termination; the final value depends on the mechanism of the transfer reaction and the nature of the end-groups assessed. The number of initiator fragments per polymer molecule and hence, the relative proportions of the termination reactions are determined by comparing the number-average degrees of polymerization obtained from end-group analysis and an independent method, e.g. osmometry (after correction for transfer). End-group concentrations are normally low and analyses are usually carried out by the radio-tracer technique. It is important to take precautions in isolating the polymer, particularly low-molecular-weight material which contains a relatively large proportion of end-groups. The necessary precautions have been described by Bevington and co-workers [33, 34]. The use of dye techniques to assess end-group concentrations has also been described [177].

A second technique directly involving initiator fragments is the coupling technique [302, 303]. This method is based on the incorporation of a functional group, such as carboxyl or hydroxyl, in the terminal unit of the polymer chain by use of a suitable primary radical to initiate polymerization. The polymer is isolated and its molecular weight is determined. The reactive end-groups are then coupled by use of a suitable difunctional reagent, and the molecular weight is redetermined. If the polymerization reaction was terminated by disproportionation then the molecular weight should not increase by more than a factor of two, while combination termination should result in much larger molecular weight increases.

We have already referred to the fact that the mode of termination affects the shape of the molecular-weight distribution and an accurate knowledge of the distribution could allow the relative proportions of the termination reactions to be assessed. The feasibility of using this approach is increasing with the development of more accurate methods for determining the distributions.

Grafting techniques have been used to detect the occurrence of combination reactions. Allen et al. [304] demonstrated that if radioactively labelled azobisisobutyronitrile is used to initiate the polymerization of methyl methacrylate in the presence of *trans*-1,4-polyisoprene no grafting occurs through the interaction of primary or propagating radicals with the polyisoprene. Hence, no activity is incorporated into the polyisoprene. Addition of a little benzoyl peroxide caused grafting to occur through the reaction between the primary radical from the peroxide and the polyisoprene. Some activity from the labelled initiator was then incorporated into the polyisoprene from combination of propagating radicals.

More recently, a kinetic technique, based on grafting and gelation, for determining the nature of the termination reaction quantitatively has been described [102, 305]. This makes use of a polymer (the prepolymer) containing polyhalogenated side-groups as part of the initiating system. Radical formation is induced by a metal carbonyl, according to the mechanism described in Sect. 3.5.6. Initiation involves the specific removal of a chlorine atom from the prepolymer, generating a radical attached to the prepolymer molecule. This radical then initiates the graft polymerization of the monomer. Combination termination results in cross-link formation between prepolymer molecules, and ultimately leads to network formation, while disproportionation yields only branches. If the weight-average degree of polymerization of the prepolymer is known the number of cross-links, and hence the number of initiations ultimately involved in combination termination, required to produce an infinite network and cause gelation of the reaction mixture can be calculated from gelation theory (Chap. 4). By measuring the time required for gelation to occur and the total rate of initiation the relative proportions of combination and disproportionation reactions can be determined. Transfer to monomer or solvent is equivalent to an increase in disproportionation. The necessary kinetic conditions and the corrections for transfer and initiator consumption have been described [102, 251] and the appropriate kinetic relations are detailed in a later chapter where network formation is considered (Chap. 4).

Model radicals can also give useful information [10], although the quantitative results could be influenced by end-group effects.

4.1.2 The results

In spite of the variety of techniques available few investigations have been undertaken. The most widely studied monomer in this connection is styrene which has been subjected to investigation by tracer techniques [96, 306, 307], comparison of molecular weights and rates of initiation

TABLE 7
Percentages of combination termination in the polymerization of methyl methacrylate

Temp. (°C)	Reference					
	312[a]	313[a]	314[a]	254[a]	305[b]	102[b]
0	40					
25	32			33	33	34
40			69			
60	15	42.5	57	27		
80			43	20	19	

[a] Radio-tracer technique; [b] gelation technique.

[308], the coupling technique [302], molecular-weight distribution studies [309], the gelation method [102], bifunctional initiation [310], and model compounds [311]. All workers appear to agree that termination takes place almost exclusively by combination over a wide range of temperatures. (See also p. 191.)

It is generally agreed that both mechanisms are operative with methyl methacrylate and that the percentage of disproportionation increases with increasing temperature, but there are discrepancies as to the relative importance of the two processes. The percentages of combination, i.e. $100k_{tc}/(k_{tc} + k_{td})$ obtained by use of the radio-tracer [254, 312—314] and gelation [102, 305] methods are summarized in Table 7. In addition the coupling technique failed to detect any combination at 100 °C [302].

Acrylonitrile undergoes combination, according to the results obtained from the dependence of molecular weights on rates of initiation [316], tracer studies [315], the coupling technique [302] and gelation studies [102].

With vinyl acetate Burnett et al. [317] obtained erratic results using the tracer technique. Subsequently Funt and Pasika [318], using the tracer technique, and Bamford and Jenkins [302], using the coupling technique, obtained results indicating disproportionation termination. In neither case was any allowance made for transfer to monomer, which is extensive, and this has the effect of overemphasizing disproportionation. The gelation technique, without correction for transfer, gave results indicating about 10 % combination at 25 °C, but after correction for transfer the results suggested 100 % combination [102].

Percentages of combination for other monomers (at 25 °C) obtained using the gelation method [102] are: methyl and ethyl acrylates, 100 %; ethyl methacrylate, 32 %; *n*-butyl methacrylate, 25 %; methacrylonitrile, 35 %. Other tracer studies [319] indicate that *p*-chlorostyrene radicals combine, while *p*-methoxystyrene radicals undergo 81 % and 53 % combination at 60 and 80 °C, respectively. The only evidence available on termination between unlike radicals is the data of Bevington et al. [312]

which suggests combination between methyl methacrylate and styrene radicals (see also Chap. 4).

At the present time there does not appear to be any absolute method for predicting the relative importance of the two termination reactions. However, we may consider the results obtained with the gelation technique which has provided comparative data on nine monomers and the results of which are generally consistent with data from other studies. All radicals which undergo 100 % combination have the structure (VI), while those which undergo disproportionation and combination have the structure (VII), viz.

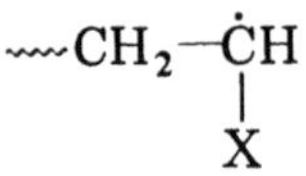

(VI)

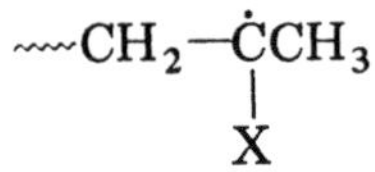

(VII)

The essential difference in the structures of these radicals is the presence of the methyl group in (VII). It could be argued that the methyl group introduces severe steric restrictions which oppose combination, and certainly molecular models show that the transition state for combination in methyl methacrylate must be under severe (compressive) strain, while that for methyl acrylate is unstrained. An argument based solely on such steric considerations does not appear valid since it is known that ethyl radicals undergo termination by both routes in the gas [320] and liquid [321] phases. Also among these monomers there is no correlation with radical reactivity parameters. The occurrence of disproportionation thus appears to be linked with the availability of transferable hydrogens. The introduction of the methyl group enhances disproportionation to a much greater extent than expected on a statistical basis and it would appear that transfer from the methyl group of the radical is much easier than from the methylene group. Indeed if it was possible for the methylene hydrogens to be abstracted it would be expected that the reactive vinyl acetate radicals would undergo disproportionation. On the basis of the available evidence, therefore, for the purposes of predicting the nature of the termination reactions, it is probably reasonable to assume that only those radicals having an α-methyl group are capable of undergoing disproportionation. The only exception to this rule appears to be *p*-methoxystyrene radicals, for which the occurrence of some disproportionation was attributed to strong polar factors [319].

If the above assumption is correct the results imply that disproportionation proceeds by reaction (29b) and probably not by (29a). There is some evidence that polymethyl methacrylate molecules possess terminal vinyl groups since such molecules can be copolymerized with acrylonitrile [322]. Degradation studies of polymers from methyl group containing monomers also support the presence of terminal vinyl groups [323].

All studies of the termination reactions of methyl methacrylate are in agreement in that they predict a negative temperature coefficient for the ratio k_{tc}/k_{td}, implying that combination and disproportionation occur through different transition states.

4.2 RATES OF BIMOLECULAR TERMINATION

Propagating polymer radicals are reactive species and the rate coefficients for bimolecular termination are high (typically 10^7 l mole^{-1} sec^{-1}), but not so high as those for the corresponding low-molecular-weight species. This is not surprising since reaction requires the close approach of large macromolecular entities and inevitably involves complex diffusive processes, often through viscous media. The first indications that restricted diffusion is important in such termination reactions came from the studies of Norrish and Smith [324], and subsequently of Tromsdorff et al. [325], who observed that, in at least some polymerizations, the rate of reaction increased rapidly after some critical conversion was reached, instead of continuing to decrease with time due to consumption of reactants. The interpretation offered was that the increase in rate is a consequence of a decrease in rate of termination, and hence, an increase in the concentration of reactive species, due to the large increase in macroscopic viscosity restricting the diffusion of the macro-radicals, the interaction changing from a chemical-controlled to a diffusion-controlled process. This view was strengthened by theoretical predictions of the critical conversions at which diffusion may control the rates of reaction [326, 327]. The auto-acceleration is a feature of many polymerizations [3], and is well known as the "gel-effect". During the period of auto-acceleration extremely high rates of polymerization, often ten times as high as the initial rates, may be observed, and measurements of individual rate coefficients for methyl methacrylate at about 30 % conversion confirmed the reduction in k_t, the value of k_p remaining unaffected [324]. The critical conversion at which the "gel-effect" becomes apparent varies from monomer to monomer, and with temperature and concentration of solvents. In bulk methyl methacrylate auto-acceleration is observed at about 12 % and 25 % conversion at 50 and 70 °C, respectively [328], and in bulk methyl acrylate at essentially zero conversion, while with styrene there is no effect until high conversions are reached. Obviously, during the auto-acceleration the kinetics are non-stationary although instantaneous relations will apply at any particular time. Lack of regard for the "gel-effect" can lead to erroneous conclusions based on crude comparative kinetic studies of polymerizations taken to finite conversions.

More recently it has been realized that diffusion-control of the termination reaction is not restricted to high conversions but also exists in many monomers at essentially zero conversion; activation energies for bimolecular termination are about the same as for diffusion. Evidence for this has appeared from studies of polymerization reactions in media of different viscosities. For acrylamide polymerizations in aqueous media, using glycerol to vary the solution viscosity, Oster et al. [265] found that the rate of polymerization was proportional to the square root of the viscosity of the solvent medium. Benson and North [329] observed that k_t for polymethyl methacrylate radicals, at low conversion, was inversely proportional to the viscosity of the solvent medium over a 100-fold variation in viscosity, while values of k_p remained unaltered; poly(olefin oxides) were used to vary the viscosity. The latter workers considered the mathematical consequences of either translational or segmental diffusion being the rate-determining process [330]. The overall termination mechanism was considered as a sequence of reversible processes, viz., diffusion of the centres of gravity of the macro-radicals to form a proximate-pair, followed by segmental rearrangement diffusion bringing the reactive sites into close proximity and finally chemical reaction. It was concluded that segmental diffusion of chain-ends is the slow step and that the effective diffusion constant of the chain-ends is dependent on the size of the macro-radicals for chain lengths less than about 100 but independent of chain length for long chains.

Following the conclusion that the segmental diffusion of chain-ends is the controlling factor it was realized that it is the local viscosity, or micro-viscosity, of the medium in the vicinity of the chain-end which is the important factor, rather than the macroscopic viscosity of the medium, and that conclusive evidence for this model cannot be derived from studies in which the viscosity is changed by added polymer, since the resulting high value of the macroscopic viscosity is partly due to chain entanglements rather than restrictions on segmental diffusion [331, 332]. Subsequently, North and Reed [333] confirmed that k_t for methyl methacrylate is inversely proportional to solvent viscosity and proportional to the mutual diffusion coefficient at zero polymer concentration, using low-molecular-weight esters as solvents, and similar evidence has been obtained for other homopolymerizations [334] and copolymerizations [335, 336].

Detailed examination of conversion—time curves for methyl methacrylate polymerizations indicated that, in the earliest stages, the rate of polymerization decreased more rapidly than could be accounted for by consumption of reagents. This decrease was attributed to an increase in the termination coefficient; subsequently the termination coefficient decreased slowly with increasing conversion [333]. The initial increase in the termination coefficient is consistent with an increase in mutual diffusion coefficient with increasing polymer concentration at low values of the latter [337], and arises from a decrease in the dimensions of the polymer coil as polymer concentration increases in a good solvent [338] under these conditions. The subsequent slow decrease in termination coefficient was attributed [333] to a slow increase in micro-viscosity with increasing polymer concentration at relatively low values of the latter; the micro-viscosity ultimately increasing more rapidly at high conversions to give rise to the "gel-effect".

An essential difficulty in predicting rate coefficients was a lack of information on the effective diffusion constant for the chain-ends. Burkhart [339] has carried out theoretical calculations of termination rate coefficients on the basis of chain-end diffusion control and in terms of solvent viscosity, chain flexibility and chain length, using the approach of Kuhn and Kuhn to describe the motion of chain-ends. The results of these calculations are that the termination coefficients are inversely proportional to solvent viscosity for long flexible chains, while deviations from this behaviour are apparent for short and stiffer chains and for very stiff chains the rate coefficient is independent of solvent viscosity. The types of deviation predicted by these calculations have been observed in polymerizations of butyl acrylate [329] and various alkyl methacrylates [334]. Calculated values of k_t were in reasonable accord with experimental observations.

It has been suggested that, since chain-end and main-chain diffusions are related, the diffusion coefficient of the whole molecule should be applied to translational motion of the chain-end [340]. While this approach can apparently predict reasonable values of termination coefficients it is considered to underestimate the flexibility of the chain-ends [341] and to be unable to describe the variations in rate coefficient with chain stiffness [342]. It now appears to be agreed that segmental diffusion of the chain-ends is the controlling factor in many of these reactions.

The influence of chain stiffness on rate coefficients has been investigated in copolymerizations of polymethyl methacrylate and maleic anhydride for which rates of copolymerization pass through a maximum as the maleic anhydride concentration is increased [336]. Incorporation of maleic anhydride residues reduces the number of C—C bonds in the polymer about which rotation can occur, and hence reduces the flexibility of the chain and decreases k_t. Chain stiffness can also be varied by changing the nature of the monomer [343] and/or the temperature of polymerization [344]. With decreasing temperature (to about −50 °C) the activation energy of termination for methyl methacrylate changes due to freezing-out of segmental motions, and, as bimolecular termination between polymer radicals decreases, primary-radical termination increases since diffusion of small entities is less affected.

Attempts have been made to extend theory to higher conversions, and Ito [345] has concluded that at all conversions the rate coefficient is proportional to the diffusion constant of the chain-end which depends on conversion but not on macroscopic viscosity. Ito has also considered primary-radical termination and concluded that the rate coefficient depends on the diffusion coefficient of the small radicals at high conversion. He has reported experimental evidence indicating a linear dependence of rate coefficient on conversion [346].

It would be of considerable interest to obtain direct information on the motions of main-chain segments and of chain-ends, possibly by the use of fluorescence techniques using appropriately labelled polymer molecules [347]. Apparently the only directly related evidence is on chain-end motion in polyacrylamide for which it has been found that the mean rotational relaxation time is proportional to T/η_0 (T = absolute temperature, η_0 = solvent viscosity) in aqueous solutions and in the presence of glycerol [348] (compare the results of Oster et al. [265]). The variation of this parameter with chain length and the influence of added polymer has been investigated [349].

Obviously, in view of the nature of the termination reaction, it is impossible to quote single values of rate coefficients for any monomer; data on termination rate coefficients is presented in Chap. 3. When carrying out comparative kinetic studies of polymerizations in different solvents, e.g. calculations of rates of initiations, rates of polymerization should be normalized to the same viscosity.

4.3 PRIMARY-RADICAL TERMINATION

Termination reactions involving primary radicals derived from the initiator are not normally considered in derivations of kinetic relations for free-radical polymerization (Sect. 2). Under normal circumstances, since most primary radicals are highly reactive and most monomers are efficient radical scavengers, neglecting such processes is usually justifiable. Consequently, primary radicals are rapidly removed from solution by reaction with monomer and their stationary concentrations are normally very low. However, there are circumstances under which primary-radical termination cannot be neglected. Under conditions of very high rates of initiation and/or very low monomer concentrations the monomer is incapable of acting as an efficient radical scavenger, hence, the concentration of primary radicals increases significantly, and the importance of primary-radical termination increases, causing certain deviations from normal kinetic behaviour. The conditions under which these effects become apparent depend on the reactivities of the various species in the system, and in a reaction involving stabilized primary radicals and unreactive monomer the effects may be observed at relatively low rates of initiation and high monomer concentration. Further, normal bimolecular termination involving interaction of polymer radicals may be restricted under conditions of high viscosity when the importance of primary-radical termination will be enhanced.

To introduce primary-radical termination into a kinetic scheme the reactions

$$R_c\cdot + R_n\cdot \xrightarrow{k_t'} P_n \tag{30}$$

$$R_c\cdot + R_c\cdot \xrightarrow{k_t''} \text{Products} \tag{31}$$

must be added to scheme (1). Reaction (31) takes place in the body of the solution and is not to be confused with geminate recombination. In addition some workers have included termination by small radicals, arising from transfer processes, within the kinetic analysis of primary-radical termination.

Exact kinetic analysis of a comprehensive reaction scheme embodying primary-radical termination leads to complex relationships and, in order to simplify calculations, it has been normal practice to adopt various simplifying assumptions [350—354]; the reader is referred to the original papers for details. The general consequences on the kinetics are that deviations from the simple relationship for rate of polymerization, eqn. (4), become apparent in that the exponent of the rate of initiation falls below 0.5 and the overall order in monomer concentration increases. These effects arise since some primary radicals are lost through termination, so that the rate of initiation does not increase linearly with the rate of radical formation, and as a result of the competition between monomer and propagating for the primary radicals the rate of initiation is dependent on [monomer].

In the original treatment, Bamford et al. [350] considered two simple cases: (i) $k_t'' = 0$, $k_t' = \lambda k_t$, implying that mutual destruction of primary radicals by reaction (31) does not occur, (ii) $k_t' = (k_t k_t'')^{1/2}$, i.e. the geometric mean hypothesis, and obtained the appropriate relations between monomer concentration, and rates of radical formation and polymerization. Both relations were applied to the polymerizations of styrene and acrylonitrile in *N,N*-dimethylformamide solution, initiated by AIBN, and gave good agreement with experimental data; k_t' was found to be about sixty times as large as k_t. It was not possible to distinguish between the validity of the two assumptions since under the experimental conditions employed the rate of reaction (31) was extremely small. Okamura and Manabe [352] also found results to be insensitive to values of k_t'' and could not distinguish between cases (i) and (ii). It is only at very high rates of initiation or very low [monomer] that reaction (31) becomes important and in most cases primary-radical termination can probably be adequately expressed in terms of eqn. (30) alone.

In addition to rates of polymerization, the molecular weights and molecular-weight distributions of the polymers are also affected and two groups of workers have considered the influence of primary-radical termination on the viscosity-average molecular weights [351, 353].

Although the use of the geometric mean hypothesis to relate the various rate coefficients of the termination reactions may be reasonable when the reactions are chemically controlled, Allen and Patrick [355] have questioned the applicability of this hypothesis to termination reactions in view of the development of ideas on diffusion-controlled termination, under which circumstances the rates of the reactions are determined by physical and not chemical parameters. These workers have considered the application of diffusion-control to all three termination reactions.

Allen and Bevington [356] have described the application of the radio-tracer technique to studies of primary-radical termination by measuring the rate of incorporation of initiator fragments into the polymer, and have applied the technique to polymerizations of styrene initiated by benzoyl peroxide. Pryor and Fiske [357] have attempted to apply this technique to polymerizations of styrene initiated by AIBN, and while they obtained strong evidence for the incorporation of initiator fragments in this way they did not obtain a quantitative interpretation due to uncertainty in the transfer coefficient to AIBN.

5. Propagation

The propagation reaction

$$P_n\cdot + M \xrightarrow{k_p} P_{(n+1)}\cdot$$

is the reaction which controls both the rate of growth and the structure of the polymer chain. The essential requirement for free-radical homopolymerization is that k_p has a finite value, i.e. the monomer is capable of adding to its conjugate radical with regeneration of the same radical species. Most substituted ethylenes of the types $CH_2{=}CHX$ (e.g. styrene, vinyl chloride, acrylonitrile, vinylacetate and acrylate esters) and $CH_2{=}CXY$ (e.g. α-methyl styrene, vinylidene chloride and methacrylate esters) are susceptible to free-radical polymerization, propagation normally taking place according to

$$R\cdot + CH_2{=}CXY \longrightarrow R{-}CH_2{-}\dot{C}XY$$

The main exceptions to the above generalization are the vinyl ethers and mono-olefins, which are not susceptible to free-radical polymerization. With 1,1-disubstituted ethylenes the second substituent Y should not be too large, since its presence makes propagation sterically more difficult. Thus, as seen from Table 8 the value of k_p for methyl methacrylate is lower than that for methyl acrylate, and while styrene is readily polymerizable, α-methyl styrene will only polymerize at low temperatures due to its low ceiling temperature (Sect. 5.2). The low values of k_p for the

TABLE 8
Values of k_p and heats of polymerization for selected monomers

Monomer	Heat of polymerization[a] (kcal mole^{-1})	k_p (l mole^{-1} sec^{-1}) at 60 °C	Ref.
Styrene	16.4	176	358
α-Methyl styrene	9.0	—	
Methyl acrylate	18.7	2090	359
Methyl methacrylate	13.0	734	25
Vinyl acetate	21.3	3700	360

[a] Taken from ref. 26.

1,1-disubstituted olefins are also associated with lower heats of polymerization than those of the corresponding monosubstituted ethylenes.

Fluorine derivatives apart, free-radical polymerization of 1,2-disubstituted ethylenes is normally considered impossible, although the homopolymerization of maleic anhydride is known [361]. Such disubstituted ethylenes are susceptible to radical reactions but k_p is zero; they can, however, often be incorporated into copolymers since the rate coefficients for cross-propagation are not zero.

In addition it is possible to use materials containing more than one double bond as monomers. These monomers may possess two vinyl groups, such as divinylbenzene or ethylene glycol dimethacrylate. Polymerization of such monomers may take place through one vinyl group, leaving the second group pendant to the polymer chain. This pendant group may subsequently be involved in polymerization, leading to branched and cross-linked structures. Finally, diene monomers, such as butadiene, isoprene or chloroprene, may be used. Such monomers may react in the same way as vinyl monomers through 1,2- or 3,4-addition, or they may polymerize through 1,4-addition yielding polymers with unsaturation in the main chain, as

$$R\cdot + CH_2{=}CH{-}CH{=}CH_2 \longrightarrow R{-}CH_2{-}CH{=}CH{-}\dot{C}H_2 \qquad (32)$$

Butadiene and isoprene are normally polymerized in emulsion since k_t is too high to lead to effective polymerization in solution.

Absolute values of k_p are determined by the reactivities of both the monomers and the radicals, which are controlled by the nature of the substituent groups. The propagating styrene radical is relatively unreactive, as a result of the stabilizing properties of the phenyl group, in contrast to the highly reactive radical derived from vinyl acetate which is much less stabilized. Normally the conjugate radical of a reactive monomer is less reactive than that derived from an unreactive monomer. Mayo and Walling [362] drew up the following order of monomer reactivities, based on the nature of the substituent groups on 1-substituted ethylenes; $-C_6H_5 > -CH{=}CH_2 > -COCH_3 > -CN > -COOR > -Cl > -CH_2Y > -OCOCH_3 > -OR$. While this order of monomer reactivities, and the reverse order for radical reactivities, holds generally, there are a large number of exceptions. A detailed discussion of reactivities is presented in a later chapter (Chap. 5). Thus the value of k_p is determined by two opposing factors, but it would appear that in many cases the radical reactivity is the predominant factor, leading to a very high value of k_p for vinyl acetate.

As seen from Table 8, polymerization is an exothermic process, and in a rapid polymerization, if the heat of polymerization is not removed from the reaction mixture, a thermal explosion may occur (p. 12).

5.1 EFFECTS OF TEMPERATURE

The rate of chain propagation is $k_p[M][R\cdot]$, and its temperature variation is determined by the activation energy for propagation E_p, which is low, about 5 kcal mole^{-1}, normally. Consequently, the rate of chain growth does not vary rapidly with changing temperature and polymerizations may be conducted at very low temperatures if a suitable source of radicals is available.

The overall effect of temperature variation, however, is more complex. Expressing the relations (4) and (5) for the rate of polymerization (ω) and kinetic chain length, respectively, in terms of activation energies we have

$$\frac{d \ln \omega}{dT} = \frac{E_p + \frac{1}{2}E_i - \frac{1}{2}E_t}{RT^2}$$

$$\frac{d \ln \nu}{dT} = \frac{E_p - \frac{1}{2}E_i - \frac{1}{2}E_t}{RT^2}$$

E_i and E_t are the activation energies for radical formation and termination, respectively.

In the case of photolytic, or radiolytic, initiation when E_i is small or zero, the temperature variations of rate and kinetic chain length are determined by $(E_p - \frac{1}{2}E_t)$ (in both cases). Since E_t is smaller than E_p, increases in temperature lead to increased rates and degrees of polymerization. For thermally sensitized polymerizations E_i is frequently greater than E_p, and increased temperatures often lead to increased rates but lower degrees of polymerization. In some cases this latter effect is accentuated by the fact that disproportionation tends to increase in importance with increasing temperature.

5.2 THE CEILING TEMPERATURE

The preceding discussion neglected the possible influence of the depropagation reaction

$$P_n\cdot \xrightarrow{k_u} P_{(n-1)}\cdot + M \qquad (33)$$

the importance of which increases rapidly with increasing temperature since the activation energy of this process is high, ~20 kcal $mole^{-1}$, typically.

Snow and Frey [363] pointed out that copolymerizations of low alkenes and sulphur dioxide would not proceed above certain temperatures characteristic of the olefin and independent of the nature of the catalyst system. These temperatures are termed "ceiling temperatures". Subsequently a number of reports have appeared describing deviations from normal kinetics (mainly at high temperatures), reactions leading to incomplete conversion of monomer and a final equilibrium concentration of monomer, the value of which is dependent only on the reaction temperature [364—368].

Dainton and Ivin [369] demonstrated that the deviations from normal kinetic behaviour and the phenomenon of ceiling temperatures could be explained by taking the depolymerization reaction (33) into account. Under such circumstances the overall rate of polymerization is

$$-\frac{d[M]}{dt} = k_p[M][R\cdot] - k_u[R\cdot]$$

(k_u is the rate coefficient for depropagation), and the ceiling temperature, T_c is the temperature at which the rate of polymerization is zero, i.e. the rates of propagation and depropagation are equal. By equating the rates of propagation and depropagation, expanding the rate coefficients and solving for T_c, we have

$$T_c = \frac{E_p - E_u}{\Delta S_p^{\ddagger} - \Delta S_u^{\ddagger} + R\ln[M]}$$

$$= \frac{\Delta H}{\Delta S_0 + R\ln[M]} \qquad (34)$$

where $\Delta S^{\ddagger}$ and E are the entropies and energies of activation, respectively, ΔS_0 is the standard entropy change on polymerization and ΔH the heat of polymerization. The driving force for propagation is the heat of polymerization, while depropagation is associated with a large entropy increase. At low temperatures ΔH is greater than $T\Delta S_0$ and propagation occurs, while at high temperatures $T\Delta S_0$ will become the dominant term and depropagation will be favoured.

From eqn. (34) it is seen that T_c is controlled by thermodynamic parameters and not the nature of the active species, consequently eqn. (34) is equally applicable to ionic as well as radical reactions. The concept of the ceiling temperature requires the presence of active centres, and in the absence of such centres polymers can exist in a thermodynamically unstable state at temperatures well above their ceiling temperatures. It is apparent that, in the presence of active centres, at any temperature there exists a concentration of monomer which is in equilibrium with polymer, as defined by eqn. (34). The existence of this equilibrium concentration prevents conversions reaching 100 %. Normally this equilibrium concentration is too low to detect, since the ceiling temperatures of most monomers are high. Perhaps the classic example is α-methyl styrene, which has a ceiling temperature for bulk monomer of 60 °C. It is this low ceiling temperature, due to the low value of ΔH, Table 8, (ΔS_0 is similar for most monomers) which normally prevents the polymerization of α-methyl styrene, except at low temperatures [370]. At 0 °C the equilibrium monomer concentration [371] is 0.76 mole l^{-1}.

Methacrylates have relatively low ceiling temperatures and direct measurements of the equilibrium concentrations of monomer in contact with polymer [365] have agreed with concentrations estimated kinetically [366]. Deviations from simple kinetic behaviour at high temperatures, observed in older data [364], can be explained by taking account of the depropagation reaction. If the effective monomer concentration (total concentration minus equilibrium concentration) is used in the normal kinetic expressions the discrepancies disappear [367], and in this way consistent values of k_p have been obtained for methyl methacrylate up to 155 °C. Bywater [367] has shown with this monomer that, at any temperature, the final equilibrium concentration of monomer is independent of the initial concentration in the temperature range 100—150 °C, and at 132.2 °C the equilibrium concentration is 0.3 mole l^{-1}.

The practical consequences of the ceiling temperature are that as the polymerization temperature is increased (at relatively low temperatures) the rate of polymerization increases but subsequently decreases as depropagation becomes important, eventually falling to zero as the ceiling temperature is approached. In addition, at high temperatures each propagating chain will undergo some depropagation reactions during its lifetime and the final length of the polymer chain will be less than that expected from simple kinetics; as the ceiling temperature is approached the degree of polymerization tends to zero. Both these effects have been observed by Dainton and Ivin [369, 372] in the formation of polysulphones by copolymerization of alkenes and sulphur dioxide.

5.3 VISCOSITY EFFECTS

We have already seen that bimolecular termination reactions are diffusion controlled even in dilute solution, and that these effects are accentuated at high conversions when the medium viscosity becomes very high. It is necessary to enquire if such effects are encountered in the propagation reaction. Intuitively it would not be expected that propagation will be greatly influenced by increases in viscosity at low conversion since it involves the diffusion of a small species; restrictions on diffusion of the propagating radical would not have so serious an effect

TABLE 9
Polymerization of vinyl acetate at 25 °C [12]

Per cent conversion	k_p (l mole^{-1} sec^{-1})	E_p (kcal mole^{-1})
4	895	4.2
23	1290	4.2
46	1980	3.6
57	550	7.4
65	87	11.4

under these conditions. However, when the viscosity becomes very high at high conversions we may anticipate some variation in k_p.

Bengough and Melville [12] carried out a study of vinyl acetate polymerizations taken to high conversion, the rate of polymerization increased with conversion to about 50 % conversion and then decreased to approximately zero at about 75 % conversion, when the reaction mixture was rigid. Absolute values of k_p and k_t, and their activation energies, were obtained from a study of non-stationary-state kinetics. Some of the data, relating to propagation, are presented in Table 9. Hayden and Melville [373] carried out a similar study for the polymerization of methyl methacrylate, where again, in characteristic fashion, the rate of polymerization increased as the conversion increased to 50 % and decreased to approximately zero at 80 % conversion. Data are presented in Table 10.

TABLE 10
Polymerization of methyl methacrylate at 25 °C [373]
(Rate of initiation 2.83 x 10^{-8} mole l^{-1} sec^{-1})

Per cent conversion	k_p [a] (l mole^{-1} sec^{-1})	E_p (kcal mole^{-1})
0	384	5.8
10	259	6.3
20	334	7.2
30	434	8.8
40	615	11.3
50	515	
60	185	
70	54	
80	6	

[a] Corrected for monomer consumption.

Both of the above sets of data show a relatively small increase in k_p as conversion increases from 0 to 50 % conversion, but too small to make a significant contribution to the large increase in rate; the increase in rate up to 50 % conversion is mainly a result of the large decrease in k_t, by a factor of 10^2 for methyl methacrylate. It may also be noted that in the conversion range 0—50 % there is an increase in E_p, the activation energy for propagation, of a few kcal mole^{-1}. This increase in E_p would give a reduction in k_p if it was not counteracted by an increase in the pre-exponential factor, A_p. The variations in both A_p and E_p are consistent with increasing viscosity of the reaction medium, the increase in E_p reflecting the restricted approach of reactants and the increase in A_p the restricted separation. Finally, k_p decreases with increasing conversion and reaction stops short of 100 % conversion. It was further observed that if the temperature was increased after the final conversion had been attained further polymerization could occur. This phenomenon, therefore, is unrelated to limiting conversions in the vicinity of the ceiling temperature.

It was suggested [373] that cessation of reaction at finite conversion was a consequence of immobilization both of monomer and of polymer radicals. Termination thus becomes a physical process of radical trapping; such radical trapping phenomena in highly viscous media are well-known [374]. Indeed, if reaction is stopped at high conversion, but below the limiting conversion, the polymerization reaction has a long after-effect [373]. At the limiting conversion unreacted monomer was considered to be immobilized in the "gelled" medium, and subsequent reaction at higher temperatures was attributed to partial release of the immobilized monomer [373].

Vaughan [326] and Robertson [375] attempted to quantify the kinetics of polymerizations at high conversion using the approach of Rabinowich [376] to diffusion-controlled processes, and critical viscosities at which the various elementary processes in a polymerization become diffusion controlled were estimated in terms of the activation energies of these processes. Thus, Vaughan [326] concluded that, in the polymerization of styrene, termination should become a diffusion-controlled process at a viscosity of 2 poise and propagation at 7×10^5 poise.

Recently, Horie et al. [377] have studied the polymerizations of methyl methacrylate and styrene at high conversions by calorimetric techniques. They obtained qualitatively the same information as the earlier workers for methyl methacrylate polymerizations, i.e. the rate passes through a maximum and then drops to zero short of 100 % conversion, and that subsequent heating of a sample to a higher temperature following the attainment of the limiting conversion brings about further polymerization. These workers calculated the glass-transition temperatures of the monomer—polymer mixtures at the limiting conversion, taking chain-end effects into account and assuming simple additivity of free volumes of unreacted monomer and of polymer. In this way it was demonstrated that the attainment of the limiting conversion corresponded to glassification of the reaction mixture. That is, reaction stops when the reaction mixture becomes a glass and large scale molecular motions, such as segmental motion of the polymer chains, effectively cease. Further heating of such a sample to a higher temperature puts that sample above its glass-transition temperature and reaction can proceed until glassification occurs once again. This interpretation seems entirely reasonable since diffusion of a monomer molecule to an active radical will require co-operative motion of segments of the polymer chain and, since this cannot occur in the glass, diffusion and hence, propagation are prevented.

In the case of styrene polymerizations Horie et al. [377] obtained 100 % conversions at temperatures above 90 °C, and it was demonstrated that the glass-transition temperature of bulk polymer, of the low molecular weights produced in these reactions, was below the reaction temperature; i.e. the reaction mixture never became a glass. Subsequently, it was shown [378] that, in anionic polymerizations carried out at lower temperatures, limiting conversions were again reached when the reaction mixture became a glass. These results indicate that, in a true glass containing large amounts of polymer, extensive polymerization is impossible. In all polymerizations taken to high conversions, especially at relatively low temperatures, the reactions are ultimately controlled by physical rather than chemical factors.

It is interesting to note that in the polymerizations of methyl methacrylate taken to high conversion, Hayden and Melville [373] observed that both k_t and k_p were somewhat dependent on the rate of initiation and therefore, on the molecular weights of the polymer formed. It may be that these variations are consistent with variations in diffusion and segmental motion as a result of variations in free volume in the reaction mixtures and glass-transition temperatures of the polymers.

5.4 EFFECTS OF HIGH PRESSURES

For an elementary process the variation in rate with pressure, p, is described by the relation

$$\frac{\mathrm{d}(\ln k)}{\mathrm{d}p} = -\frac{\Delta V^{\ddagger}}{RT}$$

where k is the rate coefficient of the process and $\Delta V^{\ddagger}$ is the difference in partial molar volumes of the transition state and the reactants. The overall effect of pressure on a polymerization reaction is complex and depends on the sensitivity of the various component reactions, including transfer, to pressure.

Thermal dissociation of initiators is associated with the extension of a valence bond and hence, a positive value of $\Delta V^{\ddagger}$. Consequently rates of thermal initiation by azo [379] and peroxide catalysts [380, 381] are decreased by increasing pressure. This reduction in rate of initiation at high pressures is counteracted by the fact that propagation is associated with a contraction. Both propagation and termination have negative values of $\Delta V^{\ddagger}$ and the rates increase with increasing pressure.

Increases in k_p have been reported for polymerizations of styrene [380, 382, 383] and α-methyl styrene [384]. At pressures above about 7000 atm k_p for α-methyl styrene decreased [384]; the reason for this is not clear but may be the result of increased viscosity or phase separation. k_t for styrene polymerization was also found, by Nicholson and Norrish [380], to decrease at high pressures, possibly as a result of increased diffusion control.

The combined effects of pressure on the various elementary processes is expected to produce an increase in degree of polymerization. In the polymerization of styrene, Merrett and Norrish [382] found that degrees of polymerization increased with pressure and then approached a limiting value at high pressures, an effect which was partly attributed to increasing transfer to monomer. More recently, Guarise et al. [385] have investigated the kinetics of styrene polymerizations under high pressures and have obtained results in agreement with the previous data of Nicholson and Norrish [380]; they calculated $\Delta V^{\ddagger}$ for the composite parameter $k_p/k_t^{1/2}$ as -18 cm^3 mole^{-1} in agreement with an earlier value [386] of -19.5 cm^3 mole^{-1}, from the same laboratory. Guarise et al. [385] also reported that the transfer constant to monomer varies little with pressure up to about 4000 atm.

The effect of pressure on the ceiling temperature of α-methyl styrene has been investigated [384] and was found to increase linearly with increasing pressure from 61 °C at 1 atm to 171 °C at 6480 atm. Effects of high pressure on polymerization processes have been reviewed by Weale [387].

5.5 SOLVENT EFFECTS

Unless solvent or monomer takes part in the mechanism of formation of primary radicals, rates of polymerization are normally considered to be independent of the nature of the solvent medium. That is, the solvent is taken to be an inert diluent and the rates of polymerization are controlled by eqn. (4). However, from time to time reports of anomalous rates in various solvents have appeared in the literature. Many of these effects have been reductions in rate and have been attributed to the simultaneous occurrence of transfer and retardation by the transfer radical; these effects are discussed in the following chapter. In addition there has been a recent debate on the significance of higher rates in the presence of solvents than would be expected on the basis of eqn. (4); that is an apparent order in monomer less than unity.

Burnett et al. [388, 389] have reported that anomalously high rates are observed for polymerizations of methyl methacrylate in various halogenated solvents, using thermal initiation by AIBN. Although the experimental data did not indicate a higher rate of decomposition of AIBN in the various solvents (see, however, Sect. 3.5.1.a), and did not show an enhanced rate of incorporation of initiator fragments, or any incorporation of solvent fragments, into the polymer, these workers attributed the high rates to increased rates of initiation in the presence of the solvents. A possible mechanism of increasing the efficiency of initiation through the participation of an initiator fragment—solvent—monomer complex, producing a monomer derived radical, was postulated. The solvent effects were correlated with reactivities of phenyl and methyl radicals.

Henrici-Olivé and Olivé [390] offered another explanation for the solvent effects, based on radical complexing. These workers pointed out that the electron affinities of typical propagating radicals are about the same as those of chloranil and tetracyanoethylene which form charge-transfer complexes with a number of common solvents used in free-radical polymerizations. They considered the situation where the propagating radicals can form complexes with both monomer and solvent and that only the monomer-complexed radical can propagate. According to their theory the rate of polymerization is given by

$$-\frac{d[M]}{dt} = k_p \left(\frac{\mathscr{I}}{k_t}\right)^{1/2} \frac{[M][M]_0}{[M] + [S]\tau_s/\tau_M} \tag{35}$$

where τ_s and τ_M are the lifetimes of the solvent and monomer-complexed radicals, respectively. Under the conditions $[M]_0 = [S] + [M]$ ($[M]_0$ is the concentration of bulk monomer), eqn. (35) predicts that if $\tau_s = \tau_M$ the solvent acts as an inert diluent, and the simple relation (4) is obeyed, while if $\tau_M > \tau_s$ higher rates than predicted by eqn. (4) are observed and if $\tau_M < \tau_s$ lower rates are predicted. A number of systems have been studied [390], the results of which appear to be in substantial agreement with the radical complex hypothesis.

Bamford and Brumby [391] pointed out that, in these previous studies, no account had been taken of effects on k_t due to diffusion-controlled termination. Measurements of absolute values of k_p and k_t showed that k_t varies only with the viscosity of the solvent system and that only part of the observed solvent effects are due to variations in k_p. The variations in k_p were generally in accord with the ideas on complexing expressed by Henrici-Olivé and Olivé.

Burnett et al. [392] have subsequently measured absolute rate coefficients for polymerizations of methyl methacrylate initiated by azo-compounds, and have obtained results generally in accord with those of Bamford and Brumby [391]; the main effect on the rate resulted from variations in k_t with solvent viscosity, some contribution from variations in k_p and a small solvent effect on thermal initiation. These workers suggest that possibly the variation in k_p arises from a variation in radical solvation rather than strong complexing of radicals and solvent. Burnett et al. [393] have also recently modified eqn. (35) to take account of variations in solvent viscosity and have calculated values of the ratios of the lifetimes of the complexed radicals in some solvents. While the modified equation appeared to be obeyed some anomalies remained and, as these workers point out, the effects of solvents on the propagation reaction are not yet understood.

5.6 THE INFLUENCE OF METAL SALTS

We have already met cases where metal derivatives influence rates of initiation and in a subsequent chapter their role in termination will be discussed. In addition, metal salts are capable of influencing polymerization reactions in other ways. If the monomer is an electrolyte and the polymer a polyelectrolyte, salting-in and salting-out effects are possible. It is also possible for the salts to complex either with monomer or radical, thereby affecting their reactivities, and in this section we are mainly concerned with this type of effect. In such cases the salt effects will not saturate until sufficient salt is added to complex the whole of the appropriate species, so that if monomer complexing alone is involved saturation will not be achieved until [salt] is of the same order as [monomer].

In many cases the influence of salts has been detected through their effect on copolymer compositions, reflecting changes in the reactivities of the reacting species. An early example of this type concerned the copolymerization of methyl methacrylate and acrylonitrile, for which it was reported that the presence of aluminium chloride increased the rates of reaction and decreased the acrylonitrile content of the polymer [394]. Lithium chloride is reported to have comparatively little effect on compositions of copolymers from acrylonitrile and vinylidene chloride [395], but some effect on rates, while it has considerable influence on copolymer compositions from acrylonitrile and *p*-vinylbenzamide [396]. In the latter case complex formation between the amide group on the monomer and lithium chloride was considered to be responsible for some of the effects; complexes of the salt with *p*-vinylbenzamide and acrylamide were prepared. Zinc chloride is reported to affect copolymer compositions from various acrylonitrile containing systems [397], and to change the reactivity ratios r_1, r_2 (defined in Chap. 4) in copolymerizations of methyl methacrylate (M_1) and vinylidene chloride (M_2), r_1 increasing and r_2 decreasing in the presence of the salt [398]. Stannic chloride affects copolymer compositions in copolymerizations of methyl methacrylate with vinylidene chloride [399] and of vinyl chloride with acrylonitrile and ethyl acrylate [400]. In a number of cases the apparent reactivity ratios vary with the relative concentrations of salt and monomer; in the stannic chloride—acrylonitrile and stannic chloride—ethyl acrylate containing systems the reactivity ratios appear to approach a limit when the ratio [salt]/[monomer] is ~ 2. Copolymerizations of the above types are by no means simple and should perhaps be considered as complex multi-component systems.

Probably zinc chloride containing systems have been studied more than any other. Imoto et al. [401] observed that the presence of zinc chloride increased rates of polymerization for various monomers containing ester and nitrile groups. 1:1 complexes of the salt with acrylonitrile and with methyl methacrylate were obtained and increased rates were attributed to enhanced propagation, the complexed monomers having a higher reactivity than free monomer. The acrylonitrile—zinc chloride 1:1 complex was obtained as a crystalline solid, and its IR spectrum indicated complexing with the nitrile group [402]; in polar solvents the complex partly dissociated. High temperature (150—200 °C) thermal polymerization of the complex produced a red-brown polymer, and both the vinyl and nitrile groups were considered to be capable of entering into polymerization. In radical-initiated polymerizations at lower temperatures, only the vinyl groups were involved.

The 1 : 1 methyl methacrylate : zinc chloride complex polymerizes to very high polymers and, during the formation of the complex, polymerization may be initiated by visible light [398]. In addition a 2 : 1 methyl methacrylate : zinc chloride complex can be formed, this complex is a colourless viscous liquid [398]. The C=C stretching frequency of complexed monomer is at longer wavelengths than in the free monomer.

Imoto et al. [398] observed that addition of zinc chloride to methyl methacrylate, in non-stoichiometric amounts, resulted in increased rates of free-radical polymerization, due to increases in the kinetic parameter $k_p/k_t^{1/2}$. The increases calculated from kinetic data and from polymer molecular weights were different, the former being the larger. It was concluded that increased rates resulted from increases in k_p and that zinc chloride acted as a transfer agent. Bamford et al. [403] refuted the latter conclusion, but supported the former; measurements of absolute rate coefficients demonstrated that k_t remained constant and, in bulk monomer, k_p increased from 2.5×10^2 l mole^{-1} sec^{-1} to about 6.1×10^2 l mole^{-1} sec^{-1} on addition of 1.42 mole l^{-1} zinc chloride. These latter workers concluded from their data and previous copolymerization data of Imoto et al. [398], that radical complexing occurs in addition to monomer complexing. It was also suggested that k_t is not reduced by the presence of the salt since the viscosity of the monomer—complex solution arises from the presence of a small concentration of a polymeric form of the complex and some support for this was obtained from vapour pressure measurements [403].

Zinc chloride is also reported to increase rates of polymerization of vinyl acetate [404] and allyl acetate [405]. In the latter case the effect was attributed to a removal of degradative chain transfer by converting the normally inactive transfer radical into a reactive complexed species.

Lithium chloride incresases rates and degrees of polymerization of

References pp. 92—103

acrylonitrile [406]; measurements of absolute rate coefficients demonstrated that this effect was due to an increase in k_p, k_t remaining unaffected. Chain transfer to carbon tetrachloride and triethylamine was increased, but to different extents. Lithium nitrate and perchlorate had less effect on acrylonitrile polymerizations and the benzoate had no effect. Lithium chloride had a small effect on methacrylonitrile polymerizations but no effect on styrene. Lithium chloride was also reported to induce the anionic polymerization of acrylonitrile at low concentrations of radical initiator. From a consideration of the absolute rate coefficients for propagation and transfer it was concluded that the effects arise from a small amount of radical complexing with both lithium and chloride ions, the lithium-complexed radical being mainly responsible for increases in transfer and the chloride-complexed radical for increases in the rate coefficient for propagation.

Aluminium chloride and bromide are reported to affect the polymerizations of methyl methacrylate [405]. At low halide concentrations ($[AlX_3]/[\text{monomer}] < 0.1$) rates of polymerization increase and degrees of polymerization decrease with increasing halide concentration. It was thought that the rate increases were due to enhanced propagation and that degrees of polymerization were reduced by increased transfer. It was concluded that complexed radicals were responsble for these effects and some supporting ESR evidence was quoted. At higher halide concentrations both rates and degrees of polymerization increased, and polymerization through organized arrays of monomer molecules was proposed.

Polymerizations of vinylpyridines in *N,N*-dimethylformamide solutions are affected by salts of non-oxidizing metals, such as zinc and cadmium compounds [407, 408], and a number of complexes with zinc salts have been prepared. The salts accelerate the polymerization of 4-vinylpyridine [408] in the order, $Cd(CH_3COO)_2 > ZnCl_2 > Zn(CH_3COO)_2 > ZnBr_2 > ZnI_2$, and a similar order was observed for accelerations of 2-vinylpyridine polymerizations [407], while the same materials reduce rates of polymerization of 2-methyl-5-vinylpyridine. The influence of complexing on monomer reactivities has been discussed in terms of IR spectra of free and complexed monomers, and of steric effects. Detailed kinetic analysis of the rate data is complicated by the fact that both physical and chemical factors can be involved. At low salt concentrations the zinc ions cause precipitation of poly(4-vinylpyridine), probably through formation of intermolecular cross-links as a result of 2 : 1 complex formation with vinylpyridine units in the polymer. At high zinc concentrations, where formation of a 1 : 1 complex is favoured, solubilization occurs as the cross-links are broken; details of these effects vary with the nature of the counter-ion. The other vinylpyridine polymers are soluble in the presence of zinc salts possibly as a result of steric hindrance preventing complex formation. However, it was concluded that a major effect on propagation in these systems stems from the nature of the metal—pyridine complexes and changes in electron density distribution in the monomer influencing the reactivity of the vinyl groups. Results of copolymerizations of the vinylpyridines with styrene supported the idea that variations in monomer reactivity are responsible for the observed effects. In the particular case of 2-vinylpyridine in the presence of zinc salts, the unusual situation $r_1 r_2 > 1$ was observed and attributed to strong steric effects.

Polymerization of 2-methylvinylimidazole is decelerated by zinc salts [260], but detailed kinetic discussion is not possible since, even at high [Zn(II)], the system remains heterogeneous as a result of the high stability constant of the 2 : 1 complex formed between monomer units in the polymer and the zinc salt. As a result of copolymerization experiments with styrene the effect was attributed to decreased reactivity of the monomer through complex formation with the imidazole ring, essentially isolating the vinyl group from its substituent.

Obviously there is now considerable evidence for the existence of monomer and/or radical complexes from polymerization data and we have referred to some examples where specific complexes have been identified. Williams [409] has recently obtained polarographic evidence for the existence of 1 : 1 methyl methacrylate complexes with $LiNO_3$, $La(NO_3)_3 \cdot 6H_2O$, $MnCl_2 \cdot 4H_2O$, $FeCl_3 \cdot 6H_2O$, $Co(NO_3)_2 \cdot 6H_2O$ and $ZnCl_2$ and 2 : 1 monomer : salt complexes with $MnCl_2 \cdot 4H_2O$, and for acrylonitrile 1 : 1 complexes with LiCl, $LiNO_3$, $Mg(NO_3)_2 \cdot 6H_2O$, $FeCl_2 \cdot 4H_2O$, $FeCl_3 \cdot 6H_2O$, $MnCl_2 \cdot 4H_2O$, $Co(NO_3)_2 \cdot 6H_2O$, $Ni(NO_3)_2 \cdot 6H_2O$ and $ZnCl_2$ and 2 : 1 complexes with $MnCl_2 \cdot 4H_2O$, $FeCl_3 \cdot 6H_2O$, $Co(NO_3)_2 \cdot 6H_2O$ and $Ni(NO_3)_2 \cdot 6H_2O$. He has observed some effects of these salts on copolymerization data.

There has been some discussion on the effects of salts on the microstructure of the polymers. Kargin et al. [410] reported that zinc chloride caused the formation of isotactic polymethyl methacrylate in contrast to the normal predominantly syndiotactic product. This observation was refuted by Bovey [411] on the basis of NMR evidence; Williams [409] did not find any influence on polymer tacticities by the salts mentioned above. Recently, Okuzawa et al. [412] prepared polymers from the 1 : 1 and 1 : 2 salt : monomer complexes of methyl methacrylate with zinc and stannic chlorides. These workers reported that, especially for materials prepared at low temperatures, the polymers had tacticities different to the normal polymers. They report that the tacticity of the polymer varies less rapidly with temperature than in the absence of the salt and that the tacticity appears to be mainly controlled by the nature of the complex. Similar results were reported by Hirai et al. [413] for polymerizations of similar complexes of methacrylonitrile. In this latter case, for the 2 : 1 monomer : salt complexes, the proportions of intra- and inter-complex propagations were estimated.

5.7 EFFECTS OF ADDED POLYMER

The addition, to a polymerization system, of polymers which interact strongly with the monomer is of obvious interest since in this way the monomer will associate with the added, or "template" polymer, increasing the local monomer concentration and possibly positioning the monomer molecules in situations suitable for "zipping-up" along the template polymer. Two groups of workers have studied the polymerization of low concentrations of acrylic acid (10^{-2}—10^{-1} mole l^{-1}) in the presence of basic polymers.

Ferguson and Shah [414, 415] carried out their polymerizations in the presence of polyvinylpyrrolidone and of polyethyleneimine in aqueous solution. While the addition of these polymers in concentrations (base mole l^{-1}) of the same order as the [monomer] certainly produced large increases in rate of polymerization, detailed kinetic interpretation of their results is impossible in view of the complexities of the system. Under most experimental conditions precipitation accompanied polymerization and this undoubtedly contributed to the increase in rate, since the addition of polyethylene oxide also produced increases in rate (although not so large as the basic polymers) and caused precipitation to occur [415]. Under conditions of pH such that the polyvinylpyrrolidone—polyacrylic acid system remained homogeneous no rate increases due to the presence of the basic polymer were observed [414]. These workers concluded that, since a maximum rate was observed when [acrylic acid] was approximately equal to the concentration of vinylpyrrolidone residues, at least some of the rate increase resulted from association of the acidic monomer with the basic polymer [414].

Bamford and Shiiki [416] investigated the acrylic acid—polyethyleneimine system at 25 °C in acetone—water mixtures in which the reaction mixture remained homogeneous during polymerization. Under their reaction conditions no polymerization was observed in the absence of the template polymer. Addition of the template caused the initial rate of polymerization to increase dramatically, pass through a maximum corresponding to about 30 % polymerization per min at base molar concentrations of template approximately equal to the initial monomer concentration, and then decrease at higher polymer concentrations. At low [template] the polymer is saturated with monomer and the rate increases with increasing [template], eventually there is insufficient monomer to saturate the polymer and further increase in [template] causes the rate to decrease as the continuity of the adsorbed monomer is interrupted. Confirmation of a "template" mechanism was provided by observations that tetramethylene pentamine did not produce increases in rate and that addition of non-polymerizable acids reduced the effects of the added polymer by occupying some of the sites on the template. The rate of polymerization in the presence of the template was described by the equation

$$\text{Rate} = C \exp(-\alpha Z^{\beta}/\boldsymbol{k}T)[\text{X}]$$

where $\boldsymbol{k}$ and T are the Boltzmann constant and the absolute temperature, respectively, [X] is the concentration of adsorbed monomer and $Z = [T]/[X]$, where [T] is the concentration of free sites on the template. α, β are constants and C is a function describing the total free radical concentration on the template molecules.

5.8 POLYMER STRUCTURES

5.8.1 Mode of propagation

It is the detailed nature of the propagation reaction which determines the structure of the polymer molecule. While the propagation reaction is usually written as

$$\sim\!\!\begin{array}{c} X \\ | \\ C\cdot \\ | \\ Y \end{array} + CH_2{=}CXY \longrightarrow \sim\!\!\begin{array}{c} X \\ | \\ C \\ | \\ Y \end{array}\!\!-CH_2-\!\!\begin{array}{c} X \\ | \\ C\cdot \\ | \\ Y \end{array}$$

the true situation is more complex. There are four possible basic modes for addition of a monomer unit to a propagating radical, viz.

$$\sim CH_2-\dot{C}XY + CH_2{=}CXY \longrightarrow \sim CH_2-CXY-CH_2-\dot{C}XY \quad (36a)$$

$$\sim CH_2-\dot{C}XY + XYC{=}CH_2 \longrightarrow \sim CH_2-CXY-CXY-\dot{C}H_2 \quad (36b)$$

$$\sim CXY-\dot{C}H_2 + XYC{=}CH_2 \longrightarrow \sim CXY-CH_2-CXY-\dot{C}H_2 \quad (36c)$$

$$\sim CXY-\dot{C}H_2 + CH_2{=}CXY \longrightarrow \sim CXY-CH_2-CH_2-\dot{C}XY \quad (36d)$$

If propagation takes place exclusively by either reaction (36a) or (36c) the polymer will have a regular head-to-tail or 1,3-structure. If reactions (36b), (36d) also contribute to propagation then head-to-head and tail-to-tail groupings will be incorporated in the final polymer. Finally, if all the reactions contribute to a significant extent the structure of the polymer will be approximately random, the relative rates of the various reactions controlling the detailed structure.

The predominance of head-to-tail structures in polystyrene was demonstrated in 1935 by Staudinger and Steinhoffer [417] who destructively distilled the polymer and found a series of products having phenyl groups on alternate carbon atoms, but no products with phenyl groups on adjacent carbon atoms. Over the years, it has generally been found that the addition polymers prepared from vinyl monomers have a predominantly 1,3-structure [26, 418—420]. This overall picture is consistent with propagation by either reaction (36a) or (36c).

The general preference for head-to-tail addition is readily understandable since head-to-head groupings will often introduce considerable steric strain into the polymer, and a radical of the type produced in eqn. (36b) may be susceptible to depropagation; in addition, polar factors in the transition state would also oppose this mode of propagation. Of reactions (36a) or (36c) which would lead to the observed structures, (36a) is preferred. This situation arises since not only is attack by the radical on the unsubstituted end of the vinyl group sterically easier but the radicals so formed will usually be stabilized by the substituent groups on the terminal carbon atom; the propagating styrene radical may be written

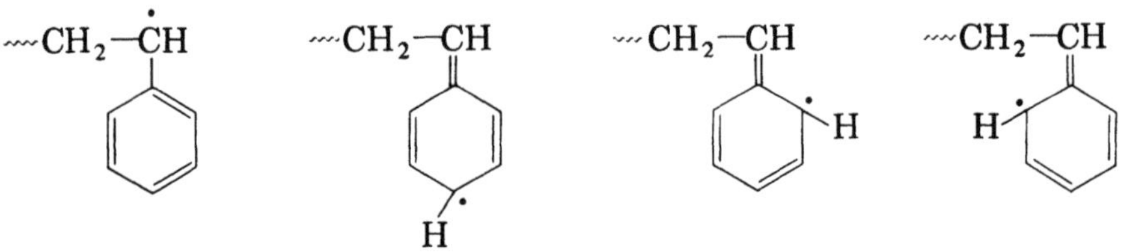

Direct experimental evidence that propagation occurs predominantly through reaction (36a) has been obtained by electron spin resonance techniques. There are many known examples of ESR studies on radicals trapped in polymers [374] as a result of precipitation of growing polymer or restricted termination in viscous media. Symons [422] demonstrated that the 9-line spectrum from radicals trapped in methyl methacrylate by a variety of methods, including polymerization to high conversion, could be interpreted in terms of preferred conformations of the propagating radical

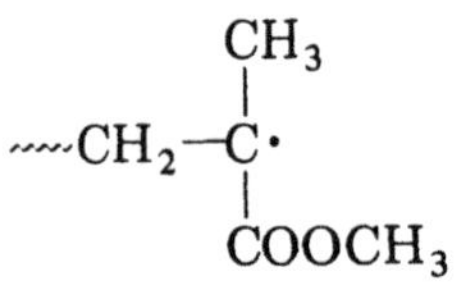

(VIII)

It has subsequently been confirmed by various methods, including the use of deuterated monomers [423], that the above radical is responsible for the spectrum obtained from solid samples. Many other workers have obtained similar spectra from related solid systems, supporting the general concept of preferred conformations [424]. Subsequently, Fischer [160, 425] and others [426, 427] obtained highly resolved ESR spectra from radicals in solution, using a flow technique, for a variety of monomers and demonstrated that the active centres could all be interpreted in terms of preferred conformations of radicals analogous to (VIII), which is taken as the normal propagating radical.

General evidence that (36a) is the predominant propagation reaction does not preclude a small amount of reaction by other processes, producing a small amount of "wrong-way" addition. While Marvel and

Denoon [418] could not detect any 1,2-diglycol units in polyvinyl alcohol by consumption of periodic acid, demonstrating a preference for 1,3-structures, Flory and Leutner [428] demonstrated the existence of 1,2-diglycol units in the polymer by using a more sensitive technique. These latter workers treated polyvinyl alcohol with periodic acid and observed a decrease in the viscosity of an aqueous solution of the polymer due to scission of the 1,2-diglycol units. It was estimated that the polymer contained about 1 % 1,2-units. This result means that in the polymerization of vinyl acetate, the polyvinyl alcohol being obtained by the hydrolysis of polyvinyl acetate, about 1 % of the monomer units react by (36b). It is perhaps not surprising that "wrong-way" addition does occur in the polymerization of vinyl acetate, since among the common monomers the vinyl acetate radical is one of the most reactive and the substituent acetate group contributes little to the stabilization of the radical.

Evidence for some contribution from 1,2-structures has been obtained from other polymers prepared by free-radical processes. Nuclear magnetic resonance studies of polyvinyl fluoride [429] and polyvinylidine fluoride [430] and their copolymers [431] have detected the presence of head-to-head structures in these materials; in polyvinylidine fluoride it was estimated that 5 % of the monomer units in a chain were reversed [430].

It is probable that (36a) is the predominant mode of addition in all vinyl free-radical polymerizations, but it should not be overlooked that the other propagation reactions can occur in some cases. We can conclude that in general the measured absolute rate coefficients for propagation are effectively those of reaction (36a), while the contribution from the alternative reactions will be small.

The preference for a radical to add to the unsubstituted end of the C=C bond of the monomer extends to the initiation reaction. ESR studies have demonstrated that $H_3C\cdot$ radicals add exclusively to the unsubstituted end of the monomer but that only 90 % of $HO\cdot$ radicals add in this way [426], 10 % adding to the α-carbon atom.

In addition it is possible for other structural irregularities to occur during chain growth in some systems. Polymethacrylonitrile, especially if it is prepared at low temperatures, can contain some ketene—imine linkages [432] due to the propagating radical reacting as

$$\sim\!\!\sim CH_2-C(CH_3)=C=N\cdot$$

It has been proposed by Grassie and McNeill [432] that the ketene—imine linkages are formed during the termination reaction.

5.8.2 Microstructure

It has long been recognized that, in a normal head-to-tail polymer prepared from a vinyl monomer, every alternate carbon atom is asymmetric. Consequently it is possible to consider the structure of the polymers in terms of the configurations of those asymmetric centres (i.e. d- or l-). Since free radicals may generally be considered as free, that is the structure of the propagating species is not subject to the effects of solvents or counterions which have a large influence on ionic polymerizations, the configurations of the asymmetric carbon atoms will be determined by steric and electronic interactions of the substituents on the monomer. Although most polymers prepared by free-radical mechanisms have regular head-to-tail structures they are amorphous and often uncrystallizable. This lack of crystallinity is attributable to a random arrangement of the asymmetric centres [433]. Following the preparation of crystalline polymers using the Ziegler-type catalysts [434, 435] a renewed interest in the detailed structure of the polymers developed. Natta et al. [436] defined three types of polymer chain; an *isotactic* chain is one in which all asymmetric carbon aroms in the chain have the same configuration (all d- or all l-), in a *syndiotactic* chain adjacent asymmetric centres are alternately d- and l-, and in an *atactic* chain the distribution of configurations is random. The isotactic (IX) and syndiotactic (X) structures may be represented as

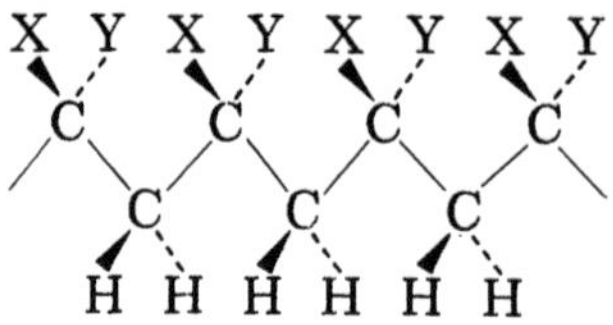

(IX)

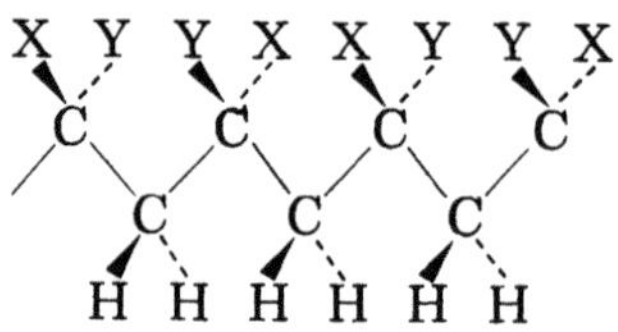

(X)

On the basis of these definitions all free-radically prepared polymers are atactic and, for a detailed discussion of structures, definitions of local structure are required. Natta [434] considered two adjacent monomer units in a chain to be isotactically bonded if their asymmetric carbon atoms are in the same configuration, and syndiotactically bonded if in the opposite configuration. This type of definition is convenient for studying the environment of the methylene groups by nuclear magnetic resonance [437]. To study structures of the α-carbon atoms by NMR it is more convenient to work in terms of triads of monomer units and Bovey and Tiers [437] have defined tacticity in this way. Accordingly, an isotactic triad is defined as one in which the asymmetric centres are ddd or lll, the central unit of such a sequence is an isotactic unit. Similarly syndiotactic sequences are dld or ldl and the remaining sequences ddl, lld; dll, ldd are defined as heterotactic. On the basis of these definitions it is possible to characterize structures of chains over the whole range from complete order to complete randomness.

Huggins [433] predicted on theoretical grounds that the stereospecificity of polymer chains should increase as the polymerization temperature is decreased. Various workers [438] have considered the effects of non-bonding interactions in determining the preferred mode of propagation, and all reach the conclusion that syndiotactic placements are energetically preferred. Thus, the activation energies for isotactic and syndiotactic placement are expected to be different and low temperature polymerizations should produce a more syndiotactic product.

Parameters have been proposed to put the structures on a quantitative basis. Coleman [439] defined α and β as the probabilities of iso- and syndiotactic placements, respectively ($\alpha + \beta = 1$). A preference for syndiotacticity implies $\beta > 1/2$. The temperature variation in β and the probabilities of forming sequences of n units with the same configuration can be calculated from the difference in enthalpy of activation for the two modes of addition [440]. Bovey and Tiers [437] defined a parameter σ (effectively the same as Coleman's α) as the probability of a unit adding to a chain in the same configuration as the preceding unit, i.e. an isotactic placement. Then, the probabilities of forming iso-, syndio- and heterotactic sequences, P_i, P_s and P_h, respectively, are given by Bernoulli trial statistics as

$$P_i = \sigma^2$$

$$P_s = (1 - \sigma)^2$$

$$P_h = 2(\sigma - \sigma^2)$$

A random polymer has $\sigma = 0.5$.

Bovey and Tiers [437, 441] analysed the NMR spectrum of polymethyl methacrylate in terms of the above probabilities and found that over a wide range of conditions the free-radical polymerization is defined by a single value of σ which is dependent only on the temperature in the range -78 °C to 100 °C. From the temperature variation in σ the difference in activation entropies and enthalpies for isotactic and syndiotactic propagation, $\Delta(\Delta H_p^\ddagger)$ and $\Delta(\Delta S_p^\ddagger)$ where

$$\Delta(\Delta H_p^\ddagger) = \Delta H_i^\ddagger - \Delta H_s^\ddagger$$

$$\Delta(\Delta S_p^\ddagger) = \Delta S_i^\ddagger - \Delta S_s^\ddagger$$

were estimated. The values obtained [411] were $\Delta(\Delta H_p^\ddagger) = 775 \pm 75$ cal mole^{-1} and $\Delta(\Delta S_p^\ddagger) = 0.0 \pm 1$ eu. Fox and Schneko [442], working over a wider range of conditions, confirmed the applicability of a single value of σ at a given polymerization temperature and found $\Delta(\Delta H_p^\ddagger) = 1.07 \pm 0.05$ kcal mole^{-1} and $\Delta(\Delta S_p^\ddagger) = 0.99 \pm 0.1$ eu.

The polymerization of menthyl methacrylate is apparently described by a single value of σ slightly greater than that for methyl methacrylate at the same temperature [443]. Vinyl acetate polymers appear to be almost random [444], $\sigma = 0.45$ or 0.55. Analysis of some polymers is more uncertain but there appears to be a general preference for syndiotactic

References pp. 92—103

addition. With methacrylic acid [445] syndiotactic addition is preferred throughout the pH range 2—10, the ratio of the rate coefficient for syndiotactic addition to that for isotactic addition increases with increasing pH, showing that the tendency for syndiotactic propagation is greater between likely charged species. Vinyl trifluoroacetate produces an almost random polymer at 65 °C, but the polymerization is not described by a single value of σ [446]. Polymerization of chlorotrifluoroethylene appears to be controlled by steric factors alone, since the probability for syndiotactic placement is twice as great as that for isotactic placement at all temperatures [447]. Fischer [448] has demonstrated that it is possible to predict polymer structures from a consideration of the preferred conformations of the propagating radical which result from hindered rotation around the terminal carbon—carbon bond, and the modes of approach of the monomer molecules.

Diene monomers have additional modes of propagation, namely, *cis*- and *trans*-1,4 addition, as in eqn. (32). With butadiene vinyl type addition proceeds to the extent of about 20 % and for isoprene both 1,2 and 3,4 addition occur to the extent of about 5 %. *Trans*-1,4 addition is the main mode of propagation with both monomers, the *cis*-1,4 content of the polymer increases with increasing temperature from about zero at —20 °C to about 25 % at 100 °C [449]. Thus in both cases the main mode of addition is *trans*-1,4, which compares with the structures of the monomers [450] where in 1,3-butadiene the *trans* configuration is the more stable and the monomer contains about 4 % of the *cis*- form, while the *cis*-configuration of isoprene is the more stable and the monomer exists mainly in this form.

REFERENCES

1 Report of the Cambridge Symposium, Trans. Faraday Soc., 32 (1936).
2 G. Gee and H. W. Melville, Trans. Faraday Soc., 40 (1944) 240.
3 C. H. Bamford, W. G. Barb, A. D. Jenkins and P. F. Onyon, The Kinetics of Vinyl Polymerization by Radical Mechanisms, Butterworths, London, 1958.
4 N. N. Semenov, Chemical Kinetics and Chain Reactions, Oxford University Press, New York, 1935.
5 C. Chachaty, M. Magat and L. T. Minassion, J. Polym. Sci., 48 (1960) 139.
6 N. S. Marans and W. D. Addy, J. Appl. Polym. Sci., 9 (1965) 3661.
7 Glacial Methacrylic Acid — Glacial Acrylic Acid, Rohm and Haas Bulletin, Philadelphia, Pa., 1960.
8 K. G. Denbigh, Trans. Faraday Soc., 43 (1947) 648.
9 A. D. Jenkins, Polymer, 1 (1960) 79.
10 G. M. Burnett, Mechanisms of Polymer Reactions, Interscience, New York, 1954.
11 F. E. Treloar, Polymer, 1 (1960) 513.
12 W. I. Bengough and H. W. Melville, Proc. R. Soc. London, Ser. A, 225 (1951) 330; 230 (1955) 429.
13 N. Grassie and H. W. Melville, Proc. R. Soc. London, Ser. A, 207 (1951) 285.

14 C. M. Burrell, T. G. Majury and H. W. Melville, Proc. R. Soc. London, Ser. A, 205 (1951) 309. T. G. Majury and H. W. Melville, Proc. R. Soc. London, Ser. A, 205 (1951) 323, 496.
15 G. M. Burnett, Trans. Faraday Soc., 46 (1950) 772.
16 C. H. Bamford and M. J. S. Dewar, Nature (London), 158 (1946) 380; Proc. R. Soc. London, Ser. A, 192 (1948) 309.
17 G. V. Schulz and G. Sing, J. Prakt. Chem., 161 (1943) 160.
18 C. H. Bamford and M. J. S. Dewar, Proc. R. Soc. London, Ser. A, 197 (1949) 356.
19 G. Dixon-Lewis, Proc. R. Soc. London, Ser. A, 198 (1949) 510.
20 D. W. E. Axford, Proc. R. Soc. London, Ser. A, 197 (1949) 374.
21 P. F. Onyon, Trans. Faraday Soc., 51 (1955) 400.
22 F. Briers, D. L. Chapman and E. Walters, J. Chem. Soc., (1926) 562.
23 W. A. Noyes and P. A. Leighton, The Photochemistry of Gases, Reinhold, New York, 1941.
24 G. M. Burnett and H. W. Melville, Proc. R. Soc. London, Ser. A, 189 (1947) 456.
25 M. S. Matheson, E. E. Auer, E. B. Bevilacqua and E. J. Hart, J. Am. Chem. Soc., 71 (1949) 497.
26 P. J. Flory, Principles of Polymer Chemistry, Cornell University Press, Ithaca, New York, 1953.
27 W. G. Burns and F. S. Dainton, Trans. Faraday Soc., 46 (1950) 411.
28 W. V. Smith and R. H. Ewart, J. Chem. Phys., 16 (1948) 592.
29 W. V. Smith, J. Am. Chem. Soc., 70 (1948) 3695.
30 M. Morton, P. Sallatiello and H. Landfield, J. Polym. Sci., 8 (1952) 111, 215, 279.
31 A. V. Tobolsky, C. E. Rogers and R. D. Brickman, J. Am. Chem. Soc., 82 (1960) 1277.
32 H. N. Friedlander, J. Polym. Sci., 58 (1962) 455.
33 J. C. Bevington, Radical Polymerization, Academic Press, New York, 1961.
34 J. C. Bevington and T. D. Lewis, Polymer, 1 (1960) 1.
35 C. H. Bamford, A. D. Jenkins and R. Johnston, Proc. R. Soc. London, Ser. A, 239 (1957) 214.
36 E. Collinson, F. S. Dainton and G. S. McNaughton, J. Chim. Phys., 52 (1955) 556.
37 R. M. Noyes, J. Am. Chem. Soc., 77 (1955) 2042.
38 R. M. Noyes, Z. Elektrochem., 64 (1960) 153; Prog. React. Kinet., 1 (1961) 129.
39 R. M. Noyes, Encyclopedia of Polymer Science and Technology, Vol. 2, Interscience, New York, 1965.
40 O. D. Trapp and G. S. Hammond, J. Am. Chem. Soc., 81 (1959) 4876.
41 G. S. Hammond, O. D. Trapp, R. T. Keys and D. L. Neff, J. Am. Chem. Soc., 81 (1959) 4878.
42 H. P. Waits and G. S. Hammond, J. Am. Chem. Soc., 86 (1964) 1911.
43 L. Hark, M. Feld and M. Szwarc, J. Am. Chem. Soc., 83 (1961) 2998.
44 C. A. Barson and J. C. Bevington, Tetrahedron, 4 (1958) 147.
45 J. C. Bevington and J. Toole, J. Polym. Sci., 28 (1958) 413.
46 J. C. Bevington, Proc. R. Soc. London, Ser. A, 239 (1957) 420.
47 R. J. Orr and H. L. Williams, J. Am. Chem. Soc., 77 (1955) 3715.
48 W. H. Atkinson, C. H. Bamford and G. C. Eastmond, Trans. Faraday Soc., 66 (1970) 1446.
49 Ch. S. Bagdasarjan, Theory of Radical Polymerization, Moscow, 1959.
50 J. C. Bevington and B. W. Malpass, J. Polym. Sci., Part A, 2 (1964) 1893.
51 C. A. Barson, J. C. Bevington and D. E. Eaves, Trans. Faraday Soc., 54 (1958) 1678.
52 J. C. Bevington, J. Toole and L. Trosarelli, Trans. Faraday Soc., 54 (1958) 863.

53 J. C. Bevington, J. Toole and L. Trosarelli, Makromol. Chem., 28 (1958) 237.
54 J. C. Bevington, J. Toole and L. Trosarelli, Makromol. Chem., 32 (1959) 57.
55 J. Pavlinic and M. Lazar, J. Polym. Sci., Part C, 22 (1968) 297.
56 M. Szwarc, J. Polym. Sci., 16 (1955) 367.
57 A. V. Tobolsky, J. Am. Chem. Soc., 80 (1958) 5927.
58 R. H. Gobran, M. B. Berenbaum and A. V. Tobolsky, J. Polym. Sci., 46 (1960) 431.
59 K. F. O'Driscoll and S. A. McArdle, J. Polym. Sci., 40 (1959) 557.
60 K. F. O'Driscoll and J. Schmidt, J. Polym. Sci., 45 (1960) 189.
61 C. B. Wooster, Macromolecules, 1 (1968) 324.
62 P. B. Ayscough, J. M. Pearson and H. E. Evans, J. Phys. Chem., 68 (1964) 3889.
63 F. M. Lewis and M. S. Matheson, J. Am. Chem. Soc., 71 (1949) 747.
64 C. G. Overberger, M. T. O'Shaughnessy and H. Shalit, J. Am. Chem. Soc., 71 (1949) 2661.
65 D. C. Blackley and H. W. Melville, Makromol. Chem., 18/19 (1956) 16.
66 M. Talat-Erben and S. Bywater, J. Am. Chem. Soc., 77 (1955) 3710, 3712.
67 L. M. Arnett, J. Am. Chem. Soc., 74 (1952) 2027.
68 J. W. Breitenbach and A. Schindler, Monatsh. Chem., 83 (1952) 724.
69 C. E. H. Bawn and S. F. Mellish, Trans. Faraday Soc., 47 (1951) 1216.
70 G. S. Hammond, J. N. Sen and C. E. Boozer, J. Am. Chem. Soc., 77 (1955) 3244.
71 J. P. van Hook and A. V. Tobolsky, J. Am. Chem. Soc., 80 (1958) 779.
72 J. Betts, F. S. Dainton and K. J. Ivin, Trans. Faraday Soc., 58 (1962) 1203.
73 R. C. Peterson, J. H. Markgraff and S. D. Ross, J. Am. Chem. Soc., 83 (1961) 3819.
74 G. Henrici-Olivé and S. Olivé, Makromol. Chem., 58 (1962) 188.
75 A. F. Moroni, Makromol. Chem., 105 (1967) 43.
76 D. Lim, Collect. Czech. Chem. Commun., 33 (1968) 1122.
77 J. C. Bevington and A. Wahid, Polymer, 4 (1963) 129.
78 K. E. Russel, J. Am. Chem. Soc., 77 (1955) 3487.
79 G. S. Hammond and R. C. Neuman, J. Am. Chem. Soc., 85 (1963) 1501.
80 S. D. Ross and M. A. Fineman, J. Am. Chem. Soc., 73 (1951) 2176.
81 C. E. H. Bawn and R. G. Halford, Trans. Faraday Soc., 51 (1955) 780.
82 C. G. Overberger and M. B. Berenbaum, J. Am. Chem. Soc., 73 (1951) 2618.
83 C. G. Overberger and A. Lebovits, J. Am. Chem. Soc., 76 (1954) 2722.
84 C. G. Overberger, W. F. Hale, M. B. Berenbaum and A. B. Finestone, J. Am. Chem. Soc., 76 (1954) 6185.
85 S. G. Cohen and C. H. Wang, J. Am. Chem. Soc., 77 (1955) 2457.
86 S. G. Cohen, S. J. Groszos and D. B. Sparrow, J. Am. Chem. Soc., 72 (1950) 3947.
87 C. G. Overberger, J. G. Lombardino, I. Tashlick and R. G. Hiskey, J. Am. Chem. Soc., 79 (1957) 2662.
88 S. G. Cohen and C. H. Wang, J. Am. Chem. Soc., 77 (1955) 3628.
89 A. T. Blomquist and I. A. Bernstein, J. Am. Chem. Soc., 73 (1951) 5546.
90 S. G. Cohen and C. H. Wang, J. Am. Chem. Soc., 75 (1953) 5504.
91 J. E. Leffler and R. A. Hubbard, J. Org. Chem., 19 (1954) 1089.
92 S. Solomon, C. H. Wang and S. G. Cohen, J. Am. Chem. Soc., 79 (1957) 4104.
93 D. L. Tuben, W. G. Bentrude and J. C. Martin, J. Am. Chem. Soc., 85 (1963) 1938.
94 J. C. Masson, in J. Brandrup and E. H. Immergut (Eds.), Polymer Handbook, New York, 1966.
95 R. Zand, Encyclopedia of Polymer Science and Technology, Vol. 2, Interscience, New York, 1965.
96 L. M. Arnett and J. H. Peterson, J. Am. Chem. Soc., 74 (1952) 2031.
97 J. C. Bevington, Trans. Faraday Soc., 51 (1955) 1392.

98 C. H. Bamford, A. D. Jenkins and R. Johnston, Trans. Faraday Soc., 58 (1962) 1212.
99 O. P. Strausz, J. W. Lown and H. E. Gunning, in C. H. Bamford and C. F. Tipper (Eds.), Comprehensive Chemical Kinetics, Vol. 5, Elsevier, Amsterdam, 1972.
100 P. Smith and A. M. Rosenberg, J. Am. Chem. Soc., 81 (1959) 2037.
101 S. Brumby, Z. Naturforsch., Teil A, 25 (1970) 12.
102 C. H. Bamford, R. W. Dyson and G. C. Eastmond, Polymer, 10 (1969) 885.
103 J. C. Bevington and H. G. Troth, Makromol. Chem., 53 (1962) 200.
104 T. Hirano, T. Miki and T. Tsuruta, Makromol. Chem., 104 (1967) 230.
105 C. H. Bamford, R. Denyer and J. Hobbs, Polymer, 8 (1967) 494.
106 C. H. Bamford, G. C. Eastmond, J. Hobbs and E. Hughes, Polymer, 10 (1969) 713.
107 A. G. Davies, Organic Peroxides, Butterworths, London, 1961.
108 J. O. Edwards (Ed.), Peroxide Reaction Mechanisms, Interscience, New York, 1962.
109 E. G. E. Hawkins, Q. Rev., 4 (1950) 251; Organic Peroxides, D. van Nostrand, Princeton, N.J., 1961.
110 A. V. Tobolsky and R. B. Mesrobian, Organic Peroxides, Interscience, New York, 1954.
111 C. G. Swain, W. H. Stockmayer and J. T. Clarke, J. Am. Chem. Soc., 72 (1950) 5426.
112 J. Ulbricht, in J. Brandrup and E. H. Immergut (Eds.), Polymer Handbook, Interscience, New York, 1966.
113 A. I. Lowell and J. R. Price, J. Polym. Sci., 43 (1960) 1.
114 H. J. Shine, J. A. Waters and D. M. Hoffman, J. Am. Chem. Soc., 85 (1963) 3613.
115 W. A. Pryor, M. D. Huston, T. R. Fiske, T. L. Pickering and E. Ciaffarin, J. Am. Chem. Soc., 86 (1964) 4237.
116 O. L. Mageli, S. D. Butaka and D. J. Bolton, Evaluation of Organic Peroxides from Half-life Data, Wallace and Tiernan, Lucidol Div., Bulletin 3030.
117 J. K. Allen and J. C. Bevington, Proc. R. Soc. London, Ser. A, 262 (1961) 271.
118 C. Walling and G. Metzger, J. Am. Chem. Soc., 81 (1959) 5365.
119 J. H. Raley, F. F. Rust and W. E. Vaughan, J. Am. Chem. Soc., 70 (1948) 1336.
120 J. A. Offenbach and A. V. Tobolsky, J. Am. Chem. Soc., 79 (1957) 278.
121 L. J. Durham and H. S. Mosher, J. Am. Chem. Soc., 82 (1960) 4537.
122 N. A. Milas and A. Golubovic, J. Am. Chem. Soc., 80 (1958) 5994.
123 W. A. Strong, Ind. Eng. Chem., Prod. Res. Dev., 3 (1964) 264.
124 P. D. Bartlett and R. E. Pincock, J. Am. Chem. Soc., 82 (1960) 1769.
125 S. G. Cohen and D. B. Sparrow, J. Am. Chem. Soc., 72 (1950) 611.
126 F. Strain, W. E. Bissinger, W. R. Dial, H. Rudoff, B. J. de Witt, H. G. Stevens and J. H. Langston, J. Am. Chem. Soc., 72 (1950) 1254.
127 V. Stannett and R. B. Mesrobian, J. Am. Chem. Soc., 72 (1950) 4125.
128 D. F. Dochnert and O. L. Mageli, Mod. Plast., 36 (1959) 142.
129 R. R. Hiatt and W. M. J. Strachan, J. Org. Chem., 28 (1963) 1893.
130 P. D. Bartlett, E. P. Benzing and R. E. Pincock, J. Am. Chem. Soc., 82 (1960) 1762.
131 A. T. Blomquist and A. F. Ferris, J. Am. Chem. Soc., 73 (1951) 3412.
132 A. T. Blomquist and A. F. Ferris, J. Am. Chem. Soc., 73 (1951) 3408.
133 E. L. O'Brien, F. M. Berringer and R. B. Mesrobian, J. Am. Chem. Soc., 81 (1959) 1506.
134 W. G. Bentrude and J. C. Martin, J. Am. Chem. Soc., 84 (1962) 1561.
135 H. Hart and D. P. Wyman, J. Am. Chem. Soc., 81 (1959) 4891.
136 P. D. Bartlett and R. R. Hiatt, J. Am. Chem. Soc., 80 (1958) 1398.
137 P. D. Bartlett and D. M. Simons, J. Am. Chem. Soc., 82 (1960) 1753.

138 M. S. Karasch, A. Fono and W. Nudenberg, J. Org. Chem., 16 (1951) 105.
139 G. S. Hammond and U. S. Nandi, J. Am. Chem. Soc., 83 (1961) 1213.
140 A. T. Blomquist and A. J. Buselli, J. Am. Chem. Soc., 73 (1951) 3883.
141 B. Barnett and W. E. Vaughan, J. Phys. Chem., 51 (1947) 926.
142 A. Conix and G. Smets, J. Polym. Sci., 10 (1953) 525.
143 G. S. Hammond, J. Am. Chem. Soc., 70 (1948) 1208.
144 D. J. Brown, J. Am. Chem. Soc., 72 (1950) 3737.
145 B. Barnett and W. E. Vaughan, J. Phys. Chem., 51 (1947) 942.
146 G. S. Hammond and L. M. Soffer, J. Am. Chem. Soc., 72 (1950) 4711.
147 H. C. Bailey and G. W. Godin, Trans. Faraday Soc., 52 (1956) 68.
148 L. E. Redington, J. Polym. Sci., 3 (1948) 503.
149 F. C. Bains, J. H. Grezlak and A. V. Tobolsky, J. Polym. Sci., Part A-1, 7 (1969) 3297.
150 J. C. Bevington and T. D. Lewis, Trans. Faraday Soc., 54 (1958) 1340.
151 H. M. Frey, Proc. Chem. Soc., (1959) 385.
152 T. F. Ferrington and A. V. Tobolsky, J. Am. Chem. Soc., 77 (1955) 4510.
153 N. Uri, Chem. Rev., 50 (1952) 375.
154 W. G. Barb, J. H. Baxendale, P. George and K. R. Hargrave, Trans. Faraday Soc., 47 (1951) 462.
155 J. H. Baxendale, M. G. Evans and G. S. Park, Trans. Faraday Soc., 42 (1946) 155.
156 F. S. Dainton and R. H. Seaman, J. Polym. Sci., 39 (1959) 279.
157 P. G. Ghosh, A. R. Mukherjee and S. R. Palit, J. Polym. Sci., Part A-2, 2 (1964) 2817.
158 P. Davies, M. G. Evans and W. C. E. Higginson, J. Chem. Soc., (1961) 2563.
159 T. Sugimura, N. Yasumoto and Y. Minoura, Kogyo Kagaku Zasshi, 68 (1965) 2513.
160 H. Fischer, Z. Naturforsch., Teil A, 19 (1964) 866.
161 J. H. Merz and W. A. Waters, Discuss. Faraday Soc., 2 (1947) 179.
162 J. W. L. Fordham and H. L. Williams, J. Am. Chem. Soc., 73 (1951) 1634, 4855.
163 I. M. Kolthoff, A. I. Medalia and H. P. Raaen, J. Am. Chem. Soc., 73 (1951) 1733.
164 C. E. H. Bawn and D. Margerison, Trans. Faraday Soc., 51 (1955) 925.
165 S. Fronaeus and C. O. Östman, Acta Chem. Scand., 10 (1956) 320.
166 D. M. Yost, J. Am. Chem. Soc., 48 (1926) 152.
167 L. B. Morgan, Trans. Faraday Soc., 42 (1946) 169.
168 K. L. Berry and J. H. Peterson, J. Am. Chem. Soc., 73 (1951) 5195.
169 S. Yuguchi and M. Hosina, Kobunshi Kagaku, 18 (1962) 381.
170 P. Ghosh, S. C. Chadha and S. R. Palit, J. Polym. Sci., Part A, 2 (1964) 4441.
171 D. H. Irvine, J. Chem. Soc., (1958) 2166; (1959) 2977.
172 M. F. Betchold, (to E. I. du Pont de Nemours and Co. Inc.) U.S. Pat. 2,972,511, (1961).
173 A. Hunyar and E. Roth, Ger. Pat., 1,052,687, (1959).
174 J. A. Price, W. M. Thomas and J. J. Padbury (to American Cyanamid Co.) U.S. Pat. 2,626,946, (1953).
175 J. M. Hamilton, Ind. Eng. Chem., 45 (1953) 1347.
176 W. K. Wilmarth and A. Hain, in J. O. Edwards (Ed.), Peroxide Reaction Mechanisms, Interscience, New York, 1962, p. 175.
177 S. R. Palit and B. M. Mandal, J. Macromol. Sci., Rev. Macromol. Chem., 2 (1968) 225.
178 E. P. Crematz, J. Polym. Sci., Part A-1, 7 (1969) 3260.
179 W. Kern, Makromol. Chem., 1 (1948) 209, 249.
180 C. S. Marvel, R. Deanin, C. G. Overberger and B. M. Kuhn, J. Polym. Sci., 3 (1948) 128.

181 C. S. Marvel, R. Deanin, B. M. Kuhn and G. B. Landes, J. Polym. Sci., 3 (1948) 433.
182 P. H. Johnson and R. C. Bebb, J. Polym. Sci., 3 (1948) 389.
183 E. B. Milovskaya and T. G. Zhuravleva, Vysokomol. Soedin., 6 (1964) 1035. E. B. Milovskaya, T. G. Zhuravleva and L. V. Zamoskaya, J. Polym. Sci., Part C, 16 (1967) 899.
184 R. J. Orr and H. L. Williams, J. Phys. Chem., 57 (1953) 925.
185 R. J. Orr and H. L. Williams, Can. J. Chem., 30 (1952) 985.
186 C. Walling, Free Radicals in Solution, John Wiley, New York, 1957.
187 F. A. Bovey, I. M. Kolthoff, A. I. Medalia and E. J. Meehan, Emulsion Polymerization, Interscience, New York, 1955.
188 G. Smets, A. Post, M. Mullier and J. P. Bex, J. Polym. Sci., 34 (1959) 287.
189 M. S. Karasch, P. Pauson and W. Nudenberg, J. Org. Chem., 18 (1953) 322.
190 L. Horner and C. Betzel, Ann., 579 (1953) 175. L. Horner, J. Polym. Sci., 18 (1955) 438.
191 L. Horner and E. Schwenk, Ann., 566 (1950) 69.
192 C. Walling and N. Indictor, J. Am. Chem. Soc., 80 (1958) 5814.
193 L. Horner and E. Schwenk, Kunstoffe, 39 (1949) 292.
194 L. Horner and E. Schwenk, Angew. Chem., 61 (1949) 411.
195 R. J. Orr and H. L. Williams, Discuss. Faraday Soc., 14 (1953) 170, 237.
196 K. Noma, O. Nishiura and A. Ichiba, Chem. High Polym. (Tokyo), 10 (1953) 231.
197 J. Lal and R. Green, J. Polym. Sci., 17 (1955) 403.
198 Y. Minoura, T. Sugimura and T. Hirahara, Kogyo Kagaku Zasshi, 70 (1967) 562.
199 G. S. Whitby, N. Wellman, V. W. Floutz and H. L. Stephens, Ind. Eng. Chem., 42 (1950) 445, 452.
200 W. H. Embree, R. Spolsky and H. L. Williams, Ind. Eng. Chem., 43 (1951) 2553.
201 I. M. Kolthoff, W. Stricks and N. Tanaka, J. Am. Chem. Soc., 77 (1955) 5215.
202 C. C. Menon and S. L. Kapur, J. Polym. Sci., 54 (1961) 45.
203 R. K. Das and S. R. Palit, J. Polym. Sci., Part C, 16 (1967) 141.
204 A. R. Mukherjee, P. G. Ghosh, S. C. Chadha and S. R. Palit, Makromol. Chem., 97 (1966) 202.
205 C. E. H. Bawn and A. G. White, J. Chem. Soc., (1951) 331, 339, 343.
206 F. S. Dainton and D. G. L. James, Trans. Faraday Soc., 54 (1958) 649.
207 B. C. Mitra, P. Ghosh and S. R. Palit, Makromol. Chem., 98 (1966) 285.
208 G. Oster and N-L. Yang, Chem. Rev., 68 (1968) 125.
209 N. Uri, Discuss. Faraday Soc., 2 (1947) 217.
210 M. G. Evans and N. Uri, Nature (London), 164 (1949) 404; J. Soc. Dyers Colour., 65 (1949) 709.
211 M. G. Evans, M. Santappa and N. Uri, J. Polym. Sci., 7 (1951) 243.
212 F. S. Dainton and W. D. Sisley, Trans. Faraday Soc., 59 (1963) 1377; A. M. Biswas and A. R. Mukherjee, Indian J. Chem., 3 (1965) 412.
213 M. K. Saha, A. R. Mukherjee, P. Ghosh and S. R. Palit, J. Polym. Sci., Part C, 16 (1966) 159.
214 E. M. Arnett and M. A. Mendelsohn, J. Am. Chem. Soc., 84 (1962) 3821.
215 E. M. Arnett and M. A. Mendelsohn, J. Am. Chem. Soc., 84 (1962) 3824.
216 E. G. Kastning, H. Naarman, H. Reis and C. Berding, Angew. Chem., Int. Ed. Engl., 4 (1965) 322.
217 K. Uehara, T. Nishi, T. Matsumura, F. Tamura and N. Murata, Kogyo Kagaku Zasshi, 70 (1967) 755.
218 (a) C. H. Bamford and D. J. Lind, Proc. R. Soc. London, Ser. A, 302 (1968) 145. (b) C. H. Bamford and A. N. Ferrar, J. Chem. Soc. D, (1970) 315.
219 C. H. Bamford, G. C. Eastmond and J. A. Rippon, Trans. Faraday Soc., 59 (1963) 2548.

220 J. Itakura, H. Tanaka and H. Ito, Kogyo Kagaku Zasshi, 70 (1967) 1227.
221 T. Otsu, Y. Nishikawa and S. Aoki, Kogyo Kagaku Zasshi, 71 (1968) 1067.
222 K. Kaeriyama, Bull. Chem. Soc. Jpn., 43 (1970) 1511.
223 Y. Nishikawa and T. Otsu, Kogyo Kagaku Zasshi, 72 (1969) 1836.
224 K. Uehara, T. Matsumura, Y. Muratu, M. Tanaka and N. Murata, Kogyo Kagaku Zasshi, 72 (1969) 1825.
225 K. Uehara, Y. Kataoka, K. Kamei, M. Tanaka and N. Murata, Kogyo Kagaku Zasshi, 72 (1969) 1831.
226 S. Aoki, S. Matsumura and T. Otsu, Bull. Chem. Soc. Jpn., 42 (1969) 2574.
227 P. E. M. Allen and T. H. Goh, Eur. Polym. J., 5 (1969) 335.
228 R. K. Freydlina and A. B. Belyavskii, Dokl. Akad. Nauk S.S.S.R., 127 (1959) 1027; Izv. Akad. Nauk S.S.S.R., Otd. Khim. Nauk, 1 (1961) 177.
229 C. H. Bamford and C. A. Finch, Proc. R. Soc. London, Ser. A, 268 (1962) 553.
230 C. H. Bamford and C. A. Finch, Trans. Faraday Soc., 59 (1963) 118.
231 C. H. Bamford, R. Denyer and G. C. Eastmond, Trans. Faraday Soc., 61 (1965) 1459.
232 C. H. Bamford, G. C. Eastmond and W. R. Maltman, Trans. Faraday Soc., 62 (1966) 2531.
233 C. H. Bamford, R. Denyer and G. C. Eastmond, Trans. Faraday Soc., 62 (1966) 688.
234 C. H. Bamford, G. C. Eastmond and W. R. Maltman, Trans. Faraday Soc., 60 (1964) 1432.
235 C. H. Bamford, G. C. Eastmond and P. Murphy, Trans. Faraday Soc., 66 (1970) 2598.
236 C. H. Bamford, and C. A. Finch, Trans. Faraday Soc., 59 (1963) 540.
237 C. H. Bamford and R. Denyer, Trans. Faraday Soc., 62 (1966) 1567.
238 C. H. Bamford, G. C. Eastmond and W. R. Maltman, Trans. Faraday Soc., 61 (1965) 267.
239 C. H. Bamford, G. C. Eastmond and K. Hargreaves, Trans. Faraday Soc., 64 (1968) 1611.
240 W. Strohmeier and P. Hartman, Z. Naturforsch., Teil B, 19 (1964) 882.
241 C. H. Bamford, R. Denyer and G. C. Eastmond, Z. Naturforsch., Teil B, 22 (1967) 580.
242 C. H. Bamford and G. C. Eastmond, Pure Appl. Chem., 12 (1966) 183.
243 C. H. Bamford and K. Hargreaves, Trans. Faraday Soc., 63 (1967) 392.
244 C. H. Bamford and K. Hargreaves, Proc. R. Soc. London, Ser. A, 297 (1967) 425.
245 C. H. Bamford, G. C. Eastmond and K. Hargreaves, Trans. Faraday Soc., 64 (1968) 175.
246 C. H. Bamford, G. C. Eastmond and F. T. Fildes, Proc. R. Soc. London, Ser. A, 326 (1972) 431.
247 C. H. Bamford, G. C. Eastmond and F. T. Fildes, Chem. Commun., (1970) 144.
248 C. H. Bamford, G. C. Eastmond and F. T. Fildes, Proc. R. Soc. London, Ser. A, 326 (1972) 453.
249 C. H. Bamford, P. A. Crowe and R. P. Wayne, Proc. R. Soc. London, Ser. A, 284 (1965) 455.
250 C. H. Bamford, P. A. Crowe, J. Hobbs and R. P. Wayne, Proc. R. Soc. London, Ser. A, 292 (1966) 153.
251 C. H. Bamford, R. W. Dyson, G. C. Eastmond and D. Whittle, Polymer, 10 (1969) 759.
252 C. H. Bamford and J. Paprotny, J. Chem. Soc. D, (1971) 140; Polymer, 13 (1972) 208.
253 C. H. Bamford, G. C. Eastmond and V. J. Robinson, Trans. Faraday Soc., 60 (1964) 751.

254 C. H. Bamford, G. C. Eastmond and D. Whittle, Polymer, 10 (1969) 771.
255 G. Henrici-Olivé and S. Olivé, Makromol. Chem., 88 (1965) 117.
256 T. Otsu and M. Yamaguchi, J. Macromol. Sci., Chem., 3 (1969) 177.
257 T. Otsu, S. Aoki, M. Nishimura, M. Yamaguchi and Y. Kusuki, J. Polym. Sci., Part A-1, 7 (1969) 3269. S. Aoki, C. Shirafuji, Y. Kusuki and T. Otsu, Makromol. Chem., 126 (1969) 8. T. Otsu and S. Kubota, Kobunshi Kagaku Zasshi, 27 (1970) 330.
258 S. Aoki, S. Kubota and T. Otsu, J. Polym. Sci., Part A-1, 7 (1969) 1567.
259 S. Tazuke and S. Okamura, J. Polym. Sci., Part A-1, 4 (1966) 141.
260 S. Tazuke and S. Okamura, J. Polym. Sci., Part A-1, 7 (1969) 851.
261 S. Tazuke and K. Imamura, private communication.
262 K. V. Vankatarao and M. Santappa, J. Polym. Sci., Part A-1, 5 (1967) 637.
263 V. Mahadeva and M. Santappa, J. Polym. Sci., 50 (1961) 361.
264 G. Oster, Nature (London), 173 (1954) 300. G. Oster and Y. Mizutani, J. Polym. Sci., 22 (1956) 177. G. K. Oster, G. Oster and J. R. Nussbaum, Polym. Prepr., Am. Chem. Soc., Div. Polym. Chem., 1 (1961) 290.
265 G. K. Oster, G. Oster and G. Prati, J. Am. Chem. Soc., 79 (1957) 595.
266 G. Delzenne, S. Toppet and G. Smets, J. Polym. Sci., 48 (1960) 347.
267 S. Toppet, G. Delzenne and G. Smets, J. Polym. Sci., Part A, 2 (1964) 1539.
268 T. Otsu, T. Sato and M. Ko, J. Polym. Sci., Part A-1, 7 (1969) 3329.
269 C. E. H. Bawn, W. H. Janes and A. M. North, J. Polym. Sci., 58 (1962) 335.
270 J. Furukawa, T. Tsuruta and Y. Takeda, Kogyo Kagaku Zasshi, 64 (1961) 1307.
271 C. S. Marvel and R. C. Woolford, J. Am. Chem Soc., 80 (1958) 830.
272 G. S. Kolesnikov and T. A. Soboleva, Khim. Nauka Prom., 2 (1957) 663. J. Furukawa, T. Tsuruta and S. Inoue, J. Polym. Sci., 26 (1957) 234.
273 J. Furukawa and T. Tsuruta, J. Polym. Sci., 28 (1958) 227.
274 C. E. H. Bawn, D. Margerison and N. M. Richardson, Proc. Chem. Soc., (1959) 397.
275 F. J. Welch, J. Polym. Sci., 61 (1962) 243.
276 Y. Ogata and Y. Minoura, Kogyo Kagaku Zasshi, 66 (1963) 1707.
277 M. Lee and Y. Minoura, Kogyo Kagaku Zasshi, 73 (1970) 125.
278 T. Otsu, M. Yamaguchi, T. Nakata, K. Murata and M. Imoto, J. Macromol. Sci., Chem., 1 (1967) 1457; and earlier papers.
279 T. Nakata, T. Otsu and M. Imoto, J. Polym. Sci., Part A, 3 (1965) 3383.
280. F. R. Mayo, J. Am. Chem. Soc., 75 (1953) 6133.
281 F. R. Mayo, Polym. Prepr., Am. Chem. Soc., Div. Polym. Chem., 2 (1961) 55; paper presented at I.U.P.A.C. Symposium, Moscow, 1961.
282 F. R. Mayo, J. Am. Chem. Soc., 90 (1968) 1289.
283 K. Kirchner, Makromol. Chem., 96 (1966) 179.
284 K. Kirchner, Makromol. Chem., 128 (1969) 150.
285 K. F. Müller, Makromol. Chem., 79 (1964) 128.
286 A. Lebovits and W. C. Teach, J. Polym. Sci., 47 (1960) 527. H. Benoit and and C. Loucheux, C. R. Acad. Sci., Paris, 251 (1960) 382.
287 W. G. Brown, Makromol. Chem., 128 (1969) 130.
288 C. Aso, T. Kunitake and Y. Tanaka, Bull. Chem. Soc. Jpn. 38 (1965) 675.
289 C. Aso, T. Kunitake, M. Shinsenji and H. Miyazaki, J. Polym. Sci., Part A-1, 7 (1969) 1497.
290 C. Aso, T. Kunitake and M. Shinsenji, J. Polym. Sci., Part A-1, 8 (1970) 797.
291 C. Walling and E. R. Briggs, J. Am. Chem. Soc., 68 (1946) 1141.
292 C. H. Bamford and M. J. S. Dewar, Discuss. Faraday Soc., 2 (1947) 310.
293 M. H. Mackey and H. W. Melville, Trans. Faraday Soc., 46 (1950) 63.
294 M. Imoto and K. Takemoto, Makromol. Chem., 98 (1966) 74; Bull. Chem. Soc., Jpn., 40 (1967) 413. K. Takemoto and M. Imoto, Cellul. Chem. Technol., 3 (1966) 347.

295 K. Takemoto, M. Sasahe and M. Imoto, Makromol. Chem., 125 (1969) 294.
296 P. A. Leeming, R. S. Lehrle and J. C. Robb, The Chemistry of Polymerization Processes, Soc. Chem. Ind. (London), Monograph, 20 (1966) 203.
297 N. C. Billingham, P. A. Leeming, R. S. Lehrle and J. C. Robb, J. Polym. Sci., Part C, 16 (1965) 3423.
298 N. C. Billingham, J. R. Eldon, R. S. Lehrle, I. L. Markham and J. C. Robb, Trans. Faraday Soc., 65 (1969) 470.
299 A. Nakano and Y. Minoura, Polymer, 10 (1960) 1.
300 N. Grassie, The Chemistry of High Polymer Degradation Processes, Butterworths, London, 1956.
301 B. Baysal and A. V. Tobolsky, J. Polym. Sci., 8 (1952) 529.
302 C. H. Bamford and A. D. Jenkins, Nature (London), 176 (1955) 78.
303 C. H. Bamford and A. D. Jenkins, Trans. Faraday Soc., 56 (1960) 907.
304 P. W. Allen, A. G. Ayrey and C. G. Moore, J. Polym. Sci., 36 (1959) 55.
305 C. H. Bamford, R. W. Dyson and G. C. Eastmond, J. Polym. Sci., Part C, 16 (1967) 2425.
306 I. M. Kolthoff, P. R. O'Connor and J. L. Hansen, J. Polym. Sci., 15 (1955) 459.
307 J. C. Bevington, H. W. Melville and R. P. Taylor, J. Polym. Sci., 12 (1954) 449.
308 F. R. Mayo, R. A. Gregg and M. S. Matheson, J. Am. Chem. Soc., 73 (1951) 1691.
309 C. A. Baker and R. J. P. Williams, J. Chem. Soc., (1956) 2352.
310 G. M. Burnett and A. M. North, Makromol. Chem., 60 (1964) 960.
311 C. G. Overberger and A. B. Finestone, J. Am. Chem. Soc., 78 (1956) 1638.
312 J. C. Bevington, H. W. Melville and R. P. Taylor, J. Polym. Sci., 14 (1954) 463.
313 G. Ayrey and C. G. Moore, J. Polym. Sci., 36 (1959) 41.
314 G. V. Schulz, G. Henrici-Olivé and S. Olivé, Makromol. Chem., 31 (1959) 88.
315 J. C. Bevington and D. E. Eaves, Trans. Faraday Soc., 55 (1959) 1777.
316 C. H. Bamford, A. D. Jenkins and R. Johnston, Trans. Faraday Soc., 55 (1959) 179.
317 G. M. Burnett, M. H. George and H. W. Melville, J. Polym. Sci., 16 (1955) 31.
318 B. L. Funt and W. Pasika, Can. J. Chem., 38 (1960) 1865.
319 G. Ayrey, F. G. Levitt and R. J. Mazza, Polymer, 6 (1966) 157.
320 A. F. Trotman-Dickenson, Gas Kinetics, Butterworths, London, 1955. J. C. Calvert, Ann. Rev. Chem., 11 (1960) 41.
321 P. S. Dixon, A. P. Stefani and M. Szwarc, J. Am. Chem. Soc., 85 (1963) 255.
322 C. H. Bamford and E. F. T. White, Trans. Faraday Soc., 54 (1958) 268.
323 G. Geuskens, E. Hellinckx and C. David, Eur. Polym. J., 7 (1971) 87. C. David and G. Geuskens, Polymer Degradation, in C. H. Bamford and C. F. H. Tipper (Eds.), Comprehensive Chemical Kinetics, Vol. 14, Elsevier, Amsterdam, 1974.
324 R. G. W. Norrish and R. R. Smith, Nature (London), 150 (1942) 336.
325 E. Tromsdorff, H. Kohle and P. Lagally, Makromol. Chem., 1 (1947) 169.
326 M. F. Vaughan, Trans. Faraday Soc., 48 (1952) 576.
327 G. V. Schulz, Z. Phys. Chem. (Frankfurt am Main), 8 (1956) 290.
328 G. V. Schulz and G. Harborth, Makromol. Chem., 1 (1947) 106.
329 S. W. Benson and A. M. North, J. Am. Chem. Soc., 81 (1959) 1339.
330 S. W. Benson and A. M. North, J. Am. Chem. Soc., 84 (1962) 935.
331 C. R. Patrick, Makromol. Chem., 43 (1961) 248.
332 A. M. North, Makromol. Chem., 49 (1961) 241.
333 A. M. North and G. A. Reed, Trans. Faraday Soc., 57 (1961) 859.
334 A. M. North and G. A. Reed, J. Polym. Sci., Part A, 1 (1963) 1311.
335 J. N. Atherton and A. M. North, Trans. Faraday Soc., 58 (1962) 2049.
336 A. M. North and D. Postlethwaite, Polymer, 5 (1964) 237.
337 L. Onsager and R. M. Fouss, J. Phys. Chem., 36 (1932) 2689.
338 T. B. Grimley, Trans. Faraday Soc., 58 (1962) 2049.

339 R. D. Burkhart, J. Polym. Sci., Part A, 3 (1965) 883.
340 P. E. M. Allen and C. R. Patrick, Makromol. Chem., 72 (1964) 106.
341 A. M. North, Makromol. Chem., 83 (1965) 15.
342 A. M. North, Q. Rev., 20 (1966) 421.
343 J. Hughes and A. M. North, Proc. Chem. Soc., (1964) 404.
344 J. Hughes and A. M. North, Trans. Faraday Soc., 60 (1964) 960.
345 K. Ito, J. Polym. Sci., Part A-1, 7 (1969) 2995.
346 K. Ito, J. Polym. Sci., Part A-1, 7 (1969) 3387.
347 Y. Nishijima, Proc. Int. Conf. Luminescence, Budapest, Hungary, 1966, p. 577.
348 Y. Nishijima, A. Teramoto, M. Yamamoto and S. Hiratsuka, J. Polym. Sci., Part A-2, 5 (1967) 23.
349 A. Teramoto, S. Hiratsuka and Y. Nishijima, J. Polym. Sci., Part A-2, 5 (1967) 37.
350 C. H. Bamford, A. D. Jenkins and R. Johnston, Trans. Faraday Soc., 55 (1959) 1451.
351 T. Manabe, T. Otsumi and S. Okamura, J. Polym. Sci., 58 (1962) 121.
352 S. Okamura and T. Manabe, Polymer, 2 (1961) 83.
353 G. Henrici-Olivé and S. Olivé, Makromol. Chem., 37 (1960) 71.
354 A. M. North, The Kinetics of Free-Radical Polymerization, Pergamon, New York, 1966.
355 P. E. M. Allen and C. R. Patrick, Makromol. Chem., 48 (1961) 89.
356 J. K. Allen and J. C. Bevington, Trans. Faraday Soc., 56 (1960) 1762.
357 W. A. Pryor and T. R. Fiske, Macromolecules, 2 (1959) 62.
358 M. S. Matheson, E. E. Auer, E. B. Bevilacqua and E. J. Hart, J. Am. Chem. Soc., 73 (1951) 1700.
359 M. S. Matheson, E. E. Auer, E. B. Bevilacqua and E. J. Hart, J. Am. Chem. Soc., 73 (1951) 5395.
360 Quoted in ref. 33.
361 J. L. Lang, W. A. Pavelick and H. D. Clarey, J. Polym. Sci., 55 (1961) 531.
362 F. R. Mayo and C. Walling, Chem. Rev., 46 (1950) 191.
363 R. D. Snow and F. E. Frey, Ind. Eng. Chem., 30 (1938) 176.
364 G. V. Schulz and F. Blaschke, Z. Phys. Chem., Abt. B, 51 (1942) 75.
365 P. A. Small, Trans. Faraday Soc., 49 (1953) 441.
366 K. J. Ivin, Trans. Faraday Soc., 51 (1955) 1273.
367 S. Bywater, Trans. Faraday Soc., 51 (1955) 1267.
368 R. E. Cook and K. J. Ivin, Trans. Faraday Soc., 53 (1957) 1132.
369 F. S. Dainton and K. J. Ivin, Q. Rev., 12 (1958) 61.
370 G. G. Lowry, J. Polym. Sci., 13 (1958) 187.
371 D. J. Worsfold and S. Bywater, J. Polym. Sci., 26 (1957) 299.
372 F. S. Dainton and K. J. Ivin, Proc. R. Soc. London, Ser. A, 212 (1952) 207.
373 P. Hayden and Sir Harry Melville, J. Polym. Sci., 43 (1960) 201.
374 See, for example, D. J. E. Ingram, Free Radicals as Studied by Electron Spin Resonance, Butterworths, London, 1958.
375 E. R. Robertson, Trans. Faraday Soc., 52 (1956) 426.
376 E. Rabinowich, Trans. Faraday Soc., 33 (1937) 1225.
377 K. Horie, I. Mita and H. Kambe, J. Polym. Sci., Part A-1, 6 (1968) 2663.
378 I. Mita, private communication.
379 A. M. Ewald, Discuss. Faraday Soc., 22 (1956) 138.
380 A. E. Nicholson and R. G. W. Norrish, Discuss. Faraday Soc., 22 (1956) 104.
381 C. Walling and J. Pellon, J. Am. Chem. Soc., 79 (1957) 4776.
382 F. M. Merrett and R. G. W. Norrish, Proc. R. Soc. London, Ser. A, 206 (1951) 309
383 A. C. Toohey and K. E. Weale, Trans. Faraday Soc., 58 (1962) 2439.
384 J. G. Kilroe and K. E. Weale, J. Chem. Soc., (1960) 3849.

385 G. B. Guarise, G. Palma, E. Siviero and G. Talamini, Polymer, 11 (1970) 613.
386 G. B. Guarise, Polymer, 7 (1966) 497.
387 K. E. Weale, Q. Rev., 16 (1962) 267.
388 D. B. Anderson, G. M. Burnett and A. C. Gowan, J. Polym. Sci., Part A, 1 (1963) 1465.
389 G. M. Burnett, W. S. Dailey and J. M. Pearson, Trans. Faraday Soc., 61 (1965) 1216.
390 G. Henrici-Olivé and S. Olivé, Makromol. Chem., 68 (1963) 219; Z. Phys. Chem. (Frankfurt am Main), 47 (1965) 286; 48 (1966) 3551; Makromol. Chem., 96 (1966) 221.
391 C. H. Bamford and S. Brumby, Makromol. Chem., 105 (1967) 122.
392 G. M. Burnett, G. G. Cameron and M. M. Zafar, Eur. Polym. J., 6 (1970) 823.
393 G. M. Burnett, G. G. Cameron and B. M. Parker, Eur. Polym. J., 5 (1969) 231.
394 H. C. Haas and E. R. Karlin, J. Polym. Sci., 9 (1952) 588.
395 M. Imoto, T. Otsu and K. Murata, Kogyo Kagaku Zasshi, 66 (1963) 1900.
396 T. Asahara, K. Ikeda and N. Yoda, J. Polym. Sci., Part A-1, 6 (1968) 2489.
397 M. Imoto, T. Otsu and B. Yamada, Kogyo Kagaku Zasshi, 68 (1965) 1132.
398 M. Imoto, T. Otsu and Y. Harada, Makromol. Chem., 65 (1963) 180.
399 M. Ibonai and T. Kato, Kogyo Kagaku Zasshi, 70 (1967) 2078.
400 B. Yamada, Y. Kusuki and T. Otsu, Kogyo Kagaku Zasshi, 72 (1969) 364.
401 M. Imoto, T. Otsu and S. Shimizu, Makromol. Chem., 65 (1963) 174.
402 M. Imoto, T. Otsu and M. Nakabayashi, Makromol. Chem., 65 (1963) 194.
403 C. H. Bamford, S. Brumby and R. P. Wayne, Nature (London), 209 (1966) 292.
404 M. Imoto, T. Otsu and T. Ito, Bull. Chem. Soc. Jpn., 36 (1963) 310.
405 V. P. Zubov, M. B. Lachenov, V. B. Golubev, V. F. Kulikova, V. A. Kabanov, L. S. Polak and V. A. Kargin, J. Polym. Sci., Part C, 23 (1968) 147.
406 C. H. Bamford, A. D. Jenkins and R. Johnston, Proc. R. Soc. London, Ser. A, 241 (1957) 364.
407 S. Tazuke, N. Sato and S. Okamura, J. Polym. Sci., Part A-1, 4 (1966) 2461.
408 S. Tazuke and S. Okamura, J. Polym. Sci., Part A-1, 5 (1967) 1085.
409 F. D. Williams, J. Macromol. Sci., Chem., 2 (1968) 459.
410 V. A. Kargin, V. A. Kabanov and V. P. Zubov, Vysokomol. Soedin., 2 (1960) 765.
411 F. A. Bovey, J. Polym. Sci., 47 (1960) 481.
412 S. Okuzawa, H. Hirai and S. Makishima, J. Polym. Sci., Part A-1, 7 (1969) 1039.
413 H. Hirai, T. Ikegami and S. Makishima, J. Polym. Sci., Part A-1, 7 (1969) 2059.
414 J. Ferguson and S. A. O. Shah, Eur. Polym. J., 4 (1968) 343.
415 J. Ferguson and S. A. O. Shah, Eur. Polym. J., 4 (1968) 611.
416 C. H. Bamford and Z. Shiiki, Polymer, 9 (1968) 595.
417 H. Staudinger and A. Steinhoffer, Liebigs Ann. Chem., 517 (1935) 35.
418 C. S. Marvel and C. E. Denoon, J. Am. Chem. Soc., 60 (1938) 1045.
419 C. S. Marvel and C. L. Levesque, J. Am. Chem. Soc., 60 (1938) 280. C. S. Marvel, J. H. Sample and M. F. Roy, J. Am. Chem. Soc., 61 (1939) 3241. C. S. Marvel, E. H. Riddle and J. O. Corner, J. Am. Chem. Soc., 64 (1942) 92. C. S. Marvel, in R. E. Burke and O. Grummitt (Eds.), The Chemistry of Large Molecules, Interscience, New York, 1943.
420 P. J. Flory, J. Am. Chem. Soc., 61 (1939) 1518; 64 (1942) 177.
421 F. T. Wall, J. Am. Chem. Soc., 62 (1940) 803; 63 (1941) 821; 64 (1942) 269.
422 M. C. R. Symons, J. Chem. Soc., (1963) 1188.
423 P. Kourim and K. Vacek, Tetrahedron Lett., (1962) 1051; paper presented at I.U.P.A.C. Symposium, Prague, 1965.
424 G. C. Eastmond, in A. D. Jenkins (Ed.), Progress in Polymer Science, Vol. 2, Pergamon, Oxford, 1971. C. H. Bamford and G. C. Eastmond, Q. Rev., 23 (1969) 271.

425 H. Fischer, J. Polym. Sci., Part B, 2 (1964) 529.
426 H. Fischer and G. Giacometti, J. Polym. Sci., Part C, 16 (1967) 2763.
427 C. Corvaja, H. Fischer and G. Giacometti, Z. Phys. Chem., 45 (1965) 1. H. Yoshida and B. Ranby, J. Polym. Sci., Part C, 16 (1967) 1333.
428 P. J. Flory and L. S. Leutner, J. Polym. Sci., 3 (1948) 880; 5 (1950) 267.
429 C. W. Wilson and E. R. Santee, Paper No. 4, Fourth OCEANS Conf. Abstracts, Session B-3, Pittsburgh, Pa., 1963. J. Polym. Sci., Part C, 8 (1964) 110.
430 R. E. Naylor and S. W. Lasoski, J. Polym. Sci., 44 (1960) 1. C. W. Wilson, J. Polym. Sci., Part A, 1 (1963) 1305.
431 R. C. Ferguson, J. Am. Chem. Soc., 82 (1960) 2416.
432 M. Talat-Erben and S. Bywater, Ric. Sci., Parte 1, Sez. A, 25 (1955) 11. N. Grassie and I. C. McNeill, J. Polym. Sci., 33 (1958) 171.
433 M. L. Huggins, J. Am. Chem. Soc., 66 (1944) 1991.
434 G. Natta, J. Polym. Sci., 16 (1955) 243.
435 G. Natta, P. Pino, P. Corradini, F. Danusso, E. Mantica, G. Mazzanli and G. Moraglio, J. Am. Chem. Soc., 77 (1955) 1708.
436 G. Natta, M. Farina and M. Peraldo, J. Polym. Sci., 43 (1960) 289.
437 F. A. Bovey and G. V. D. Tiers, J. Polym. Sci., 44 (1960) 173.
438 J. W. L. Fordham, J. Polym. Sci., 39 (1959) 321, 355. L. L. Ferstandig and F. C. Goodrich, J. Polym. Sci., 43 (1960) 373. D. J. Cram and K. R. Kopecky, J. Am. Chem. Soc., 81 (1959) 2748. D. J. Cram, J. Chem. Educ., 37 (1960) 317. C. E. H. Bawn, W. R. Janes and A. M. North, J. Polym. Sci., Part C, 4 (1963) 427.
439 B. D. Coleman, J. Polym. Sci., 31 (1958) 155.
440 C. E. H. Bawn and A. Ledwith, Q. Rev., 16 (1962) 361.
441 F. A. Bovey, J. Polym. Sci., 46 (1960) 59.
442 T. G. Fox and H. W. Schneko, Polymer, 3 (1962) 575.
443 H. Sobue, K. Matsuzaki and S. Nakano, J. Polym. Sci., Part A, 2 (1964) 3339.
444 F. A. Bovey, E. W. Anderson, D. C. Douglas and J. A. Manson, J. Chem. Phys., 39 (1963) 1199.
445 G. V. D. Tiers and F. A. Bovey, J. Polym. Sci., 47 (1960) 479. W. L. Miller, W. S. Brey and G. B. Butler, J. Polym. Sci., 54 (1961) 329.
446 K. C. Ramey and N. D. Field, J. Polym. Sci., Part B, 3 (1965) 63.
447 G. V. D. Tiers and F. A. Bovey, J. Polym. Sci., Part A, 1 (1963) 833.
448 H. Fischer, Kolloid Z., 206 (1965) 131.
449 W. S. Richardson, J. Polym. Sci., 13 (1954) 229. V. N. Nikitin and T. V. Yakovleva, J. Phys. Chem. (Moscow), 28 (1954) 697.
450 J. G. Aston, G. J. Szasz, H. W. Woolley and F. G. Brickwedde, J. Chem. Phys., 14 (1946) 67. R. G. Parr and R. S. Mulliken, J. Chem. Phys., 18 (1950) 1338. C. M. Richards and J. R. Nielson, J. Opt. Soc. Am., 40 (1950) 438.

Chapter 2

Chain Transfer, Inhibition and Retardation

G. C. EASTMOND

1. Introduction

The preceding chapter considered the essential processes involved in any free-radical polymerization; namely, chain initiation, chain growth and termination by mutual destruction of propagating radicals. This chapter describes reactions which are frequently encountered but are not essential ingredients of every polymerization reaction; namely, chain transfer, retardation and inhibition. The distinguishing characteristic of these processes is that they interrupt the normal growth of the polymer chains. Often the processes under consideration occur through reaction of the propagating (or initiating) radicals with an added substrate. However, in some cases, such as transfer to monomer, the process may be inherent in a given polymerization performed in the absence of an added substrate. Transfer to monomer and polymer are considered in this chapter, rather than in the preceding chapter, for convenience in discussion.

When a reactive substrate is added to a polymerization system its mode of reaction may, in principle, be identified by the influence of the substrate on the polymerization kinetics over and above that which would occur through simple dilution of the reactants on addition of an inert diluent. The simplest and usual method of classifying substrates through their effects on polymerization reactions is as follows. Chain transfer has no effect on the rate of polymerization (apart from dilution effects) but causes a decrease in the degree of polymerization of the resulting polymer. Inhibition corresponds to a complete suppression of polymerization by the added substrate, the inhibitor, while a retarder reduces both the rate and degree of polymerization. On the basis of these simple qualitative definitons it is often difficult to differentiate clearly between the various modes of behaviour. For example, an additive present at low concentration may reduce the rate of polymerization (i.e. act as a retarder) but may completely suppress the polymerization at a higher concentration (i.e. act as an inhibitor). Thus, the distinction between inhibition and retardation might be considered one of degree. Further, the same additive may act in different ways under different conditions, e.g. in the polymerization of different monomers.

Wherever possible, it is better to differentiate between the types of behaviour in terms of mode of reaction and specific influence on kinetic parameters. The most tenuous distinction is that between retardation and

inhibition. An inhibitor may be defined as a substance which destroys initiating, or primary, radicals (or reduces the rate of formation of initiating radicals) and a retarder as one which destroys the propagating, or chain-carrying radicals. We shall see later that these definitons allow a clear distinction to be drawn between inhibition and retardation in terms of kinetic parameters. These definitions may be applicable when the chemical natures of the initiating and propagating radicals are sufficiently different that a clear distinction can be drawn between their reactivities towards the additive. However, in the majority of polymerizations the nature of initiating and propagating radicals is sufficiently similar that such a clear distinction is not possible and we must resort to the more tenuous differentiation based on the relative efficiencies of monomer and additive as radical scavengers. If the additive is so reactive that at "low" concentration it effectively competes with monomer for radicals and completely suppresses polymerization it may be defined as an inhibitor, while if the additive is not so reactive and only reduces the rate we may define it as a retarder.

2. Chain transfer

2.1 RATES OF POLYMERIZATION

The essential mechanism of a chain-transfer process in free-radical polymerization is described by the equations

$$R_n\cdot + A{-}B \longrightarrow P_n + B\cdot \tag{1a}$$

$$B\cdot + M \longrightarrow BM\cdot\ (R_1\cdot) \tag{1b}$$

where $R_n\cdot$ is a propagating radical containing n monomer units. AB is the transfer agent and A is a labile atom (usually hydrogen, chlorine or bromine) or group of atoms which is abstracted from the transfer agent to form dead polymer P_n and a radical $B\cdot$. Radical $B\cdot$ is termed a transfer radical and reinitiates polymerization by addition of monomer M. Hence, the reactive site is transferred from the original to a new propagating chain. The occurrence of such transfer reactions was first suggested by Flory [1] and it has subsequently been realized that the transfer agent may be monomer, initiator, solvent, added substrate or polymer produced during the polymerization reaction.

The detailed kinetic consequences of the transfer reaction depend on the reactivity of $B\cdot$. If $B\cdot$ has a reactivity at least comparable to that of the propagating radical then $B\cdot$ rapidly reinitiates polymerization, through addition of monomer, regenerating the propagating species and maintaining the rate of polymerization. In such cases AB is a conventional transfer agent. If reinitiation by $B\cdot$ is relatively slow the concentration of B· in the stationary state will be relatively high and these radicals will take part in termination reactions, either with themselves or with propagating radicals. The consequence of this is a reduction in the concentration of reactive radicals compared with the corresponding polymerization in the absence of the transfer agent and hence, the rate of polymerization is decreased by the presence of the substrate. Such processes are termed degradative chain transfer and the kinetic consequences are detailed in the discussion of retarded polymerizations where reinitiation is allowed for (Sect. 3.3.2). Finally, if B· is stable and will not reinitiate polymerization AB is a conventional retarder; this case is also discussed under retardation (Sect. 3.3.1).

A formal kinetic treatment of a polymerization reaction including transfer to monomer and solvent (or added substrate) can be represented by the reaction scheme (2)

Initiation

$$\mathrm{C} \longrightarrow 2\mathrm{R_c}\cdot$$

$$\mathrm{R_c}\cdot + \mathrm{M} \xrightarrow{k_i} \mathrm{R_1}\cdot$$

Propagation

$$\mathrm{R}_n\cdot + \mathrm{M} \xrightarrow{k_p} \mathrm{R}^{\cdot}_{(n+1)}$$

Transfer to monomer

$$\mathrm{R}_n\cdot + \mathrm{M} \xrightarrow{k_{fm}} \mathrm{P}_n + \mathrm{M}\cdot \qquad (2)$$

Transfer to solvent or substrate

$$\mathrm{R}_n\cdot + \mathrm{S} \xrightarrow{k_{fs}} \mathrm{P}_n + \mathrm{S}\cdot$$

Reinitiation by transfer radicals

$$\mathrm{M}\cdot + \mathrm{M} \xrightarrow{k_{pf}} \mathrm{R_2}\cdot$$

$$\mathrm{S}\cdot + \mathrm{M} \xrightarrow{k_{ps}} \mathrm{R_1}\cdot$$

Termination

$$\mathrm{R}_n\cdot + \mathrm{R}_m\cdot \xrightarrow{k_{tc}} \mathrm{P}_{(n+m)}$$

$$\mathrm{R}_n\cdot + \mathrm{R}_m\cdot \xrightarrow{k_{td}} \mathrm{P}_n + \mathrm{P}_m$$

where C and M are initiator and monomer, respectively, and S is solvent or added substrate. Denoting the rate of formation of $\mathrm{R_1}\cdot$ in the initiation process (not including reinitiation) as $\mathcal{I}$ and applying the usual stationary-state approximation (Chap. 1), equations can be written to describe the stationary-state concentrations of various radical species. Writing $[\mathrm{R}\cdot] = \Sigma_n\ [\mathrm{R}_n\cdot]$ we have, for the total concentration of propagating radicals

$$\frac{\mathrm{d}[\mathrm{R}\cdot]}{\mathrm{d}t} = \mathscr{I} - k_{\mathrm{fm}}[\mathrm{M}][\mathrm{R}\cdot] - k_{\mathrm{fs}}[\mathrm{S}][\mathrm{R}\cdot] + k_{\mathrm{pf}}[\mathrm{M}][\mathrm{M}\cdot] + k_{\mathrm{ps}}[\mathrm{M}][\mathrm{S}\cdot] - (k_{\mathrm{tc}} + k_{\mathrm{td}})[\mathrm{R}\cdot]^2 = 0 \tag{3}$$

and for the various transfer radicals

$$\frac{\mathrm{d}[\mathrm{M}\cdot]}{\mathrm{d}t} = k_{\mathrm{fm}}[\mathrm{M}][\mathrm{R}\cdot] - k_{\mathrm{pm}}[\mathrm{M}][\mathrm{M}\cdot] = 0 \tag{4}$$

$$\frac{\mathrm{d}[\mathrm{S}\cdot]}{\mathrm{d}t} = k_{\mathrm{fs}}[\mathrm{S}][\mathrm{R}\cdot] - k_{\mathrm{ps}}[\mathrm{M}][\mathrm{S}\cdot] = 0 \tag{5}$$

From eqns. (3)—(5) we see that

$$\mathscr{I} = (k_{\mathrm{tc}} + k_{\mathrm{td}})[\mathrm{R}\cdot]^2 \tag{6}$$

which is identical with the relation applying to simple polymerizations in the absence of a transfer agent (Chap. 1). Continuing to make the assumptions usually applied to simple polymerizations, and assuming that reinitiation is rapid but that transfer does not occur to such an extent that a significant fraction of monomer is used in the reinitiation step of the transfer processes, the rate of polymerization, for long chains, is given by

$$\omega = -\frac{\mathrm{d}[\mathrm{M}]}{\mathrm{d}t} = k_{\mathrm{p}}[\mathrm{M}][\mathrm{R}\cdot] = k_{\mathrm{p}}[\mathrm{M}]\left(\frac{\mathscr{I}}{k_{\mathrm{tc}} + k_{\mathrm{td}}}\right)^{1/2} \tag{7}$$

Equation (7) is again identical to that appertaining to simple free-radical polymerizations and, from eqn. (7), it is seen that simple transfer processes have no influence on overall rates of polymerization.

The foregoing derivation implicitly assumes that transfer radicals take no part in the termination reactions. Generally this assumption is valid since the concentration of transfer radicals is much smaller than $[\mathrm{R}\cdot]$. This can be seen by rearranging, say, eqn. (5) in the form

$$[\mathrm{S}\cdot]/[\mathrm{R}\cdot] = k_{\mathrm{fs}}[\mathrm{S}]/k_{\mathrm{ps}}[\mathrm{M}]$$

For simple transfer k_{ps} is at least comparable in magnitude with k_{p} and for most transfer agents $k_{\mathrm{fs}}/k_{\mathrm{p}}$ is small and, since the ratio $[\mathrm{S}]/[\mathrm{M}]$ is not normally excessive, $[\mathrm{S}\cdot]$ is small compared with $[\mathrm{R}\cdot]$.

Circumstances can arise in which $[\mathrm{S}\cdot]$ is not negligible compared with $[\mathrm{R}\cdot]$. For example, if k_{fs} is high or k_{ps} is low, compared with k_{p}, especially if, simultaneously, $[\mathrm{S}]/[\mathrm{M}]$ is not small, $[\mathrm{S}\cdot]$ cannot be neglected. Such conditions arise in telomerization reactions where high concentrations of active transfer agents are used specifically to produce low-molecular-weight polymers possessing end-groups derived from the transfer agent, i.e. telomers; or, if k_{ps} is low and reinitiation is slow. Under such circumstances it is necessary to introduce into the kinetic scheme (2) the termination reactions

$$S\cdot + R_n\cdot \xrightarrow{k_{tsr}} P_n \quad (8a)$$

$$S\cdot + S\cdot \xrightarrow{k_{tss}} \text{Products} \quad (8b)$$

Corresponding equations can be introduced for terminations involving transfer radicals derived from the monomer. Combining eqns. (8) with scheme (2), the stationary-state equations for various radical species become

$$\frac{d[R\cdot]}{dt} = \mathscr{I} - k_{fm}[M][R\cdot] - k_{fs}[S][R\cdot] + k_p[M][M\cdot] + k_{ps}[M][S\cdot] - (k_{tc} + k_{td})[R\cdot]^2 - k_{tsr}[S\cdot][R\cdot] = 0 \quad (9)$$

$$\frac{d[M\cdot]}{dt} = k_{fm}[M][R\cdot] - k_{pm}[M][M\cdot] = 0 \quad (10)$$

$$\frac{d[S\cdot]}{dt} = k_{fs}[S][R\cdot] - k_{ps}[M][S\cdot] - k_{tsr}[S\cdot][R\cdot] - k_{tss}[S\cdot]^2 = 0 \quad (11)$$

These equations do not yield useful simple kinetic relations. Qualitatively, inclusion of reactions (8) causes deviations from the usual simple proportionality between rate of polymerization and $\mathscr{I}^{1/2}$ (eqn. (7)). The transfer agent reduces the rate of polymerization to a greater extent than expected on the basis of simple dilution effects. The magnitude of the reduction in rate naturally depends on the nature of S· and its reactivity and hence, the effect will vary from one transfer agent to another and from one monomer to another. Such deviations from simple kinetic behaviour can be used to indicate the occurrence of reactions such as those indicated in eqns. (8).

Kinetic scheme (2) may be further complicated by the introduction of transfer to initiator or polymer. As in the previous discussion, if the transfer radicals are reactive, such processes have no influence on the overall rates of polymerization. In the case of transfer to initiator, although the instantaneous rate of polymerization remains unaffected, kinetic complications may arise in consequence of the transfer mechanism (Sect. 2.5).

Manabe et al. [2] considered a reaction scheme involving transfer to monomer, solvent and initiator. These workers classified the transfer radicals as primary radicals, along with primary radicals $R_c\cdot$ derived directly from the initiator, and included all primary radicals in termination reactions of the type (8a, b). They then assessed the importance of primary-radical termination (Chap. 1) by deriving an expression for the ratio of the concentrations of primary to propagating radicals and concluded that, under conditions normally encountered, primary-radical termination is insignificant.

2.2 INFLUENCE OF TRANSFER ON DEGREES OF POLYMERIZATION

The predominant influence of transfer reactions on polymerization processes is on the molecular weight of the polymer formed. For kinetic scheme (2) it is easy to derive an expression illustrating how the number-average degree of polymerization DP is influenced by the presence of transfer processes. DP is simply the average number of monomer molecules consumed per inactive polymer molecule formed, or, the rate of polymerization divided by the rate of formation of polymer molecules. Thus, DP is given by the expression

$$\mathrm{DP} = \left(-\frac{\mathrm{d[M]}}{\mathrm{d}t}\right) \Big/ \left(\frac{\mathrm{d[P]}}{\mathrm{d}t}\right)$$

$$= \frac{k_\mathrm{p}[\mathrm{M}][\mathrm{R}\cdot]}{(k_\mathrm{td} + k_\mathrm{tc}/2)[\mathrm{R}\cdot]^2 + k_\mathrm{fm}[\mathrm{M}][\mathrm{R}\cdot] + k_\mathrm{fs}[\mathrm{S}][\mathrm{R}\cdot]}$$

which can be written in the form

$$\frac{1}{\mathrm{DP}} = \frac{(k_\mathrm{td} + k_\mathrm{tc}/2)[\mathrm{R}\cdot]}{k_\mathrm{p}[\mathrm{M}]} + \frac{k_\mathrm{fm}}{k_\mathrm{p}} + \frac{k_\mathrm{fs}[\mathrm{S}]}{k_\mathrm{p}[\mathrm{M}]}$$

or, by use of eqn. (6)

$$\frac{1}{\mathrm{DP}} = \frac{(k_\mathrm{td} + k_\mathrm{tc}/2)\mathscr{I}^{1/2}}{(k_\mathrm{td} + k_\mathrm{tc})^{1/2} k_\mathrm{p}[\mathrm{M}]} + \frac{k_\mathrm{fm}}{k_\mathrm{p}} + \frac{k_\mathrm{fs}[\mathrm{S}]}{k_\mathrm{p}[\mathrm{M}]} \tag{12}$$

Equation (12) shows how various factors influence the reciprocal of the degree of polymerization. The first term on the right-hand side of this equation is identical with the expression derived in Chap. 1 for polymerizations in which chain-transfer processes are absent. The second and third terms describe the effects of transfer to monomer and to solvent (or other transfer agent), respectively.

Naturally, in cases where chain transfer to monomer does occur this process is inherent in the particular polymerization and will always influence the DP of the polymer. For such reactions it is usual to write the DP of the polymer formed in the absence of any other transfer reaction as DP_0, so that

$$\frac{1}{\mathrm{DP}_0} = \frac{(k_\mathrm{td} + k_\mathrm{tc}/2)\mathscr{I}^{1/2}}{(k_\mathrm{td} + k_\mathrm{tc})^{1/2} k_\mathrm{p}[\mathrm{M}]} + \frac{k_\mathrm{fm}}{k_\mathrm{p}} \tag{13}$$

From eqns. (13) and (14) we have

$$\frac{1}{\mathrm{DP}} = \frac{1}{\mathrm{DP}_0} + \frac{k_\mathrm{fs}[\mathrm{S}]}{k_\mathrm{p}[\mathrm{M}]} \tag{14}$$

an equation first developed by Mayo [3] and usually referred to as the Mayo equation. The applicability of the Mayo equation to retarded polymerizations is discussed subsequently (Sect. 3.3).

Chain-transfer processes not only affect the number-average degree of polymerization but may also modify the molecular-weight distribution of the final polymer. For polymerizations which terminate by disproportionation the polymers prepared in the presence and absence of transfer have a most-probable distribution with the ratio of weight- to -number-average molecular weights equal to 2. If bimolecular termination of propagating radicals proceeds by combination a different situation arises. In the absence of transfer the molecular-weight distribution of the polymer is narrower than the most probable; the ratio of weight-to-number-average molecular weights being equal to 1.5. However, those polymer molecules produced as dead polymer as a result of transfer reactions have a most-probable distribution. Consequently, since the whole polymer produced in such a polymerization contains chains terminated by both mechanisms, the molecular-weight distribution of the polymer is the superposition of two distributions appropriately weighted according to the proportions of chains produced by the two chain-stopping reactions, and hence, depends on the extent of the transfer reaction. A number of individual cases have been considered and molecular-weight distributions calculated [4]. The subject of molecular-weight distributions is considered in a subsequent chapter.

Although it was stated in Sect. 2.1 that, classically, transfer has no effect on rates of polymerization, the rate of polymerization does affect the extent of transfer. For high rates of polymerization low-molecular-weight polymers are produced and the first term on the right-hand side of eqn. (13) is large so that for transfer to have a significant influence on DP the transfer terms in eqn. (13) must be relatively large. On the other hand, for low rates of polymerization high-molecular-weight polymers are produced, the first term on the right-hand side of eqn. (13) will be small and the transfer terms can have a comparatively large effect on DP. Quite simply this effect arises from the fact that for low rates of polymerization radical lifetimes are relatively long and the probability that any propagating radical undergoes transfer is comparatively high.

2.3 TRANSFER CONSTANTS

It is apparent, from eqn. (12), that the kinetic parameters which are important in describing the magnitudes of the effects of transfer are the ratios of the rate coefficients for the particular transfer process to that for propagation. These ratios, which quantify the relative efficiencies of transfer agent and monomer for reaction with the propagating radical, are termed transfer constants. The transfer constant for monomer denoted C_M, equals k_{fm}/k_p and measures the relative rates of reaction of monomer with the propagating radical for the two mechanisms involved. Similarly, $C_S (= k_{fs}/k_p)$ is the transfer constant for solvent or added substrate. A corresponding term for transfer to initiator can be defined and inclusion of this process simply involves the addition of another term in eqns. (12) and (14). If two solvents or substrates, S and S′ are present, and both are active in transfer, their effects on the reciprocal of DP are additive in the sense that eqn. (14) becomes

$$\frac{1}{\mathrm{DP}} = \frac{1}{\mathrm{DP}_0} + C_\mathrm{s}\frac{[\mathrm{S}]}{[\mathrm{M}]} + C_{\mathrm{s}'}\frac{[\mathrm{S}']}{[\mathrm{M}]}$$

In certain specific cases the total effect of transfer from a mixture of transfer agents cannot be predicted from a knowledge of the individual transfer constants. For example, water appears to reduce the reactivity of *N,N*-dimethylformamide as a transfer agent in acrylonitrile polymerization [5], presumably through formation of a less-reactive *N,N*-dimethylformamide hydrate.

Transfer constants have two major applications. First, a knowledge of transfer constants allows prediction and control of molecular weights and molecular-weight distributions in a polymer. Second, they provide the essential data necessary for theoretical discussions of radical and substrate reactivity (Chap. 5).

In principle, transfer constants can be determined quite readily from degrees of polymerization of polymers prepared under various conditions. To obtain a specific transfer constant accurately, certain prerequisites must be met to ensure the Mayo equation applies; these conditions are described in subsequent sections dealing with individual classes of transfer reactions. Since transfer may affect both the molecular-weight distribution and degree of polymerization of the polymer, values of DP should be obtained, wherever possible, by an absolute method rather than by viscometry. (Viscosity—molecular-weight relations are strictly only applicable to polymers having a particular molecular-weight distribution.) A general comment on determination of transfer constants is that the Mayo equation only applies to instantaneous degrees of polymerization and can, therefore, only be applied under conditions where reactant concentrations remain sensibly constant during the polymerization.

A second general approach to calculation of transfer constants is based on the mechanism of the reaction. Referring to eqn. (1), which considers the general case where AB is the transfer agent, fragment A is incorporated into the dead polymer molecule as an end-group (1a). Transfer radical B· is then incorporated as an end-group in the reinitiation step (1b). Overall one molecule of transfer agent is incorporated into the polymer for each act of transfer. If sensitive analytical methods are available, e.g. radio-tracer techniques and radiation activation analysis, transfer constants can be determined by studying the incorporation of transfer agent. Practical considerations related to the radio-tracer method have been discussed in detail by Bevington [6]. Errors may arise if the transfer agent can also be incorporated into the polymer through a copolymerization reaction, or if a radioactive polymerizable impurity is present in the radio-tracer [7], or if the mechanism of the transfer reaction is such that the transfer agent is not actually incorporated into the polymer [3]. The most prevalent mechanism in chain transfer is the transfer of a hydrogen atom from the substrate to the propagating radical. Naturally, such reactions are subject to isotope effects and, unless due account is taken of this, the use of deuterated transfer agents, to aid polymer analysis, can lead to errors in the derivation of transfer constants [8].

2.4 TRANSFER TO MONOMER

In polymerizations involving monomer transfer a monomer molecule can either add to a propagating radical or undergo transfer with it (usually by transfer of a hydrogen atom). The transfer constant C_M $(=k_{fm}/k_p)$ measures the relative tendencies of the two processes and the contribution of monomer transfer in eqn. (12) is independent of monomer concentration. Determination of C_M relies on the fact that $1/DP_0$ is dependent on the polymerization conditions, according to the first term of eqn. (13). For catalysed polymerizations it is preferable to carry out the reaction in the absence of solvent or other transfer agent and to use an initiator which is not susceptible to transfer. Under these conditions, writing eqn. (13) in terms of rate of polymerization $\omega(= -d[M]/dt)$ we have

$$\frac{1}{DP_0} = \frac{(k_{tc}/2 + k_{td})\omega}{k_p^2[M]^2} + C_M \tag{15}$$

and a plot of $1/DP_0$ against ω is linear with intercept C_M. This case is illustrated in Fig. 1. If it is necessary to have a solvent present during

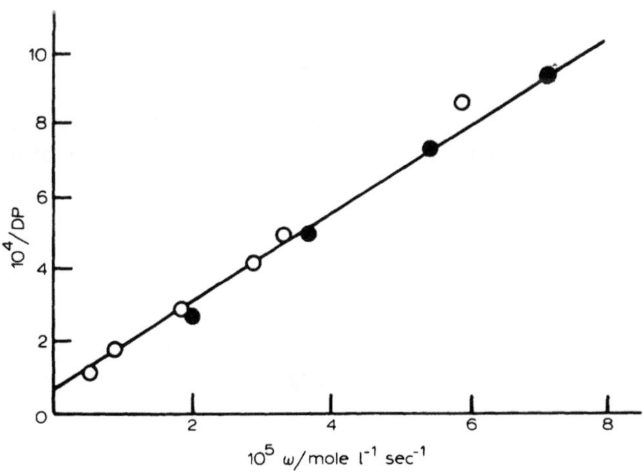

Fig. 1. 1/DP vs. ω for polymerizations of styrene at 60 °C initiated by ●, azo-bis-isobutyronitrile; ○, benzoyl peroxide (ref. 9, courtesy J. Am. Chem. Soc.).

TABLE 1
Selected monomer transfer constants

Monomer	Temp. (°C)	$10^5 C_M$	Ref.
Acrylonitrile	60	2.6	10
Allyl acetate	80	7×10^3	11
Allyl chloride	80	16×10^3	11
Methyl acrylate	60	0.7—4.0	12
Methyl methacrylate	60	1.0	13
Styrene	60	6.0	14
Vinyl acetate	50	200	15
Vinyl chloride	50	45	16

polymerization (e.g. to maintain homogeneity) a separate experiment must be performed to determine if the solvent is also a transfer agent when the intercept would also include a contribution from transfer to solvent.

It is interesting to note that if ω is proportional to $[M]^2$, i.e. if $\mathscr{I}$ is proportional to $[M]^2$, as might be the case under certain conditions, $1/DP_0$ is completely independent of [M].

Representative values of C_M for various monomers are given in Table 1. It is immediately apparent from the table that allylic monomers are the most susceptible to transfer and, in addition, reinitiation by their transfer radicals is very inefficient. Bartlett and Altshul [11] found that, as a result of these effects, the polymerization of allyl acetate is first-order in [initiator] and that the molecular weights of the polymers are low and independent of the rate of polymerization. By use of deuterated monomers, and observing isotope effects, Bartlett and Tate [17] demonstrated that the transfer reaction proceeds according to the equation

$$R_n + CH_2{=}CHCH_2O\cdot CO\cdot CH_3 \longrightarrow P_n + CH_2{=}CH\dot{C}HO\cdot CO\cdot CH_3$$

and that polymerization of the deuterated monomer

$$CH_2{=}CHCD_2O\cdot CO\cdot CH_3$$

proceeds more rapidly and yields a higher molecular-weight polymer. Lack of reinitiation by transfer radicals is a consequence of resonance stabilization between the structures

$$CH_2{=}CH{-}\dot{C}HX \quad \text{and} \quad \dot{C}H_2{-}CH{=}CHX$$

Effects of pressure on these transfer processes have been discussed by Walling [18].

Monomers containing α-methyl group substituents are potentially susceptible to transfer by abstraction of a hydrogen atom from the α-methyl group to give the stabilized structures

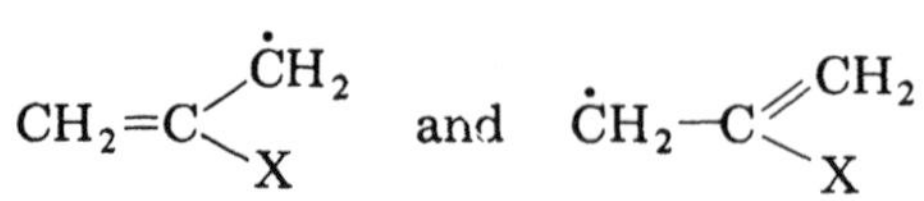

In accordance with this idea isopropenyl acetate has been found to undergo degradative chain transfer [19, 20]. Other monomers of similar structure, e.g. methyl methacrylate, show no evidence of retardation accompanying monomer transfer, possibly as a result of higher monomer reactivity.

Transfer to vinyl acetate proceeds by hydrogen abstraction from the acetoxy methyl group [21], producing the radical

$$CH_2{=}CH{\cdot}OCO\dot{C}H_2$$

Initiation by this radical produces a polymer molecule with a reactive unsaturated end-group which can at some stage be involved in a propagation reaction leading to the formation of a branched polymer. In the specific case of vinyl acetate polymerization branches produced by this type of mechanism may be removed by hydrolysis, and are therefore not present in poly(vinyl alcohol) produced by hydrolysis of poly(vinyl acetate).

2.5 TRANSFER TO INITIATOR

For polymerizations involving transfer to initiator, reaction (16), viz.

$$R_n\cdot + C \xrightarrow{k_{fc}} P_n + \dot{C} \tag{16}$$

a further term $k_{fc}[C]/k_p[M]$ must be added to the right-hand side of the preceding equations for 1/DP; [C] is the concentration of initiator. Since [C], $\mathscr{I}$ and ω are related it is usual to express 1/DP in terms of ω. The exact relationship depends on the number of radicals produced by decomposition of a molecule of initiator and on the initiator efficiency (Chap. 1). Assuming $\mathscr{I}$ is proportional to [C] we can write $\mathscr{I} = K[C]$, where K is related to the rate coefficient for initiator decomposition and takes into account the details of the initiation mechanism. Then by use of eqns. (7) and (15) and introducing the transfer to initiator term we have, in the absence of transfer to solvent,

$$\frac{1}{DP} = \frac{(k_{tc}/2 + k_{td})\omega}{k_p^2[M]^2} + \frac{k_{fm}}{k_p} + \frac{k_{fc}(k_{tc} + k_{td})\omega^2}{Kk_p^3[M]^3} \tag{17}$$

The occurrence of transfer to initiator can be identified by deviations from linearity when 1/DP is plotted against ω (Fig. 2).

Equation (17) can be recast in the form

$$\frac{1}{DP} - \frac{k_{fm}}{k_p} = \frac{(k_{tc}/2 + k_{td})K[C]}{(k_{tc} + k_{td})\omega} + \frac{k_{fc}[C]}{k_p[M]}$$

from which it is apparent that a plot of $(1/DP - k_{fm}/k_p)/[C]$ against $1/\omega$ should be linear with intercept $k_{fc}/k_p[M]$; hence allowing evaluation of the transfer constant to initiator $C_I(= k_{fc}/k_p)$.

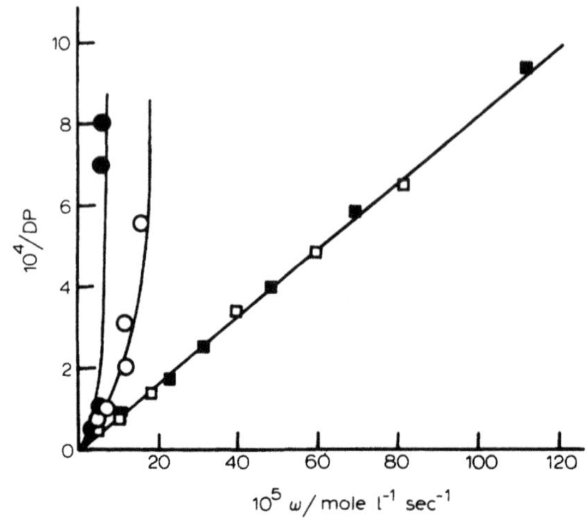

Fig. 2. 1/DP vs. ω for polymerizations of methyl methacrylate at 60 °C, initiated by ■, azo-bis-isobutyronitrile; □, benzoyl peroxide; ○, cumene hydroperoxide and ●, *t*-butylhydroperoxide (ref. 13, courtesy J. Polym. Sci.).

In general, transfer to initiator can lead to a variety of complications according to the specific mechanism of the transfer reaction. Considering reaction (16), in general a fragment of initiator will be incorporated into the dead polymer molecule and the second fragment C will be incorporated into a new chain on reinitiation. C· is used to denote the transfer radical derived from initiator since there is no reason to suppose that it will always be identical with the primary radical R_c· which normally initiates polymerization.

If reaction (16) destroys the reactive moiety of the initiator, for example the —O—O— linkage in a peroxide, then essentially "unreactive" fragments are incorporated into the polymer molecules. Such a mechanism destroys the initiator. Consequently, as polymerization proceeds over a finite period of time [C] and, hence, $\mathscr{I}$ decrease more rapidly than in the absence of transfer, this reaction constitutes induced decomposition of initiator by the propagating radical.

Assuming transfer to diaroyl peroxides follows the usual mechanism for induced decomposition of these compounds the process can be represented by

$$R_n\cdot + RCO{\cdot}O{-}O{\cdot}OCR \longrightarrow P_nCO{\cdot}OR + RCO{\cdot}O\cdot \qquad (18)$$

Reaction (18) incorporates an aroyloxy group into the dead polymer and produces an aroyloxy radical, identical with that produced as a primary radical on decomposition of the initiator. Reinitiation may then occur by direct addition of monomer to the aroyloxy radical or by addition to the aryl radical formed by loss of carbon dioxide (as discussed in Chap. 1).

Bevington et al. [22] developed an isotope technique for determining rates of initiation and transfer constants to initiator and described its application to transfer to diaroyl peroxides. The object of the experiment is to measure the rate of incorporation of initiator fragments into the polymer. In cases where decomposition of primary and transfer radicals may occur prior to initiation or reinitiation, the initiator must be labelled in such a position that unambiguous data are obtained. Each act of initiation incorporates one initiator fragment into the polymer (rate of incorporation = $\mathscr{I}$) and each act of transfer incorporates two fragments (rate of incorporation = $2k_{fc}[\mathrm{R}\cdot][\mathrm{C}]$). Thus, we have, for initiator fragments

$$\text{Total rate of incorporation} = \mathscr{I} + 2k_{fc}[\mathrm{R}\cdot][\mathrm{C}]$$

and from eqn. (7)

$$\text{Total rate of incorporation} = \mathscr{I} + \frac{2k_{fc}[\mathrm{C}]\,\omega}{k_p[\mathrm{M}]} \tag{19}$$

Experimentally, ω is varied, keeping [C] constant, by addition of a second unlabelled initiator. This results in a variation in the total rate of initiation but does not affect the rate of incorporation of labelled initator fragments in the initiation process. A plot of total rate of incorporation of labelled fragments against ω allows determination of $\mathscr{I}$ from the intercept and C_I from the slope. Use of suitably labelled initiators in experiments of this type allows the proportion of radicals which undergo some decomposition prior to reinitiation to be determined.

A method analogous to the mixed initiator technique, based on molecular-weight measurements, has been proposed by Laita [23]. This method is suitable for use with photoinitiators where the rate of initiation can be varied by altering the light intensity or the initiator concentration. Equation (12), including a term for transfer to initiator, becomes

$$\frac{1}{\mathrm{DP}} = \frac{(k_{tc}/2 + k_{td})\mathscr{I}^{1/2}}{(k_{tc} + k_{td})^{1/2}k_p[\mathrm{M}]} + \frac{k_{fm}}{k_p} + \frac{k_{fs}[\mathrm{S}]}{k_p[\mathrm{M}]} + \frac{k_{fc}[\mathrm{C}]}{k_p[\mathrm{M}]} \tag{20}$$

If the rate of initiation is varied by changing the light intensity (maintaining [C] constant) it is apparent from eqn. (20) that the intercept of a plot of 1/DP against $\mathscr{I}^{1/2}$, at zero $\mathscr{I}$, gives a value for

$$\frac{k_{fm}}{k_p} + \frac{k_{fs}[\mathrm{S}]}{k_p[\mathrm{M}]} + \frac{k_{fc}[\mathrm{C}]}{k_p[\mathrm{M}]}$$

while if [C] is varied ($\mathscr{I}$ varies with [C]) and the light intensity is kept constant the intercept of a plot of 1/DP against [C], at [C] = 0, gives

$$\frac{k_{fm}}{k_p} + \frac{k_{fs}[\mathrm{S}]}{k_p[\mathrm{M}]}$$

Comparison of these two intercepts allows evaluation of the transfer constant to initiator.

The preceding discussion has assumed that transfer to initiator results in destruction of the "initiating moiety". This is not necessarily so. Consider an initiator X—X which normally generates initiating radicals by scission of the weak link between the X residues. In principle, it is possible that transfer could occur by removal of a hydrogen atom from one of the X residues to leave the transfer radical X—X•. This radical could then reinitiate polymerization, incorporating the weak link at the end of the new polymer chain, ultimately giving rise to a polymer molecule $X{-}XM_n$ (assuming disproportionation termination). Such transfer processes do not reduce the concentration of initiating centres and would not cause reductions in rates of polymerizations continued over finite periods of time in excess of those arising from normal decomposition of the initiator. Subsequent decomposition of the terminal weak link would generate $M_nX\cdot$ as a primary radical which could initiate polymerization in the usual way to produce a polymer molecule M_nXM_m. This process would extend the length of the initial polymer chain and offset the reduction in molecular weight produced by the initial transfer process. Instances of such reactions are relatively rare in practice and variable in nature.

In the absence of efficient radical scavengers diaroyl peroxides are very sensitive to induced decomposition by reactive radicals. Processes of this type are often suppressed to some extent in polymerization reactions. Some typical transfer constants for disubstituted benzoyl peroxides in styrene polymerizations, determined by Cooper [24], are given in Table 2.

The presence of electron-withdrawing substituents increases susceptibility to transfer and also reduces the rate of thermal decomposition, necessitating the use of relatively high initiator concentrations to achieve a given rate of polymerization. A given peroxide often shows marked differences in activity with different propagating radicals. Cooper [24] demonstrated that bis(2,3,4,5-tetrachlorobenzoyl)peroxide is quite susceptible to transfer with the polystyryl radical but less reactive towards

TABLE 2

Transfer constants for substituted dibenzoyl peroxides in polymerizations of styrene at 70 °C [24]

Para substituent	C_I	*Meta* substituent	C_I	*Ortho* substituent	C_I
Tert-butyl	Approx. 0	Fluorine	0.246	Methyl	0.175
Methyl	0.003	Chlorine	0.346	Fluorine	0.40
Methoxy	0.074	Bromine	0.465	Chlorine	1.91
(Hydrogen)	(0.075)	Iodine	0.262	Bromine	2.17
Fluorine	0.219	Cyanide	0.804		
Chlorine	0.216	Nitro	6.2		
Bromine	0.193				
Iodine	0.293				

radicals derived from methyl methacrylate and vinyl acetate. Bevington et al. [22] obtained similar results for bis(4-methoxy-3,5-dibromobenzoyl)peroxide, from which results they concluded that polar factors are important in the transition state for transfer (see Chap. 5).

Hydroperoxides are susceptible to transfer, as illustrated by the results of Baysal and Tobolsky [13] for *tert*-butyl and cumene hydroperoxides in Fig. 2. Walling and Chang [25] have proposed that transfer to hydroperoxides involves scission of the O—H bond rather than the O—O bond.

As a class of initiators azo-nitriles have been considered to be unreactive in chain transfer and for this reason azo-bis-isobutyronitrile has been a preferred initiator for kinetic studies. Recent data, however, indicate that chain transfer to this initiator is by no means insignificant (Chap. 3). Laita and Machacek [26] obtained evidence that with azo-bis-isopropane transfer occurs by hydrogen abstraction rather than by dissociation of the azo-linkage and simultaneous elimination of nitrogen, which is the mechanism of thermal and photolytic decomposition of these initiators.

Many sulphur-containing initiators are efficient in initiation but lead to degradative chain transfer. This must arise from the generation of different radicals in the two processes. Ferrington and Tobolsky [27] proposed that, while initiation by the disulphide $(C_6H_5S)_2$ produces reactive $C_6H_5S\cdot$ radicals, transfer proceeds according to the equation

$$R_n\cdot + (C_6H_5S)_2 \longrightarrow P_nC_6H_5 + C_6H_5S{-}S\cdot$$

and that the disulphide radical is unreactive. Similarly, the same workers suggested that simple dissociation of tetra-alkylthiuram disulphides gives the radical $R_2N{-}CS{-}S\cdot$ but that in transfer the reaction is

$$R_n\cdot + (R_2N{-}CS{-}S)_2 \longrightarrow P_nCS{-}NR_2 + R_2N{-}CS{-}S{-}S\cdot$$

with the transfer radical unreactive towards monomer.

Other initiators which are reported to undergo degradative chain transfer include diazothioethers [28] and diazoaminobenzene [29].

2.6 TRANSFER TO POLYMER

The possibility of transfer to polymer exists, in principle, in all polymerizations. This process produces a polymer molecule with a radical site at some point along the polymer chain, viz.

$$R_n\cdot + \sim\!CH_2{-}CHX\!\sim \longrightarrow P_n + \sim\!CH_2\dot{C}X\!\sim$$

Reinitiation by the polymeric transfer radical leads to the formation of branched polymeric species. If bimolecular radical termination proceeds by disproportionation then the polymer molecule formed is a simple branched species. On the other hand, if termination is by combination then termination of the branched species with a linear propagating radical yields a longer branch and if two branched species terminate mutually a cross-link is formed between the two initially dead polymer molecules. Transfer to polymer increases the molecular weight of the dead polymer and affects the molecular-weight distribution [30, 31].

In favourable cases, by introducing dead polymer into a polymerization medium, use may be made of transfer reactions to produce branched structures. If the dead polymer is derived from a monomer different to that in the polymerizing system then graft polymers can be produced. These processes, however, are generally inefficient in producing multicomponent species and simultaneously produce large proportions of homopolymer.

Transfer to polymer obviously increases in importance as the concentration of polymer increases and, in most conventional homopolymerizations, transfer to dead polymer is insignificant in the early stages of polymerization when the polymer concentration is minimal.

Chain transfer constants for polymers can be estimated from determinations of the extent of chain branching [32], by the use of model compounds [32] and by isotope techniques [33, 34]. In general it appears that transfer constants to monomer and polymer are similar in magnitude, but differences will be expected especially when transfer to monomer gives, for example, a highly stabilized radical which cannot be generated from the polymer.

Polyethylene produced by free-radical polymerization is highly branched. Long-chain branches can arise by conventional chain transfer to polymer and reinitiation by the polymeric transfer radical; the branch length is comparable with the initial chain length. In addition, the presence of large numbers of short-chain branches, about four carbon atoms long, has been established [35—38]. It is believed that these short-chain branches arise as a result of a "back-biting" type of process involving intramolecular hydrogen transfer through formation of a transient six-membered ring, viz.

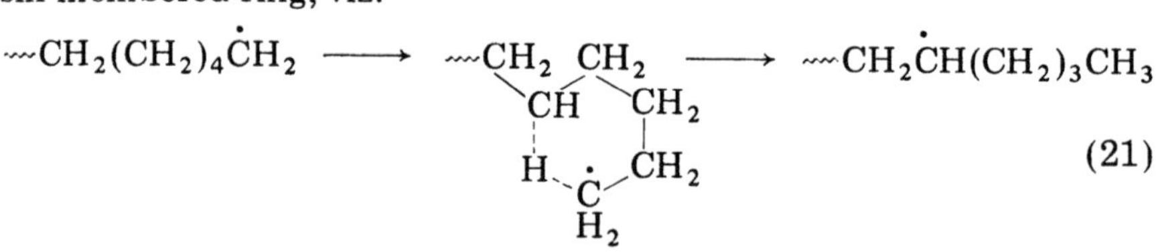

(21)

Chain growth then continues by addition of monomer to the transfer radical leaving the original chain-end as a short branch. When the transfer radical has added monomer there is a high probability of repeating the process, resulting in the formation of clusters of two-carbon-atom side chains [39—41]. Clusters of short-chain branches could also occur following transfer to the middle of a chain. Intramolecular chain transfer of this type is not dependent on the presence of dead polymer and has essentially the same probability of occurrence at all stages of chain growth.

A characteristic feature of vinyl acetate polymerizations is the formation of highly branched polymer. Many of the branches may be removed by hydrolysis. Undoubtedly a large proportion of the hydrolysable branches arise from transfer to monomer, as discussed earlier (Sect. 2.4), since the transfer constant for monomer is high and, at least in the early stages, the concentration of unreacted monomer is much higher than that of polymerized monomer units. Long-chain hydrolysable branches can also result from conventional transfer to polymer by transfer of a hydrogen atom from the acetoxy methyl group in the polymer and reinitiation in the normal way [42]. Long-chain non-hydrolysable branches can be formed by transfer of the α-proton on the polymer chain, viz.

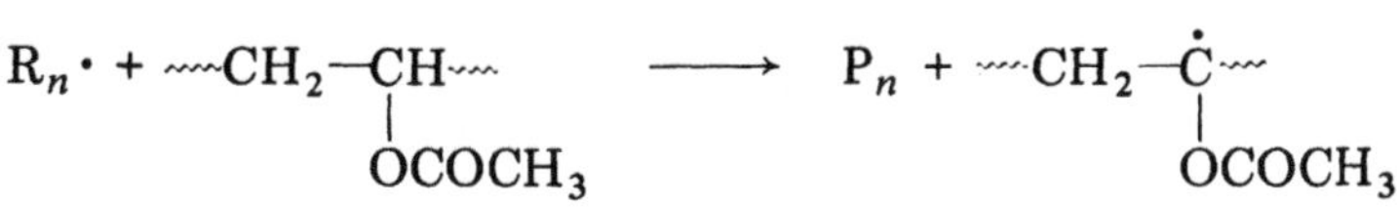

Clarke et al. [43] estimated the relative rates of formation of hydrolysable and non-hydrolysable branches by determining transfer constants to the following compounds

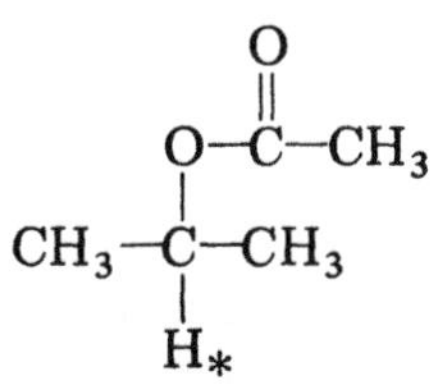

(I)

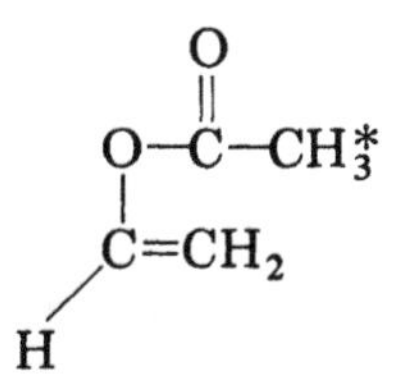

(II)

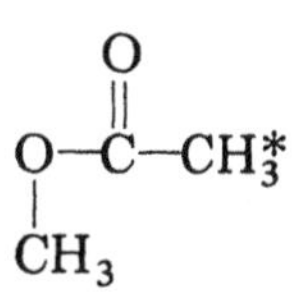

(III)

The measured transfer constants for the hydrogen atoms marked with an asterisk in vinyl acetate polymerizations at 60 °C were

Compound	(I)	(II)	(III)
Transfer constant	0.0008	0.00025	0.00025

On the basis of these data we would expect transfer to polymer to produce a higher rate of formation of non-hydrolysable than hydrolysable branches.

It has also been suggested [42, 44, 45] that short-chain branches, similar to those formed in polyethylene by reaction (21), can be produced

by a corresponding mechanism, as in

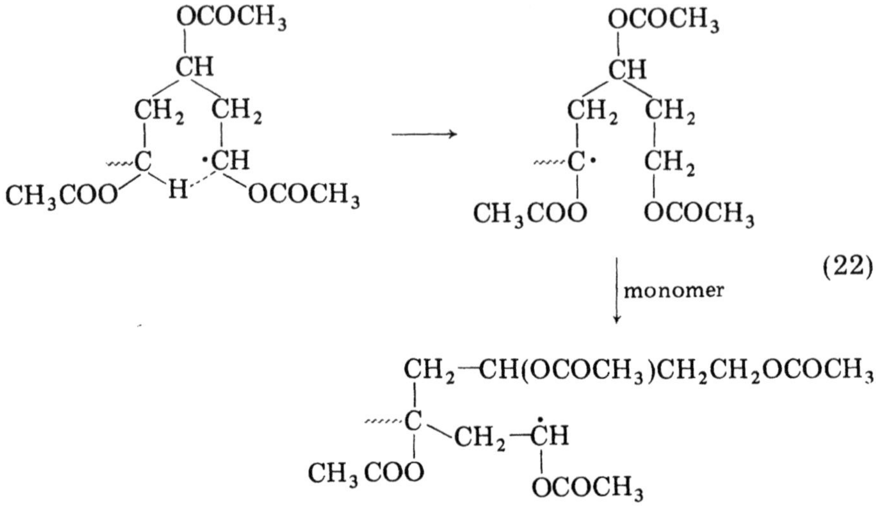

(22)

In addition to transfer to the main-chain components of the polymer, rates of transfer to polymer, and to added dead polymer, may be high if the polymer chains contain reactive end-groups derived from, for example, an initiator, transfer agent or disproportionation termination [46—49], and the rates of such processes are dependent on the molecular weight of the dead polymer.

2.7 TRANSFER TO SOLVENT OR TRANSFER AGENT

Chain transfer constants for solvents and added substrates are usually determined by application of the Mayo equation (14). To obtain accurate values of C_s certain conditions must be fulfilled. Determination of C_s from a plot of 1/DP against [S]/[M] requires that $1/DP_0$ is constant throughout the series of experiments. This condition implies that $\mathcal{I}^{1/2}/[M]$ must remain constant. In the simplest systems this requirement is achieved by carrying out the polymerizations in an inert solvent (i.e. a solvent to which transfer is known to be negligible) when [S]/[M] can be varied by replacing inert solvent by transfer agent, maintaining all other concentrations constant. If no suitable unreactive solvent is available [S]/[M] must be varied by dilution of monomer with transfer agent in which case the initiator concentration must be adjusted to maintain $\mathcal{I}^{1/2}/[M]$ constant. When extremely reactive transfer agents are used, so that large variations in DP are produced by very low concentrations of transfer agent, [M] and $\mathcal{I}^{1/2}/[M]$ may be considered constant.

The Mayo equation can only be applied strictly in the absence of degradative chain transfer. Under conditions such that retardation accompanies chain transfer the situation is more complex and is considered in a subsequent section (Sect. 3.3). In addition, the transfer agent should only act as a simple diluent and not cause other kinetic complications, such as precipitation of polymer. Other complications, such as transfer to initiator, should be avoided as far as possible. Copolymerization with the transfer agent should not occur. Conversions should be kept low, especially if the transfer constant is high, so that there is effectively no consumption of reagents during the polymerization. It is known that variations in the viscosity of the solvent medium can influence rates and degrees of polymerization through effects on the rate coefficients for termination (Chap. 1). Yokoto et al. [50] measured the transfer constant to *N,N*-dimethylaniline in methyl methacrylate polymerizations in different solvents using the normal Mayo equation and observed variations in C_s with the solvent used. However, if viscosity effects were taken into account by writing the Mayo equation in the form

$$\frac{1}{\mathrm{DP}} = \frac{k_t^0 \omega}{k_p^2 [\mathrm{M}]^2 \eta} + C_\mathrm{M} + C_s \frac{[\mathrm{S}]}{[\mathrm{M}]}$$

(where η is the viscosity of the solvent medium and k_t^0 is the value of $(k_{tc}/2 + k_{td})$ at a standard viscosity) the values of C_s were independent of the solvent.

It follows from reaction scheme (2) that

$$\left(-\frac{\mathrm{d[S]}}{\mathrm{d}t}\right) \Big/ \left(-\frac{\mathrm{d[M]}}{\mathrm{d}t}\right) = \frac{k_{fs}[\mathrm{S}][\mathrm{R}\cdot]}{k_p[\mathrm{M}][\mathrm{R}\cdot]} = \frac{k_{fs}[\mathrm{S}]}{k_p[\mathrm{M}]}$$

or

$$\frac{\mathrm{d}\log[\mathrm{S}]}{\mathrm{d}\log[\mathrm{M}]} = C_s \tag{23}$$

Thus, transfer constants can in principle be determined by measuring the relative rates of consumption of transfer agent and monomer, and this method has been successfully applied in some cases [51—54]. This approach can only be satisfactorily employed for transfer agents of fairly high reactivity. Derivation of eqn. (23) assumes that the polymer chains are long, so that consumption of monomer in the transfer process is negligible. More accurately, taking reinitiation into account but assuming termination by transfer radicals is unimportant

$$\left(-\frac{\mathrm{d[M]}}{\mathrm{d}t}\right) \Big/ \left(-\frac{\mathrm{d[S]}}{\mathrm{d}t}\right) = \frac{k_p[\mathrm{M}][\mathrm{R}\cdot] + k_{ps}[\mathrm{M}][\mathrm{S}\cdot]}{k_{fs}[\mathrm{S}][\mathrm{R}\cdot]}$$

$$= \frac{k_p[\mathrm{M}][\mathrm{R}\cdot] + k_{fs}[\mathrm{S}][\mathrm{R}\cdot]}{k_{fs}[\mathrm{S}][\mathrm{R}\cdot]}$$

$$= \frac{k_p[\mathrm{M}]}{k_{fs}[\mathrm{S}]} + 1$$

Over the years numerous transfer constants have been reported with values for different systems covering six orders of magnitude, reflecting a tremendous range of reactivities. Unfortunately the accuracy of these data leaves a lot to be desired; not infrequently transfer constants for the same process vary by an order of magnitude. In spite of numerical uncertainties, obvious correlations can be drawn between the magnitudes of transfer constants and the monomer and radical reactivities. While reactivities are discussed in a subsequent chapter a few general comments may be made at this juncture.

In a series of early studies of chain transfer processes Gregg and Mayo [55, 56] investigated chain transfer in styrene polymerizations and produced general correlations between transfer constants and the structures of the transfer agents. Using a series of aromatic hydrocarbons they concluded that benzylic hydrogen atoms are susceptible to transfer and that their activity increases with the degree of substitution. Thus, as seen from Table 3, benzene is relatively inert at 60 °C, the introduction of benzylic hydrogen increases the reactivity in the order toluene $<$ ethyl benzene $<$ isopropyl benzene, but *tert*-butyl benzene is relatively unreactive [55]. Transfer constants with a given radical also increase with increased stabilization of the transfer radical [55], viz. diphenylmethane, triphenylmethane and fluorene (Table 3).

TABLE 3
Some transfer constants with polystyryl radicals at 60 °C [55, 56]

Substrate	$10^5 C_s$
Benzene	0.18
Toluene	1.25
Ethyl benzene	6.7
Isopropyl benzene	8.2
tert-Butyl benzene	0.6
n-Butyl chloride	0.4
n-Butyl bromide	0.6
n-Butyl iodide	18.5
Methylene chloride	1.5
Chloroform	5
Carbon tetrachloride	920
Carbon tetrachloride	136,000
n-Butyl mercaptan	2.2×10^6 [a]
tert-Butyl mercaptan	360,000[b]
n-Dodecyl mercaptan	1.9×10^6 [b]
Ethyl thioglycolate	5.8×10^6 [b]

[a] Ref. 57; [b] ref. 53.

A secod major series of transfer agents are halogen-containing materials, and their activity generally increases with the degree of substitution and the polarizability of the halogen which is transferred [56] (Table 3).

Another major class of active transfer agents are mercaptans which normally transfer a hydrogen atom to the propagating radical, as shown by isotope effects [58], and a few representative data are given in Table 3. Disulphides are even more reactive than the monosulphides, transfer apparently occurring through attack on the S—S bond. This view is supported by the observation that cyclic disulphides are incorporated into the polymer chain by copolymerization [59, 60], presumably through the reaction

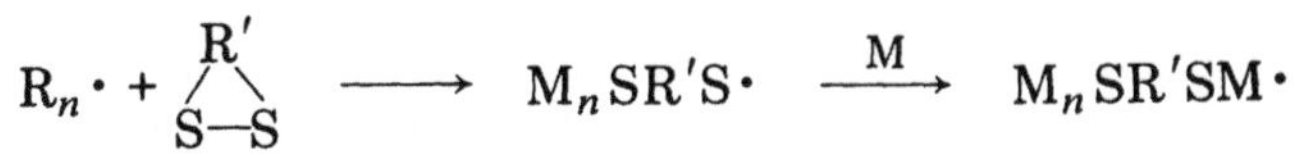

The relative reactivities of substrates towards polystyryl radicals cannot be universally used to predict relative reactivities in polymerizations of other monomers. For example, carbon tetrachloride is much more reactive as a transfer agent than benzene in polymerizations of styrene, but the reverse order of reactivity is observed in acrylonitrile polymerizations; quoted values of transfer constants for benzene and carbon tetrachloride in the polymerization of acrylonitrile at 60 °C include [61] 2.46×10^{-4} and 0.85×10^{-4}, respectively. Various schemes have been put forward to explain such variations in behaviour and these are discussed in Chap. 5.

3. Inhibition and retardation

3.1 GENERAL CONCEPTS OF INHIBITION AND RETARDATION

In simple free-radical polymerizations the termination or chain-stopping reaction is the mutual bimolecular destruction of propagating radicals (Chap. 1, Sect. 4). Termination may also proceed through reaction of reactive radicals with some other species present in the reaction mixture, species classified as inhibitors and retarders. We have already discussed briefly (Sect. 1) the distinction between inhibition and retardation. Inhibitors and retarders are substrates Z which react with active radicals R· to produce inactive species, as depicted in eqn. (24), viz.

$$R\cdot + Z \longrightarrow \text{Inactive products} \qquad (24)$$

Reaction (24) is in direct competition with the chain-growth process (eqn. (25))

$$R\cdot + M \longrightarrow RM\cdot \qquad (25)$$

In any particular reaction system the detailed consequences of the competition between monomer and Z for the active radicals depend on the relative values of the rate coefficients for the two processes, the concentrations of reagents and the properties of the so-called "inactive products". The important common factor in inhibition and retardation is that the concentration of active radicals capable of supporting polymerization is reduced by the presence of the substrate Z and rates of polymerization are reduced.

Radical R· may be any active radical species in the reaction system. If R· is a primary (initiating) radical or a very short-chain propagating radical it is probable that the kinetics of the reaction will conform with those of an inhibited reaction, while if R· is a longer propagating radical kinetics of a retarded polymerization will be observed.

Substrate Z may itself be a free radical which is too stable to initiate polymerization by forming a propagating radical through reaction with monomer. Reaction between R· and Z may then be the simple addition of the two species, forming a covalent bond between R and Z and producing an inert species totally incapable of supporting further polymerization.

Alternatively Z may be a neutral molecule which may add to R· according to the reaction

$$R\cdot + Z \longrightarrow RZ\cdot \tag{26}$$

to produce an "inactive radical" which may subsequently terminate another propagating chain. *Z* may also react with an active radical by a transfer mechanism which may be represented as

$$R\cdot + Z \longrightarrow R + Z\cdot \tag{27}$$

Again, Z· may participate further by terminating another active radical.

A further possibility occurs if Z is a transition metal salt when reaction (24) can involve charge transfer and a change in the oxidation state of the metal. For example, Z may be ferric chloride [62], in which case reaction (24) can be written

$$R\cdot + FeCl_3 \longrightarrow RCl + FeCl_2 \tag{28}$$

Ferrous ions may be a constituent of an initiating system, such as the ferrous plus hydrogen peroxide system (Chap. 1), when ferric is a product of radical formation. Subsequent chain termination by ferric through reaction (28) regenerates ferrous which may then be involved in further initiation.

Reaction products of both eqns. (26) and (27) include a radical derived from the substrate Z. Since Z is, by definition, an inhibitor or retarder the radicals so formed are inactive. That is, they are much less reactive than radicals responsible for polymerization and, ideally, should be completely inert as regards supporting further polymerization. In practice the radicals derived from Z may not be completely inert but may slowly reinitiate further polymerization by addition of monomer. Under such conditions, reaction of R· with Z does not change the instantaneous total concentration of radicals capable of partaking in polymerization but does reduce the concentration of reactive radicals present in the system, thereby reducing the rate of polymerization and causing retardation. Under such circumstances reaction (26) is a copolymerization reaction in which Z acts as a co-monomer and reaction (27) is a transfer reaction with Z as the transfer agent. Simultaneous occurrence of transfer and retardation is termed degradative chain transfer. The extent to which such polymerizations are retarded depends on the reactivity of Z towards a particular active radical involved in reaction and also on the reactivity of the radical derived from Z towards monomer.

Over the whole field of free-radical polymerization a wide variety of monomer and radical reactivities is encountered. In addition, it is known that a wide range of substances are capable of inhibiting or retarding at least some polymerizations. Coupled with the fact that various modes of reaction are possible, it is obvious that the variety of detailed possible consequences is enormous. Indeed, the general field of inhibition and retardation is probably the most neglected and least well-understood aspect of free-radical polymerization. In subsequent sections we shall consider a number of specific cases and their kinetic consequences.

3.2 KINETICS OF INHIBITED POLYMERIZATIONS

We have already made mention (Sect. 1) of the fact that the distinction between inhibition and retardation may be considered as one of degree rather than of type. However, there are cases in which reaction may be specifically referred to as inhibition and we shall proceed to consider the kinetics of these in some detail.

Consider the possible reaction scheme

$$\mathrm{C} \longrightarrow \mathrm{R_c}\cdot \tag{29a}$$

$$\mathrm{R_c}\cdot + \mathrm{M} \xrightarrow{k_i} \mathrm{R_1}\cdot \tag{29b}$$

$$\mathrm{R_c}\cdot + \mathrm{Z} \xrightarrow{k_{zc}} \text{Inactive products} \tag{29c}$$

$$\mathrm{R}_n\cdot + \mathrm{M} \xrightarrow{k_p} \mathrm{R}^{\cdot}_{(n+1)} \tag{29d}$$

$$\mathrm{R}_n\cdot + \mathrm{Z} \xrightarrow{k_z} \text{Inactive products} \tag{29e}$$

$$\mathrm{R}_n\cdot + \mathrm{R_m}\cdot \xrightarrow{k_t} \text{Polymer} \tag{29f}$$

Scheme (29) represents the situation where the inhibitor Z can react either with primary radicals $\mathrm{R_c}\cdot$, derived from the initiator C, or with propagating radicals $\mathrm{R}_n\cdot$ to give products completely incapable of further reaction. In the absence of inhibitor scheme (29) reduces to the simple scheme for free-radical polymerizations when, applying the usual stationary-state assumptions (Chap. 1), the rate of polymerization ω_0 is given by

$$\omega_0 = -\left(\frac{d[M]}{dt}\right)_0 = k_p[M][R\cdot]_0 = \frac{k_p}{k_t^{1/2}}[M]\mathscr{I}_0^{1/2}$$

where $[R\cdot]_0$ is the total stationary concentration of propagating radicals (subscript 0 is used here to indicate the values of parameters in the absence of inhibitor). $\mathscr{I}_0$ is the rate of formation of $R_c\cdot$ and the uninhibited rate of initiation or formation of propagating radicals $R_1\cdot$. k_p and k_t are, respectively, the rate coefficients for propagation and termination (for simplicity no distinction is drawn between combination and disproportionation termination).

Inclusion of reactions between substrate Z and various radicals reduces the concentration of active radicals in the reaction system. Assuming that Z in no way interferes with the extent of geminate recombination, i.e. that the rates of formation of primary radicals $R_c\cdot$ in the absence and presence of Z are identical, the rate of formation of $R_c\cdot$ in the presence of Z is $\mathscr{I}_0$. Avoiding situations where the rates of change of ω or radical concentration are large, stationary-state kinetics are applicable when, for primary and propagating radicals, we have

$$\frac{d[R_c\cdot]}{dt} = \mathscr{I}_0 - k_{zc}[Z][R_c\cdot] - k_i[M][R_c\cdot] = 0 \tag{30}$$

$$\frac{d[R\cdot]}{dt} = k_i[M][R_c\cdot] - k_z[Z][R\cdot] - k_t[R\cdot]^2 = 0 \tag{31}$$

From eqn. (30)

$$[R_c\cdot] = \frac{\mathscr{I}_0}{k_{zc}[Z] + k_i[M]} \tag{32}$$

The rate of formation of propagating radicals in the presence of inhibitor, or rate of initiation $\mathscr{I}$, is

$$\mathscr{I} = k_i[M][R_c\cdot]$$

and by substituting from eqns. (30) and (32)

$$\mathscr{I} = \mathscr{I}_0\left\{1 - \frac{k_{zc}[Z]}{k_{zc}[Z] + k_i[M]}\right\} \tag{33}$$

Under conditions such that $k_{zc}[Z] \gg k_i[M]$, $\mathscr{I}$ will be very small compared with $\mathscr{I}_0$ and may be effectively zero. When $\mathscr{I}$ is very small but finite, Z will compete with monomer for the propagating radicals at each stage of growth (competition between (29d) and (29e)). Inclusion of reaction (29e) reduces the total stationary concentration of propagating radicals $[R\cdot]$ below that which would normally exist in a simple polymerization with rate of initiation $\mathscr{I}$, as seen from the presence of the term Z in eqn.

(31). If $k_{zc}[Z] \gg k_i[M]$, then it is probable that $k_z[Z] \gg k_p[M]$ also, so that chains will mainly be terminated in the early stages of growth and both rates and degrees of polymerization will be low. Under these conditions it is almost inevitable that all radicals will be terminated by inhibitor and normal bimolecular termination will be absent. Since the rate of polymerization ω is always given by

$$\omega = \frac{-d[M]}{dt} = k_p[M][R\cdot] \tag{34}$$

substituting from eqn. (31)

$$\omega = \frac{k_i k_p [R_c\cdot][M]^2}{k_z[Z]} = \frac{k_p[M]\mathscr{I}}{k_z[Z]}$$

$$= \frac{k_p[M]\mathscr{I}_0}{k_z[Z]} \left\{1 - \frac{k_{zc}[Z]}{k_{zc}[Z] + k_i[M]}\right\} \tag{35}$$

The degree of polymerization DP, or the average number of monomer molecules consumed per polymer molecule formed, in the presence of the inhibitor is

$$DP = \omega / k_z[Z][R\cdot]$$

$$= \frac{k_p[M]}{k_z[Z]} \tag{36}$$

Thus, under conditions $k_{zc}[Z] \gg k_i[M]$ and $k_z[Z] \gg k_p[M]$ it can be seen from eqns. (35) and (36) that any polymerization which does occur proceeds at a very low rate and yields only very low-molecular-weight polymer. Since chain lengths are very short the consumption of monomer in the initiation process (29b) will not be negligible compared with consumption of monomer in propagation so that more accurately, eqn. (35) should be written

$$DP = \left(\frac{-d[M]}{dt}\right) \Big/ k_z[Z][R\cdot] = \frac{k_p[M][R\cdot] + k_i[M][R_c\cdot]}{k_z[Z][R\cdot]}$$

$$= \frac{k_p[M]}{k_z[Z]} + 1$$

In many systems containing a powerful inhibitor (k_z large) the rate of polymerization is, for all practical purposes, zero, even at low [Z], although by use of sufficiently sensitive techniques a low rate of monomer consumption may be detected [63]. These conditions previal until sufficient inhibitor has been consumed so that Z can no longer effectively compete with monomer for all active radicals. The rate of polymerization then increases as the radical concentration increases until, when all Z has been consumed, it reaches its normal value for a simple polymerization. As seen from Fig. 3, an inhibited polymerization is characterized by

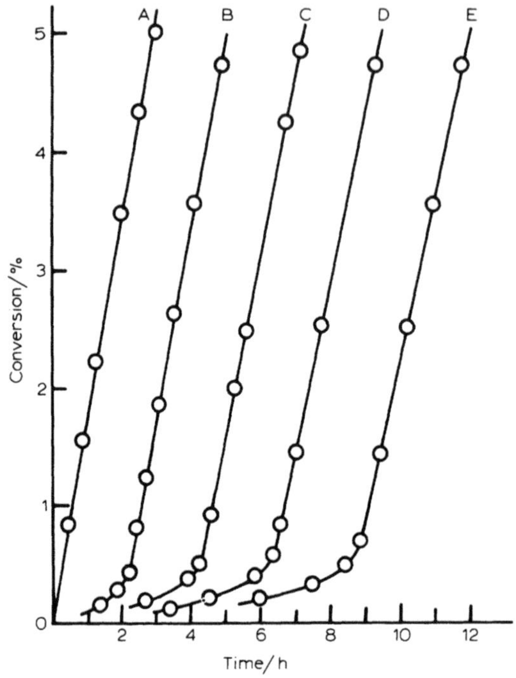

Fig. 3. Conversion—time curves for polymerizations of styrene at 60 °C initiated by azo-bis-isobutyronitrile (0.5 g l^{-1}) in the presence of various concentrations of *p*-benzoquinone, as indicated: A, 0 g l^{-1}; B, 0.050 g l^{-1}; C, 0.075 g l^{-1}; D, 0.123 g l^{-1}; E, 0.157 g l^{-1} (ref. 63, courtesy J. Chem. Soc.).

an initial period during which the rate of polymerization is zero (i.e. an inhibition period), followed by a transitional region when the rate of polymerization increases, finally adopting its normal rate commensurate with consumption of initiator during the inhibition period. The length of the transitional period depends on the activity of Z as a radical scavenger and is shortest for the most reactive inhibitors. During this period the reaction has something of the character of a retarded polymerization but, where [Z] is very low, [R·] and ω are generally changing too rapidly with time to justify use of the stationary-state approximation.

So long as the consumption of monomer during the transitional period is small compared with the final conversion, polymer formed after the inhibition period is produced under conditions of normal simple kinetics and $k_p/k_t^{1/2}$ calculated (see Chap. 1) from the degree of polymerization and the true rate of polymerization (i.e. not the average over the whole reaction period) will have its normal value.

The rate of consumption of inhibitor is

$$-\frac{d[Z]}{dt} = k_{zc}[Z][R_c\cdot] + k_z[Z][R\cdot] \qquad (37)$$

and under conditions such that bimolecular termination of propagating radicals (29f) is completely suppressed (i.e. during the inhibition period) we have, from eqns. (30), (31) and (37), that

$$-\frac{d[Z]}{dt} = \mathscr{I}_0$$

Consequently, the rate of initiation in the absence of inhibitor can be calculated if the rate of consumption of initiator can be measured (see also Chap. 1). It also follows that the length of the inhibition period is equal to the initial concentration of Z divided by $\mathscr{I}_0$. This assumes that inhibition is ideal in that the inhibition period can be defined accurately, say, by extrapolation of the monomer conversion—time curve to zero conversion (see also subsequent discussion in Sect. 3.3) and that a molecule of inhibitor reacts with only one radical to produce completely inert products. Allowance can be made for a molecule of inhibitor terminating more than one chain.

In the whole of the preceding discussion it has been assumed that Z is an efficient radical scavenger, capable of reacting with and destroying any active radical species. Cases can be visualized, at least in principle, where Z reacts with primary radicals only. This corresponds to removing reaction (e) from scheme (29). Equations (30), (32) and (33) remain identical while eqn. (31) reduces to

$$\frac{d[R\cdot]}{dt} = k_i[M][R_c\cdot] - k_t[R\cdot]^2 = 0$$

under stationary-state conditions. Thus, primary radicals which escape reaction with Z initiate polymerization which proceeds normally but at a reduced rate of initiation given by eqn. (33). Although under such circumstances $\mathscr{I}$ will be less than $\mathscr{I}_0$, $\mathscr{I}$ may be comparable with $\mathscr{I}_0$ and an inhibition period will not necessarily be observed. Indeed, rates of polymerization may only be slightly reduced by the presence of Z which is still classified as an inhibitor since it destroys primary active species. Values of $k_p/k_t^{1/2}$ calculated from rates and degrees of polymerization under these conditions will be identical with normal values calculated for uninhibited polymerizations. This is to be contrasted with modified values of $k_p/k_t^{1/2}$ which characterize polymerizations whose rates are reduced by the presence of a retarder.

A similar situation arises when Z interferes with the mechanism of radical formation, reducing the rate of formation of primary radicals and hence, of initiation, but not destroying any radical species which are formed. Again, the polymerization will proceed normally but with a reduced rate of initiation and calculated values of $k_p/k_t^{1/2}$ will be normal. Perhaps one of the most convenient methods of classifying a substrate Z as an inhibitor in free-radical polymerizations is if it reduces the rate of polymerization, or even completely suppresses it for a time, but that any polymerization which does occur proceeds with a normal value of the composite parameter $k_p/k_t^{1/2}$.

3.3 KINETICS OF RETARDED POLYMERIZATIONS

It will be appreciated, from the discussion in Sect. 3.1, that a variety of reaction schemes can be devised to describe retarded polymerizations. The retarder Z may add to (eqn. (26)) or undergo transfer with (eqn. (27)) the propagating radical. If the retarder is a simple molecule, one species formed by interaction of the retarder with the propagating radical will be a relatively stable radical, which may, however, reinitiate slowly. Additional complications might arise through the occurrence of such processes as primary-radical termination and interaction of the radical derived from Z with active radicals. A variety of possible schemes have been examined by various workers and we shall consider a number of these.

3.3.1 Reaction schemes excluding reinitiation

The simplest case to consider is that in which the retarder reacts with the propagating radicals to produce completely inert products. For such a system we may write the reaction scheme

$$\mathrm{C} \longrightarrow \mathrm{R_c\cdot} \tag{38a}$$

$$\mathrm{R_c\cdot + M} \longrightarrow \mathrm{R_1\cdot} \tag{38b}$$

$$\mathrm{R}_n\cdot + \mathrm{M} \xrightarrow{k_p} \mathrm{R}^{\cdot}_{(n+1)} \tag{38c}$$

$$\mathrm{R}_n\cdot + \mathrm{Z} \xrightarrow{k_z} \text{Inert products} \tag{38d}$$

$$\mathrm{R}_n\cdot + \mathrm{R}_m\cdot \xrightarrow{k_t} \text{Polymer} \tag{38e}$$

The inert products formed in (38d) include a dead polymer P_n.

Retarders are, in general, similar in operation to, but less reactive than, inhibitors. Since the concentration of primary radicals $R_c\cdot$ derived from the initiator C will be much less than the total concentration of propagating radicals we may assume that the retarder does not react with primary radicals. Thus, the rate of initiation $\mathscr{I}$ (i.e. the rate of formation of $R_1\cdot$) in the presence and absence of the retarder are equal. Scheme (38) is therefore a simplified form of scheme (29). In the absence of the retarder, scheme (38) again reduces to the simple scheme for free-radical polymerization (see Chap. 1) when, applying the stationary-state approximation,

$$\frac{d[R\cdot]}{dt} = \mathscr{I} - k_t[R\cdot]_0^2 = 0$$

and

$$\omega_0 = k_p[M][R\cdot]_0 = \frac{k_p}{k_t^{1/2}}[M]\mathscr{I}^{1/2}$$

where $[R\cdot] = \sum_n [R_n\cdot]$ and subscript 0 denotes the absence of retardation.

In the presence of retarder

$$\frac{d[R\cdot]}{dt} = \mathscr{I} - k_z[Z][R\cdot] - k_t[R\cdot]^2 \tag{39}$$

The reduction in concentration of active radicals by the retarder reduces the rate of polymerization, which may be very low initially. For relatively low initial retarder concentrations, as retarder is consumed the rate gradually increases to a maximum rate equal to that which would hold had no retarder been present initially. Under such conditions the rate of polymerization changes continuously during a retarded polymerization but application of the stationary-state approximation is still justifiable if the rate in the presence of the retarder is essentially constant over a period of time (e.g. in the presence of a high concentration of weak retarder) or if $d[R\cdot]/dt$ is very small compared with the rate of formation or destruction of propagating radicals [4]. In general, it is not justifiable to apply the stationary-state approximation during the period in which the rate of polymerization is changing significantly with time.

Assuming the stationary-state approximation is valid in the initial stages of a retarded polymerization $d[R\cdot]/dt = 0$, so that eqn. (39) becomes

$$\mathscr{I} = k_z[Z][R\cdot] + k_t[R\cdot]^2 \tag{40}$$

and the rate of polymerization is again given by eqn. (34). Also from scheme (38)

$$-\frac{d[Z]}{dt} = k_z[Z][R\cdot] \tag{41}$$

The fraction of total termination proceeding by mutual bimolecular termination of propagating radicals is

$$\left(-\frac{d[M]}{dt}\right)^2 \Big/ \left(-\frac{d[M]}{dt}\right)_0^2 = \frac{k_t[R\cdot]^2}{k_t[R\cdot]_0^2} = \frac{k_t[R\cdot]^2}{\mathscr{I}}$$

It has been pointed out [64] that if the retarded rate of polymerization is less than 20 % of the unretarded rate then less than 4 % of the propagating radicals undergo mutual termination. Consequently, in the presence of an efficient retarder (or sufficiently high concentration of weak retarder) bimolecular termination of propagating radicals can be neglected when

$$\mathcal{I} = k_z[Z][R\cdot] = -d[Z]/dt \tag{42}$$

and knowledge of the rate of disappearance of Z allows $\mathcal{I}$ to be estimated. Under such conditions, from eqns. (34) and (42)

$$\omega = \frac{k_p[M]}{k_z[Z]}\mathcal{I} \tag{43}$$

and the rate of polymerization is proportional to $\mathcal{I}$, rather than to $\mathcal{I}^{1/2}$ as in a simple free-radical polymerization. The kinetic chain length ν or, in the absence of transfer, the degree of polymerization of polymer formed under such conditions (assuming chains are not very short, cf. eqn. (36) and subsequent discussion) is given by

$$\nu = \mathrm{DP} = k_p[M]/k_z[Z]$$

It also follows from eqns. (34) and (42) that, even without stationary-state assumptions,

$$\frac{d\log[M]}{d\log[Z]} = \frac{k_z}{k_p}$$

allowing the parameter k_z/k_p to be evaluated if variations in [M] and [Z] can be determined accurately. k_z/k_p is analogous to the transfer constant in simple cases and is often used as a measure of the effectiveness of a retarder.

If Z is a simple molecule a product of reaction (38d) may be a radical which, although incapable of supporting polymerization, can subsequently terminate an additional propagating radical and if one molecule of Z destroys y chains the rate of disappearance of Z will be $\mathcal{I}/y$. Bartlett and Kwart [64] described methods for calculating y, $\mathcal{I}$ and k_z/k_p from accurate dilatometric data obtained under conditions where the retarded rate is much less than the unretarded rate but not equal to zero; values of k_z/k_p were obtained for various retarders for vinyl acetate polymerizations.

It cannot always be assumed that reaction (38d) is the only termination process. For example, if Z is a weak retarder the simultaneous occurrence of (38e) must be considered. [R·] is obtained by solving the quadratic eqn. (40), when

$$[R\cdot] = \frac{1}{2k_t^{1/2}}\left\{\left(\frac{k_z^2[Z]^2}{k_t} + 4\mathcal{I}\right)^{1/2} - \frac{k_z[Z]}{k_t^{1/2}}\right\} \tag{44}$$

and the rates of polymerization and consumption of retarder are obtained by substituting [R·] into eqns. (34) and (41), respectively.

Bamford et al. [62] demonstrated that ferric chloride is a very powerful retarder for the polymerization of styrene, that the polymerization kinetics are entirely in accord with scheme (38) and that the stationary-state approximation is valid throughout the course of the reaction. These workers differentiated eqn. (40) to obtain

$$\frac{d[Z]}{dt} = -\left(\frac{\mathscr{I} + k_t[R\cdot]^2}{k_z[R\cdot]^2}\right)\frac{d[R\cdot]}{dt}$$

and, by substituting with the aid of eqns. (34) and (40), rearranging and integrating they obtained the relation

$$-\frac{[R\cdot]_{max}}{[R\cdot]} + \log\left(\frac{[R\cdot]_{max} + [R\cdot]}{[R\cdot]_{max} - [R\cdot]}\right) = k_z[R\cdot]_{max}t + A \qquad (45)$$

where A is a constant. A "reduced rate" β_t was defined as the ratio of the rate of polymerization at time t to that when $[Z] = 0$ (i.e. $\beta_t = [R\cdot]/[R\cdot]_{max}$) so that, if [M] is effectively constant

$$\frac{1}{\beta'} - \frac{1}{\beta_t} + \log\left(\frac{1-\beta'}{1+\beta'}\cdot\frac{1+\beta_t}{1-\beta_t}\right) = k_z[R\cdot]_{max}t \qquad (46)$$

where β' is the value of β_t at $t = 0$ and equals $(\mathscr{I}k_t)^{1/2}/(k_z[Z])$. If $\beta' \ll 1$ eqn. (46) may be rewritten as

$$\frac{1}{\beta'} - \frac{1}{\beta_t} + \log\left(\frac{1+\beta_t}{1-\beta_t}\right) = k_z[R\cdot]_{max}t \qquad (47)$$

Then, by choosing a time $t = \tau$ such that

$$-\frac{1}{\beta_t} + \log\left(\frac{1+\beta_t}{1-\beta_t}\right) = 0 \qquad (48)$$

eqn. (47) becomes

$$\frac{1}{\beta'} = k_z[R\cdot]_{max}\tau$$

and

$$\tau = [Z]_0/\mathscr{I} \qquad (49)$$

where $[Z]_0$ is the initial concentration of Z. The solution of eqn. (48) is $\beta_\tau = 0.648$. This is a general result for retarded reactions consistent with scheme (38) provided that (a) the initial rate of polymerization is sufficiently small, (b) the products of reaction (38d) do not react further and (c) [M] and $\mathscr{I}$ remain effectively constant throughout the reaction. Determination of τ, the time at which the rate of polymerization is 0.648 of the maximum rate, allows $\mathscr{I}$ to be calculated from eqn. (49). τ is fundamentally a more satisfactory parameter than the induction period obtained by back-extrapolation of the linear portion of a conversion—time plot to its intercept on the time axis.

Bamford et al. [62] also wrote eqn. (45) in the form

$$-\frac{1}{\beta_t} + \log\left(\frac{1+\beta_t}{1-\beta_t}\right) = k_z[\mathrm{R}\cdot]_{\max} t + A \qquad (50)$$

By plotting β_t against $(k_z[\mathrm{R}\cdot]_{\max} t + A)$ and fitting experimental values of β_t to the curve the value of $k_z[\mathrm{R}\cdot]_{\max} = k_z(\mathscr{I}/k_t^{1/2})$ may be determined An example of such a plot is shown in Fig. 4. Knowledge of $\mathscr{I}$ from eqn. (49) allows calculation of k_z/k_p (or $\mathscr{I}/k_t^{1/2}$).

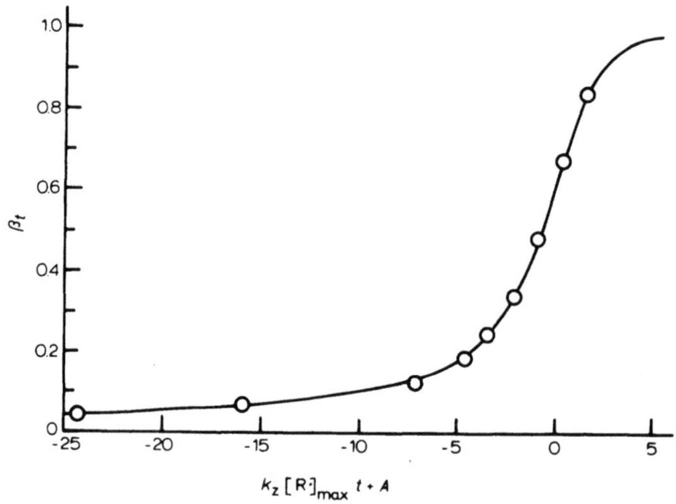

Fig. 4. Dependence of the reduced rate β_t on the reaction time t, showing calculated curve (eqn. (50)) and experimental data for the polymerization of styrene (2.61 mole l^{-1}) in *N,N*-dimethylformamide solution at 60 °C, initiated by azo-bis-isobutyronitrile (7.34×10^{-2} mole l^{-1} initially) in the presence of $FeCl_3$ (7.91×10^{-3} mole l^{-1} initially) (ref. 60, courtesy Proc. R. Soc. London).

Kinetically it is of no consequence if reaction (38d) proceeds by addition of Z to the propagating radical or by transfer. However, if a product of (38d) is a radical which can terminate a second chain then the molecular weight of the polymer may vary according to whether or not radical combination occurs. Bonsall et al. [65] made use of this effect in attempting to determine the nature of the bimolecular termination reaction in unretarded polymerizations.

Kinetic scheme (38) has been extended to include depolymerization processes [66].

3.3.2 *Reaction schemes including reinitiation*

As stated in the earlier discussion (Sect. 3.1), if a stable radical is a product of the retardation reaction there is a real possibility that the radical may be sufficiently reactive to reinitiate polymerization slowly. This possibility has been realized by several authors who have included such a reaction in kinetic schemes describing retarded polymerizations. One of the earliest such schemes, proposed by Flory [67], was essentially

$$\text{Catalyst} \longrightarrow \text{R}_\text{c}\cdot \tag{51a}$$

$$\text{R}_\text{c}\cdot + \text{M} \longrightarrow \text{M}_1\cdot \tag{51b}$$

$$\text{R}_n\cdot + \text{M} \xrightarrow{k_\text{p}} \text{R}^{\bullet}_{(n+1)} \tag{51c}$$

$$\text{R}_n\cdot + \text{Z} \xrightarrow{k_\text{z}} \text{Z}\cdot \tag{51d}$$

$$\text{Z}\cdot + \text{M} \xrightarrow{k_\text{pz}} \text{R}^{\bullet}_{(n+1)} \tag{51e}$$

$$\text{Z}\cdot + \text{Z}\cdot \xrightarrow{k_\text{tz}} \text{Polymer} \tag{51f}$$

In (51d) Z· is a stable radical formed by the addition of Z to a propagating radical but the reaction could equally have been written as a transfer process (products of (51d) would then include dead polymer and the reinitiation reaction (51e) would produce the radical $\text{R}_1\cdot$). The molecular weight of the final polymer will depend on the mechanism of reaction (51d), otherwise the processes are kinetically equivalent. The only true termination reaction included in scheme (51) is (51f), on the assumption [67] that, since Z· is, by definition, relatively unreactive, $[\text{Z}\cdot] \gg [\text{R}\cdot]$, so that termination between Z· and $\text{R}_n\cdot$ and between two $\text{R}_n\cdot$ radicals may be neglected (see later). Assuming the stationary-state approximation applies we have

$$\frac{\text{d}[\text{R}\cdot]}{\text{d}t} = \mathscr{I} - k_\text{z}[\text{Z}][\text{R}\cdot] + k_\text{pz}[\text{M}][\text{Z}\cdot] = 0$$

$$\frac{\text{d}[\text{Z}\cdot]}{\text{d}t} = k_\text{z}[\text{Z}][\text{R}\cdot] - k_\text{pz}[\text{M}][\text{Z}\cdot] - k_\text{tz}[\text{Z}\cdot]^2 = 0$$

From the above equations

$$[\text{R}\cdot] = \{\mathscr{I} + k_\text{pz}(\mathscr{I}/k_\text{tz})^{1/2}[\text{M}]\}/k_\text{z}[\text{Z}]$$

$$[\text{Z}\cdot] = (\mathscr{I}/k_\text{tz})^{1/2}$$

It also follows that

$$-\frac{\text{d}[\text{Z}]}{\text{d}t} = k_\text{z}[\text{Z}][\text{R}\cdot] = \mathscr{I} + k_\text{pz}(\mathscr{I}/k_\text{tz})^{1/2}[\text{M}]$$

and the rate of consumption of Z is independent of [Z]. Since eqn. (34) holds, the ratio k_z/k_p is given by

$$\frac{k_z}{k_p} = \frac{[M]}{[Z]}\frac{d[Z]}{d[M]}$$

The extension of reaction scheme (51), by inclusion of the termination reactions

$$Z\cdot + R_n\cdot \xrightarrow{k_{tzr}} \text{Polymer} \tag{52}$$

$$R_n\cdot + R_n\cdot \xrightarrow{k_t} \text{Polymer} \tag{53}$$

has been considered by Kice [68]. In the course of a comprehensive treatment of the kinetic scheme (51) with reactions (52) and (53), Kice defined two parameters

$$\mu = \frac{\text{Retarded rate of polymerization}}{\text{Unretarded rate of polymerization}} = \frac{\omega}{\omega_0}$$

and

$$\phi = k_{tzr}/(k_t k_{tz})^{1/2}$$

and derived the relation

$$\frac{\mu^2[Z]}{1-\mu^2}\left[1 + \left\{1 + \frac{(1-\mu^2)}{\phi^2\mu^2}\right\}^{1/2}\right] = \frac{k_t\omega}{k_p k_z}\left\{1 + \frac{(1-\mu^2)}{\phi^2\mu^2}\right\}^{1/2} + \frac{k_{pz}k_t[M]}{k_{tzr}k_z} \tag{54}$$

The parameter ϕ is often used in kinetics of copolymerization and its significance is discussed in Chap. 4.

Under conditions such that the initial rate is sufficiently large to be determined accurately and [Z] remains effectively constant for a sufficient time to allow the retarded rate of polymerization ω to be determined a plot of

$$\frac{\mu^2[Z]}{1-\mu^2}\left[1 + \left\{1 + \frac{(1-\mu^2)}{\phi^2\mu^2}\right\}^{1/2}\right] \text{ vs. } \omega\left\{1 + \frac{(1-\mu^2)}{\phi^2\mu^2}\right\}^{1/2}$$

should give a straight line. The slope of the straight line is, from eqn. (54), $k_t/k_p k_z$, so that, from a knowledge of k_p/k_t (Chap. 1), k_z may be determined. The intercept is $k_{pz}k_t[M]/k_{tzr}k_z$ so that if k_t is known (Chap. 1), using k_z from the slope, a value of k_{pz}/k_{tzr} may be estimated. In order to obtain the plot described above it is necessary to assume a value of ϕ, Kice suggested trying values of ϕ covering a wide range and adopting the value which gives the least deviation from a straight line. This treatment was applied to polymerizations of methyl methacrylate [68] and methyl acrylate [69] in the presence of a variety of retarders.

Kice [68, 69] found that for a number of retarders the best values of ϕ

were so large that the second term in braces in eqn. (54) was negligible compared with unity. Under such conditions eqn. (54) simplifies to

$$\frac{2\mu^2[\mathrm{Z}]}{1-\mu^2} = \frac{k_t \omega}{k_p k_z} + \frac{k_{pz} k_t [\mathrm{M}]}{k_{tzr} k_z}$$

The definition of μ implicitly assumes that values of [M] in the retarded and unretarded polymerizations are identical. Jenkins [70] repeated Kice's analysis allowing for the possibility that [M] may be lower in the retarded polymerizations if it is necessary to use relatively high concentrations of weak retarder to attain sufficient difference between the retarded and unretarded rates. Special cases were also considered [70, 71] for $\phi = 1$, $\phi \to \infty$ and $\phi \to 0$ when simplified relations were obtained and it was demonstrated that in each case ω is proportional to $\mathscr{I}$ raised to a power between 0.5 and 1.0, the absolute value of the exponent depending on the relative values of $k_z[\mathrm{Z}]$ and $(\mathscr{I}k_t)^{1/2}$.

Allen et al. [72] used a scheme similar to that of Kice with the assumption that

$$k_t = k_{tzr}/\sigma = k_{tz}/\sigma^2$$

where σ is a constant for a given system. Again, it was shown that the order with respect to $\mathscr{I}$ lay between 0.5 and 1.0, the absolute value depending on the efficiency of reinitiation. These authors investigated the polymerization of vinyl acetate retarded by dihydromyrcene and by isopropylbenzene and were unable to distinguish between retardation through transfer or through copolymerization on the basis of kinetic data.

Application of the kinetic treatments of Kice and others requires a knowledge of the unretarded rate of polymerization ω_0. Such information cannot always be obtained, for example, in the solution polymerization of vinyl chloride the propagating radical is highly reactive and readily undergoes transfer which is always accompanied to some extent by retardation. Even if it is possible to obtain a value of ω_0 in the absence of retardation using a particular solvent, the same value may not be applicable to polymerizations in other solvents, which act as retarders, in view of possible solvent effects on k_p and on k_t (Chap. 1). To overcome these difficulties Bamford and co-workers [7] performed an alternative comprehensive analysis for retarded reactions in which they considered the reactions in scheme (50) together with termination reactions (51) and (52) but excluding reaction (50f) on the grounds that mutual destruction of transfer radicals is unlikely except under unusual conditions [68, 73]; this is equivalent to assuming $\phi = \infty$. Application of the stationary-state approximation then allows equations to be written for [R·] and [Z·], and by eliminating [Z·] we have

$$(k_{pz}[\mathrm{M}] + k_{tzr}[\mathrm{R}\cdot])(\mathscr{I} - k_t[\mathrm{R}\cdot]^2) - 2k_{tzr}k_z[\mathrm{Z}][\mathrm{R}\cdot]^2 = 0 \qquad (55)$$

Since eqn. (34) holds at any time, eqn. (55) can be written in terms of the rate of polymerization ω and becomes

$$\frac{b}{a}\omega^3 + \left(1 + 2abc\frac{[S]}{[M]}\right)\frac{\omega^2}{a^2} - \frac{b}{a}\mathscr{I}\omega - \mathscr{I} = 0 \qquad (56)$$

where

$$a = k_p[M]/k_t^{1/2};\ b = k_{tzr}/k_{zp}k_t^{1/2}[M];\ c = k_z/k_p \qquad (57)$$

a is the usual proportionality factor between rate of polymerization and $\mathscr{I}^{1/2}$ for simple polymerizations while c measures the relative tendencies for substrate and monomer to react with the propagating radical so that, for rapid reinitiation by transfer radical, c is the conventional transfer constant.

When plotted as log ω against log $\mathscr{I}$ eqn. (56) gives a sigmoid plot, as shown schematically in Fig. 5, and comparison of the experimental sigmoid curve with eqn. (56) allows evaluation of a, b and c separately. The sigmoid curve in Fig. 5 has certain properties which aid the determination of a, b and c.

At sufficiently high [R·] (or $\mathscr{I}$)

$$k_{tzr}[R\cdot] \gg k_{pz}[M], \quad k_t[R\cdot] \gg k_z[Z]$$

when, from eqn. (55)

$$[R\cdot] = (\mathscr{I}/k_t)^{1/2}$$

and

$$\omega = k_p[M](\mathscr{I}/k_t)^{1/2} \qquad (58)$$

which is the simple rate expression for unretarded polymerizations.

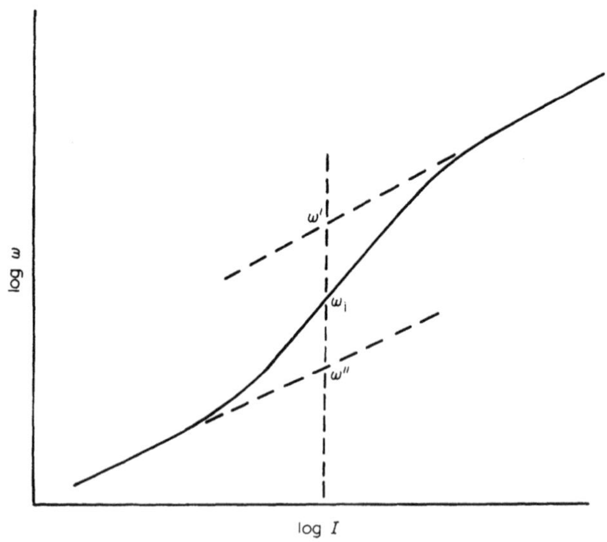

Fig. 5. Diagrammatic representation of the sigmoid curve corresponding to eqn. (56) (ref. 7, courtesy Trans. Faraday Soc.).

At sufficiently low [R·] (or $\mathscr{I}$), unless $k_{pz} = 0$,

$$k_{tzr}[\mathrm{R}\cdot] \ll k_{pz}[\mathrm{M}]$$

when

$$\omega = k_p[\mathrm{M}]\left(1 + 2\mathscr{R}\frac{[\mathrm{S}]}{[\mathrm{M}]}\right)^{-1/2}\left(\frac{\mathscr{I}}{k_t}\right)^{1/2} \tag{59}$$

where

$$\mathscr{R} = abc$$

It was suggested $\mathscr{R}$ be termed the intrinsic retardation of the system.

From eqns. (58) and (59) it is seen that the slopes of the sigmoid curve at both extreme values of $\mathscr{I}$ are 0.5. At the lowest values of $\mathscr{I}$ the reaction differs from a simple free-radical polymerization in that the order in [M] is greater than unity (except when there is no retardation and $\mathscr{R} = 0$).

If asymptotes are drawn to the sigmoid curve at high and low $\mathscr{I}$, corresponding to eqns. (58) and (59), respectively (as shown in Fig. 5) the vertical distance between the asymptotes, i.e. $\log \omega' - \log \omega''$ in Fig. 5, is given by

$$\tfrac{1}{2}\log\left(1 + 2\mathscr{R}\frac{[\mathrm{S}]}{[\mathrm{M}]}\right)$$

Using $\mathscr{I}_i$ and ω_i to denote, respectively, rates of initiation and polymerization at the point of inflexion of the curve in Fig. 5.

$$\omega_i = \frac{a}{b}\left(1 + 2\mathscr{R}\frac{[\mathrm{S}]}{[\mathrm{M}]}\right)^{1/2}$$

and, if ω', ω'' are the rates of polymerization on the asymptotes, at $\mathscr{I}_i$

$$\omega_i^2 = \omega'\omega''$$

so that the point of inflexion is midway between the points corresponding to ω' and ω''.

The gradient of the sigmoid curve at the point of inflexion is

$$\{1 + (1 + 2\mathscr{R}[\mathrm{S}]/[\mathrm{M}])^{1/2}\}/\{3 + (1 + 2\mathscr{R}[\mathrm{S}]/[\mathrm{M}])^{1/2}\} \tag{60}$$

The numerical value of this gradient is the maximum value of the exponent for $\mathscr{I}$ and reduces to 0.5 when [Z] = 0 or $\mathscr{R} = 0$, when there is no retardation. Expression (60) only has the value unity if $\mathscr{R} = \infty$, which in practice, means no reinitiation by Z, i.e. $k_{pz} = 0$, $b = \infty$, where there is no point of inflexion and the gradient increases asymptotically to unity as $\mathscr{I}$ decreases (cf. eqn. (43)).

The usefulness of the above approach depends on a knowledge of absolute rates of initiation and rates of polymerization over a wide enough range of values of $\mathscr{I}$ to establish the full sigmoid curve. Known values of $\mathscr{I}$

can often be obtained using established photoinitiators (Chap. 1). It may be necessary to measure accurately very low rates of polymerization ($\omega \sim 10^{-7}$ mole l^{-1} sec^{-1}) and for this purpose Bamford and co-workers [7] recommend a dilatometric technique using a "dummy" dilatometer to allow corrections for thermal fluctuations. At the highest values of $\mathscr{I}$ it is necessary to use an initiating system which produces radicals which are sufficiently reactive that primary-radical termination can be neglected. The 2-cyano-isopropyl radical from azo-bis-isobutyronitrile becomes involved in primary-radical termination in vinyl chloride polymerizations if $\mathscr{I} > 10^{-7}$ mole l^{-1} sec^{-1}.

An additional check on the acceptability of the parameters a, b and c derived from kinetic data can be obtained from the molecular weights of the polymers formed in the retarded polymerizations [7]. For polymerization in which retardation occurs through transfer to the retarder the number-average degree of polymerization is given by

$$\text{DP} = \frac{\omega}{\text{d[P]}/\text{d}t} = \left\{c\,\frac{[\text{S}]}{[\text{M}]}\left(1 + \frac{b\omega}{a + b\omega}\right) + \frac{1}{2}\,\frac{\omega}{a^2}\right\}^{-1} \tag{61}$$

The factor 1/2 has been included in eqn. (61) on the assumption that propagating radicals terminate by combination. Allowance for simultaneous transfer to monomer, not involving retardation, can be made by the inclusion of the usual transfer constant C_m (Sect. 2.4) as an additional term in the braces.

Discussion of molecular weights of polymers produced in retarded polymerizations involving transfer has centred on the applicability of the Mayo equation (14) and it is often concluded that this equation is valid [10, 72]. However, an alternative expression to eqn. (61) for DP, cast in the form of the Mayo equation is

$$\frac{1}{\text{DP}} = \frac{\mathscr{I}}{2\omega} + C_m + c\,\frac{[\text{S}]}{[\text{M}]} \tag{62}$$

from which it is seen that the Mayo equation is not strictly applicable unless $\mathscr{I}/\omega$ is maintained constant as [S]/[M] is varied.

4. Some chemical aspects of inhibition and retardation

In addition to assessing the relative reactivities of substrates by determining rate coefficients for the interactions of propagating radicals with retarders or inhibitors, it is often necessary to know the mechanisms of the termination processes. For example, in the case of retarded reactions in which reinitiation occurs it is important to know whether reaction proceeds through transfer or copolymerization, since the mechanism directly influences the molecular weight of the polymer obtained.

When an inhibitor is used to assess rates of initiation, it is essential to know the number of radicals terminated by each molecule of inhibitor.

From information in the literature it is apparent that the chemistry of many of these termination reactions is complex and a given substrate may react differently in different situations. In addition, an extremely large number of substances are known to act as inhibitors or retarders in various polymerization systems and the majority of these reactions have only been subjected to a cursory investigation. Consequently, it is difficult to give a general comprehensive account of these reactions and we shall present only a brief account of some of the more well-known systems.

4.1 STABLE RADICAL INHIBITORS

In principle it might be anticipated that the simplest systems would be those in which a stable free radical is used to terminate a primary radical or a propagating chain. One of the most well-known stable radicals is triphenylmethyl which acts as an effective inhibitor and gives rise to inhibition periods. However, under certain conditions it may also initiate polymerization slowly [74].

The most widely-used radical inhibitor is 2,2-diphenyl-1-picrylhydrazil (DPPH). While the structure of DPPH is conventionally written as (IV),

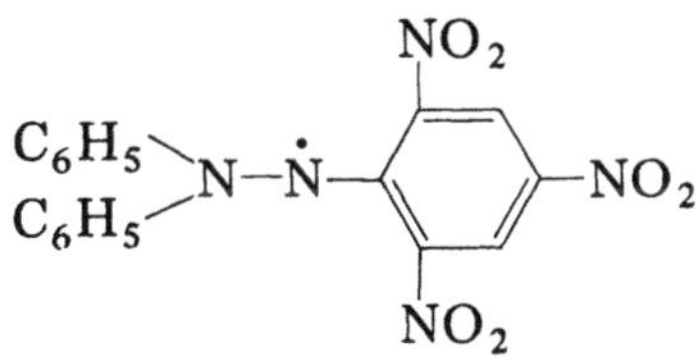

(IV)

the stability of the radical arises from considerable delocalization of the unpaired electron; electron spin resonance spectroscopy has shown that the electron densities on the two nitrogen atoms are equal [75]. DPPH has a high extinction coefficient in the visible spectrum which allows spectrophotometric monitoring of low concentrations during inhibition periods. In polymerizations of methyl methacrylate the disappearance of DPPH colour and the onset of polymerization are coincident, but complications arise in other systems. When vinyl acetate is used as the monomer, the rate of consumption of monomer continues to be insignificant for some time after the disappearance of the DPPH colour [76], while for styrene polymerizations rates of polymerization following the inhibition period are lower than would be anticipated on the basis of the kinetics of retarded polymerizations [77]. These effects are understandable if the products of inhibition are capable of acting as inhibitors or retarders in some reactions [76], and the effect probably arises from the presence of nitro-groups in the DPPH; it is known that aromatic nitro-compounds can act as chain terminators and are more reactive towards vinyl acetate radicals than towards methyl methacrylate radicals [69]. In many cases, rates of initiation can be determined by measuring the rate of loss of DPPH assuming one radical reacts with one DPPH molecule, if certain precautions are taken [78]. However, DPPH readily abstracts hydrogen from many organic compounds and, consequently, cannot be used to measure rates of initiation produced by, for example, photo-dissociation of azo-isopropane [26].

In catalysed polymerizations, inhibitors often react with primary radicals derived from the initiator. One of the most common initiators is azo-bis-isobutyronitrile and reservations have been expressed concerning the suitability of DPPH in "counting" the isobutyronitrile radicals derived from this initiator [18, 79, 80]. The reaction between DPPH and isobutyronitrile radicals is complex. It has been suggested that the radical becomes attached to a phenyl group of the DPPH [81] and a possible product of a complex reaction has been suggested [82]. DPPH residues also become incorporated into polymer formed in the latter stages of reactions carried out in the presence of this inhibitor [83]. The reaction appears to be further complicated if the azo-bis-isobutyronitrile is decomposed in the presence of oxygen [84].

4.2 QUINONE RETARDERS

Probably the most important and widely investigated class of retarders are the quinones and their derivatives. *p*-Benzoquinone and its derivatives are effective inhibitors or retarders and inhibition periods are generally proportional to the initial concentration of the added substrate. Although hydroquinone and its alkyl ether derivatives are often used as monomer stabilizers they are not in themselves inhibitors but rely on the presence of oxygen to produce quinones.

The mechanism of inhibition and retardation by quinones has been the subject of some debate and a variety of mechanisms have been proposed to account for various observations [85]. Although some details are not yet fully understood, it is now accepted that the main mode of reaction is addition of the propagating (rather than primary) radical to the oxygen of the quinone, as in reaction (63)

(63)

The relatively unreactive radical so formed may terminate a second chain, as in eqn. (64)

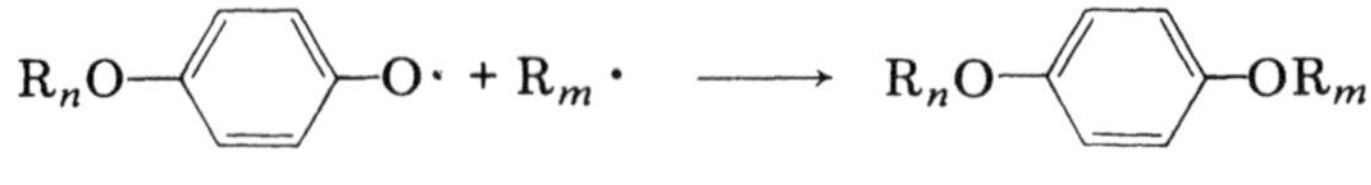

(64)

This mechanism is supported by certain studies on small radical reactions [86]. Consequently, the effectiveness of quinones as retarders varies with the nature of both quinone and monomer concerned. *p*-Benzoquinone may be classified as an inhibitor for styrene polymerizations but only as a strong retarder of polymerization of methyl methacrylate. While most workers favour reaction between propagating radicals and quinone as the major reaction [87, 88], Funt and Williams [89] postulated the formation of an adduct between *p*-benzoquinone and primary radicals produced by decomposition of azo-bis-isobutryronitrile.

The mechanism represented by eqns. (63) and (64) is consistent with radio-tracer studies by Bevington et al. [88] on polymers of methyl methacrylate prepared in the presence of *p*-benzoquinone, when it was found that the polymers formed contained two initiator fragments and one quinone residue per molecule. For styrene polymerizations carried out in the presence of *p*-benzoquinone, radio-tracer studies showed that the polymer may contain more than one quinone residue per chain [63]. This evidence, and the fact that large numbers of quinone molecules are consumed for every polystyrene chain terminated [74, 90—92], led to the view that the so-called unreactive radical formed in eqn. (63) is sufficiently reactive to reinitiate the polymerization of styrene by adding to a monomer molecule. In this way the quinone may be considered to be a co-monomer. These ideas are supported by considerable reductions in molecular weight of polystyrene prepared in the presence of quinone when the polymer is treated with a reagent which will sever ether linkages [63]. Activation energies for reaction (63) and for reinitiation of styrene polymerizations have been estimated by Tűdős [93]. Using chloranil as retarder, evidence for copolymerization of the retarder has been obtained for polymerizations of styrene [94—96] and of methyl methacrylate [68]. In the former case the product was a low-molecular-weight polymer containing one chloranil residue per two styrene residues [96]. Hauser and Zutty [97] have recently prepared a number of copolymers containing large proportions of quinone residues.

4.3 AROMATIC NITRO-COMPOUNDS

A further important class of retarders, generally less effective than quinones, are the aromatic nitro-compounds. The materials will inhibit polymerizations of vinyl acetate [64, 92], retard polymerizations of styrene [68] but have very little effect on polymerizations of methyl methacrylate [68] and of methyl acrylate [69]. These differences in behaviour have been attributed to polar factors [68]. The effectiveness of these substances as retarders increases with the number of nitro-groups per molecule.

Various mechanisms have been proposed to explain the reactions of nitro-compounds with propagating radicals [18] and it is possible that no

References pp. 149—152

single mechanism is universally applicable. Price et al. [98—100] proposed an initial attack by the propagating radical on the aromatic ring, according to

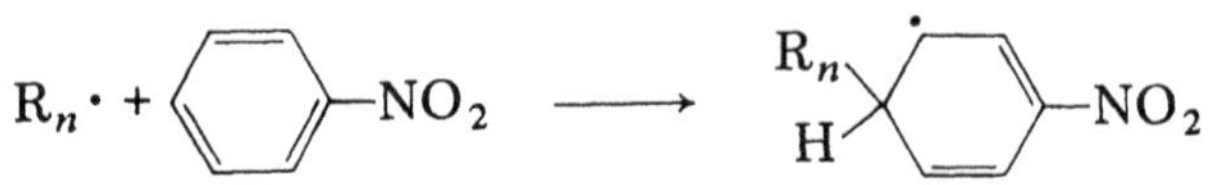

followed by

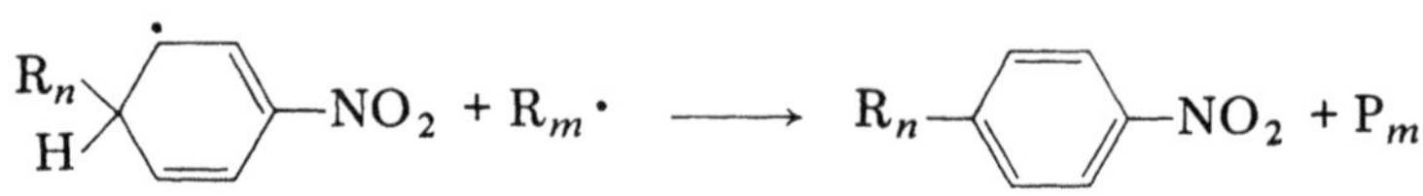

and possibly reinitiation through

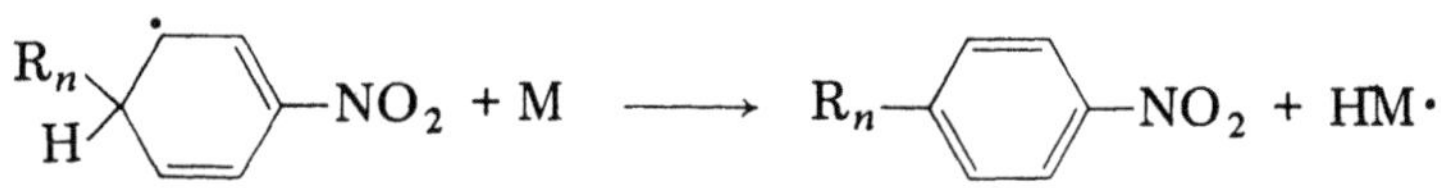

In contrast, Bartlett et al. [64, 101, 102] obtained evidence supporting the view that initial reaction occurs between the propagating radical and the nitro-group, as in reaction (65), viz.

(65)

(V)

Other workers [103—107] support attack on the nitro group as the primary reaction in retardation but propose different reactions for the subsequent fate of radical (V) and its analogues. Hammond and Bartlett [101] proposed reaction (66), viz.

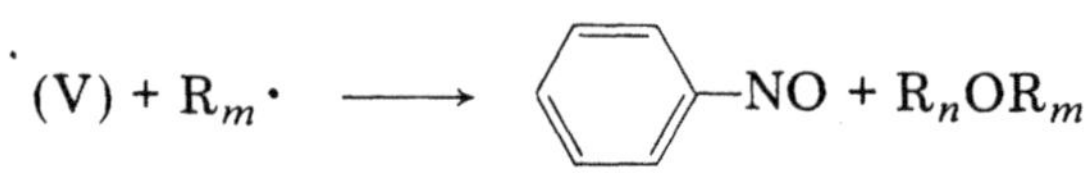

(66)

while Inamoto and Simamura [103] proposed

followed by

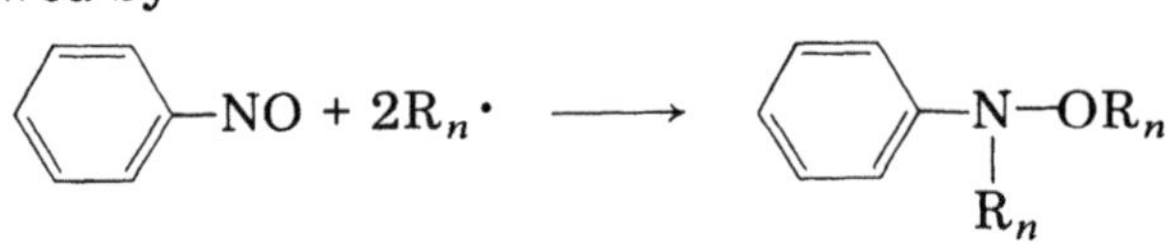

Tűdős suggested reaction (67)

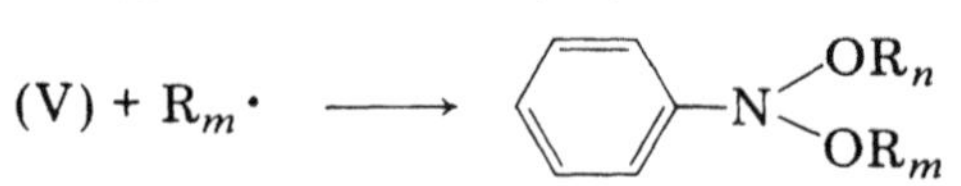

(67)

should be considered.

For styrene polymerizations Bevington and Ghanem [106] found that a mechanism in which

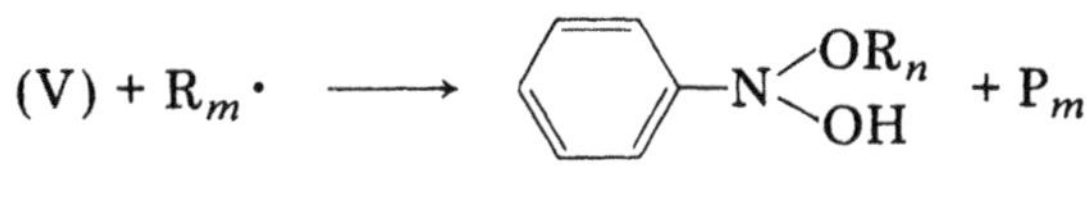

was the predominant reaction and

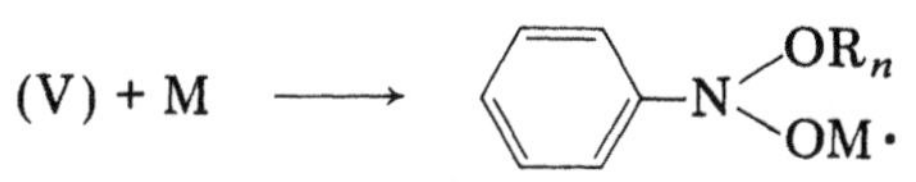

was a minor reaction would explain their observations.

4.4 METAL SALTS AS RETARDERS

A number of metal salts are known to act as inhibitors or retarders under various conditions. The termination reactions are redox processes in which the radical may either be oxidized or reduced. Naturally, the outcome of an interaction between an ion and a radical will depend on the redox potentials appertaining to the system in question.

Early work in this area concentrated on reactions in aqueous solution and it was established that the propagating radicals derived from acrylamide could be terminated by Fe(III), Cu(II), Ce(IV), Hg(II), Tl(III) and Ag(I); kinetic parameters for a number of these processes were determined [108, 109]. The VO_2^{2+} [110] and UO_2^{2+} [111] ions were also shown to terminate polymerizations of acrylamide and kinetic parameters were determined, but it was found that UO_2^+ ion did not terminate the polymerizations of methyl methacrylate, methyl acrylate or acrylonitrile in aqueous solution [112]. It has been reported [113] that Ti(III), Mo(III), U(III), W(III), Cr(II), U(II) and Eu(II) will terminate polymerizations of acrylonitrile.

In non-aqueous systems the most well-known retarder of this type is ferric chloride and in those cases studied the metal salt acts as an ideal retarder [62] with no reinitiation by products of the termination reaction (Sect. 3.3.1). Values of the rate coefficients for the termination reaction are in the order styrene > acrylonitrile > methacrylonitrile. The reaction involves reduction of Fe(III) to Fe(II) and formation of Fe(II) can be used to monitor the process and derive rates of initiation. For acrylonitrile polymerizations, for example, the termination reaction could be written as

$$\sim\!CH_2\!-\!\dot{C}HCN + FeCl_3 \longrightarrow \sim\!CH_2\!-\!CHClCN + FeCl_2 \qquad (68)$$

or as

$$\sim\!CH_2\!-\!\dot{C}HCN + FeCl_3 \longrightarrow \sim\!CH{=}CHCN + HCl + FeCl_2$$

Since no evidence of unsaturation was found in polyacrylonitrile prepared in the presence of $FeCl_3$, eqn. (68) was proposed as the preferred mode of reaction. It was also observed that the rate of polymerization of acrylonitrile in the presence of low concentrations of $FeCl_3$ increased markedly when low concentrations of *N,N*-dimethylformamide were added as solvent and this effect was attributed to the formation of a complex between the solvent and $FeCl_3$ which is less reactive towards the propagating radical than is $FeCl_3$ itself [62]. Reactivity parameters, including a polar factor, were determined for $FeCl_3$ in *N,N*-dimethylformamide solutions [114]. The influence of a number of other soluble Fe(III) salts on the polymerization of methyl methacrylate in *N,N*-dimethylformamide solution have been investigated, the magnitude of the effects varying from complete inhibition to zero retardation [115].

4.5 MISCELLANEOUS RETARDERS

Numerous cases have been reported of substances retarding polymerizations of various monomers. Polymerizations of those monomers which yield the most reactive radicals (e.g. vinyl acetate) are likely to be retarded, even if only slightly, by the presence of any substance which transfers with or adds to the propagating radical since the transfer or co-monomer radicals are likely to be less reactive than the propagating radical. In the case of vinyl chloride polymerizations virtually every solvent for poly(vinyl chloride) retards the polymerization [7].

One of the most well-known retarders of free-radical polymerizations is molecular oxygen. The literature on the effects of oxygen is vast and, at first sight, conflicting [116]. Oxygen has been reported to have effects from catalytic to inhibitory. A number of organic compounds, including monomers, can peroxidize and the peroxides subsequently act as initiators of thermal polymerization. However, it is well known that oxygen adds to many organic radicals to form peroxy radicals and in a polymerization the reaction may be represented as

$$R_n\cdot + O_2 \longrightarrow R_n{-}O{-}O\cdot \qquad (69)$$

The peroxy radical may then reinitiate polymerization in the process

$$R_n{-}O{-}O\cdot + M \longrightarrow R_nOOM\cdot \qquad (70)$$

so that oxygen is a potential co-monomer. In many cases the reinitiation, reaction (70), is slower than reaction (69) and oxygen acts as a retarder. In other cases, where homopolymerization is slow or non-existent, copolymerization may proceed quite rapidly [117]. Some workers have succeeded in isolating low-molecular-weight polyperoxides formed by copolymerization of oxygen with monomers [118, 119].

The peroxy linkages formed in polyperoxides as a result of copolymerization of oxygen with a monomer may subsequently initiate the thermal or photopolymerization of a monomer. Consequently the kinetics of polymerizations performed in the presence of oxygen may be complex. Initial rates of polymerization may be extremely low during the period in which molecular oxygen is acting as a retarder, or even as an inhibitor, but subsequent decomposition of the polyperoxides (e.g. at the end of an inhibition period) may produce rates of polymerization which are greater than if no retarder had been present initially.

Sulphur can retard or inhibit free-radical polymerizations by adding to propagating radicals to produce $R_n(S)_m\cdot$ radicals which are relatively inactive in reinitiation but take part in termination reactions [120]. However, subsequent rupture of S—S bonds can yield reactive radicals capable of initiating polymerization.

General references to retardation produced by other types of compounds, including thiols, aromatic amines and nitroso-compounds, have been given elsewhere [6, 85, 90, 115, 121].

REFERENCES

1 P. J. Flory, J. Am. Chem. Soc., 59 (1937) 241.
2 T. Manabe, T. Otsumi and S. Okamura, J. Polym. Sci., 58 (1962) 121.
3 F. R. Mayo, J. Am. Chem. Soc., 65 (1943) 2324.
4 C. H. Bamford, W. G. Barb, A. D. Jenkins and P. F. Onyon, The Kinetics of Vinyl Polymerisation by Radical Mechanisms, Butterworths, London, 1958.
5 A. D. Jenkins and R. Johnston, J. Polym. Sci., 39 (1959) 81.
6 J. C. Bevington, Radical Polymerisation, Academic Press, New York, 1961.
7 W. H. Atkinson, C. H. Bamford and G. C. Eastmond, Trans. Faraday Soc., 66 (1970) 1446.
8 J. C. Bevington and H. G. Troth, Trans. Faraday Soc., 58 (1962) 2005.
9 D. H. Johnson and A. V. Tobolsky, J. Am. Chem. Soc., 74 (1952) 938.
10 C. H. Bamford and E. F. T. White, Trans. Faraday Soc., 52 (1956) 716.
11 P. D. Bartlett and R. Altshul, J. Am. Chem. Soc., 67 (1945) 816.
12 V. Mahadevan and M. Santhappa, Makromol. Chem., 16 (1955) 119.
13 B. Baysal and A. V. Tobolsky, J. Polym. Sci., 8 (1952) 529.
14 F. R. Mayo, R. A. Gregg and M. S. Matheson, J. Am. Chem. Soc., 73 (1951) 1691.
15 M. S. Matheson, E. E. Auer, E. B. Bevilacqua and E. J. Hart, J. Am. Chem. Soc., 71 (1949) 2610.
16 G. Vidotto, A. Crosato-Arnaldi and G. Talamini, Makromol. Chem., 114 (1968) 217.
17 P. D. Bartlett and F. A. Tate, J. Am. Chem. Soc., 75 (1953) 91.
18 C. Walling, Free Radicals in Solution, J. Wiley, New York, 1957.
19 R. Hart and G. Smets, J. Polym. Sci., 5 (1950) 55.
20 N. G. Gaylord and F. R. Eirich, J. Polym. Sci., 5 (1950) 743.
21 H. W. Melville and P. R. Sewell, Makromol. Chem., 32 (1959) 139.
22 J. C. Bevington, J. Toole and L. Trossarelli, Makromol. Chem., 32 (1959) 57.
23 Z. Laita, J. Polym. Sci., 38 (1959) 247.
24 W. Cooper, J. Chem. Soc., (1952) 2408.
25 C. Walling and Y. W. Chang, J. Am. Chem. Soc., 76 (1954) 4878.
26 Z. Laita and Z. Mačhacek, J. Polym. Sci., 38 (1959) 459.

27 T. Ferrington and A. V. Tobolsky, J. Am. Chem. Soc., 80 (1958) 3215.
28 W. B. Reynolds and E. W. Cotton, Ind. Eng. Chem., 42 (1950) 1905.
29 R. N. Haward and W. Simpson, Trans. Faraday Soc., 47 (1951) 212.
30 C. H. Bamford and H. Tompa, J. Polym. Sci., 10 (1953) 345.
31 J. K. Beasley, J. Am. Chem. Soc., 75 (1953) 6123.
32 D. Lim and O. Wichterle, J. Polym. Sci., 29 (1958) 579.
33 J. C. Bevington, G. M. Guzman and H. W. Melville, Proc. R. Soc. London, Ser. A, 221 (1954) 437, 453.
34 E. H. Gleason, M. L. Miller and G. F. Sheats, J. Polym. Sci., 38 (1959) 133.
35 W. M. D. Bryant, J. Polym. Sci., 2 (1947) 547.
36 M. J. Roedel, J. Am. Chem. Soc., 75 (1953) 6110.
37 F. M. Rugg, J. J. Smith and L. H. Wartman, Ann. N.Y. Acad. Sci., 57 (1953) 398.
38 A. G. Morrell, Discuss. Faraday Soc., 22 (1956) 152.
39 A. H. Willbourn, J. Polym. Sci., 34 (1959) 569.
40 K. Casey, C. T. Elston and M. K. Phibbs, J. Polym. Sci., Part B, 2 (1964) 1053.
41 J. C. Woodbrey and P. E. Ehrlich, J. Am. Chem. Soc., 85 (1963) 1580.
42 H. W. Melville and P. R. Sewell, Makromol. Chem., 32 (1959) 139.
43 J. T. Clarke, R. O. Howard and W. H. Stockmayer, Makromol Chem., 44—46 (1961) 427.
44 I. Sakurada, Y. Sakaguchi and Z. Shiiki, Kobunshi Kagaku, 21 (1964) 289.
45 S. Imoto, J. Uchida and T. Kominami, Kobunshi Kagaku, 14 (1957) 101.
46 C. H. Bamford and E. F. T. White, J. Polym. Sci., 54 (1958) 268.
47 M. S. Glukman, M. J. Kampf, J. L. O'Brien, T. G. Fox and R. K. Graham, J. Polym. Sci., 37 (1959) 411.
48 S. Okamura and K. Katagiri, Makromol. Chem., 28 (1958) 177.
49 G. V. Schulz, G. Henrici and S. Olivé, J. Polym. Sci., 17 (1955) 45.
50 K. Yokota, H. Tomioka and A. Tazumi, J. Macromol. Sci., Chem., 6 (1972) 1337.
51 W. V. Smith., J. Am. Chem. Soc., 68 (1946) 2059.
52 R. A. Gregg and F. R. Mayo, J. Am. Chem. Soc., 70 (1948) 2373.
53 R. A. Gregg, D. M. Alderman and F. R. Mayo, J. Am. Chem. Soc., 70 (1948) 3740.
54 J. L. O'Brien and F. Gornick, J. Am. Chem. Soc., 77 (1955) 4757.
55 R. A. Gregg and F. R. Mayo, Discuss. Faraday Soc., 2 (1947) 328.
56 R. A. Gregg and F. R. Mayo, J. Am. Chem. Soc., 75 (1953) 3530.
57 C. Walling, J. Am. Chem. Soc., 70 (1948) 2561.
58 L. A. Wall and D. W. Brown, J. Polym. Sci., 14 (1954) 513.
59 W. A. Stockmayer, R. O. Howard and J. T. Clarke, J. Am. Chem. Soc., 75 (1953) 1756.
60 A. V. Tobolsky and B. Baysal, J. Am. Chem. Soc., 75 (1953) 1757.
61 S. K. Das, S. R. Chatterjee and S. R. Palit, Proc. R. Soc. London, Ser. A, 227 (1955) 252.
62 C. H. Bamford, A. D. Jenkins and R. Johnston, Proc. R. Soc. London, Ser. A, 239 (1957) 214.
63 J. C. Bevington, N. A. Ghanem and H. W. Melville, J. Chem. Soc., (1955) 2822.
64 P. D. Bartlett and H. Kwart, J. Am. Chem. Soc., 72 (1950) 1051.
65 E. P. Bonsall, L. Valentine and H. W. Melville, Trans. Faraday Soc., 49 (1953) 686.
66 G. M. Burnett and P. R. E. J. Cowley, Trans. Faraday Soc., 49 (1953) 1490.
67 P. J. Flory, Principles of Polymer Chemistry, Cornell University Press, Ithaca, New York, 1953.
68 J. L. Kice, J. Am. Chem. Soc., 76 (1954) 6274.
69 J. L. Kice, J. Polym. Sci., 19 (1956) 123.

70 A. D. Jenkins, Trans. Faraday Soc., 54 (1958) 1895.
71 A. D. Jenkins, Trans. Faraday Soc., 54 (1958) 1885.
72 P. W. Allen, F. M. Merrett and J. Scanlan, Trans. Faraday Soc., 51 (1955) 95.
73 L. H. Peebles, J. T. Clarke and W. H. Stockmayer, J. Am. Chem. Soc., 82 (1960) 4730.
74 F. R. Mayo and R. A. Gregg, J. Am. Chem. Soc., 70 (1948) 1284.
75 D. J. E. Ingram, Free Radicals as Studied by Electron Spin Resonance, Butterworths, London, 1958.
76 W. I. Bengough, Chem. Ind. (London), (1955) 599.
77 M. S. Matheson, E. E. Auer, E. B. Bevilacqua and E. J. Hart, J. Am. Chem. Soc., 73 (1951) 1700.
78 J. A. Lyons and W. F. Watson, J. Polym. Sci., 18 (1955) 141.
79 G. S. Hammond, J. N. Sen and C. E. Boozer, J. Am. Chem. Soc., 77 (1955) 3244.
80 F. Tűdős, T. F. Berezhnikh and M. Azori, Acta Acad. Sci. Hung., 24 (1960) 91.
81 R. H. Poirier, E. J. Kahler and F. Bennington, J. Org. Chem., 17 (1952) 1437.
82 C. E. H. Bawn and D. Verdin, Trans. Faraday Soc., 51 (1960) 925.
83 J. C. Bevington, J. Chem. Soc., (1956) 1127.
84 D. Verdin, Trans. Faraday Soc., 56 (1960) 823.
85 G. Goldfinger, W. Yee and R. D. Gilbert, in H. F. Mark, N. G. Gaylord and N. M. Bikales (Eds.), Encyclopedia of Polymer Science and Technology, Vol. 7, Interscience, New York, 1967.
86 A. F. Bickel and W. A. Waters, Rec. Trav. Chim. Pays-Bas, 69 (1950) 1490.
87 S. G. Cohen, J. Polym. Sci., 2 (1947) 511.
88 J. C. Bevington, N. A. Ghanem and H. W. Melville, Trans. Faraday Soc., 51 (1955) 346.
89 B. L. Funt and F. D. Williams, J. Polym. Sci., 46 (1960) 139.
90 S. G. Foord, J. Chem. Soc., (1940) 48.
91 G. Goldfinger, I. Skeist and H. Mark, J. Phys. Chem., 47 (1943) 578.
92 P. D. Bartlett, G. S. Hammond and H. Kwart, Discuss. Faraday Soc., 2 (1947) 342.
93 F. Tűdős, J. Polym. Sci., 30 (1958) 343.
94 J. W. Breitenbach and H. L. Breitenbach, Z. Phys. Chem., Abt. A, 190 (1942) 361.
95 J. W. Breitenbach and A. J. Renner, Can. J. Res., Sect. B, 28 (1950) 507.
96 F. A. Bovey and I. M. Kolthoff, J. Polym. Sci., 5 (1950) 569.
97 C. F. Hauser and N. L. Zutty, J. Polym. Sci., Part A-1, 8 (1970) 1385.
98 C. C. Price, J. Am. Chem. Soc., 65 (1943) 2380.
99 C. C. Price and D. A. Durham, J. Am. Chem. Soc., 65 (1943) 757.
100 C. C. Price and D. H. Read, J. Polym. Sci., 1 (1946) 44.
101 G. S. Hammond and P. D. Bartlett, J. Polym. Sci., 6 (1950) 617.
102 P. D. Bartlett and H. Kwart, J. Am. Chem. Soc., 74 (1952) 3969.
103 N. Inamoto and M. Simamura, J. Org. Chem., 23 (1958) 408.
104 W. P. Norris, J. Am. Chem. Soc., 81 (1959) 4239.
105 B. A. Gringas and W. A. Waters, J. Chem. Soc., (1954) 1920.
106 J. C. Bevington and N. A. Ghanem, J. Chem. Soc., (1959) 2071.
107 F. Tűdős, I. Kende and M. Azori, J. Polym. Sci., 53 (1961) 17.
108 E. Collinson, F. S. Dainton and G. S. McNaughton, J. Chim. Phys., 52 (1955) 556.
109 E. Collinson, F. S. Dainton, D. R. Smith, G. J. Trudel and S. Tazuke, Discuss. Faraday Soc., 29 (1960) 188.
110 E. A. S. Cavell and A. C. Meeks, Makromol. Chem., 78 (1964) 178.
111 E. A. S. Cavell and A. C. Meeks, Polymer, 8 (1967) 79.
112 Y. Mahadevan and M. Santhappa, J. Polym. Sci., 50 (1961) 361.

113 F. S. Dainton and D. G. L. James, Trans. Faraday Soc., 54 (1958) 649.
114 C. H. Bamford, A. D. Jenkins and R. Johnston, Trans. Faraday Soc., 55 (1959) 418.
115 E. R. Entwistle, Trans. Faraday Soc., 56 (1960) 284.
116 F. A. Bovey and I. M. Kolthoff, Chem. Rev., 42 (1948) 491.
117 F. R. Mayo, A. A. Meller and G. A. Russell, J. Am. Chem. Soc., 80 (1958) 2500.
118 F. A. Bovey and I. M. Kolthoff, J. Am. Chem. Soc., 69 (1947) 2143.
119 C. F. Barnes, R. M. Elofson and G. D. Jones, J. Am. Chem. Soc., 72 (1950) 210.
120 P. D. Bartlett and D. S. Triffan, J. Polym. Sci., 20 (1956) 457.
121 K. K. Georgieff, J. Appl. Polym. Sci., 9 (1965) 2009.

Chapter 3

Kinetic Data for Homogeneous Free-Radical Polymerizations of Various Monomers

G. C. EASTMOND

1. Introduction

The essential elementary processes which contribute to the mechanism of free-radical polymerizations and their kinetic consequences were established in the 1930s. Since that time the study of free-radical polymerization kinetics has remained an active field of investigation and literally thousands of papers, on various aspects of this field, have been published. These papers have achieved a variety of objectives. They have provided evidence to substantiate the basic mechanism of the polymerization processes. They have described the development of techniques for obtaining useful kinetic parameters, such as absolute rate coefficients, and have provided such data for a number of systems. They have enabled side reactions, such as primary-radical termination, chain transfer and retardation, to be detected, their consequences to be understood and parameters which assess the importance of such reactions (e.g. transfer constants) to be defined and measured. In consequence the mechanisms of polymerization processes have been refined. Finally, they have provided a means by which rates and mechanisms of new radical-forming processes have been studied by using rates of polymerization to monitor rates of radical formation.

In spite of the total effort applied to this field, examination of the literature reveals that surprisingly few definitive data are available for all but a few of the simplest systems relating to the most common monomers. Various factors have contributed to this situation. The first is historical. Accurate techniques for measuring rates of polymerization (e.g. dilatometry) have long been available. However, determination of absolute rate coefficients requires, in addition to measurements of stationary and non-stationary rates, a knowledge of molecular weights or degrees of polymerization of polymers. In the early days there was a shortage of accurate and reliable methods for determining absolute molecular weights; inevitably this led to inaccuracies in viscosity—molecular weight relationships used by many workers to calculate degrees of polymerization. Naturally, such inaccuracies led to uncertainties in absolute values of rate coefficients for propagation and termination. Gradually the situation has improved, especially since the advent of rapid osmometric techniques

References pp. 276—285

which allow more accurate determination of number-average degrees of polymerization, the most useful parameter as an adjunct to rate measurements.

Secondly, as in any topic, there has been a gradual improvement in the understanding of the detailed processes involved. Various complicating features have been recognized and due allowance for these permits the determination of more meaningful absolute parameters. For example, calculation of absolute rate coefficients for the polymerization of any monomer requires a knowledge of the relative rates of combination and disproportionation termination appertaining to that particular system. Initially incorrect assumptions on this point have resulted in some old data being superseded.

Thirdly, concepts which have been developed through studies on one particular system have not been applied generally. An example of this is diffusion-controlled termination, which was shown to be an important feature of methyl methacrylate polymerizations and for which system it has been established that the rate coefficient for termination is inversely proportional to viscosity of the solvent medium (see Chap. 1, Sect. 4.2). The physical concepts involved in diffusion-controlled termination are applicable generally. But, while it is now known that viscosity effects are important in polymerizations of styrene [1] and chloroprene [2], there has been no general application of these ideas. Unless due account is taken of such physical effects, experimentally observed solvent effects may be incorrectly attributed to changes in chemical reactivity.

Chain transfer processes are frequently encountered in free-radical polymerizations. The extent to which transfer occurs and the reactivity of, say, solvent (or other transfer agent) are usually assessed through application of the Mayo equation

$$\frac{1}{\mathrm{DP}} = \frac{1}{\mathrm{DP}_0} + C_s \frac{[\mathrm{S}]}{[\mathrm{M}]} \qquad (1)$$

DP, DP_0 are, respectively, the number-average degrees of polymerization in the presence and absence of transfer, M and S are monomer and solvent (or transfer agent) and C_s is the appropriate transfer constant (the ratio of the rate coefficient for transfer to that for propagation). While values of C_s can be obtained from plots of 1/DP vs. [S]/[M] determination of true values of C_s requires DP_0 to remain constant throughout the series of experiments. As explained in the discussion of transfer processes (see Chap. 2, Sect. 2.7), DP_0 may change as [S]/[M] is varied unless adequate precautions are taken. Many workers fail to provide evidence that such precautions have been taken and, for this reason, many reported values of C_s should be treated with caution. Simple application of the Mayo equation to processes in which retardation accompanies chain transfer yields only apparent values of C_s. In addition, as [S]/[M] is varied DP_0 may vary as a result of changes in viscosity of the solvent medium which affects the kinetic chain length through changes in the termination rate coefficient. In at least one case it has been shown that if kinetic data obtained for polymerizations of methyl methacrylate in viscous and non-viscous systems are normalized to a given viscosity the transfer constant to *N,N*-dimethylaniline is identical for different solvents, whereas if no correction is made apparently different values are found [3]. Consequently, effects which arise from physical phenomena may be incorrectly attributed to changes in chemical reactivity.

For the simplest cases the rate of polymerization of a monomer M in the presence of an added initiator is given by

$$-\frac{\mathrm{d}[\mathrm{M}]}{\mathrm{d}t} = \frac{k_\mathrm{p}}{k_\mathrm{t}^{1/2}} [\mathrm{M}] [\mathscr{I}]^{1/2} \tag{2}$$

where k_p and k_t are, respectively, the rate coefficients for propagation and termination, and $\mathscr{I}$ is the rate of initiation. Since $\mathscr{I}$ is often proportional to [initiator], eqn. (2) may be written

$$-\frac{\mathrm{d}[\mathrm{M}]}{\mathrm{d}t} = \frac{k k_\mathrm{p}}{k_\mathrm{t}^{1/2}} [\mathrm{M}] [\mathrm{initiator}]^{1/2} \tag{3}$$

where k is an effective rate coefficient describing the rate of formation of propagating radicals. In turn, eqn. (3) may be simplified to

$$-\frac{\mathrm{d}[\mathrm{M}]}{\mathrm{d}t} = K[\mathrm{M}] [\mathrm{initiator}]^{1/2} \tag{4}$$

where the composite parameter K is an overall rate coefficient for polymerization. In the general case eqn. (4) may be written

$$-\frac{\mathrm{d}[\mathrm{M}]}{\mathrm{d}t} = K[\mathrm{M}]^a [\mathrm{initiator}]^b \tag{5}$$

Numerous kinetic studies have been concerned with the effects of additives on polymerization systems and with the study of new initiators. While some of these investigations have involved a detailed study of the influence of additives on polymerization reactions and others have used polymerization reactions to carry out detailed analyses of mechanisms of radical formation from new initiators, the major proportion of the studies have been very cursory. Many authors have been content to obtain values of the parameters K, a and b and the order with respect to any additive and, possibly, overall activation energies. Some authors have simply reported effects of additives on degrees of polymerization. In some cases from a knowledge of rates and degrees of polymerization apparent values of $k_\mathrm{p}/k_\mathrm{t}^{1/2}$. have been obtained. Many of these cursory investigations are embodied in short papers which may identify deviations from normal kinetic behaviour appertaining to simple polymerizations but are not sufficiently comprehensive to decide whether the observed effects arise from complications in the mechanism of initiation or from modifications to the polymerization processes, e.g. from retardation or specific effects on k_p or k_t. Although such investigations constitute a large proportion of the literature on free-radical polymerization kinetics they provide very little insight into the mechanisms of the reactions.

Any attempt to produce an up-to-date and comprehensive compilation of kinetic data obviously requires the examination of a vast amount of information. Inevitably there must be some selection of data. An attempt has been made to restrict data in the present compilation to homogeneous systems, although it is not always clear from the original literature whether or not polymerizations remain homogeneous to the maximum conversions employed. Some rate coefficients obtained from emulsion polymerization of the more common monomers have been retained for comparative purposes since in a few cases the only indication of the magnitude of the rate coefficients comes from emulsion polymerizations. Values of transfer constants obtained from emulsion polymerizations are not included. Composite parameters determined in the course of cursory investigations, referred to above, have not been included since they are often ambiguous and may only apply under a limited range of conditions.

An extremely useful parameter in practical investigations is the composite parameter $k_p/k_t^{1/2}$. Numerous values of this parameter have been quoted in the literature for a variety of systems but are not included in this compilation since many are only apparent values and only applicable under specific circumstances which are often not well-defined.

The present compilation is restricted to certain specific parameters, namely, absolute values of k_p and k_t (from which values of $k_p/k_t^{1/2}$ can be calculated) together with their activation energies, transfer constants and values of the parameter k_z/k_p (k_z is the rate coefficient for chain termination in retarded polymerizations). Values of k_t are overall values of the termination rate coefficient. Specific reaction conditions (e.g. solvents and pressures) which have been identified have been noted as comments under the relevant references. Where viscosity effects have been considered they have been referred to and, where available, solution viscosities have been given to allow examination of relationships between k_t and solvent viscosity. Other data, such as rate coefficients for decompositions of initiators and relative proportions of combination and disproportionation termination, have been considered in some detail in discussions of these topics (see Chap. 1).

Examination of quoted values of transfer constants reveals considerable scatter of values for some processes. The scatter may reflect the use of incorrect values of degrees of polymerization in some cases, or the existence of unidentified effects which have not been allowed for, or the incorrect use of the Mayo equation as referred to above. Many quoted transfer constants are apparent values resulting from simple application of the Mayo equation to retarded polymerizations. Whether such values are meaningful and generally useful and whether they should be included as transfer constants or as values of k_z/k_p may be debatable and must depend on the extent to which retardation occurs. In any case, values quoted in the Tables should be used with caution.

When analysing kinetic data relating to polymerization processes it is customary, and normally justifiable, to assume that rate coefficients are independent of the chain length of the propagating chain. However, in some cases, notably chain transfer with highly active transfer agents, there is good evidence that chain-transfer constants vary with degree of polymerization, especially for short chains. The investigation of such highly efficient transfer processes is usually associated with the study of telomerization reactions (the formation of telomers or very short chain species carrying end-groups derived from the transfer agent) and Starks [4] has recently compiled a large volume of such data.

In producing the present compilation of data considerable use has been made of previous compilations, particularly the Polymer Handbook [5], as sources of older data. In the absence of any specific reason to the contrary the accuracy of such compilations has been accepted (citations to Chemical Abstracts included with the more obscure references in the older compilations have not been included here). With regard to the more recent literature the more popular journals have been scanned for relevant information. However, in view of the vast amount of polymerization data published and the relatively small proportion which contains quotable data this has proved to be a relatively unprofitable task. Since it has not been possible to read all the literature in detail some data have probably been overlooked.

In the present compilation, data relating to the polymerization of the major monomers are presented in individual sections and those relating to derivatives of the major monomers (e.g. styrene derivatives) are given in subsequent sections. Finally, a section on the polymerizations of miscellaneous monomers is included. It is obvious from these sections that very few monomers have received any extensive investigation. The text associated with each section is minimal and only serves to summarize the major characteristic features of the polymerizations of the monomer concerned. Further details of these particular aspects and a number of relevant references can be found in the more detailed discussions of these topics (Chaps. 1 and 2).

2. Styrene

Of all vinyl monomers styrene is the most well-known and most extensively investigated. Styrene is a relatively reactive monomer and yields a comparatively unreactive (stabilized) propagating radical. Kinetically, free-radical polymerizations of styrene in bulk and in solution are relatively simple and these reactions have often been used as a base-line

TABLE 1
Absolute rate coefficients and activation energies for free-radical polymerizations of styrene

Temp. (°C)	k_p (l $mole^{-1}$ sec^{-1})	E_p (kcal $mole^{-1}$)	$10^{-7}k_t$ (l $mole^{-1}$ sec^{-1})	E_t (kcal $mole^{-1}$)	$10^6 k_p/k_t$	Comments	Ref.
	($k_p = 2.4 \times 10^8 \exp\{-(9000 \pm 400)/RT\}$ l $mole^{-1}$ sec^{-1})[b]					a	7
0	6.91	6.5 ± 1	0.183	2.8 ± 1		c	8
	13.2	7.76	3.32	2.37		d	9
5	22	7.4, 8.4				e	10
10					0.583	d	9
15	2.92	6.3	0.555	1.9		f	11
					0.65, 0.68	f	12
					2.0	f	13
	40 ± 20[g]		8.0 ± 4.0[g]		0.5	d	14
20	24		1.4			d	15
					0.55	d	16
25	18.7		0.279			c	8
	39.5		0.596			f	11
	44		4.75			d	9
	25 ± 3		0.9 ± 0.15			d, dodecane (40 % v/v)	17
	66.5					a	7
	35 ± 7		1.9 ± 0.4			d,i, η = 11.1 mP	1
	35 ± 5		1.9 ± 0.3			d,i, bromobenzene (20 % v/v), η = 10.8 mP	1
	29 ± 8		1.6 ± 0.3			d,i, bromobenzene (50 % v/v), η = 10.5 mP	1

	29 ± 7		1.5 ± 0.4		d,i, diethyl malonate (20 % v/v), η = 12.1 mP	1
	23 ± 8		1.1 ± 0.2		d,i, diethyl phthalate (20 % v/v), η = 15.2 mP	1
	20 ± 5		0.8 ± 0.2		d,i, diethyl phthalate (50 % v/v), η = 26.1 mP	1
	27 ± 4		0.31 ± 0.04		d,i, dinonyl phthalate (60 % v/v), η = 58.2 mP	1
30	51.9		1.05	4.93	d	18
	55		5.05	1.089	d	9
				2.36	d	16
				2.06, 2.28	f	19
	72.5		6.65		d	20
	106		10.8		d	21
	46.4		3.36		d, 1 atm	22
	56.2		2.63		d, 250 atm	22
	66.6		2.24		d, 500 atm	22
	78.8		2.02		d, 750 atm	22
	96.1		1.99		d, 1000 atm	22
30.5	51.0	17.6			e	23
35				2.50	f	19
	63	13.0, 14.1			e	24
40	139				e	10
				2.77	f	19
	96				e	24
40.5	120.5				e	23

TABLE 1—*continued*

Temp. (°C)	k_p (l mole^{-1} sec^{-1})	E_p (kcal mole^{-1})	$10^{-7} k_t$ (l mole^{-1} sec^{-1})	E_t (kcal mole^{-1})	$10^6 k_p/k_t$	Comments	Ref.
45	151					[e]	24
	164					[e]	25
50	123				1.895	[d]	9
	223					[e]	24
	209	5.9	11.5	0.5		[h]	26
	206					[e]	25
50.2	300.3 (291.0)					[e]	23
55	270					[e]	24
60	176		7.2			[d]	9
	376					[e]	24
	186 ± 8		7.2[j]			[i, k], η = 4.74 mP	1
	183 ± 8		6.7[l]			[i, k], bromobenzene (20 % v/v), η = 5.09 mP	1
	167 ± 8		6.2[l]			[i, k], bromobenzene (40 % v/v), η = 5.48 mP	1
	165 ± 8		6.0[l]			[i, k], bromobenzene (50 % v/v), η = 5.71 mP	1
	178 ± 8		6.4[l]			[i, k], diethyl malonate (20 % v/v), η = 5.34 mP	1
	133 ± 8		5.2[l]			[i, k], diethyl malonate (50 % v/v), η = 6.56 mP	1
	174 ± 8		5.1[l]			[i, k], diethyl phthalate (20 % v/v), η = 6.63 mP	1

	163 ± 8	3.8[l]	[i,k], diethyl phthalate (40 % v/v), η = 8.95 mP	1
	162 ± 8	3.0[l]	[i,k], diethyl phthalate (50 % v/v), η = 11.53 mP	1
	186 ± 8	6.6[l]	[i,k], benzonitrile (20 % v/v), η = 5.14 mP	1
	158 ± 8	6.1[l]	[i,k], benzonitrile (40 % v/v), η = 5.58 mP	1
	134 ± 8	5.8[l]	[i,k], benzonitrile (50 % v/v), η = 5.86 mP	1
70.9	357.9		[e]	23

[a] Direct determination of radical concentration by ESR.
[b] Units of R undefined in text, probably cal mole^{-1} deg^{-1}.
[c] Viscosity method.
[d] Rotating-sector technique.
[e] Emulsion polymerization.
[f] Direct study of pre- or after-effect.
[g] Units are kg mole^{-1} sec^{-1}.
[h] Calculated from data of ref. 9.
[i] Diffusion controlled termination considered
[j] Value of k_t quoted in ref. 9 assumed to hold.
[k] From steady-state kinetics assuming a value of k_t.
[l] Value of k_t calculated on the assumption that k_t in ref. 9 holds for bulk monomer and that k_t is inversely proportional to solvent viscosity.

TABLE 2
Monomer transfer constants (C_M) for free-radical polymerization of styrene

Temp. (°C)	$10^4 C_M$	Solvent	Ref.
	$C_M = 0.2 \exp(-5400/RT)$[a]		27
0	~0		27
	0.108		8, 28
25	0.279		18, 28
	0.358		8
27	0.31		29
30	0.2		27, 30
	0.32		18
45	0.3		27
50	0.4	Benzene	30, 31
	0.6		29
	0.62		29
	0.78		32
60	0.595		33
	0.6		27, 29, 30, 34, 35—37
	0.79		29
	0.85		29
	1.1		38
67.8	1.0		34
70	0	Ethyl acetate	39
	0.6		27, 40
	0.96		41
	1.16		29
	1.35		29
	2.0	Benzene	39
75	0	Ethyl acetate	39
	1.6		39
	5.0	Benzene	39
	5.00	Ethyl methyl ketone	39
80	0.7		42
	0.75		27
	1.00		30
	1.604		43
80.3	4.0		34

TABLE 2—*continued*

Temp. (°C)	$10^4 C_M$	Solvent	Ref.
90	0.85		27
	1.25		29
	1.47		29
	1.79		29
	2.147		43
99	1.5		42
100	1.72		29
	1.8		29
	1.83		29
110	2.80		29
117	1.40		30
132	2.45		29
	3.0		29
	3.4		29
	5.33		29

[a] Units of R, cal mole^{-1} deg^{-1}.

TABLE 3
Transfer constants to polymer (C_p) relating to free-radical homopolymerizations of various monomers

Monomer	Temp. (°C)	$10^4 C_p$	Comments	Ref.
Acrylonitrile	50	4.7	$MgClO_4$ soln.	44
	60	3.5	[a]	45
1,3-Butadiene	50	11		46
N,N-Dimethylacrylamide	50	0.61		47
Methyl acrylate	60	0.5	[a]	48
		1.0	[a]	49
Methyl methacrylate	50	0.22	[b]	50
		1.5	[b]	31, 51
		350	[c]	51
		360	[c]	31
		1000	[c]	50

TABLE 3—*continued*

Monomer	Temp. (°C)	$10^4 C_p$	Comments	Ref.
	60	0.1		49
		2.1		52
		1.5	b	51
		360	c	51
	80	2.48		52
	90	0.22	b	50
		1000	c	50
Styrene	50	1.9		26, 53
		4.5		31
		14.0		31
		16.6		54
	55	15		55
	60	1.9		56
		3.1		49
		16.6		54
		15.4	c, d	57
		15.8	c, d	57
		0.8	a	49
	90	5.8		31
	100	2.0	b	58
	110	9.2		55
		10.8		54
	130	1.8		59
	154	1.5		59
Vinyl acetate	−15	0.36		60
	0	0.5		60
		1.7		61
		2.48	e	62
		0.52	f	62
	11	2	f	29
	21	4	f	29
	31	16	f	29
	40	11.2		63
		30.9		59
		32.0		64

TABLE 3—*continued*

Monomer	Temp. (°C)	$10^4 C_p$	Comments	Ref.
	50	3		59
		10.2		63
		0.15	1 kg cm^{-2}	65
		0.11	500 kg cm^{-2}	65
		0.06	1000 kg cm^{-2}	65
	60	1.4		66
		1.8		67
		1.9		68
		2.5		69
		3.0		70
		4.0		49
		6.8		63
		8.0		49
		47.0		64
		3.03	b	62
		1.27	c	62
	60—70	3.5		61
	70	0.25		71
		2	e	72
		4	f	72
Vinyl chloride	50	5	a	73

[a] Estimated from studies on model compounds.
[b] For non-terminal units.
[c] For end-groups.
[d] —CBr_3 end-groups.
[e] Transfer to main chain.
[f] Transfer to acetoxy —CH_3.

TABLE 4
Transfer constants to initiators (C_I) for free-radical polymerization of styrene

Initiator	Temp. (°C)	C_I	Comments	Ref.
Acetyl peroxide	70	0		74, 75
p-Acetoxybenzoyl peroxide	70	0.187		74

TABLE 4—*continued*

Initiator	Temp. (°C)	C_I	Comments	Ref.
2,2′-Azo-bis-isobutyronitrile	50	~0	Benzene	31
	60	0		30, 35
		0.012	[a]	76
		0.0164	[b]	76
		~0.16		77
Benzoyl peroxide	22	~0.1		78
	50	0.13		79
	60	0.048		35
		0.055		37
	70	0		75
		0.075		74
		0.12		41
		0.18		78, 79
	80	0.13		78, 79
2-Bromobenzoyl peroxide	50	1.0		78, 79
	70	2.17		74
3-Bromobenzoyl peroxide	70	0.465		74
4-Bromobenzoyl peroxide	70	0.193		74
2-Butanone peroxide	70	0.0667	Benzene	39
		0.1250	Ethyl acetate	39
		0.1670	Ethyl methyl ketone	39
	75	0.1250	Benzene	39
		0.1670	Ethyl acetate	39
		0.2000	Ethyl methyl ketone	39
		0.243		39
t-Butoxymaleoyl peroxide	70	1.52		74
t-Butoxyphthaloyl peroxide	70	0.018		74
p-*t*-Butylbenzoyl peroxide	70	0		74
t-Butyldeuteroperoxide	70	0.039		80
		0.039	Benzene	80
t-Butyl hydroperoxide	60	0.035		35
	70	0.060		81
		0.064		80
		0.063	*o*-Dichlorobenzene	80
		0.066	Benzene, chlorobenzene	80

TABLE 4—*continued*

Initiator	Temp. (°C)	C_I	Comments	Ref.
	80	0.0033		82
Butyl peroxide	60	0.00076	1,4-Dioxan	83
		0.00092	Benzene	83
	80	0.0029	1,4-Dioxan	83
sec-Butyl peroxide	60	0.0004	Benzene	84
	80	0.0021	Benzene	84
t-Butyl peroxide	60	0.00086		85
		0.0013		86
		0.0003		87
		0.0006	Benzene, heptane	87
		0.000234	1,4-Dioxan	87
	70	0.064		80
		0.066	Benzene, chlorobenzene	80
		0.063	*o*-Dichlorobenzene	80
	80	0.0033		82
		0.0022	Benzene	87
Butyroyl peroxide	70	0.018		74
2-Chlorobenzoyl peroxide	22	~2.0		78
	70	1.91		74
3-Chlorobenzoyl peroxide	70	0.346		74
4-Chlorobenzoyl peroxide	70	0.216		75
Cinnamoyl peroxide	70	1.10		74
Crotonyl peroxide	70	0.146		74
4-Cyanobenzoyl peroxide	70	0.804		74
2-Cyano-2-propylazoformamide	100	0.17		88
Cyclohexanone peroxide	60	0.062		89
1,1-Di-(*t*-butylperoxy)butane	80	0.00077		43
	90	0.0140		43
2,2-Di-(*t*-butylperoxy)butane	80	0.00072		43
	90	0.0115		43
1,1-Di-(*t*-butylperoxy)isobutane	80	0.00083		43
	90	0.0155		43
2,2-Di-(*t*-butylperoxy)propane	80	0.00040		43
	90	0.0105		43
2,4-Dichlorobenzoyl peroxide	60	2.9		75

TABLE 4—*continued*

Initiator	Temp. (°C)	C_I	Comments	Ref.
	70	2.6		75
α,α-Dimethylbenzyl hydroperoxide	60	0.063		35
	70	0.10		81
Ethyl peroxide	60	0.00066	Benzene	84
	80	0.0024	Benzene	84
2-Fluorobenzoyl peroxide	70	0.40		74
3-Fluorobenzoyl peroxide	70	0.246		74
4-Fluorobenzoyl peroxide	70	0.219		74
Furoyl peroxide	70	0.23		74
Hexanoyl peroxide	70	0.166		74
Isopropyl peroxide	60	0.0003	Benzene	84
	80	0.0015	Benzene	84
3-Iodobenzoyl peroxide	70	0.262		74
4-Iodobenzoyl peroxide	70	0.293		74
Lauroyl peroxide	70	0		75
		0.024		74
	84	0		75
p-Methoxycarbonyloxybenzoyl peroxide	70	0.208		74
Methoxyphenyl peroxide	70	0.074		74
Myristoyl peroxide	70	0		75
		0.116		74
2-Naphthoyl peroxide	70	0.178		74
3-Nitrobenzoyl peroxide	70	6.2		74
4-Nitrobenzoyl peroxide	70	7.4		74
Octanoyl peroxide	70	0		75
		0.098		74
Oleoyl peroxide	70	0.154		74
Palmitoyl peroxide	70	0.142		74
5-Phenyl-2,4-pentadienoyl peroxide	70	5.24		74
Propyl peroxide	60	0.00084		90
Sorboyl peroxide	70	1.19		74
Stearoyl peroxide	70	0.154		74
2-Thenoyl peroxide	50	0.23		78

TABLE 4—*continued*

Initiator	Temp. (°C)	C_I	Comments	Ref.
	70	0.38		78
2-Toluoyl peroxide	70	0.175		74
4-Toluoyl peroxide	50	0.17		78
	70	0.19		78
		0.003		74
9-Undecanoyl peroxide	70	0.065		74

[a] Apparent value, affected by primary-radical termination.
[b] Average value after allowing for primary-radical termination.

TABLE 5
Transfer constants (C_s) to various solvents and transfer agents for free-radical polymerization of styrene

Solvent	Temp. (°C)	$10^4 C_s$	Comments	Ref.
Acetaldehyde	60	8.5		91
Acetic acid	60	2.0		82
		5.42		92
Acetic anhydride	60	0.7		93
Acetone	60	<0.5		93
		4.1		94
Acetone oxime	60	2.2		93
Acetonitrile	60	0.44		93
Acetyl bromide	60	8600		95
Allyl alcohol	60	1.5		93
Allyl chloride	70	15.1	a	96
Allyl stearimide	90	5.82	a, b	97
		8.30	a, c	97
Aminobenzene	50	20	Benzene	99
	60	2.0		93
4-Aminotoluene	50	78	Benzene	99
Benzaldehyde	60	5.5		91
	100	2.6		100

TABLE 5—*continued*

Solvent	Temp. (°C)	$10^4 C_s$	Comments	Ref.
Benzene	50	0.01		31
	60	0.018		101
		0.023		102
		0.028		103
		0.04	Data from ref. 101	29
	70	5.50		39
	75	6.67		39
	80	0.121		103
		0.156		104
		0.061	Data from ref. 101	105
	100	0.184		101
		0.23	Data from ref. 101	29
		0.31		106
		0.42		106
	132	0.81		106
		0.89		101
		1.5	Data from ref. 101	29
2-Benzimidazothiol	99	2100		107
Benzoin	60	40		93
2-Benzothiazolethiol	50	300	Pyridine	108
	99	2600		107
Benzyl methyl ether	68	6.0		109
4-Benzyloxyphenol	60	290	[a]	110, 111
4-Benzyloxyphenol-*d*	60	10		110
Bis-(2-aminophenyl)disulphide	25	38000		108
	50	30000		108
		34000		108
	75	30000		108
Bis-(4-aminophenyl)disulphide	50	2400		112
Bis-[2-(bromomethyl)phenyl] disulphide	50	10000		108
2,2-Bis(bromomethyl)-1,3-propanediol diacetate	60	40500		113
Bis-[4-(*N*-carboethoxy)amino phenyl]disulphide	50	2400		112
Bis-(2-carboethoxyphenyl) disulphide	50	<50		112
Bis-(4-carboethoxyphenyl) disulphide	50	1100		108, 112
Bis-(2-carbomethoxyphenyl) disulphide	99	2000		107
Bis-(β-carbopropoxyethyl) disulphide	50	<50		108

TABLE 5—*continued*

Solvent	Temp. (°C)	$10^4 C_s$	Comments	Ref.
Bis-(4-carboxy-2-aminophenyl) disulphide	50	30000		112
Bis-(4-carboxymethylphenyl) disulphide	50	2400	Pyridine	112
Bis-(2-carboxyphenyl) disulphide	50	100		112
Bis-(4-carboxyphenyl) disulphide	50	1100	Pyridine	108
		1700		112
Bis-(β-chloroacetoxyethyl) disulphide	50	<50		108
Bis-(chlorobenzyl)disulphide	50	<50		108
Bis-(2-chloroethoxy)methane disulphide	80	6.0		95
Bis-(2-chloroethyl)disulphide	50	100		108
Bis-[2-(chloromethyl)phenyl] disulphide	50	13000		108
Bis-[4-(*N*-*n*-dodecylamino)-phenyl]disulphide	50	2100		112
Bis-(2-ethoxyphenyl)disulphide	99	750		107
Bis-(4-ethoxyphenyl)disulphide	99	3300		107
Bis-(2-ethylhexyl)disulphide	50	<50		108
Bis-(*N*-ethyl-*N*-phenylthio-carbamoyl)disulphide	60	17500		114
Bis-(ethyl-xanthogen)disulphide	60	7240		114
Bis-(2-hydroxyethyl)disulphide	50	<50		108
Bis-(2-hydroxymethylphenyl) disulphide	50	5800		112
Bis-(4-hydroxymethylphenyl) disulphide	50	900		108
Bis-(isopropyl-xanthogen) disulphide	50	53000		108
	99	75000		107
Bis-(4-methoxyphenyl) disulphide	50	1800		112
Bis-(4-methoxytolyl)disulphide	99	210		112
Bis-(1-methylheptyl)disulphide	99	104		112
Bis-(2-methylquinoyl) disulphide	50	400		112

TABLE 5—*continued*

Solvent	Temp. (°C)	$10^4 C_s$	Comments	Ref.
Bis-(methyl-xanthogen) disulphide	60	11100		114
	70	136		40
Bis-(*N*-morpholine)disulphide	50	<50		108
Bis-(morpholinothiocarbonyl) disulphide	60	61000		114
Bis-(1-naphthylmethyl) disulphide	99	330		107
Bis-(1-phenylethyl)disulphide	50	<50		112
Bis-(2-pyridyl-β-ethyl) disulphide	50	300		108
Bis-(2,3,5,6-tetramethylphenyl) disulphide	50	7300		108
Bis-(2,4,6-triisopropylphenyl) disulphide	50	1200		112
Bis-[3-(trimethoxysilyl)propyl] disulphide	79.5	118		115
Bis-(2,4,6-trimethylphenyl) disulphide	50	6900		112
Bis-(2,6-xylyl)disulphide	50	6900		108
Bromoacetic acid	60	430		95
	68	300		109
Bromobenzene	60	1.78		94
	155	3		116
4-Bromobenzyl methyl ether	68	6.0		109
1-Bromobutane	60	0.06		93
	100	0.35		93
4-Bromo-1-isopropylbenzene	60	7.57		117
4-Bromophenylacetylene	60	158	a, h, j	118
		219	a, i, j	118
Bromotrichloromethane	30	330000		119
	60—80	$(65 \pm 4) \times 10^4$		120
	80	9450	Data from ref. 101	121
Bromotrifluoromethane	60	0.7	d	122
		2.0	d	122
α-Bromostyrene	70	10000		123
β-Bromostyrene	70	2000		123
1-Butanethiol	25	132000		124
	60	210000		125
		220000		28, 126
		250000		127

TABLE 5—*continued*

Solvent	Temp. (°C)	$10^4 C_s$	Comments	Ref.
	70	150000		125
	80	170000		125
	99	154000		107
1-Butanethiol-*d*	60	52000		125
	70	40000		125
	80	70000		125
2-Butanone	60	4.98		94
	70	8.60		39
	75	12.00		39
1-Butene	100	2.6		128
2-Butene	100	2.0		128
Butyl alcohol	60	0.06		93
		1.6		82, 129
		10.66		92
sec-Butyl alcohol	60	5.62		92
t-Butyl alcohol	60	0.22		33
		6.65		82
	100	0.55		130
	130	1.0		130
Butylamine	60	0.5		93
		7.00		92
t-Butylbenzene	60	0.06		101
		0.04	Data from ref. 101	29
	80	0.193	Data from ref. 101	121
	100	0.55		101
1,4-Butylene glycol	60	3.2		33
	80	5.6		33
	100	9.3		33
4-*t*-Butyl-1-isopropylbenzene	60	3.46	Benzene	117
Butyl mercaptoacetate	50	260000		107
4-*t*-Butylphenol	60	26		111
4-*t*-Butylpyrocatechol	60	3600	a	110, 111
4-*t*-Butylpyrocatechol-*d*	60	370		110
α-*t*-Butylstyrene	110	$<10^{-3}$	a	131
Butyraldehyde	60	5.7		93
	100	11.0		93
Carbon disulphide	60	6.55		132
Carbon tetrabromide[g]	40	18000	Benzene	133
	45–80	$(42 \pm 6) \times 10^5$	e	134
	60	17800		57
		136000		93

TABLE 5—*continued*

Solvent	Temp. (°C)	$10^4 C_s$	Comments	Ref.
		22000	Benzene	133
		25000	Acetonitrile	133
	70	18000		135
	80	23000	Benzene	133
	90	25100		57
	100	23500		93
Carbon tetrachloride	60	87		94, 95
		92		93, 136, 137
		98		138
		110		36
		122	f	139
		148	f	140
		144.08		92
		84	Pressure 56000 psi	138
		90	Pressure 28400 psi	138
	76	117		141
	80	133	Data from ref. 101	121
	100	185		93, 136
	132	304		136
	140	300		79
Carbonyl-bis-(thioglycolic acid)	50	3600	Pyridine	108
Chloroacetic acid	60	28.60		92
	68	200		109
Chloroacetyl chloride	60	3300		95
4-Chlorobenzaldehyde	100	5.6		100
Chlorobenzene	60	0.133		103
		1.50		94
	80	0.235		103
		0.874	Data from ref. 101	121
	100	0.54		106
	140	0.6		79
4-Chlorobenzyl methyl ether	68	4.0		109
1-Chlorobutane	60	0.04		93
	100	0.37		93
2-Chlorobutane	60	1.2		93
	100	0.3		93
Chloroform	60	0.5		93, 137
		0.566		103
		3.40		94
	68	4.0		109
	80	0.50	Data from ref. 101	121
		0.916		103

TABLE 5—*continued*

Solvent	Temp. (°C)	$10^4 C_s$	Comments	Ref.
4-Chloro-1-isopropylbenzene	60	6.90	Benzene	117
1-Chloro-2-methylpropane	60	1.4		93
	100	3.0		93
3-Chloro-2-methyl propene	60	24.0		93
1-Chloropentane	60	0.49		93
2-Chlorophenol	60	6.0		142
4-Chlorophenol	60	−11		111
4-Chlorophenylacetylene	60	168	a, h, j	118
		154	a, i, j	118
α-Chlorotoluene	60	1.56		93
2-Chlorotoluene	70	0.62		41
	140	1.8		41
Chloro(trimethyl)silane	50	12.5	Benzene	143
4-Cyanobenzyl methyl ether	68	20.0		109
4-Cyano-1-isopropylbenzene	60	18.6	Benzene	117
Cyclohexane	60	0.024		101
		0.04	Data from ref. 101	29
		0.031		102
		0.063		103
	80	0.066	Data from ref. 101	121
		0.083		103
		0.156		104
	100	0.16		101
		0.23	Data from ref. 101	29
		0.31		106
	132	0.81		106
		0.87		101
		1.5	Data from ref. 101	29
Cyclohexanone	60	7.90		94
Cyclohexene	80	0.083		103
Cyclopentanone	60	3.30		94
Decahydronaphthalene	60	0.4		101
1-Decanethiol	60	17100		34
6-Deoxy-6-dipropylamino-methyl-β,*D*-glucoside	100	22		144
6-Deoxy-6-mercaptomethyl-α,*D*-glucoside	100	55000		144
Di-2-benzothiazyl disulphide	25	23000		108
	50	21000		108
		23000		108
	75	24000		108
	99	27300		107

TABLE 5—*continued*

Solvent	Temp. (°C)	$10^4 C_s$	Comments	Ref.
Di-2-benzothiazyl sulphide	50	<50		107
Dibenzoyl disulphide	50	<50		112
	60	107		86
	99	1100		107
Dibenzyl diselenide	60	20000		93
Dibenzyl disulphide	50	200		112
		300		108
	60	100		93
	99	110		107
Di-*O*-benzyl-methyl-α,*D*-glucoside	100	62		144
Dibenzyl sulphone	50	0.052		145
α,α′-Dibromodibenzyl	60	3020		95
1,2-Dibromoethane	60	0.988		103
	80	1.914		103
1,2-Dibromoethylbenzene	60	1950		95
Dibromomethane	60	110		95
Dibromostilbene	60	3020		95
Dibromostyrene	60	1950		95
α,α′-Dibromo-*p*-xylene	60	150		95
Dibutylcadmium	100	1170		146
Dibutyl disulphide	60	24		86
	99	68		107
	150	220		147
Di-*sec*-butyl disulphide	50	<50		112
Di-*t*-butyl disulphide	50	<50		112
	60	1.4		86
Di-*t*-butyl ether	60	2.6	Benzene	84
	80	1.0	Benzene	84
2,6-Di-*t*-butylphenol	64	49		111
Dibutylphosphine	100	20800		148
Dibutyl sulphide	60	22		86
Di-*n*-butyl sulphide	55	15.4		149
Di-*t*-butyl sulphide	60	250		86
Dichloroacetic acid	60	35.00		92
2-Dichlorobenzene	60	3.4		94
3-Dichlorobenzene	140	0.2		79
		1.4		41
4-Dichlorobenzene	60	2.6		94

TABLE 5—*continued*

Solvent	Temp. (°C)	$10^4 C_s$	Comments	Ref.
1,4-Dichloro-2-butene	80	51		95
Dichlorodimethylsilane	50	17.8		143
1,2-Dichloroethane	60	0.333		103
		4.12		94
	70	1.1		41
	80	1.137		103
		9.8		95
	100	3.84		106
Dichloromethane	60	0.15		93
	80	9.5		95
	100	11.8		93
2,4-Dichloro-1-naphthol	60	490	a	111
α,α-Dichlorotoluene	60	50.0		93
Diethylamine	60	25.61		92
Diethylbenzene	100	3.35	Mixed isomers	106
		6.33	Mixed isomers	106
	132	5.13	Mixed isomers	106
Diethyl bromomalonate	60	700		93
	100	1200		93
Diethyl dibromomalonate	60	12000		93
Diethyl dichloromalonate	60	30.0		93
	100	62.0		93
Diethyl disulphide	99	45		107
Diethyl dithiodiacetate	60	150		91
Diethyl ether	60	5.64		150
Diethyl malonate	60	0.47		93
		2.70		92
Diethyl 3-mercaptopropyl ether	60	210000		28
		141000		151
	100	137000		28
Diethylmercury	100	0.335		146
Diethylphosphine	100	13500		148
Diethylzinc	100	3660		146
Dihexyl disulphide	99	104		107
Dihydroanthracene	50	610		152
Diiodomethane	60	710		95
Diisobutyl disulphide	60	20		86
p-Diisopropylbenzene	60	3.30	Benzene	117

TABLE 5—*continued*

Solvent	Temp. (°C)	$10^4 C_s$	Comments	Ref.
Diisopropyl disulphide	60	6.6		86
2,6-Diisopropylphenol	60	310	a	111
Dilauryl disulphide	60	2.3		93
N,N-Dimethylacetamide	50	0.743		153
	60	4.6		154
N,N-Dimethylaminobenzene	50	12	Benzene	99
		53		155
N,N-Dimethyl-4-aminotoluene	50	16	Benzene	99
2,2-Dimethylbutane	60	0.43		86
2,3-Dimethyl-2-butene	100	5.4		128
Dimethyl disulphide	60	94		86
Dimethyl dithiodiacetate	99	1000		107
N,N-Dimethylformamide	50	0.869		153
	60	4.0		154
	100	1.08		156
2,6-Dimethyl-2,6-octadiene	60	2.0		157, 158
2,2-Dimethylpropylstyrene	110	<10	a	131
Dimethylsulphoxide	50	0.548		153
	60	0.48		159
Di-1-naphthoyl disulphide	50	3400		112
Di-1-naphthyl disulphide	99	15700		107
Di-2-naphthyl disulphide	25	1700		108
	50	1900		108
	75	2900		108
	99	3600		107
Dioctadecyl disulphide	99	240		107
1,4-Dioxan	60	2.28		150
		2.75		94
	100	0.8		144
Diphenylamine-*t*	60	0.9	Benzene	160
2,4-Diphenylbutene-1	110	70		131
Diphenyl disulphide	50	600		108
	60	1470		161
1,1-Diphenylethylene	110	<400		131
Diphenylmethane	60	2.3		101
	100	4.2		101
2,4-Diphenylpentene-1	110	170		131
1,3-Diphenyl-1,3-propanedione	60	7.0		93

TABLE 5—*continued*

Solvent	Temp. (°C)	$10^4 C_s$	Comments	Ref.
Diphenyl sulphide	50	560		145
Diphenyl sulphone	50	210		145
Diphenyl sulphoxide	50	240		145
Dipropyl disulphide	60	23.4		86
Di-2-pyridyl disulphide	50	100		112
Di-2-quinolyl disulphide	50	500		112
Dithiodiacetic acid	50	<50		108
	99	2000		107
Di-2-tolyl disulphide	25	2200		108
	50	2300		108
		2800		108
	75	3200		108
Di-4-tolyl disulphide	50	1100		108
	99	1500		107
N,N-Divinylaminobenzene	60	130		162
2,2,3,3,4,4,5,5,6,6,7,7-Dodecafluoro-*n*-heptyl alcohol	60	13.33		163
Dodecamethylpentasiloxane	79.5	0.285		115
1-Dodecanethiol	60	148000		28, 151
		190000		164
	100	130000		146
	110	260000		146
Dodecanol	60	10.92		92
Dodecyl vinyl ether	60	3.32		150
		4.11		150
Ethanethiol	50	171000		165
2-Ethoxybenzenethiol	99	251000		107
4-Ethoxybenzenethiol	99	1300		107
Ethyl acetate	60	5.70		92
	70	5.5		39
	75	6.67		39
Ethyl alcohol	60	1.32		166
		16.11		92
	80	2.60		166
	100	2.6		166
Ethylbenzene	60	0.67		101
		0.83	Data from ref. 101	29
		0.70		102
		0.710		103
		2.7		167

TABLE 5—*continued*

Solvent	Temp. (°C)	$10^4 C_s$	Comments	Ref.
	80	1.07	Data from ref. 101	121
		1.113		103
	100	1.38		106
		1.62		101
		2.2	Data from ref. 101	29
		2.33		106
	132	2.31		106
		2.9		101
		4.9	Data from ref. 101	29
Ethyl dibromoacetate	90	2700		57
Ethyl dichloroacetate	60	1.3		93
Ethyl 2,4-diphenylbutyl sulphide	50	300000		165
Ethylene glycol	60	1.36		33
		13.31		92
	80	2.70		33
	100	4.70		33
Ethyl mercaptoacetate	60	580000		151
Ethyl phenethyl sulphide	50	71500		165
α-Ethylstyrene	110	<10	a	131
Ethyl tribromoacetate	90	105000		57
Ethyl trichloroacetate	60	65.0		93
	90	90		57
	100	145.0		93
Fluorobenzene	60	0.0372		168
Fluorene	60	75.0		101
	100	124.0		101
4-Fluorophenol	60	54		111
Glycerol	60	28.64		92
Heptane	60	0.42		101
	100	0.95		101
1-Heptanethiol	99	151000		107
1-Heptene	100	2.7		128
2-Heptene	100	3.2		128
Hexaethyl ethanehexacarboxylate	50	<50		112
Hexamethyldisiloxane	79.5	0.387		115
1,1,3,3,5,5-Hexamethyl-1-hexanethiol	25	32000		108
	50	29000		108
	75	24000		108

TABLE 5—*continued*

Solvent	Temp. (°C)	$10^4 C_s$	Comments	Ref.
Hexane	100	0.9		128
1-Hexanethiol	99	153000		107
1-Hexene	100	2.5		128
2-Hexene	100	3.6		128
Hydrogen chloride	100	~0		169
Hydrogen sulphide	60	19900		170
	70	50000		29
Hydroisobutylaluminium	100	269000		146
		280000		171
Hydroquinone	60	3.6		93
4-Hydroxybutyric acid γ-lactone	50	0.409		153
2-Hydroxytoluene	50	33		172, 173
	60	43		111
3-Hydroxytoluene	50	11		172, 173
4-Hydroxytoluene	50	22.1		172, 173
	60	39		111
Iodoacetic acid	68	8000		109
1-Iodobutane	60	1.85		93
	100	5.5		93
Isobutyl alcohol	60	0.17		82
Isobutyl alcohol-*d*	100	2.9		130
	130	7.8		130
α-Isobutylstyrene	110	<10	a	131
Isobutyric acid	60	2.5		82
		4.6		82
Isobutyronitrile	100	2.7		130
	130	3.5		130
2-Isopropenylnaphthalene	80	56		42
		70		42
	99	69		42
		86		42
Isopropyl alcohol	60	3.05		166
	80	4.00		166
	100	1.7		130
		6.00		166
	130	2.7		130
Isopropyl alcohol-*d*	100	1.6		130
Isopropyl-1-*d* alcohol-*d*	100	0.78		130

References pp. 276—285

TABLE 5—*continued*

Solvent	Temp. (°C)	$10^4 C_s$	Comments	Ref.
Isopropylbenzene	60	0.8		157
		0.82		101
		1.04	Data from ref. 101	29
		3.88	Benzene	117
	80	1.31	Data from ref. 101	121
	100	2.00		101
		2.90	Data from ref. 101	29
Mercaptoacetic acid	99	>140000	a	107
2-Mercaptobenzoic acid	99	>147000		107
4-Mercaptomethylmethoxy-benzene	99	260000		107
1-Mercaptomethylnaphthalene	25	127000		108
	50	183000		108
	75	157000		108
	99	246000		107
3-Mercaptopropionic acid	50	60000		108
4-Methoxy-1-isopropylbenzene	60	3.23	Benzene	117
2-Methoxynaphthalene	60	<5.0		111
2-Methoxyphenol	60	43		142
3-Methoxyphenol	60	<5		111
4-Methoxyphenol	60	260	a	111
4-Methoxyphenylacetylene	60	66.7	a, h, j	118
		53.4	a, i, j	118
Methyl alcohol	60	0.74		166
		2.96		92
	80	1.10		166
	100	1.22		166
N-Methylaminobenzene	50	13	Benzene	99
N-Methylaminotoluene	50	11	Benzene	99
3-Methyl-2-butanone	60	2.41		168
2-Methyl-1-butene	100	3.1		128
3-Methyl-1-butene	100	6.9		128
2-Methyl-2-butene	100	2.9		128
Methyl chloroacetate	60	0.3		93
	100	0.75		93
4-Methyl-2,4-diphenyl-pentene-1	110	2900		131
Methyl mercaptoacetate	60	6300		129
	99	276000		107
Methyl 2-mercaptobenzoate	99	170000		107

TABLE 5—*continued*

Solvent	Temp. (°C)	$10^4 C_s$	Comments	Ref.
Methyl oleate	60	3.15 ± 0.09	a	174
		3.52 ± 0.09	a	174
	90	6.64 ± 0.19	a	174
4-Methyl-2-pentanone	60	0.767		168
2-Methyl-2-pentene	100	6.3		128
4-Methyl-2-pentene	100	6.9		128
4-Methylphenylacetylene	60	88.8	a, h, j	118
		55.3	a, i, j	118
2-Methyl-2-propanethiol	60	31000		130
		37000		28, 151
		46000		164
	100	18000		130
		23000		28, 151
2-Methyl-1-propanol	60	4.97		92
2-Methyl-2-propanol	60	24.76		92
2-Methyl-1-propene	100	1.7		128
Methyl stearate	60	1.06 ± 0.17	a	174
	90	0.676 ± 0.09	a	174
α-Methylstyrene	74	95		175
	80	3.2		42
		4.9		42
	99	5.6		42
		8.5		42
	110	~8		131
Methyltrichlorosilane	50	19.2	Benzene	143
Naphthalene	60	11		176
1-Naphthalenethiol	99	1500		107
2-Naphthalenethiol	99	1800		107
1-Naphthol	60	480		110, 111
1-Naphthol-*d*	60	75		110, 111
2-Naphthol	60	77		142
Nitromethane	60	10		91
4-Nitrophenylacetylene	60	2310	a, h, j	118
		3950	a, i, j	118
1-Octadecanethiol	90	147000		107
2,2,3,3,4,4,5,5-Octafluoro-*n*-pentyl alcohol	60	11.36		163
Octamethyltrisiloxane	79.5	6.9		115
2-Octanethiol	99	32000		107

TABLE 5—*continued*

Solvent	Temp. (°C)	$10^4 C_s$	Comments	Ref.
Octanoic acid	60	13.26		92
2-Octene	100	2.8		128
Octylphosphine	60	36000		177
1-Oxa-4,5-dithiacycloheptane	150	570		147
3-Pentanone	60	2.6		82
Pentaphenylethane	60	20000		101
1-Pentene	100	2.3		128
2-Pentene	100	4.2		128
Pentyl acetate	60	17.83		34
Phenol	50	8.1		172, 173
	60	14		142
Phenylacetic acid	60	6.0		93
Phenylacetylene	60	98.3	a, h, j	118
		94.3	a, i, j	118
α-Phenyl-4-hydroxytoluene	60	<5		111
2-Phenylphenol	60	<5		111
3-Phosphinopropionitrile	60	50000		177
3,3′-Phosphinylidenedipropionitrile	60	50000		177
Phosphorous trichloride	57	250		178
Piperidine	60	1.0		93
Pivalonitrile	60	0.38		76
1-Propanethiol	60	21800		92
Propionic acid	60	0.05		93
		4.3		82
		4.5		129
		4.65		82
n-Propyl alcohol	60	2.00		166
	80	3.14		166
	100	3.6		166
1,2-Propylene glycol	60	2.08		33
	80	3.90		33
	100	6.80		33
2-Propyn-1-ol	60	7.0		93
Pyridine	60	0.6		93
Pyrocatechol	60	1340	a	110, 111
Pyrocatechol-*d*	60	260		110
Pyrogallol	60	10400	a	110, 111

TABLE 5—*continued*

Solvent	Temp. (°C)	$10^4 C_s$	Comments	Ref.
Pyrogallol-*d*	60	1600		110
2,3,4,6-Tetra-*O*-acetylmethyl-α,D-glucoside	100	2.0		144
1,1,1,3-Tetrabromo-3-phenylpropane	90	36500		57
Tetrabutyltin	100	3.71		146
1,1,2,2-Tetrachloroethane	100	10.8		106
Tetrachlorosilane	50	20	Benzene	143
Tetraethyl lead	100	1.24		146
Tetraethylsilane	100	8.12		146
Tetrahydrofuran	60	1.44	d	168
		1.62	d	168
2,3,4,6-Tetramethylphenol	60	580	a	110, 111
2,3,4,6-Tetramethylphenol-*d*	60	20		110
Tetramethylsilane	50	3.1	Benzene	143
Tetramethylsuccinodinitrile	60	0.37		76
Tetraphenylsuccinonitrile	50	28000		112
Thioacetic acid	99	>147000	a	107
Thiobenzoic acid	99	62300	a	107
Thiocarbonyl-bis-(thioglycolic acid)	50	2100	Pyridine	108
Toluene	60	0.105		103
		0.121		102, 179, 180
		0.125		101
		0.16	Data from ref. 101	29
		0.134		181
		0.21		30
		1.10		167
		2.05		167
	80	0.15		82
		0.298	Data from ref. 101	121
		0.3		82
		0.308		103
		0.310		182, 183
		0.313		104, 181
	100	0.53		106
		0.55		82
		0.645		101
		0.8	Data from ref. 101	29
		0.72		106
	132	1.12		106

TABLE 5—*continued*

Solvent	Temp. (°C)	$10^4 C_s$	Comments	Ref.
6-*O*-(*p*-Toluenesulphonyl)-methyl-α,*D*-glucoside	100	2.0		144
4-Toluenethiol	99	700		107
α-Toluenethiol	99	255000		107
2,3,4-Tri-*O*-acetyl-6-deoxy-6-iodomethyl-α,*D*-glucoside	100	50		144
Tribromoacetic acid	90	24000		57
1,1,1-Tribromopropane	90	24100		57
Tributoxyborane	100	<0.1		146
Tributylborane	100	34.8		146
Tributylphosphate	100	<0.1		146
Tributylphosphine	100	24.4		146
Tributylstibine	100	58.0		146
Trichloroacetic acid	60	66.0		93
		100.0		92
α,α,α-Trichlorotoluene	60	57.5		93
Triethoxyaluminium	100	<0.1		146, 171
Triethylaluminium	100	80500		146
		170000		171
Triethylamine	60	7.1		180
		7.5		184, 185
		17.17		92
		3.0	Pressure 26600 psi	185
		1.4	Pressure 64600 psi	185
Triethylindium	100	17600		146
Triethylsilane	70	33.4		186
	80	36.8		186
Triisobutoxyaluminium	100	<1.0		171
Triisobutylaluminium	110	285000		171

TABLE 5—*continued*

Solvent	Temp. (°C)	$10^4 C_s$	Comments	Ref.
Trimethoxymethylsilane	79.5	0.23		115
3-Trimethoxysilylpropane-thiol	79.5	59000		115
2,2,4-Trimethylpentane	100	<10		171
Trimethyl-s-triazine	60	0.468		187
2,4,6-Trinitroaminobenzene	50	118000		188
2,4,6-Trinitrochlorobenzene	50	585000		188
2,4,6-Trinitromethoxybenzene	50	203000		188
2,4,6-Trinitrophenol	50	2110000		188
2,4,6-Trinitrotoluene	50	146000		188
Triphenylmethane	60	3.5		101
	80	6.0		189
	100	8.0		101
6-*O*-Triphenylmethyl methyl-α,*D*-glucoside	100	21		144
2,4,4-Triphenylpentene-1	110	2600		131
Triphenylsilane	70	2.44		186
	80	1.3		186
Tripropylamine	100	24.2		146
Water	60	0.06		92
2,6-Xylenol	60	110		111

[a] Apparent value, retardation occurs.
[b] Thermal (uncatalysed) initiation.
[c] Initiator used.
[d] Same experimental data, different methods of calculation.
[e] Very low conversions used.
[f] Corrected for loss of low-molecular-weight polymer.
[g] A recent determination indicates a value of $C_s = 50.9 \pm 1.8$ at 60°C (C. H. Bamford, private communication).
[h] From degrees of polymerization.
[i] From rates of polymerization.
[j] Copolymerization considered.

TABLE 6
Kinetic parameters for inhibition of free-radical polymerization of styrene

Inhibitor	Temp. (°C)	k_z/k_p	k_z ($l\ mole^{-1}\ sec^{-1}$)	Comments	Ref.
Anthracene	44.4	2	170		190
p-Benzoquinone	50	518 ± 25			191
	60	227			192
	90	560			193
3,4-Benzpyrene	44.4	14	1360		190
Chloro-*p*-benzoquinone	50	720 ± 20			191
Copper(II) chloride	50	~11000		DMF	194
	60	~10000	1.81×10^6	DMF	194
	70	~9000		DMF	194
2,3,7,8-Dibenzphenanthrene	44.4	13	1260		190
1,2,3,4-Dibenzpyrene	44.4	6	550		190
1,2,7,8-Dibenztetracene	44.4	13	1260		190
1,2,9,10-Dibenztetracene	44.4	13	1260		190
2,5-Dichloro-*p*-benzoquinone	50	>0.0002			195
2,6-Dichloro-*p*-benzoquinone	50	$>10^{-4}$			195
2,3-Dimethyl-*p*-benzoquinone	50	120 ± 20			191
2,5-Dimethyl-*p*-benzoquinone	40	160			196
	50	82 ± 10			191
	60	61			196
	90	43			193

o-Dinitrobenzene	50	2.82			197
m-Dinitrobenzene	50	5.17			197
p-Dinitrobenzene	50	13.52			197
2,4-Dinitrotoluene	50	1.543			197
Ethyl *p*-nitrobenzoate	50	1.68			197
Ethyl 1,3,5-trinitrobenzoate	50	57.2			188
Iron(III) chloride	60	536	94000	DMF	198, 199
			848	[a,b] $[Fe(DMF)_6^{3+}]$, DMF	200
			41500	[a,b] $[Fe(DMF)_5Cl^{2+}]$, DMF	200
			65500	[a,b] $[Fe(DMF)_4Cl_2^+]$, DMF	200
			31400	[a,b] $[FeCl_4^-]$, DMF	200
Methoxy-*p*-benzoquinone	50	193 ± 10			191
1-Methoxy-4-nitrobenzene	50	0.035			197
2-Methoxy-1,3,5-trinitrobenzene	50	20.3			188
Methyl-*p*-benzoquinone	50	266 ± 15			191
2-Methyl-*p*-benzonquinone	80	210			193
2′,3′-Naphtho-3,4-pyrene	44.4	240	22000		190
Nitrobenzene	50	0.326			197
1-Nitro-4-chlorobenzene	50	0.364			197
2-Nitrotoluene	50	0.055			197
4-Nitrotoluene	50	0.203			197

TABLE 6—*continued*

Inhibitor	Temp. (°C)	k_z/k_p	k_z ($l\ mole^{-1}\ sec^{-1}$)	Comments	Ref.
Oxygen	50	14600	10^6—10^7		201
Picramide	50	11.8			188
Picric acid	50	211			188
Picryl chloride	50	58.5			188
Tetrabromo-*p*-benzoquinone	50	618 ± 10			195
Tetrachloro-*p*-benzoquinone	50	2040			195
	80	950			193
Tetraiodo-*p*-benzoquinone	50	2740 ± 30			195
Tetramethyl-*p*-benzoquinone	90	0.68			193
Trichloro-*p*-benzoquinone	50	$>10^4$			195
Trimethyl-*p*-benzoquinone	50	25 ± 5			191
	80	26			193
1,3,5-Trinitrobenzene	40	94.8			188
	50	64.2			188
	60	35.1			188
1,3,5-Trinitroluene	50	14.6			188

[a] Kice treatment applied, reinitiation considered.
[b] Species considered to terminate chains [ ].
DMF, *N*,*N*-Dimethylformamide solution.

for investigating various complications which are more difficult to study in other polymerization processes (e.g. primary-radical termination).

The thermal (uncatalysed) polymerization is facile and represents perhaps the most serious complication in detailed kinetic studies of this monomer. However, the effects of thermal initiation may be minimized using conventional rates of initiation (employing an added initiator) and working at comparatively low temperatures (<100 °C).

Propagation may be simply represented by the equation

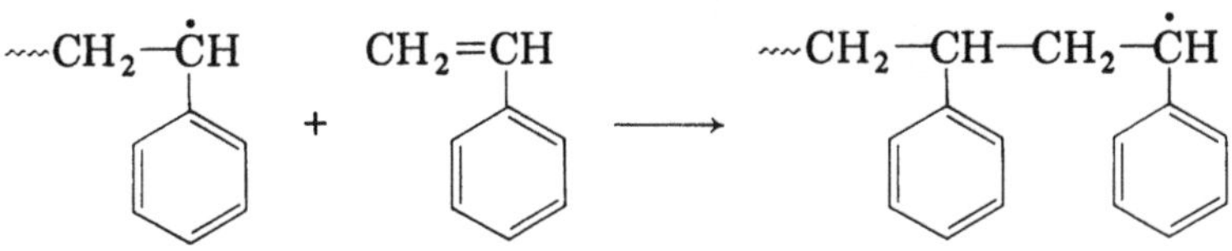

in which monomer adds to the propagating radical to give a regular head-to-tail structure and in such a way as to give the most highly stabilized propagating radical possible. Evidence has been quoted [6] to indicate that "wrong" addition of monomer can occur occasionally (about 1 in 10^4 monomer units) probably through addition of monomer to one of the canonical forms of the styrene radical, viz.

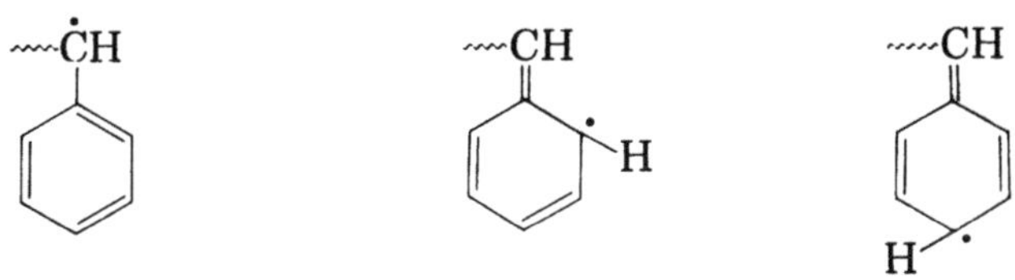

It is generally accepted that bimolecular termination of the propagating radicals proceeds almost exclusively by combination at all temperatures normally used in polymerization studies*. Styrene polymerizations are uncomplicated by the occurrence of a "gel-effect" or Norrish—Tromsdorff effect (see Chap. 1, Sect. 4.2) unless taken to high conversion, i.e., marked increases in rate of polymerization with increasing conversion are not observed at conversions normally used in kinetic studies. Although gel-effects are unimportant with this monomer it has been established recently that the normal termination reaction is a diffusion-controlled process [1].

Values of various kinetic parameters are given in the accompanying Tables. Table 1 contains values of rate coefficients for propagation and termination relating to various experimental conditions. Table 2 presents transfer constants to monomer for the homopolymerization of styrene, while Table 3 contains values of transfer constants to polystyrene for the styrene propagating radical along with corresponding figures

* Some workers maintain that disproportionation termination in styrene polymerizations has been underestimated (e.g. K. C. Berger, Makromol. Chem., 176 (1975) 3575).

TABLE 7
Absolute rate coefficients and activation energies for free-radical polymerizations of various styrene derivatives

Derivative	Temp. (°C)	k_p (l mole^{-1} sec^{-1})	E_p (kcal mole^{-1})	$10^{-7}k_t$ (l mole^{-1} sec^{-1})	Comments	Ref.
p-Chloro	30	150		7.7	[a]	202
p-Cyano	30	219		3.5	[a], dimethylacetamide	202
p-Methoxy	0	2.92		0.106	[b]	203
	30	71		3.3	[a]	21
o-Methyl	50	60	13.9		[c]	23
p-Methyl	30	84		6.6	[a]	21
	50	140	7.7		[c]	23

[a] Rotating-sector technique.
[b] Viscosity method.
[c] Emulsion polymerization.

relating to the homopolymerizations of other monomers. Tables 4 and 5 list quoted values of transfer constants to initiators and to solvents and other transfer agents. Table 6 presents data on inhibition and retardation of styrene polymerizations by various reagents.

3. Substituted styrenes

In comparison with the volume of data on styrene polymerizations, few quantitative data exist on the kinetics of polymerization of styrene derivatives. The most notable monomer of this class is α-methyl styrene, often considered to be non-polymerizable by radical mechanisms. α-Methyl styrene provides a classical example of the ceiling-temperature phenomenon. For most monomers, at normal polymerization temperatures, the depropagation reaction (loss of a monomer unit from a propagating radical) is much slower than propagation (addition of monomer to the propagating radical) but is associated with a higher activation energy. Thus, at some elevated temperature (the ceiling-temperature) the rates of propagation and depropagation become equal.

TABLE 8
Monomer transfer constants (C_M) for free-radical polymerization of substituted styrenes

Monomer	Temp. (°C)	$10^4 C_M$	Ref.
p-Bromostyrene	50	23	204
p-Iodostyrene[a]	50	1.9	205
p-Methoxystyrene	0	0.198	203
	60	0.74	38

[a] Benzene solution.

For α-methyl styrene the ceiling-temperature is about 60 °C. Consequently, α-methyl styrene can only be polymerized at relatively low temperatures and the kinetics are complicated by the simultaneous occurrence of propagation and depropagation reactions (see also Chap. 1, Sect. 5.2).

Kinetic data for various styrene derivatives are given in Table 7 (absolute rate coefficients), Table 8 (transfer constants to monomer) and Table 9 (transfer constants to solvents and other transfer agents).

TABLE 9

Transfer constants (C_S) to solvents and transfer agents for free-radical polymerizations of substituted styrenes

Monomer	Solvent	Temp. (°C)	$10^4 C_S$	Comments	Ref.
4-Chlorostyrene	Benzaldehyde	100	1.9		100
	4-Bromo-1-isopropylbenzene	60	5.71	Benzene	117
	4-*t*-Butyl-1-isopropylbenzene	60	3.52	Benzene	117
	Carbontetrabromide	60	52000	Benzene	133
	4-Chlorobenzaldehyde	100	9.7		100
	4-Chloro-1-isopropylbenzene	60	4.97	Benzene	117
	4-Cyano-1-isopropylbenzene	60	8.84	Benzene	117
	1,4-Diisopropylbenzene	60	3.62	Benzene	117
	Isopropylbenzene	60	3.44	Benzene	117
	4-Methoxy-1-isopropylbenzene	60	3.86	Benzene	117
4-Iodostyrene	Benzene	50	0.2		205
4-Methoxystyrene	4-Bromo-1-isopropylbenzene	60	11.8	Benzene	117
	4-*t*-Butyl-1-isopropylbenzene	60	3.64	Benzene	117
	4-Chloro-1-isopropylbenzene	60	8.52	Benzene	117
	4-Cyano-1-isopropylbenzene	60	40.9	Benzene	117
	1-Methoxy-1-isopropylbenzene	60	3.40	Benzene	117
4-Methylstyrene	4-Bromo-1-isopropylbenzene	60	9.23	Benzene	117
	4-*t*-Butyl-1-isopropylbenzene	60	3.59	Benzene	117

	4-Chloro-1-isopropylbenzene	60	7.67	Benzene	117
	4-Cyano-1-isopropylbenzene	60	26.0	Benzene	117
	1,4-Diisopropylbenzene	60	3.67	Benzene	117
	Isopropylbenzene	60	4.12	Benzene	117
	4-Methoxy-1-isopropylbenzene	60	3.27	Benzene	117
Pentafluorostyrene	Ethylbenzene	60	5.86	[a,b]	168
			5.35	[a]	168
	Fluorobenzene	60	0.117		168
	Isopropylbenzene	60	6.51	[a,b], THF	168
			6.76	[a], THF	168
			6.82	Fluorobenzene	168
	Methylisobutyl ketone	60	0.287	[a,b]	168
			0.33	[a]	168
	Methylisopropyl ketone	60	1.14	[a,b]	168
			1.09	[a]	168
	Tetrahydrofuran	60	1.54	[a,b]	168
			1.53	[a]	168
	Toluene	60	2.25	[a,b]	168
			1.20	[a]	168

[a] Same experimental data, different methods of calculation.
[b] Using Mayo equation.
THF, Tetrahydrofuran solution.

4. Methyl methacrylate

After the polymerization of styrene, the free-radical polymerization of methyl methacrylate (MMA) is the most widely investigated and thoroughly understood vinyl polymerization. Rates of the uncatalysed thermal polymerizations (much slower than those of styrene) may vary from sample to sample [206] and depend to some extent on the history of the monomer [207—209]. Catalysed polymerizations of pure MMA are well-behaved and yield reproducible kinetic data. The fundamental kinetic parameters for MMA polymerizations are now well-documented and rates of polymerization are often used to monitor rates of radical-forming processes.

Although, under standard conditions, rates of polymerization are reproducible, polymerizations of MMA are associated with a number of complicating features which provide an insight into some general aspects of free-radical polymerization. A number of the phenomena are associated with the termination reaction. Bimolecular termination of the propagating radicals proceeds simultaneously by combination and disproportionation, the relative rates of the two processes depending on the polymerization temperature. Disproportionation has the higher activation energy and

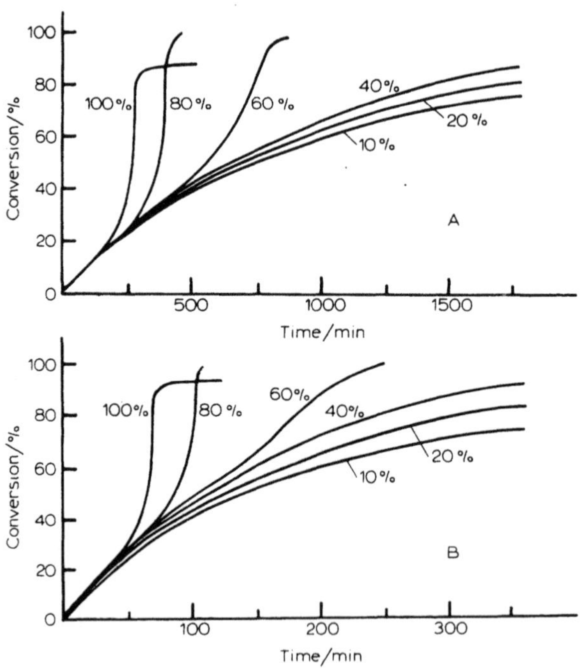

Fig. 1. Conversion—time curves for polymerizations of methyl methacrylate in benzene solution initiated by benzoyl peroxide (10 g l^{-1}) at A, 50 °C and B, 70 °C. Monomer concentrations indicated on curves (ref. 210, courtesy Makromol. Chem.).

percentages of termination by combination are approximately 30 % at 25 °C and 10 % at 80 °C (for further details see Chap. 1, Sect. 4.1.2).

Rates of polymerization are influenced by physical effects arising from viscosity of the reaction medium. The first such effect to be observed, and the most dramatic, is the so-called "gel"- or Norrish—Tromsdorff-effect (see Chap. 1, Sect. 4.2). This effect is observed as an auto-acceleration which commences at finite conversions in bulk and solution polymerizations of the monomer, and arises in consequence of reduced rates of termination in very viscous systems. The magnitude of the effect and its point of appearance are dependent on both temperature and initial monomer concentration [210], as shown in Fig. 1. The second, more subtle effect, occurs in relatively non-viscous systems and at zero conversion. Under these conditions the rate of termination is influenced by the viscosity of the solvent medium (see discussion of Diffusion-controlled Termination, Chap. 1, Sect. 4.2); the rate coefficient for termination is inversely proportional to the viscosity of the solution in the absence of any polymer.

Solvent effects on rates of polymerization of MMA have appeared in the literature from time to time. Early workers reported effects of solvents on the parameter $k_p/k_t^{1/2}$. Subsequently it was realized that the effects were partly due to changes in viscosity influencing k_t but, by making allowance for this effect, it has been realized that values of k_p are also solvent dependent. Detailed discussions of these effects can be found in the references cited in Table 10, which details values of absolute rate coefficients.

Transfer constants to monomer, initiators and other transfer agents are given in Tables 11—13 and to polymer in Table 3. Kinetic parameters relating to retardation and inhibition are given in Table 14.

5. Methyl acrylate

Kinetic studies of methyl acrylate polymerizations have been reported by a number of workers but the polymerizations of this monomer have not received the same extensive study as those of the related methyl methacrylate. Difficulties in achieving reproducible data have been reported and values of $k_p/k_t^{1/2}$ quoted in the literature show considerable disagreement.

One of the problems with methyl acrylate polymerizations is the phenomenon of the "gel-effect" which is very marked, and for polymerizations of bulk monomer is apparent at virtually zero conversion. The gel-effect may be overcome by carrying out polymerizations in solution.

Few studies of the mechanism of bimolecular termination have been reported with some disagreement between various workers. However, it now appears that termination is essentially by combination of propagating radicals, at least at about 25 °C (see Chap. 2, Sect. 4.1.2).

TABLE 10
Absolute rate coefficients and activation energies for free-radical polymerizations of methyl methacrylate

Temp. (°C)	k_p (l mole^{-1} sec^{-1})	E_p (kcal mole^{-1})	$10^{-6} k_t$ (l mole^{-1} sec^{-1})	E_t (kcal mole^{-1})	$10^6 k_p/k_t$	Comments	Ref.
	($k_p = 2.5 \times 10^6 \exp\{-(5400 \pm 60)/RT\}$ l mole^{-1} sec^{-1}; $k_t = 7.2 \times 10^8$ l mole^{-1} sec^{-1}; $E_t = 0$)[b,a]						211
−45	11.1		0.76			c	212
	5.43		0.184			c	212
			0.0136			c	212
			0.00375			c	212
			0.0045			d	212
−30	13.2		0.488			c	212
			0.0475			c	212
			0.045			d	212
−20	24		0.97			c	212
			0.062			c	212
			0.12			d	212
0	41.6		2.69			[e], biradical initiation assumed	208
	220 ± 12 %		35 ± 20 %			[d,f], η = 0.74 cP	213
	170 ± 12 %		38 ± 20 %			[d,f], methyl formate, η = 0.47 cP	213
	170 ± 12 %		21 ± 20 %			[d,f], benzene, η = 0.82 cP	213
	230 ± 12 %		13 ± 20 %			[d,f], diethyl phthalate η = 1.9 cP	213
	270 ± 12 %		13 ± 20 %			[d,f], diethyl malonate, η = 2.4 cP	213

	250 ± 12 %		6.6 ± 20 %			d,f, diethyl phthalate, η = 4.2 cP	213
	290 ± 12 %		3.1 ± 20 %			d,f, dimethylglycolphthalate, η = 8.1 cP	213
	260 ± 12 %		2.9 ± 20 %			d,f, diethyl phthalate, η = 9.5 cP	213
	255 ± 12 %		0.80 ± 20 %			d,f, dimethylglycolphthalate, η = 30.5 cP	213
	290 ± 12 %		0.35 ± 20 %			d,f, dimethylglycolphthalate, η = 80 cP	213
5					7.78	d	214
15					6.75	c	13
					9.8	c	215
22	128		8.44		15.2	c	216
22.5	384	5.8	44	0.5	8.7	c	217
23.6	310 ± 20	4.4	66 ± 4	0		d	218
25	512.6		46.6			d	219
	390 ± 40		42 ± 4			d	17
	410 ± 40		29 ± 3			d, methanol (50 % v/v)	17
	410 ± 40		26 ± 3			d, *n*-octane (5 % v/v) +*n*-nonane (20 % v/v)	17
	260		21			d,f, benzene	220
	270		21			d,f, fluorobenzene	220
	280		19.5			d,f, chlorobenzene	220
	285		17.5			d,f, anisole	220
	310		17			d,f, bromobenzene	220
	330		17			d,f, benzonitrile	220

TABLE 10—*continued*

Temp. (°C)	k_p (l mole^{-1} sec^{-1})	E_p (kcal mole^{-1})	$10^{-6} k_t$ (l mole^{-1} sec^{-1})	E_t (kcal mole^{-1})	$10^6 k_p/k_t$	Comments	Ref.
	410		42.7			d,f	221
	450		41.1			d,f, 1-fluoronaphthalene (45 % v/v)	221
	480		48.2			d,f, 1-chloronaphthalene (15 % v/v)	221
	420		30.4			d,f, 1-chloronaphthalene (30 % v/v)	221
	440		26.9			d,f, 1-chloronaphthalene (45 % v/v)	221
	500		30.5			d,f, 1-chloronaphthalene (60 % v/v)	221
	460		26.1			d,f, 1-bromonaphthalene	221
	370		38.9			d,f, toluene	221
	340		17			d,f, methylbenzoate	222
	240		11.5			d,f, methyl phenyl acetate	222
	335		9			d,f, dimethyl phthalate	222
	270		22			d,f, dimethyl carbonate	222
	330		16			d,f, diethyl oxalate	222
	250		29			d,f, methyl formate	222
	240		25			d,f, methyl propionate	222
	280		14			d,f, diethyl succinate	222
	320		28			d,f, acetonitrile	222
30	143	6.31	12.2	2.84		d	214

	251	5.0 ± 0.2	21	4.0 ± 0.5[g] 0 ± 0.5[h]		Calculated from data of ref. 214	223
	248		22.7		10.9	d	224
					13.6	c	215
					7.41	c	19
					14	d	225
					40	c	226
					14	c, ethylacetate	226
					46	c, diethyl ether	226
			33			d,f	227
	141		11.5			d, benzene	228
	283		45.5			d, 1 bar	229
	348		23.5			d, 250 bar	229
	441		19.7			d, 500 bar	229
	495		16.7			d, 750 bar	229
	616		15.4			d, 1000 bar	229
32	106		5.7			c	230
35					8.08	c	19
35.9	410 ± 50		68 ± 12			d	218
40					8.92	c	19
	404		17.6		23	c, ethylacetate	226
	322		22.5			Calc. from data of ref. 214	223
45	224					i	25
50					16.06	d	214
	410		24			Calc. from data of ref. 214	223
					19	d	225

TABLE 10—*continued*

Temp. (°C)	k_p (l mole^{-1} sec^{-1})	E_p (kcal mole^{-1})	$10^{-6} k_t$ (l mole^{-1} sec^{-1})	E_t (kcal mole^{-1})	$10^6 k_p/k_t$	Comments	Ref.
50.5	580 ± 60		69 ± 10			[d]	218
60	367		18.7			[d]	214
	515		25.5			Calc. from data of ref. 214	223
	573		{ 11.9[g] 2.0[h]			Calc. from data of refs. 214 and 232	231
	640		25.6[j]			[f,k]	221, 233
	680		21.5[l]			[f,k], 1-chloronaphthalene (20 % v/v)	221
	710		17.4[l]			[f,k], 1-chloronaphthalene (40 % v/v)	221
	700		17.4[l]			[f,k], 1-chloronaphthalene (40 % v/v)	221
	720		13.4[l]			[f,k], 1-chloronaphthalene (60 % v/v)	221
	710		13.4[l]			[f,k], 1-chloronaphthalene (60 % v/v)	221
	690		9.4[l]			[f,k], 1-chloronaphthalene (80 % v/v)	221
	740		21.3[l]			[f,k], 1-bromonaphthalene (20 % v/v)	221
	800		17.0[l]			[f,k], 1-bromonaphthalene (40 % v/v)	221
	840		17.0[l]			[f,k], 1-bromonaphthalene (40 % v/v)	221
	820		12.8[l]			[f,k], 1-bromonaphthalene (60 % v/v)	221

References pp. 276—285

	660		18.4^{l}	[f,k], 1-fluoronaphthalene (40 % v/v)	221
	630		18.4^{l}	[f,k], 1-fluoronaphthalene (40 % v/v)	221
	640		14.9^{l}	[f,k], 1-fluoronaphthalene (60 % v/v)	221
	660		36.2^{l}	[f,k], diethyl ether	233
	640		18.9^{l}	[f,k], methyl benzoate	233
	650		17.0^{l}	[f,k], phenyl ethyl bromide	233
	690		14.8^{l}	[f,k], *p,p*-dichlorodiphenylether	233
	740		13.9^{l}	[f,k], diethyl malonate	233
	700		10.3^{l}	[f,k], phenyl salicylate	233
	690		8.64^{l}	[f,k], diethyl phthalate	233
	650		5.96^{l}	[f,k], dioctyl phthalate	233
	600		3.31^{l}	[f,k], dimethylglycolphthalate	233
70	640		27.5	Calc. from data of ref. 214	223
80	800		30.5	Calc. from data of ref. 214	223

[a] Direct determination of radical concentration by ESR.
[b] Units of R undefined in text, probably cal K^{-1} $mole^{-1}$.
[c] Direct study of pre- or after-effect.
[d] Rotating-sector technique.
[e] Viscosity method.
[f] Diffusion-controlled termination considered.
[g] Disproportionation termination assumed.
[h] Combination termination assumed.
[i] Emulsion polymerization.
[j] Assuming value of k_t quoted in ref. 223.
[k] From steady-state kinetics assuming a value of k_t.
[l] k_t calculated assuming k_t in ref. 223 holds in bulk monomer and that k_t in solution is inversely proportional to solution viscosity.

TABLE 11
Monomer transfer constants (C_M) for free-radical polymerization of methyl methacrylate

Temp. (°C)	$10^4 C_M$	Solvent	Ref.
0	0.128		28
	0.148		208
30	0.117		28
	0.260		214
	0.15[a]	Benzene, dimethylphthalate	3
50	0.10	Benzene	31, 50
	0.15		29
	0.477		214
	0.85		28
60	0.07		231
	0.10		34, 234, 235
	0.103		236
	0.18		29
65	0.20		39
70	0.17	Ethyl acetate	39
	0.23		29
	0.29	Toluene	39
	0.30		39
	0.45		40
	0.807		214
75	0.27	Ethyl acetate	39
	0.33		39
	0.60	Ethyl methyl ketone	39
	0.70	Benzene	39
80	0.25		29
	0.40		39
	0.157		237
90	0.10		50
100	0.38		29
120	0.58		29

[a] Diffusion-controlled termination considered.

Few data are available on values of absolute rate coefficients for propagation and termination and these are presented in Table 15. Tables 16 and 17 contain, respectively, transfer constants to monomer and initiators for methyl acrylate polymerizations. Transfer constants to solvents and other transfer agents are given in Table 18, along with data for other esters of acrylic and methacrylic acids.

Values of k_p for methyl acrylate polymerizations are high and polymerizations are susceptible to retardation (see Chap. 2, Sect. 3). Data relating to retardation and inhibition are given in Table 19.

TABLE 12
Transfer constants to initiator (C_I) for free-radical polymerization of methyl methacrylate

Initiator	Temp. (°C)	C_I	Comments	Ref.
2,2′-Azo-bis-2-ethylbutyronitrile	60	0		235
2,2′-Azo-bis-isobutyronitrile	50	~0	Benzene	31
	60	0	Ethyl acetate	34, 234, 235
		0.02		238
2,2′-Azo-bis-2-methylbutyronitrile	60	0		235
2,2′-Azo-bis-2-methylvaleronitrile	60	0		235
2,2′-Azo-bis-2,3,3-trimethylbutyronitrile	60	0		235
1,1′-Azo-dicyclohexanecarbonitrile	60	0		235
Benzoyl peroxide	50	0.01		29
	60	~0		34
		0.02	Data of ref. 34	29
2-Butanone peroxide	65	0.00698		39
		0.0025	Benzene	39
	70	0.0092		39
		0.0033	Toluene, ethyl acetate	39
		0.0040	Ethyl methyl ketone	39
	75	0.0089		39
		0.00553	Ethyl methyl ketone	39
		0.00667	Benzene	39
		0.0071	Ethyl acetate	39
	80	0.0128		39
		0.0111	Toluene	39
t-Butyl hydroperoxide	60	1.27		34

TABLE 12—*continued*

Initiator	Temp. (°C)	C_I	Comments	Ref.
2-Chlorobenzoyl peroxide	60	0.019		236
		0.35	Ethyl acetate	234
		0.8	Data of ref. 234	29
3-Chlorobenzoyl peroxide	60	0.003		236
4-Chlorobenzoyl peroxide	60	0.009		236
Cinnamoyl peroxide	60	0.009		236
α,α-Dimethylbenzyl hydroperoxide	60	0.33		34
Hydrogen peroxide	60	0.46	Ethyl methyl ketone	234
4-Methyoxyphenyl peroxide	60	0.037		236
3-Nitrobenzoyl peroxide	60	0.012		236
4-Nitrobenzoyl peroxide	60	0.144		236
Palmitoyl peroxide	60	~0		236
		0.16	Data of ref. 236	29
o-Toluyl peroxide	60	0.046		236
		0.06	Data of ref. 236	29

TABLE 13
Transfer constants (C_s) to various solvents and transfer agents for free radical polymerization of methyl methacrylate

Solvent	Temp. (°C)	$10^4 C_s$	Comments	Ref.
Acetaldehyde	60	6.5		91
Acetic acid	80	0.24		239
Acetone	30	0.36		240
	60	0.195		241
	80	0.275		241
		0.225		239
N-Allyl stearimide	90	3.01	a, b	97
Aluminium trichloride	25	130	a	242
Aminobenzene	60	4.2		243
	80	6.3		243
	100	9.0		243
Anthracene	50	0		115
Benzaldehyde	60	2.5		91
Benzene	30	0.01	f	3
	50	0.036		31, 50
	52	0.027		241
	60	0.040		241
		0.83		245
	75	0.33		39
	80	0.075		239
		0.24	Data of ref. 239	29
		0.080		241
	90	0.036		50
Benzenethiol	45	47000		246
	60	27000		231

TABLE 13—*continued*

Solvent	Temp. (°C)	$10^4 C_s$	Comments	Ref.
Benzyl acetate	30	0.32		240
Bis-(4-bromophenyl)disulphide	50	46		161
	60	98		161
Bis-(4-chlorophenyl)disulphide	50	72		161
	60	117		161
Bis-(dimethylthiocarbamoyl)disulphide	70	115		40
Bis-(4-methoxyphenyl)disulphide	50	44		161
	60	52		161
Bis-(2-nitrophenyl)disulphide	50	176		161
	60	508		161
Bis-(4-nitrophenyl)disulphide	50	127		161
	60	193		161
Bis-[3-(trimethoxysilyl)propyl]disulphide	79.5	25.8		115
4-Bromoisopropylbenzene	60	3.71	[a], benzene	247
Bromotrichloromethane	30	450 ± 30		244
		12000		248
Bromotrinitromethane	45	10000		249
3-Bromobenzenethiol	45	38000		246
4-Bromobenzenethiol	45	56000		246
1-Butanethiol	60	6600		231
		6700		28, 126
2-Butanone	60	0.45		234
	70	0.56		39
	75	0.83		39

	80	0.70		239
1-Butene	40	3.1		250
	50	5.1		250
cis-2-Butene	40	3.2		250
	50	4.9		250
trans-2-Butene	40	3.0		250
	50	5.2		250
n-Butyl alcohol	80	0.25		239
sec-Butyl alcohol	80	0.85		239
t-Butyl alcohol	60	0.085		241
	80	0.100		239
		0.152		241
t-Butylbenzene	80	0.260		239
1,4-Butylene glycol	60	0.61		237
	80	1.07		237
4-*t*-Butylisopropylbenzene	60	2.74	[a], benzene	247
Butyl mercaptodithiodiacetate	50	6000		29
4-*t*-Butylpyrocatechol	45	9		246
Carbon disulphide	60	0.68		132
Carbon tetrabromide	60	2700	Benzene	133
	80	3300	Benzene	133
	100	4600	Benzene	133
Carbon tetrachloride	60	0.925		241
		2.40		245
		5		91
	80	2.393		239
		2.421		241
		3.3	Data of ref. 239	29

TABLE 13—*continued*

Solvent	Temp. (°C)	$10^4 C_S$	Comments	Ref.
Chlorobenzene	60	0.074		241
	80	0.200		239
		0.207		241
3-Chlorobenzenethiol	45	35000		246
4-Chlorobenzenethiol	45	46000		246
1-Chlorobutane	80	1.20		239
Chloroform	60	0.454		241
		1.77		245
	80	1.129		241
		1.400		239
		1.9	Data of ref. 239	29
4-Chloroisopropylbenzene	60	3.07	[a], benzene	247
α-Chlorotoluene	60	4.17		245
4-Cyanoisopropylbenzene	60	4.99	[a], benzene	247
Cyclohexane	60	12		91
	80	0.10		239
Dibutyl ether	60	0.8		91
Dibenzoyl sulphone	50	40		145
1,2-Dichloroethane	60	0.35		241
	80	0.756		241
Dichloromethane	60	0.100		241
	80	0.217		241
1,2-Dichloropropane	80	0.675		239
Diethyl disulphide	60	1.3		251

Diethyl dithiodiacetate	60	6.5		91
Diethyl mercury	60	0.898		146
Dimethoxyphenylmethylsilane	79.5	0.2		115
N,*N*-Dimethylaminobenzene	30	2.45	f	3
	50	430		155
	60	11.3		243
	70	10.8		252
	80	17.0		243
	100	20.0		243
Dimethyldichlorosilane	50	2.45	Benzene	143
2,6-Dimethyl-2,6-octadiene	60	6.7		158
		8.0		157
Dimethyl phthalate	30	0.2	f	3
Dimethyl sulphoxide	60	0.071	c	159
1,1-Dimethyl-2,2,2-trinitroethyl acetate	45	520		249
4,4-Dimethyl-5,5,5-trinitro-2-pentanone	45	400		249
m-Dinitrobenzene	50	52		253
1,1-Dinitropropane	45	68		249
2,2-Dinitropropane	45	15		249
1,4-Dioxan	80	0.222		239
Diphenylamine	44.1	~0		254
Diphenylamine-*t*	60	0.3	Benzene	160
Diphenyl disulphide	50	38		161
	60	7		251
		85		161
Diphenyl sulphide	50	64		145
Diphenyl sulphone	50	6		145

TABLE 13—*continued*

Solvent	Temp. (°C)	$10^4 C_s$	Comments	Ref.
Diphenyl sulphoxide	50	13		145
Di-4-tolyl disulphide	50	31		161
	60	44		161
Di(trimethylsiloxy)phenyl methyl silane	79.5	0.232		115
N,N-Divinylaminobenzene	60	340		162
2,2,3,3,4,4,5,5,6,6,7,7-Dodecafluoro-*n*-heptyl alcohol	60	2.80		163
Dodecamethylpentasiloxane	79.5	0.145		115
Ethyl acetate	60	0.100		236
		0.13		234, 236
		0.132		236
		0.134		234
		0.155		234
		0.156		234
		0.46	Data of ref. 234	29
	70	0.55		39
	75	0.83		39
	80	0.240		239
Ethyl alcohol	60	0.400		166
	80	0.625		166
	100	0.80	d	166, 255
		0.72—0.75	e	166, 255
Ethylbenzene	52	0.501		241
	60	0.766		241
	80	1.311		241
		1.350		239
		2.1	Data of ref. 239	29

Ethylene glycol	60	0.28		237
	80	0.60		237
Ethyl mercaptoacetate	60	6300		231
Furfurylidene malononitrile	44.1	12000		254
Heptane	50	1.8		256
Hexamethyldisiloxane	79.5	0.104		115
Hydrogen sulphide	60	810		170
Hydroquinone	45	7.0		246
	131	100		206
Isobutyl alcohol	60	0.10		241
	80	0.229		241
		0.250		239
4-Isobutylisopropylbenzene	60	5.72	[a], benzene	247
Isobutyric acid	80	0.900		239
Isopropyl alcohol	60	0.583		241
	80	1.09		166
		1.907		241
	100	3.00	[d]	166, 255
		2.71—2.85	[e]	166, 255
Isopropylbenzene	60	1.9		157
		2.56	[a], benzene	247
	80	1.9		239
		2.4	Data of ref. 239	29
3′-Mercaptoacetophenone	45	42000		246
4′-Mercaptoacetophenone	45	26000		246
2-Mercaptoethyl alcohol	50	4430	[d]	257
		4300	[e]	257
	60	6200		231
	70	4600	[d]	257
		4000	[e]	257

TABLE 13—*continued*

Solvent	Temp. (°C)	$10^4 C_s$	Comments	Ref.
2-Mercaptoethylamine hydrochloride	50	1690	[d]	257
		1660	[e]	257
	70	1580	[d]	257
		1500	[e]	257
3-Mercapto-*n*-propyl methyldimethoxybismuthane	79.5	1285		115
3-Mercapto-*n*-propyltrimethoxysilane	79.5	6930		115
4-Methoxyisopropylbenzene	60	3.46	[a], benzene	247
4-Methoxyphenol	45	<5.0		246
Methyl alcohol	60	0.2		166
	80	0.333		166
	100	0.45	[d]	166, 255
		0.32—0.36	[e]	166, 255
Methylaminobenzene	60	7.0		243
	80	10.0		243
	100	13.3		243
Methylcyclohexane	80	0.195		239
Methyldichlorosilane	79.5	15.2		115
Methyldi(trimethylsiloxy)silane	79.5	0.284		115
Methyl isobutyrate	60	0.26		258
Methyl octadecanoate	60	0.282 ± 0.03	[a]	174
Methyl oleate	60	1.68 ± 0.17	[a]	174
4-Methyl-2-pentanone	80	0.700		239
2-Methyl-2-propanethiol	60	1800		231
2-Methyl-1-propane	40	2.5		250
	50	4.4		250

Methyltrichlorosilane	50	2.7	Benzene	143
Methyltriisopropoxysilane	79.5	0.687		115
Methyltrimethoxysilane	79.5	0.331		115
2-Naphthalenethiol	45	13000		246
	60	31000		231
1-Naphthol	45	<5.0		246
N-Nitrobutylamine	40	0		249
Nitroethane	45	2.0		249
Nitromethane	60	2.0		91
1-Nitropropane	45	5		249
1-Nitrosopiperidine	45	8.2		249
4-Nitrotoluene	44.1	~0		254
2,2,3,3,4,4,5,5-Octafluoro-*n*-pentyl alcohol	60	2.36		163
Octamethylcyclotetrasiloxane	79.5	0.08		115
Octamethyltrisiloxane	79.5	0.0316		115
Octylphosphine	60	23000		177
3-Pentanone	60	0.833		241
	80	1.729		241
		1.775		239
		2.7	Data of ref. 239	29
Phenyldimethoxytrimethylsiloxysilane	79.5	0.032		115
Phenyldi(trimethylsiloxy)silane	79.5	1.84		115
β-Phenylethyl acetate	30	0.18		240
3-Phosphinopropionitrile	60	14000		177
3,3′-Phosphinylidenedipropionitrile	60	13000		177
2-Propanethiol	60	0.38		231

TABLE 13—*continued*

Solvent	Temp. (°C)	$10^4 C_s$	Comments	Ref.
n-Propyl alcohol	60	0.692		166
	80	0.84		166
	100	1.25	d	166, 255
		1.147—1.293	e	166, 255
1,2-Propyleneglycol	60	0.48		237
	80	0.86		237
n-Propylmethyldi(trimethylsiloxy)silane	79.5	0.189		115
Pyrocatechol	45	<5		246
Pyrogallol	45	26		246
Tetrabutyltin	60	1.32		146
1,1,2,2-Tetrachloroethane	60	0.155		241
	80	0.200		239
		0.235		241
Tetrachlorosilane	50	3.0	Benzene	143
Tetraethylsilane	60	5.75		146
2,2,3,3-Tetrafluoro-*n*-propyl alcohol	60	1.86		163
2,3,4,6-Tetramethylphenol	45	11.0		246
Thioglycolic acid	50	3960	d	257
		3900	e	257
	70	3480	d	257
		3800	e	257
Toluene	30	0.62		240
	52	0.84		241
	60	0.170		179, 241
		0.190		236
		0.202		235, 236, 259

		0.250		236
		0.26		258
		0.400		236
	70	0.567		39
	80	0.292		259
		0.303		241
		0.525		239
		0.52	Data of ref. 239	29
		0.91		39
3-Toluenethiol	45	47000		246
4-Toluenethiol	45	74000		246
Tributylamine	70	25.7		252
Tributylborane	60	7.45		146
Tributylphosphine	60	30.6		146
Tributylstibine	60	<1.0		146
1,1,1-Trichloroethane	80	0.600		239
Triethylaluminium	60	1240		146
Triethylamine	60	8.3	a	180
		1900		260
Triethylindium	60	332		146
3,3,3-Trifluoro-*n*-propylmethyldi(trimethylsiloxy)-silane	79.5	0.205		115
Trimethoxyphenylsilane	79.5	0.053		115
4-Trimethoxysilyl-α-bromotoluene	79.5	2.98		115
Trimethylchlorosilane	50	2.2	Benzene	143
2,2,4-Trimethylpentane	50	1.2		256
1,1,1-Trinitrobutane	45	8300		249

TABLE 13—*continued*

Solvent	Temp. (°C)	$10^4 C_s$	Comments	Ref.
1,1,1-Trinitroethane	45	1400		249
2,2,2-Trinitroethyl propionate	45	3000		249
Trinitromethane	45	5400		249
Triphenylmethane	60	4.0		189
Tripropylamine	60	14.6		146
Water	60	0		261

[a] Apparent value due to retardation.
[b] Some copolymerization.
[c] Mayo equation not strictly applicable under experimental conditions.
[d] From Mayo equation.
[e] From end-group analysis.
[f] Viscosity effects allowed for.

TABLE 14
Kinetic parameters for inhibition of free-radical polymerization of methyl methacrylate

Inhibitor	Temp. (°C)	k_z/k_p	k_z ($l\ mole^{-1}\ sec^{-1}$)	Comments	Ref.
p-Benzoquinone	44.1	5.5	2400		254
	50	5.7			253
	60	4.5			192
N,N'-Bis(*p*-chlorophenyl)thiourea	50	0.35	197	[a]	266
N,N'-Bis(*p*-dimethylamino)thiourea	50	0.09	48	[a]	266
N,N'-Bis(*m*-methoxyphenyl)thiourea	50	0.33	185	[a]	266
N,N'-Bis(*m*-nitrophenyl)thiourea	50	0.55	307	[a]	266
N,N'-Bis(*m*-tolyl)thiourea	50	0.21	115	[a]	266
N,N'-Bis(*p*-tolyl)thiourea	50	0.13	72	[a]	266
Cobalt(III) nitrate	15	0.178		[b] $[Co^{3+}]$, aq. HNO_3	262
	20	0.261		[b] $[Co^{3+}]$, aq. HNO_3	262
Cobalt(III) perchlorate	15	0.237		[b] $[Co^{3+}]$, aq. $HClO_4$	262
	20	0.264		[b] $[Co^{3+}]$, aq. $HClO_4$	262
	25	0.231		[b] $[Co^{3+}]$, aq. $HClO_4$	262
Cobalt(III) sulphate	15	0.096		[b] $[Co^{3+}]$, aq. H_2SO_4	262
	20	0.187		[b] $[Co^{3+}]$, aq. H_2SO_4	262
Copper(II) bromide	($k_z = 3.3 \times 10^{12} \exp(-40000/RT)\ l\ mole^{-1}\ sec^{-1}$)[c]			DMF	263
	40	1870		DMF	263
	50	1980		DMF	263
	60	2190	1.61×10^6	DMF	263
	70	2942		DMF	263

TABLE 14—*continued*

Inhibitor	Temp. (°C)	k_z/k_p	k_z (l mole^{-1} sec^{-1})	Comments	Ref.
Copper(II) chloride	60	1027		DMF	264
2,5-Dichloro-*p*-benzoquinone	44.1		5500		265
2,6-Dichloro-*p*-benzoquinone	44.1		16500		265
m-Dinitrobenzene	44.1	0.0048	2.2		254
Diphenylpicrylhydrazil	44.1	2000			254
Diphenyl thiourea	50	0.22	121	a	266
Furfurylidene malonitrile	44.1	1.2	550		254
Iron(III) chloride	60		5000	DMF	198, 199
Oxygen	50	33000	10^7		267
Tetrachloro-*p*-benzoquinone	44.1	0.26	120		254
1,3,5-Trinitrotoluene	44.1	0.05	23		254

[a] Parameters related to reinitiation and cross-termination quoted.
[b] Species considered responsible for chain termination.
[c] R in J K^{-1} mole^{-1}.
DMF, *N,N*-dimethylformamide solution.

TABLE 15

Absolute rate coefficients and activation energies for free-radical polymerizations of acrylic acid esters

Ester	Temp. (°C)	k_p ($l\ mole^{-1}\ sec^{-1}$)	E_t ($kcal\ mole^{-1}$)	$10^{-6} k_t$ ($l\ mole^{-1}\ sec^{-1}$)	E_t ($kcal\ mole^{-1}$)	$10^6 k_p/k_t$	Comments	Ref.
Butyl	15					820	[a]	12
						680	[a]	12
	25	13	2.1	0.018	0		[b]	268
		2100	12.5	330	17.6		[a], at 20 % conversion	269
						59	[a]	215
	30					1250	[a]	226
						1410	[a], benzene	226
						833	[a], ethyl acetate	226
						840	[a], *n*-butyl propionate	226
		1237		6.18			[b], 1 bar	270
		1452		4.63			[b], 250 bar	270
		1977		3.84			[b], 500 bar	270
		2687		3.73			[b], 750 bar	270
		3387		3.50			[b], 1000 bar	270
	35	14.5		0.018			[b]	268
Methyl	15	1300		75		17	[b]	271
		880		260		3.4	[b]	272
						12	[a]	12
	25	1580		55		28	[b]	273
		580		6.5			[a]	275
	30	720	7.1	4.3	5.3		[b]	276
	50	1000	4.7	3.55	~0	283	[b]	277
	60	2090		9.5			[b]	276

[a] Direct use of pre- or after-effect.

[b] Rotating-sector technique.

TABLE 16
Monomer transfer constants (C_M) for free-radical polymerization of methyl acrylate

Temp. (°C)	$10^4 C_M$	Solvent	Ref.
	($C_M = 0.4 \exp[-6219/RT]$)[a]		278
55	0.275		278
60	0.036		278
	0.325		278
65	0.10		278
	0.11		278
	0.37		278
	0.11	Ethyl acetate	39
70	0.072		278
	0.16		278
	0.18		278
	0.405		278
	0.01	Ethyl acetate	279
75	0.224		278
	0.25		278
	0.25	Ethyl acetate	39

[a] Units of R in cal K^{-1} $mole^{-1}$.

TABLE 17
Transfer constants to initiator (C_I) for free-radical polymerizations of methyl acrylate

Initiator	Temp. (°C)	C_I	Comments	Ref.
Benzoyl peroxide	55	0.0143		278
	60	0.0246		278
	65	0.0375		278
	70	0.05		278
		~0.01	Ethyl acetate	279
2-Butanone peroxide	65	0.05		278
		0.05	Ethyl acetate	39
	70	0.077		278
	75	0.113		278
		0.113	Ethyl acetate	39
t-Butyl hydroperoxide	60	0.01		278
	70	0.0266		278
t-Butyl peroxide	65	0.00047		278
	70	0.00082		278
	75	0.00111		278

TABLE 18
Transfer constants (C_s) to solvents and transfer agents for free-radical polymerizations of esters of acrylic and methacrylic acids (excluding methyl methacrylate)

Monomer	Solvent	Temp. (°C)	$10^4 C_s$	Ref.
Butyl acrylate	*N*,*N*-Dimethylaminobenzene	50	380	155
Butyl methacrylate	Benzene	60	0.158	280
Ethyl acrylate	Acetic Acid	50	0.176	281
		80	0.537	282
		100	1.05	281
	Acetone	40	0.27	281
		60	0.27	283
		80	1.10	282
		100	2.3	281
	Acetonitrile	50	0.158	281
		60	0.245	281
		80	0.550	282
		100	1.43	281
	Allyl chloride	30	1.5[a, b]	97
	Benzene	40	0.016	281
		60	0.22	281
			0.27	285
			0.45	283
		80	0.525	282
		100	2.2	281
	Bromobenzene	60	0.163	281
		80	0.685	282
			0.688	281
		100	3.34	281
	2-Butanone	60	0.151	283
	n-Butyl alcohol	40	1.28	281
		60	2.91	281
		80	5.85	282
		100	12.6	281
	sec-Butyl alcohol	45	10.6	281
		63	18.5	281
		80	22.2	282
		100	31.5	281
	t-Butyl alcohol	40	0.068	281
		60	0.17	281
		80	0.712	282
		100	1.64	282
	n-Butyric acid	80	0.855	282
	Carbon tetrachloride	40	0.332	281
		60	0.90	281
		70	1.13	283
		80	1.55	282
		100	2.8	281

TABLE 18—*continued*

Monomer	Solvent	Temp. (°C)	$10^4 C_s$	Ref.
	Chlorobenzene	40	0.054	281
		80	0.370	282
		100	1.68	281
	Chloroform	40	0.195	281
		60	0.89	283
		70	1.57	283
		80	1.489	282
		100	4.74	281
	Cyclohexane	50	0.289	281
		60	0.48	281
			0.61	283
		80	1.220	282
		100	3.08	281
	Ethyl acetate	50	0.298	281
		60	0.448	281
			0.69	283
		80	0.89	282
		100	1.82	281
	Ethyl benzene	40	0.688	281
		60	1.16	281
		80	16.80	282
		100	28.7	281
	Ethyl methyl ketone	50	1.44	281
		60	1.92	281
		80	3.29	282
		90	4.45	281
	Formic acid	80	0.046	282
	n-Hexane	50	0.524	281
		60	0.593	281
		80	0.970	282
		100	1.46	281
	Isobutyl alcohol	45	2.10	281
		63	3.31	281
		80	4.65	282
		100	8.06	281
	Isopropylbenzene	50	11.7	281
		60	13.8	281
		80	22.2	282
		90	26.0	281
	Methyl alcohol	60	0.32	283
	2-Propyl alcohol	—	20	286
	Toluene	40	0.88	281
		60	0.32	283
			1.7	281
		70	1.84	283

TABLE 18—*continued*

Monomer	Solvent	Temp. (°C)	$10^4 C_s$	Ref.
		80	2.60	282
			2.9	281
		100	6.8	281
Ethyl methacrylate	Acetone	80	0.102	287
	Acetophenone	80	0.281	287
	Benzene	80	0.081	287
	2-Butanone	80	0.252	287
	Butyl alcohol	80	0.454	287
	sec-Butyl alcohol	80	1.604	287
	t-Butyl alcohol	80	0.417	287
	Carbon tetrachloride	80	0.901	287
			5.640	288
	Chlorobenzene	80	0.436	287
	Chloroform	80	0.703	287
			2.360	288
	Cyclohexane	80	0.928	287
	1,2-Dichloroethane	80	1.821	288
	Ethyl acetate	80	0.919	287
	Ethyl alcohol	80	0.429	287
	Ethylbenzene	80	1.428	287
	Heptane	80	0.865	287
	2-Heptanone	80	0.702	287
	Isobutyl alcohol	80	0.445	287
	Isopropylbenzene	80	2.067	287
	2,4-Pentanedione	80	0.236	287
	1,1,2,2-Tetrachloroethane	80	0.311	287
			1.820	288
	1,1,1-Trichloroethane	80	0.536	287
	Toluene	80	0.436	287
Hexadecyl methacrylate	Carbon tetrachloride	70	0.983	289
	Isopropylbenzene	70	2.05	289
Isobutyl methacrylate	Benzene	60	0.165	280
	Carbon tetrachloride	80	1.971	288
	Chloroform	80	1.110	288
	1,2-Dichloroethane	80	0.510	288
	1,1,2,2-Tetrachloroethane	80	0.510	288

TABLE 18—*continued*

Monomer	Solvent	Temp. (°C)	$10^4 C_s$	Ref.
Methyl acrylate	Acetone	80	0.622	290
			1.1	291
	Acetophenone	50	<2.5[a]	253
	Benzene	80	0.326	290
			0.45	291
	Benzonitrile	50	<1.0[a]	253
	Benzoyl chloride	50	<1.5[a]	253
	Biphenyl	50	<2.0	253
	1-Butanethiol	30	15300	126
		60	16900	28, 126
	2-Butanone	80	3.238	290
			3.61	291
	Butyl alcohol	80	2.747	290
	sec-Butyl alcohol	80	14.14	290
	t-Butyl alcohol	80	0.389	290
	Carbon tetrabromide	40	3500[c]	133
		50	3500[c]	133
		60	4100[c]	133
	Carbon tetrachloride	80	1.25	291
			1.266	288
			1.323	290
			1.55	288
	Chlorobenzene	80	0.52	291
			0.986	290
	Chloroform	80	2.100	288
			2.144	290
			2.333	288
			2.5	291
	Cyclohexane	80	0.027	290
			1.2	291
	Cyclohexanone	80	5.5	291
	o-Dichlorobenzene	80	0.71	291
	1,2-Dichloroethane	80	0.82	288
			1.00	288
	Dimethyl 2,4-dimethyl-glutarate	60	0.45	48
	2,6-Dimethyl-2,6-octadiene	60	42	158
	Diphenylamine-*t*	60	300[c]	160
	Ethylbenzene	80	6.056	290
	Isobutyl alcohol	80	2.496	290

TABLE 18—*continued*

Monomer	Solvent	Temp. (°C)	$10^4 C_s$	Ref.
	Isopropylbenzene	80	6.966	290
			16.2	291
	Methyl dodecanoate	60	0.751 ± 0.18[a]	174
	Methyl isobutyrate	60	1.4	48
	Methyl oleate	60	3.66 ± 0.52[a]	174
	Naphthalene	50	<2.0	253
	Phenanthrene	50	26	253
	Phosphorous trichloride	40	100	178
	2-Propanethiol	—	6000	292
	1,1,2,2-Tetrachloroethane	80	0.932	290
			1.55	288
			1.561	288
	Tetraethylsilane	60	33.1[c]	146
	Toluene	60	2.7	179, 180
		80	1.775	290
			2.7	291
	Tributylphosphine	60	1890[c]	146
	1,1,1-Trichloroethane	80	0.574	290
	Triethylamine	60	400	180
	Trimethyl 2,4,6-heptane-tricarboxylate	60	0.54	48
	Tripropylamine	60	470[c]	146

[a] Apparent value due to retardation.
[b] Calculated from data of ref. 284.
[c] Benzene solution.

TABLE 19
Kinetic parameters for inhibition of free-radical polymerization of methyl acrylate

Inhibitor	Temp. (°C)	k_z/k_p	k_z(l mole^{-1} sec^{-1})	Ref.
Acetophenone	50		<0.25	293
Aniline	50	0.0001	<0.1	253, 293
Anthracene	50	0.098	100	293
Benzonitrile	50		<0.1	253
Benzophenone	50		<0.15	253

TABLE 19—*continued*

Inhibitor	Temp. (°C)	k_z/k_p	k_z(l mole^{-1} sec^{-1})	Ref.
p-Benzoquinone	44.1		1200	265
Benzoyl chloride	50		<0.15	253
Cobalt(III) nitrate[a]	10	0.133		294
	15	0.222		294
Cobalt(III) perchlorate[b]	10	0.125		294
	15	0.206		294
Diazoaminobenzene	50	0.00419		293
2,5-Dichloro-*p*-benzoquinone	44.1		10200	265
2,6-Dichloro-*p*-benzoquinone	44.1		16700	265
m-Dinitrobenzene	44.1		58	265
	44.1	0.035		295
o-Dinitrobenzene	50	0.018		293
1,5-Dinitronaphthalene	50	0.0151		293
2,4-Dinitrophenol	50	0.0649		293
2,4-Dinitrotoluene	50	0.0188		293
Diphenyl	50		<0.2	253
Fluorene	50	0.00602		293
Furfurylidene malonitrile	44.1	1.2	550	254
Iron(III) chloride	60		6800[c]	198, 199
1-Methoxy-4-nitrobenzene	50	0.00727	7.2	253, 293
Naphthalene	50		<0.2	253
α-Naphthol	50	0.00496		293
2-Nitroaniline	50	0.00341		293
3-Nitroaniline	50	0.00421	4.21	253, 293
4-Nitroaniline	50	0.00267	2.67	253, 293
Nitrobenzene	44.1	0.00642		295
	50	0.00464		253, 293
p-Nitrobenzoic acid	50	0.0107	10.7	253, 293
Nitrodeuterobenzene	50	0.00468		293
Nitrodiphenyl	50	0.00604		293
2-Nitro-α-naphthol	50	0.0111		293
1-Nitro-β-naphthol	50	0.0154		293
2-Nitrophenol	50	0.0108		293
3-Nitrophenol	50	0.00562	5.62	253, 293

TABLE 19—*continued*

Inhibitor	Temp. (°C)	k_z/k_p	k_z(l mole^{-1} sec^{-1})	Ref.
4-Nitrophenol	50	0.00426	4.26	253, 293
2-Nitrotoluene	50	0.00323		293
3-Nitrotoluene	50	0.00412	4.12	253, 293
4-Nitrotoluene	50	0.00486	4.86	253, 293
Phenanthrene	50	0.0026	2.6	253, 293
Phenol	50	0.0002	<0.2	253, 293
Picric acid	50	0.319		293
Stilbene	50	0.00196		293
Sulphur	44.1		1100	265
Tetrachloro-*p*-benzoquinone	44.1		2000	265
1,3,5-Trinitrobenzene	50	0.204	204	253, 293
1,3,5-Trinitrotoluene	44.1		105	265
	50	0.0596		293
Triphenylmethane	50	0.00283		293
$(C_6H_5)_3CH_{0.3}D_{0.7}$	50	0.00144		293

[a] Termination by Co(III) in nitric acid solution.
[b] Termination by Co(III) in perchloric acid solution.
[c] In *N,N*-dimethylformamide solution.

6. Miscellaneous esters of acrylic and methacrylic acids

Very little detailed kinetic information is available on the free-radical polymerizations of esters of acrylic and methacrylic acids other than for the methyl esters. Apparently absolute rate coefficients for propagation and termination in acrylate polymerizations are only available for the butyl ester (Table 15) although limited data are available for a wider range of methacrylates (Table 20) and diffusion-controlled termination has been considered for polymerizations of benzyl and tetrahydrofurfuryl methacrylates. The bulk polymerization of all the esters show "gel-effects".

From the limited data available it appears that probably all acrylates terminate by combination of propagating radicals and methacrylates terminate by both combination and disproportionation.

Transfer constants to monomer are given in Table 21 and to other transfer agents and solvents in Table 18.

TABLE 20
Absolute rate coefficients for free-radical polymerizations of methacrylic acid esters (excluding methyl methacrylate)

Ester	Temp. (°C)	k_p ($l\ mole^{-1}\ sec^{-1}$)	E_p ($kcal\ mole^{-1}$)	$10^{-6} k_t$ ($l\ mole^{-1}\ sec^{-1}$)	E_t ($kcal\ mole^{-1}$)	Comments	Ref.
Benzyl	−20	144		0.83		[a], η = 7.6 mPa s	296
		158		0.20		[a], diethyl phthalate, η = 78 mPa s	296
	0	246		5.94		[a], ethyl acetate, η = 0.87 mPa s	296
		290		6.69		[a], ethyl acetate, η = 1.05 mPa s	296
		257		1.66		[a], diethyl malonate, η = 3.7 mPa s	296
		307		1.79		[a], η = 4.6 mPa s	296
		283		0.36		[a], diethyl phthalate, η = 23.5 mPa s	296
		330		0.13		[a], dimethylglycol phthalate, η = 84 mPa s	296
	20	510		2.87		[a], η = 2.7 mPa s	296
		561		0.81		[a], diethyl phthalate, η = 8.6 mPa s	296
	30	1410		41.9		Benzyl butyrate	297
		1250		41.9		Benzene	228
Butyl	30	372		28.8		1 bar	298
		461		22.0		250 bar	298
		573		18.0		500 bar	298
		768		16.9		750 bar	298

		925		13.4		1000 bar	298
n-Butyl	30	369		10.2			274
t-Butyl	25	350	4.4	14	1.1		299
β-Chloroethyl	30	254		6.71		Benzene	228
β-Cyclohexyl	30	1190		32.8		Benzene	228
Ethyl	30	126		7.35		Benzene	228
Isopropyl	30	121		4.52		Benzene	228
β-Methoxyethyl	30	249		9.30		Benzene	228
Phenyl	30	411		11.9		Benzene	228
β-Phenylethyl	30	228		1.88		Benzene	228
γ-Phenylpropyl	30	149		0.813		Benzene	228
n-Propyl	30	467		45.1			300
Tetrahydrofurfuryl	60	1211		75.2		[a], η = 0.425 cP	301
		1094		72.1		[a], chloroform, η = 0.445 cP	301
		1111		69.3		[a], benzene, η = 0.454 cP	301
		1246		65.8		[a], bromobenzene, η = 0.430 cP	301
		1286		61.3		[a], methyl benzoate, η = 0.531 cP	301
		1312		58.5		[a], α-bromonaphthalene, η = 0.546 cP	301
		1288		59.6		[a], 1,2-dimethylethylene phthalate, η = 0.547 cP	301

All data obtained using rotating-sector technique.
[a] Diffusion-controlled termination considered.

TABLE 21
Monomer transfer constants (C_M) for free-radical polymerization of esters of acrylic and methacrylic acids (excluding methyl esters)

Monomer	Temp. (°C)	$10^4 C_M$	Ref.
Benzyl acrylate	55	0.883	302
	60	0.905	302
	65	0.955	302
Bornyl methacrylate	60	2.85	303
Butyl methacrylate	60	0.14	280
Ethyl methacrylate	[C_M = 6.44 exp(−4120/RT)][a]		304
	45	0.248	304
	60	0.259	304
	80	0.456	304
	90	0.442	304
Isobornyl methacrylate	60	1.85	303
Isobutyl methacrylate	[C_M = 4.77 exp(−3540/RT)][a]		304
	35	0.189	304
	50	0.179	304
	60	0.14	280
		0.165[b]	280
		0.224	304
	80	0.301	304
	100	0.382	304

[a] Units of R in cal K^{-1} $mole^{-1}$.
[b] Benzene solution.

7. Acrylonitrile

Polyacrylonitrile is insoluble in its monomer and in most common solvents and polymerizations performed in such media are heterogeneous processes. It is only by carrying out polymerizations in solvents such as *N,N*-dimethylformamide or dimethyl sulphoxide or in aqueous solutions of certain salts (acrylonitrile is readily soluble in water but the polymer is not) that homogeneous polymerization can be achieved. Pure acrylonitrile does not undergo spontaneous thermal polymerization. Kinetic data on homogeneous polymerizations of acrylonitrile are limited. Values of rate coefficients for propagation and termination are included in Table 22. Termination of propagating radicals appears to proceed by combination (see Chap. 1, Sect. 4.1.2).

Transfer constants to monomer are given in Table 23 and to polymer in Table 3. The meagre data on transfer to initiator are included in Table 36, along with data for miscellaneous monomers. Transfer constants to various solvents and transfer agents are presented in Table 24 and data relating to inhibition and retardation in Table 25.

TABLE 22
Absolute rate coefficients and activation energies for free-radical polymerizations of acrylonitrile

Temp. (°C)	k_p (l mole^{-1} sec^{-1})	E_p (kcal mole^{-1})	$10^{-7}k_t$ (l mole^{-1} sec^{-1})	E_t (kcal mole^{-1})	$10^6 k_p/k_t$	Comments	Ref.
0	178	3.88	2.49	3.70		DMF, $FeCl_3$	305
	3000—5000					[a]	306
15	23000	4.1	280	5.4		[b], aq.	307
25	127		1.22		10.4	[c]	308
	14500		200			[b], aq.	309
	28000		370			[b], aq.	307
					3.4	[b], DMF	310
	52		0.5			[c]	311
	1910		29			[b], DMSO	312
	382 ± 230		4.76 ± 2.2			[b], DMF	305
	51	6.4	1.8	~0	29	[c], DMF	313
30	32500		440			[b], aq.	307
					58.8	[b], aq.	314
40	20000					[a]	306
60	1960		78.2			[a], DMF	327
	11600		1900			[a], DMSO	312
66	660		8.65			[a], DMF	305

[a] Emulsion polymerization; [b] Rotating-sector technique; [c] Direct use of pre- or after-effect; DMF, *N,N*-Dimethylformamide solution; DMSO, Dimethylsulphoxide solution.

TABLE 23
Monomer transfer constants (C_M) for free-radical polymerization of acrylonitrile

Temp. (°C)	$10^4 C_M$	Solvent	Ref.
20	0.18		315
25	0.105		316
30	1.5	DMSO	317
40	0.17		316
50	0.050	Zinc chloride, aqueous	318
	0.27	Magnesium perchlorate, aqueous	44
	8.2	Sulphur dioxide	319
	~0		320
60	0.26		321
	0.3		322
	~0		320

DMSO, Dimethylsulphoxide.

TABLE 24
Transfer constants (C_s) to solvents or other transfer agents for free-radical polymerization of acrylonitrile

Solvent	Temp. (°C)	$10^4 C_s$	Comments	Ref.
Acetonitrile	60	2.0	Succinonitrile	91
Allyl chloride	30	5.950	[a], calc. from data of ref. 323	97
Anthracene	50	18000		253
Arabinose	60	13.0	DMSO	324
1,4-Benzoquinone	50	13000		253
γ-Butyrolactone	50	0.658		153
		0.74		325
Carbon tetrabromide	50	400	[a]	326
	60	500		327
		1900	[a], DMF	180
Carbon tetrachloride	80	1.13		328
Chloroform	80	5.9		328
Copper(II) chloride	35	18×10^4	0.01 M NaCl	329
		19×10^4	0.1 M NaCl	329
		32×10^4	1.0 M NaCl	329

TABLE 24—*continued*

Solvent	Temp. (°C)	$10^4 C_s$	Comments	Ref.
Copper(II) sulphate	35	1900	[a], aqueous	329
		2800	[a], 10^{-4} M NaBr	329
		10700	[a], 0.001 M NaBr	329
		39300	[a], 0.01 M NaBr	329
		21 x 10^4	[a], 0.1 M NaBr	329
		3000	[a], 0.01 M NaCl	329
		13500	[a], 0.1 M NaCl	329
		136000	[a], 1.0 M NaCl	329
6-Deoxy-6-dipropylaminomethyl-β,*D*-glucoside	60	1100	DMF	144
6-Deoxy-6-iodomethyl-α,*D*-glucoside	60	50	DMF	144
6-Deoxy-6-mercaptomethyl-α,*D*-glucoside	60	1300	DMF	144
6-Deoxy-6-phthalimidomethyl-α,*D*-glucoside	60	50	DMF	144
2,3-Di-*O*-benzylmethyl-α,*D*-glucoside	60	90	DMF	144
Dibutyl cadmium	60	55000	DMF	146
Diethyl mercury	60	72.2	DMF	146
Diethyl zinc	50	16000	DMF	146
N,N-Dimethylacetamide	50	4.945		153
		5.05		330
N,N-Dimethylformamide	25	[b]	[a]	331
	50	1.0		332
		2.70		325
		2.78		330
		2.83		153
		10		333
	60	2.412		332
		4.494		334
		5.0	Succinonitrile	91
2,4-Dimethylglutaronitrile	50	0.6	$MgClO_4$ soln.	44
2,6-Dimethyl-2,6-octadiene	60	450		335
Dimethylsulphoxide	30	0.47		317
	50	0.11		312
		0.29		330
		0.795		153
Diphenylamine-*t*	60	700	DMF	135
Dodecyl vinyl ether	50	4.95		150
Erythritol	60	12.8	DMSO	324

TABLE 24—*continued*

Solvent	Temp. (°C)	10^4C_s	Comments	Ref.
Ethylene carbonate	50	0.073		336
		0.33		330
		0.39		325
		0.474		153
		0.5		312
		1.0		333
	60	0.128		336
N-Ethylpiperidine	60	3300	[a]	180
Glucose	60	6.9	DMSO	324
Glyceraldehyde	60	32.1	DMSO	324
Glycerol	60	23.5	DMSO	324
2,4,6-Heptanetricarbonitrile	50	1.0	$MgClO_4$ soln.	44
Hydrodiisobutyl aluminium	60	3940	DMF	146
Iron(III) chloride	60	33300	DMF	198
Isobutyronitrile	50	1.3	$MgClO_4$ soln.	44
	60	3.5		45
		1.8	Calc. from data of ref. 45	44
		4.0	Succinonitrile	91
Magnesium perchlorate	50	<0.05		44
Methyl-α,*D*-glucoside	60	20	DMF	144
Methyl-β,*D*-glucoside	60	20	DMF	144
N-Methylpiperidine	60	2300	[a]	180
Nitromethane	60	6.0	Succinonitrile	91
Sorbitol	60	6.1	DMSO	324
Succinonitrile	60	0.6		91
Sulphur dioxide	50	0		319
2,3,4,6-Tetra-*O*-acetylmethyl-α,*D*-glucoside	60	30	DMF	144
Tetrabutyltin	60	80.8	DMF	146
Tetraethyllead	60	243	DMF	146
Tetraethylsilane	60	21.0	DMF	146
6-*O*-(*p*-Toluenesulphonyl)-methyl-α,*D*-glucoside	60	10	DMF	144
Tributylamine	60	6700	[a]	180
Tributylborane	60	6470	DMF	146
Tributylstilbene	60	111000	DMF	146
Triethyl aluminium	60	590	DMF	146

TABLE 24—*continued*

Solvent	Temp. (°C)	$10^4 C_s$	Comments	Ref.
Triethylamine	60	3800		334
		1900		327
		1700	0.4 M $LiNO_3$ soln.	327
		5900	[a], DMF	180
		6600	[a]	180
Triethylindium	60	2220	DMF	146
Trimethylamine	60	790	[a]	180
6-*O*-Triphenylmethylmethyl-α,*D*-glucoside	60	80	DMF	144
Tripropylamine	60	10500	[a]	180
		4280	DMF	146
Vinyl benzoate	65	1400		337
Zinc chloride	50	0.006	Aqueous	318

[a] Apparent value, some retardation observed.
[b] Rate coefficient for transfer quoted as 0.19 l $mole^{-1}$ sec^{-1}.
DMSO, Dimethylsulphoxide solution.
DMF, *N,N*-Dimethylformamide solution.

TABLE 25
Kinetic parameters for inhibition of free-radical polymerization of acrylonitrile

Inhibitor	Temp. (°C)	k_z/k_p	k_z (l $mole^{-1}$ sec^{-1})	Comments	Ref.
Anthracene	50	2.67	2670		338
p-Benzoquinone	50	0.91	910		338
Cobalt(III) nitrate	15	0.66		Aq. HNO_3	294
Cobalt(III) sulphate	15	0.202		Aq. H_2SO_4	294
	20	0.84		Aq. H_2SO_4	294
Cobalt(III) perchlorate	15	0.855		Aq. $HClO_4$	294
	20	1.25		Aq. $HClO_4$	294
	25	0.86		Aq. $HClO_4$	294
Copper(II) chloride	60	100		DMF	339
Iron(III) chloride	0		8200 ± 300	DMF	305
	25	4.85		DMF	340
			980 ± 70	DMF	331
	60	3.33	6500	DMF	198, 199
		3.42	6600	DMF	340

DMF, *N,N*-Dimethylformamide solution.

TABLE 26
Absolute rate coefficients and activation energies for free-radical polymerizations of vinyl acetate

Temp. (°C)	k_p (l mole^{-1} sec^{-1})	E_p (kcal mole^{-1})	$10^{-7}k_t$ (l mole^{-1} sec^{-1})	E_t (kcal mole^{-1})	$10^6 k_p/k_t$	Comments	Ref.
($k_p = 2.0 \times 10^6 \exp\{-(4700 \pm 700)/RT\}$ l mole^{-1} sec^{-1}; $k_t = 2.75 \times 10^7$, $E_t = 0$)[b]						a	211
−15	2000	3.2	22	0		c	60
0	2800		22			c	60
15					3.1 (3.5)	d	12
					20	d	13
	910 (680)	4.2	5.7 (3.2)	0	16 (19)	d	341
15.9	586 (556)	4.4	304 (268)	0		e	342
	670		250			e	343
	700		260			e, n-hexane	344
16					26	d	215
20	559		5.18		10.8	d	345

25	1100		3.0			e	346
	1012	7.32	5.88	5.24	17.2	e	347
	1000		5.9			e	348
	895	4.2	2.4	<1	37.2	d, 4 % conversion	349
30					12.1	d	19
35					13.4	d	19
40					14.4	d	19
50	2640		11.68		22.6	e	347
60	9500—19000		38—76			Calc. from data of refs. 347 and 348	69

[a] Determination of radical concentration by ESR.
[b] Units of R undefined in text, probably cal K^{-1} $mole^{-1}$.
[c] Viscosity method.
[d] Direct use of pre- or after-effect.
[e] Rotating-sector technique.

TABLE 27
Monomer transfer constants (C_M) for free-radical polymerization of vinyl acetate

Temp. (°C)	$10^4 C_M$	Comments	Ref.
0	0.90	Data of ref. 60	91
20	0.94		68
25	2.4		348
	1.3	Data of ref. 348	91
	1.45	Data of ref. 348	91
	10.7		28
40	1.29		63
	1.32		68
50	1.29		63
	4.55		350
	20		347
	0.25		351
	0.41	1 kg cm^{-2}	65
	0.54	500 kg cm^{-2}	65
	0.61	1000 kg cm^{-2}	65
60	1.75		63
	1.91		68
	1.93		67
	2.0		352
	2.1	Data of ref. 353	29
	2.4		354
	2.5		66, 91
	2.6		66
	2.8		355
	1.91		356
	1.8		351
	2.15	Overall	62
	2.26	Acetyl group	62
65	2.1		351
70	2.4		72
	2.9	Data of ref. 72	91
	0.4		71

TABLE 28
Transfer constants to initiator (C_I) for free-radical polymerization of vinyl acetate at 60 °C

Initiator	C_I	Ref.
Benzoyl peroxide	0.09	352
	0.15	355
	0.032	356
2-Bromobenzoyl peroxide	0.25	355
	3.5[a]	29
3-Bromobenzoyl peroxide	0.24	355
	0.6[a]	29
4-Bromobenzoyl peroxide	0.17	355
2-Chlorobenzoyl peroxide	0.17[a]	29
Lauroyl peroxide	0.10	355
Palmitoyl peroxide	0.10	355
	0.17[a]	29

[a] Calculated from data of ref. 355.

8. Vinyl acetate

A number of difficulties are encountered in the study of the kinetics of polymerization of vinyl acetate. The monomer is relatively difficult to purify to a sufficient extent to obtain reproducible kinetic data and the polymerization of bulk monomer shows a "gel-effect" at about 20 % conversion. Thus, meaningful kinetic data can only be obtained from carefully purified monomer and at low conversion. Even then the polymerizations are complicated by extensive transfer as a result of the high reactivity of the propagating radical. Transfer to monomer occurs readily to produce polymer with unsaturated end-groups which can subsequently be incorporated into growing chains to form long branches (see Chap. 2, Sect. 2.4). Transfer to polymer can occur intra- or intermolecularly to give polymers containing long and short branches (see Chap. 2, Sect. 2.6). According to the transfer mechanism, branches may be hydrolyzable. Transfer to solvents also occurs readily and, because the propagating radical is highly reactive, the transfer radicals are usually less reactive and transfer is often associated with retardation. These problems make the study of the details of the polymerization mechanism difficult. For example, in determining the mechanism of bimolecular termination accurate allowance must be made for the effects of transfer; it now appears that propagating radicals terminate largely, if not exclusively, by combination.

TABLE 29
Transfer constants (C_s) to solvents and other transfer agents for free-radical polymerization of vinyl acetate

Solvent	Temp. (°C)	$10^4 C_s$	Comments	Ref.
Acetaldehyde	30	400		357
	45	530	[a]	91
	60	220		358
		200	Data of ref. 358	29
		660	[a]	91
		570	Benzene	91
	75	700	[a]	91
Acetaldehyde diacetate	60	40	[a]	91
Acetic acid	50	0.180		351
	60	1.13		359
		1.0	Data of ref. 359	29
		10	[a], benzene	91
		0.200		351
		0.166		356
	65	0.170		351
Acetic anhydride	60	8.0	[a]	91
Acetone	60	1.5	[a]	91
		11.70		359
		12.0		29
	70	25.6		360
	75	42		361
Acetonitrile	60	10	[a]	91
Acetophenone	50	91.5	[a]	253
	60	100	[a]	91
	70	62.0		360
Acetyl disulphide	60	2900	[a]	91
Allyl acetate	60	94	[a]	91
		85	Benzene	91

Allyl chloride	60	3100		379
N-Allyl stearimide	90	620.1	a,b	379
Aniline	50	149	a	253
		210	Benzene	99
Benzaldehyde	60	230	a	29, 362
		540	a	91
		460	Benzene	91
	70	421	a	360
	75	600	a	91
Benzene	60	1.07		64
		1.2		354
		2.4		355
		2.96		359
		1.2	a	91
		20		354
	70	5.27		360
	75	1.40		64
		3.6		361
		1.4	a	91
Benzoic acid	60	50	a	91
Benzoic anhydride	60	130	a	91
Benzoin	60	800	a	91
Benzoyl chloride	50	366		253
	70	300		363
Benzyl acetate	60	80	a, benzene	91
Benzyl alcohol	70	556		360
Benzyl methyl ether	60	280	a, benzene	91
Biphenyl	50	263		253
	60	6.4		364
Bis-(2-chloroethyl)ether	70	245		363

TABLE 29—*continued*

Solvent	Temp. (°C)	$10^4 C_s$	Comments	Ref.
Bromoacetic acid	60	489	[a]	365
	70	4450		363
Bromobenzene	50	18.9		253
	60	10.0	1 kg cm^{-2}	65
		9.2	1000 kg cm^{-2}	65
		8.0	2000 kg cm^{-2}	65
	70	134.2		363
	75	25.2		361
1-Bromobutane	60	50	[a]	91
	70	1100		363
2-Bromo-2-methylpropane	60	150	[a]	91
4-Bromotoluene	50	20.6		253
Bromotrichloromethane	60	6000		366
		$>4 \times 10^5$		366
	70	6303		363
2,3-Butanedione	60	670	[a], benzene	91
1-Butanethiol	60	480000		29, 126
2-Butanone	60	73.80		359
	70	63.6		360
	75	165		361
1-Buten-3-yne	60	~0	[a]	358
N-*n*-Butylacetamide	60	40	[a]	91
Butyl acetate	50	13.2		367
sec-Butyl acetate	50	4.4		367
	60	8.0	[a]	91

t-Butyl acetate	50	1.5		367
	60	6.2		358
Butyl alcohol	60	20.39		359
		20.0	Data of ref. 359	29
	70	29.1		360
sec-Butyl alcohol	60	31.74		359
	70	6.21		360
	75	95.0		361
t-Butyl alcohol	60	0.46		359
		1.3	a	91
		0.5	Data of ref. 359	29
	70	12.1		360
t-Butylbenzene	60	3.61		359
Butyldisulphide	60	10000	a	91, 368
Butyl sulphide	60	260	a	368
Butyraldehyde	40	670	1 kg cm^{-2}	369
		590	1000 kg cm^{-2}	369
		510	2000 kg cm^{-2}	369
		440	4000 kg cm^{-2}	369
	60	650		358, 362
		1000	a	91
	70	388		360
Carbon tetrabromide	60	390000	Benzene	133
	70	28740		363
Carbon tetrachloride	0	1500		357
	20	4700		370
	45	7600	a	91
	60	10000		137
		10700		371
		9600	a	91
		7300	Benzene	91

References pp. 276—285

TABLE 29—*continued*

Solvent	Temp. (°C)	$10^4 C_s$	Comments	Ref.
	70	2023		363
	75	10500	[a]	91
Chloral	60	5000	[a]	91
	70	4927		363
Chloral hydrate	70	4312		363
Chloroacetic acid	70	2550		363
2-Chlorobenzaldehyde	60	390	[a]	91
3-Chlorobenzaldehyde	60	860	[a]	91
4-Chlorobenzaldehyde	60	340	[a]	91
Chlorobenzene	60	8.35		359
		80	[a]	91
		8.0	1 kg cm^{-2}	65
		6.8	1000 kg cm^{-2}	65
		5.6	2000 kg cm^{-2}	65
	70	2.61		363
	75	12.7		361
1-Chlorobutane	60	10	[a]	91
Chloroform	—	100		372
	30	140		357
	60	125.18		359
		130	Data of ref. 359	29
		160		137
		150	[a]	91
		170	Benzene	91
	70	554		363
3-Chloro-2-methyl-1-propane	60	150	[a]	91
3-Chloro-2-methyl-1-propene	60	400	[a]	91

α-Chlorotoluene	60	450	[a]	91
	70	584		363
	75	118		361
2-Chlorotoluene	70	92.86		363
4-Chlorotoluene	70	195		363
Crotonaldehyde	60	1800	[a]	362
3-Cyanobenzaldehyde	60	1070	[a]	91
4-Cyanobenzaldehyde	60	610	[a]	91
Cyclohexane	60	6.59		359
		7.0	Data of ref. 359	29
		100	[a], benzene	91
Cyclohexanol	70	127		360
Cyclohexanone	60	180	[a], benzene	91
	75	670		361
Cyclohexene	60	620	[a]	91
		1600	Benzene	91
	75	770	[a]	91
Decahydronaphthalene	60	48	[a]	91
1,1-Dibromoethane	60	1100	[a]	91
1,2-Dibromoethane	70	134		363
Di-*n*-butylether	60	76	[a]	91
o-Dichlorobenzene	75	42		361
p-Dichlorobenzene	70	11.8		363
1,1-Dichloroethane	60	65	[a]	91
1,2-Dichloroethane	60	5	[a]	91
		7.18		359
		7	Data of ref. 359	29
	70	10.2		363

References pp. 276—285

TABLE 29—*continued*

Solvent	Temp. (°C)	$10^4 C_s$	Comments	Ref.
Dichloromethane	60	4	[a]	91
Diethyl dithiodiacetate	60	15000		368
		14100		91
Diethyleneglycol	70	85.3		360
Diethyl ether	60	45.3		150
Diethyloxalate	60	4.0	[a]	91
Diethyl succinate	60	41 ± 4.1		373
Di-*n*-hexyl succinate	60	40 ± 4.0		373
3,4-Dimethoxybenzaldehyde	60	550	[a]	91
N,N-Dimethylaniline	50	260	Benzene	99
5,5-Dimethyl-1,3-cyclohexane-dione	70	5580		360
N,N-Dimethylformamide	60	50	[a]	91
Dimethyl malonate	60	17	[a]	91
2,6-Dimethyl-2,6-octadiene	60	430		158
		700		157
Dimethyl oxalate	60	1.0	[a]	91
		2.0		29, 358
Dimethyl succinate	60	31 ± 3.1		373
N,N-Dimethyl-*p*-toluidine	50	750	[b], benzene	99
	50	380	[b], benzene	99
1,4-Dioxan	60	20	[a], benzene	91
	70	49.1		360
Diphenylamine	50	138		253
	60	240		160

		170	c	160
		230	d	160
2,2,3,3,4,4,5,5,6,6,7,7-Dodecafluoro-*n*-heptyl alcohol	60	33.3		63
Dodecyl vinyl ether	60	57.2		150
		73.5		150
Ethanediol diacetate	60	40		91
Ethyl acetate	20	1.52		68
	40	2.11		68
	50	2.9		367
		12		350
	60	1.07		359
		1.25		355
		2.6		29, 358
		2.96		68
		3.3	a	91
		3.4 ± 0.34		373
	70	7.8		360
Ethyl acetoacetate	70	80.4		360
Ethyl alcohol	60	25	a, benzene	91
	70	26.3		360
Ethyl benzene	60	55.15		359
		100	a	91
Ethyl benzoate	60	26	a	91
Ethyl butyrate	50	45		350
Ethyl dichloroacetate	60	210	a	91
Ethyleneglycol	70	83.0	a	360
Ethyl 2-ethylhexanoate	50	65		350
Ethyl formate	50	22		350
Ethyl isobutyrate	50	160		350

TABLE 29—*continued*

Solvent	Temp. (°C)	$10^4 C_s$	Comments	Ref.
Ethyl lactate	60	700	a	91
Ethyl laurate	50	105		350
Ethyl nonanoate	50	80		350
Ethyl octadecanoate	50	140		350
Ethyl octanoate	50	70		350
Ethyl propionate	50	40		350
Ethyl trichloroacetate	60	4400	a	91
Ethyl trifluoroacetate	50	30		350
Fluorene	60	4700	a	91
	70	3610	a	360
2-Furaldehyde	60	15000	a	91
Furfuryl alcohol	70	2520	a	360
Heptane	50	17.0		367
Hexachloroethane	70	1210		363
1,5-Hexadien-3-yne	60	~0		358
1-Iodobutane	60	800	a	91
Iodomethane	70	1230		363
Isobutyl acetate	50	9.1		367
Isobutyl alcohol	60	21.75		359
	70	32.4		360
Isobutyric acid	60	5.02		359
Isobutyronitrile	60	100	a	91

Isopropyl acetate	50	3.5		367
	60	3.1		29, 358
		8.0	a	91
	67.5	9.0	a	91
	75	10	a	91
Isopropyl alcohol	70	44.6		360
4-Isopropylbenzaldehyde	60	540	a	91
Isopropylbenzene	60	89.9		359
		100		91, 157
	70	139		360
	75	356		361
p-Mentha-1,8-diene	60	1900	a	91
Methanol	10	3.4		105
	60	2.26		29, 352
		4.3		374
		6.0	a	91
	70	5.5		374
		9.0—10.5		375
Methallyl chloride	60	400	a, e	379
2-Methoxybenzaldehyde	60	420	a	91
3-Methoxybenzaldehyde	60	2500	a	91
4-Methoxybenzaldehyde	60	370	a	91
Methoxybenzene	60	10	a, benzene	91
Methyl acetate	60	1.56	a	29, 352
		1.6		29, 352
		2.5	a	91
N-Methylaniline	50	360	Benzene	99
3-Methyl-2-butanone	60	118.16		359
2-Methyl-3-butyne-2-ol	60	400	a	91

TABLE 29—*continued*

Solvent	Temp. (°C)	$10^4 C_s$	Comments	Ref.
Methyl butyrate	60	18		29, 358
		19	[a]	91
Methyl cyanoacetate	60	5000	[a]	91
Methyl cyclohexane	60	11.75		359
		24	[a]	91
Methyl formate	60	3	[a]	91
Methyl glycolate	60	300	[a]	91
Methyl isobutyrate	60	58		29, 358
		71 ± 7.1		373
		86	[a]	91
Methyl lactate	60	640	[a]	91
Methyl 9-octadecenoate	60	1000	[a]	91
4-Methyl-2-pentanone	60	34.52		359
Methyl propionate	60	23		358
		26 ± 2.6		373
N-Methyl toluidine	50	830	Benzene	99
Nitromethane	60	2300	[a], benzene	91
	75	2600	[a]	91
2,2,3,3,4,4,5,5-Octafluoro-*n*-pentyl alcohol	60	26.1		163
1-Oxa-4,5-dithiacycloheptane	60	2500—25000		368
Paraldehyde	70	136		360
Pentachloroethane	70	1348		363
2,4-Pentanedione	60	10	[a]	91

3-Pentanone	60	114.39		359
		10.0	Data of ref. 359	29
Pentyl acetate	70	7.2		360
	75	87.0		361
Pentyl alcohol	75	56		361
Phenol	50	120		253
	60	600	a	91
Phenyl acetate	60	400	a	91
9-Phenylfluorene	70	16240		360
Phosphorous trichloride	60	15000		178
Propionaldehyde	60	950	a	91
		1000	Benzene	91
	70	457		360
Propyl acetate	50	6.2		367
	60	3.4		29, 358
Salicylic acid	70	296		360
1,1,2,2-Tetrabromoethane	60	6000	a	91
Tetrabromoethylene	70	2800		363
1,1,2,2-Tetrachloroethane	60	107.03		359
		160		91
	70	67.72		363
Tetrachloroethylene	70	465		363
1,1,3,3-Tetrachloropropyl acetate	70	423.2		363
2,2,3,3-Tetrafluoro-*n*-propyl alcohol	60	7.11		163
Tribromoethylene	70	34720		363
Tribromomethane	70	34760		363
Tributylborane	10	9000		376

TABLE 29—*continued*

Solvent	Temp. (°C)	$10^4 C_s$	Comments	Ref.
1,1,1-Trichloroethane	60	71.11		359
1,1,2-Trichloroethane	60	35.98		359
Trichloroethylene	70	3810		363
Trichlorophenyl acetate	70	1445		363
Triethoxymethane	—	7.6		372
	30	8.0		357
Triethylamine	50	360	Benzene	99
	60	370	a	29, 180
2,2,4-Trimethylpentane	50	8.0		367
Triphenylmethane	60	700		377
m-Tolualdehyde	60	570	a	91
p-Tolualdehyde	60	440	a	91

Toluene	50	12.0		253
		14.9		378
	60	17.8		378
		20.75		355
		20.89		164, 179, 359
		21.6		378
		34	a	91
		35	Benzene	91
		69.0	1 kg cm^{-2}	65
		49.0	1000 kg cm^{-2}	65
	70	21.1		378
		21.8		360
		29.2		378
	75	66		361
	80	91.6		104
α-Toluenethiol	70	8850		360
Tolunitrile	60	2100	a	91
Xylene	50	14.9		253
	70	278		360
	75	166		361

[a] Apparent value, retardation observed.
[b] Different monomer concentrations used.
[c] Partially deuterated transfer agent.
[d] Tritiated in ring.
[e] Some copolymerization observed.

TABLE 30
Kinetic parameters for inhibition of free-radical polymerization of vinyl acetate

Inhibitor	Temp. (°C)	k_z/k_p	k_z ($l\ mole^{-1}\ sec^{-1}$)	Ref.
Anthanthrene	$k_z = 4.5 \times 10^6 \exp(-1950/RT)$ $l\ mole^{-1}\ sec^{-1}$ [d,e]			380
	40	156.6[f]		380, 381
	50	129.3[f]		380
	60	105.4[f]		380
Anthracene	$k_z = 4.3 \times 10^{-5} \exp(-1200/RT)$ $l\ mole^{-1}\ sec^{-1}$ [d,e]			380
	20	64.8[f]		380
	30	60.3[f]		380
	40	45.6[f]		380
	50	36.5[f]		380, 381
		20.9	36000	253, 382
	60	27.8[f]		380
1,2-Benzanthracene	50	11.7[f]		380, 381
Benzonitrile	50	0.0041	7.1	253, 382
Benzophenone	50	0.030	50	253, 382
Benzoyl chloride	50	0.037	64	253, 382
1,12-Benzperylene	50	1.820		381
1,2-Benzpyrene	50	0.840		381
3,4-Benzpyrene	50	30.6[f]		380, 381
3,4-Benztetracene	50	28.3[f]		380, 381
1-Bromo-3-nitrobenzene	50	27.59 ± 0.23		383
1-Bromo-4-nitrobenzene	50	17.16 ± 0.47		383
α-Bromo-3-nitrotoluene	50	18.01 ± 0.34		384
α-Bromo-4-nitrotoluene	50	19.21 ± 0.54		384
1-*t*-Butyl-3-nitrobenzene	50	10.03 ± 0.19		384
1-*t*-Butyl-4-nitrobenzene	50	9.51 ± 0.18		384
1-Chloro-3-nitrobenzene	50	29.29 ± 0.81		383
1-Chloro-4-nitrobenzene	50	17.26 ± 0.67		383
α-Chloro-3-nitrotoluene	50	15.50 ± 0.39		384
α-Chloro-4-nitrotoluene	50	17.67 ± 0.56		384
3-Chlorophenol	45	0.027[b]		385
		0.014[c,g]		385
(Deuterated)		0.007[b]		385
		0.007[c,g]		385
4-Chlorophenol	45	0.046[b]		385
		0.034[c,g]		385
(Deuterated)		0.010[b]		385
		0.006[c,g]		385
Chrysene	50	0.336		381

TABLE 30—*continued*

Inhibitor	Temp. (°C)	k_z/k_p	k_z (l mole^{-1} sec^{-1})	Ref.
3-Cyanophenol	45	0.069[b]		385
		0.095[c,g]		385
(Deuterated)		0.067[b]		385
		0.094[c,g]		385
1,2,5,6-Dibenzanthracene	50	8.7		381
1,2,3,4-Dibenzpyrene	50	73.6[f]		380, 381
1,2,4,5-Dibenzpyrene	50	13.5[f]		380, 381
2,5-Di-*t*-butylhydroquinone	50	3.8		386
1,4-Dihydroxy-2,3,5,6-tetramethylbenzene	50	7.4 ± 2.0		386
2,6-Dimethylhydroquinone	50	2.9		386
N,N-Dimethyl-3-nitroaniline	50	8.69 ± 0.14		383
N,N-Dimethyl-4-nitroaniline	50	3.99 ± 0.20		383
o-Dinitrobenzene	45	96		380, 381
m-Dinitrobenzene	45	105		387, 388
	50	108.60 ± 1.60		383
		66.0	113000	253, 382
p-Dinitrobenzene	45	267		387, 388
	50	68.5	116000	253, 382
		154.00 ± 3.90		384
Diphenyl	50	0.027	46	253, 382
Ethyl 3-nitrobenzene sulphonate	50	55.94 ± 0.77		384
Ethyl 4-nitrobenzene sulphonate	50	74.97 ± 0.37		384
Ethyl 3-nitrobenzoate	50	23.71 ± 2.07		383
Ethyl 4-nitrobenzoate	50	39.52 ± 3.60		383
1-Fluoro-3-nitrobenzene	50	30.04 ± 1.00		383
1-Fluoro-4-nitrobenzene	50	12.79 ± 0.32		383
Hydroquinone	50	0.7		386
α-Hydroxy-3-nitrotoluene	50	12.26 ± 0.36		383
α-Hydroxy-4-nitrotoluene	50	10.20 ± 0.63		383
3-Hydroxyphenyl acetate	45	0.160[b]		385
		0.121[c,g]		385
(Deuterated)		0.137[b]		385
		0.121[c,g]		385
3-Hydroxytoluene	45	0.046[b]		385
		0.029[c,g]		385
(Deuterated)		0.012[b]		385
		0.005[c,g]		385

TABLE 30—*continued*

Inhibitor	Temp. (°C)	k_z/k_p	k_z (l $mole^{-1}$ sec^{-1})	Ref.
4-Hydroxytoluene	45	0.071[b]		385
		0.071[c,g]		385
		0.015[b]		385
		0.011[c,g]		385
1-Iodo-3-nitrobenzene	50	31.55 ± 0.25		383
1-Iodo-4-nitrobenzene	50	23.21 ± 1.07		383
Iron(III) chloride[a]	60		2.35×10^6	198, 199
1-Methoxy-3-nitrobenzene	50	13.90 ± 0.60		383
1-Methoxy-4-nitrobenzene	50	5.07 ± 0.40		383
Methyl 3-nitrobenzene sulphinate	50	53.51 ± 1.38		384
Methyl 4-nitrobenzene sulphinate	50	65.91 ± 1.65		384
Naphthalene	50	0.115		381
		0.144	25.5	253, 382
	70	0.172		360
3-Nitroacetanilide	50	14.88 ± 0.82		383
4-Nitroacetanilide	50	8.22 ± 0.80		383
3-Nitroacetophenone	50	23.91 ± 0.41		383
4-Nitroacetophenone	50	45.44 ± 1.50		
3-Nitroaniline	50	9.34 ± 0.43		383
4-Nitroaniline	50	4.97	8500	253, 382
		3.82 ± 0.01		383
3-Nitrobenzaldehyde	50	28.27 ± 0.72		383
4-Nitrobenzaldehyde	50	58.90 ± 1.56		383
3-Nitrobenzamide	50	22.90 ± 2.35		383
4-Nitrobenzamide	50	31.36 ± 0.33		383
Nitrobenzene	45	19		387
	50	11.2	19300	253, 382
		12.04 ± 0.27		384
3-Nitrobenzene sulphonamide	50	43.73 ± 2.28		384
4-Nitrobenzene sulphonamide	50	50.61 ± 0.78		384
3-Nitrobenzene sulphonyl chloride	50	110.90 ± 1.46		384
4-Nitrobenzene sulphonyl chloride	50	165.30 ± 2.47		384
3-Nitrobenzoic acid	50	20.50 ± 1.00		383
4-Nitrobenzoic acid	50	37.91 ± 0.76		383
		24.9	43000	253, 382
3-Nitrobenzonitrile	50	43.56 ± 1.07		383
4-Nitrobenzonitrile	50	70.55 ± 0.81		383
3-Nitrobenzophenone	50	23.11 ± 1.21		383

TABLE 30—*continued*

Inhibitor	Temp. (°C)	k_z/k_p	k_z (l mole^{-1} sec^{-1})	Ref.
4-Nitrobenzophenone	50	34.12 ± 1.21		383
3-Nitrobenzyl acetate	50	14.61 ± 0.92		384
4-Nitrobenzyl acetate	50	15.17 ± 1.25		384
3-Nitrobenzyl phenyl ether	50	16.54 ± 0.34		383
4-Nitrobenzyl phenyl ether	50	13.63 ± 0.79		383
(3-Nitrobenzyl)phenyl sulphone	50	17.91 ± 0.55		384
(4-Nitrobenzyl)phenyl sulphone	50	18.50 ± 0.41		384
3-Nitrophenol	50	13.08 ± 0.14		383
4-Nitrophenol	50	9.07	15500	253, 382
		5.08 ± 0.05		383
3-Nitrophenyl acetate	50	21.68 ± 1.35		383
4-Nitrophenyl acetate	50	17.50 ± 1.57		383
3-Nitrophenylacetonitrile	50	18.47 ± 0.66		384
4-Nitrophenylacetonitrile	50	19.56 ± 0.58		384
3-Nitrophenylnitromethane	50	23.01 ± 0.42		384
4-Nitrophenylnitromethane	50	24.77 ± 0.59		384
(3-Nitrophenyl)phenylmethane	50	10.55 ± 0.55		384
(4-Nitrophenyl)phenylmethane	50	10.34 ± 0.55		384
3-Nitrotoluene	50	8.92 ± 0.47		383
4-Nitrotoluene	45	20		387
	50	10.8	18600	253, 293
		7.28 ± 0.24		383
4-*t*-Octylpyrocatechol	50	1.1		386
Perylene	50	28.9[f]		380
		28.3		381
Phenanthrene	50	0.087		381
		0.57	98.0	253, 382
	70	0.34		91
Phenol	45	0.019[b]		385
		0.024[c,g]		385
	50	0.012	21	253, 382
	60	0.06		91
(Deuterated)	45	0.002[b]		385
		0.005[c,g]		385
9-Phenylanthracene	50	27.3[f]		380, 381
Phenyl 3-nitrobenzenesulphonate	50	54.72 ± 0.47		384
Phenyl 4-nitrobenzenesulphonate	50	70.18 ± 1.36		384
Pyrene	50	1.15		381

TABLE 30—*continued*

Inhibitor	Temp. (°C)	k_z/k_p	k_z (l mole^{-1} sec^{-1})	Ref.
Pyrogallol	50	5.0 ± 1.0		386
Tetracene	50	862.0		381
Tetramethylbenzoquinone	45	90		193
1,3,5-Tetramethyldinitrobenzene	45	1.3		387
2,3,4,6-Tetramethylphenol	45	0.94[b]		385
		1.32[c,g]		385
(Deuterated)		0.075[b]		385
		0.085[c,g]		385
Tolunitrile	50	0.0039	6.7	253, 382
2,4,6-Tri-*t*-butylphenol	45	0.017 ± 0.004[b]		389
		0.021 ± 0.004[c]		389
2,4,6-Trimethylphenol	50	0.5		386
1,3,5-Trinitrobenzene	45	890		388
	50	404	760000	253, 382
1,3,5-Trinitrotoluene	45	890		388
Triphenylene	50	0.016		381
p-Xylene	50	0.0015	2.6	253, 382

[a] *N,N*-Dimethylformamide solution.
[b] From molecular weight data.
[c] From rate measurements.
[d] Assuming $k_p = 3.2 \times 10^7 \exp(-6300/RT)$ l mole^{-1} sec^{-1}.
[e] R in cal K^{-1} mole^{-1}.
[f] Average value.
[g] Assuming disproportionation termination.

Values of absolute rate coefficients for propagation and termination are presented in Table 26 (p. 238). Transfer constants to monomer are given in Table 27 (p. 240) and to polymer in Table 3 (p. 163). Table 28 (p. 241) gives transfer constants to initiator and transfer constants to solvents and other reagents are given in Table 29 (p. 242) (many of these values are apparent due to the occurrence of retardation). Data relating to retardation and inhibition are presented in Table 30 (p. 256).

9. Vinyl chloride

Poly(vinyl chloride) is insoluble in its monomer and most common solvents. Consequently, many kinetic studies of vinyl chloride polymerization have been concerned with heterogeneous systems. Homogeneous reaction can be achieved in certain solvents such as 1,2-dichloroethane, chlorobenzene and tetrahydrofuran, especially if reactions are only taken to low conversion.

Various groups of workers have studied homogeneous polymerizations of vinyl chloride, often with conflicting results including the observation of different dependences on monomer and initiator concentrations. These observations are a consequence of the high reactivity of the propagating radical and low reactivity of the monomer. Thus, chain transfer occurs readily and the transfer radicals reinitiate polymerization slowly and the polymerizations are very susceptible to retardation (see Chap. 2, Sect. 3). Additionally, the low reactivity of the monomer makes the polymerizations susceptible to primary-radical termination unless highly reactive radicals are used to initiate polymerization [390].

The rather meagre data available on rate coefficients for propagation and termination and for inhibition are included in Tables 34 and 38 which give data for miscellaneous monomers. Transfer constants to monomer are given in Table 31, and to solvents and other transfer agents in Table 32.

TABLE 31
Monomer transfer constants (C_M) for free-radical polymerization of vinyl chloride

Temp. (°C)	$10^4 C_M$	Solvent	Ref.
($C_M = 125 \exp[-7300/RT]$)[a]			79
25	3.2	Diethyl oxalate	391
	3.2	Ethylene dichloride	390
30	6.25		79
40	50		392
50	8.5		393
	11		394
	13.5		79
	6.4	Butyl acetate	393
	7.8	Butyl chloride	393
60	12.3		393
	10.8	Butyl acetate	393
	12.8	Butyl chloride	393
70	23.8		79

[a] Units of R in cal K^{-1} mole^{-1}.

TABLE 32
Transfer constants (C_s) of solvents and other transfer agents for free-radical polymerization of vinyl chloride

Solvent	Temp. (°C)	$10^2 C_s$	Comments	Ref.
Acetaldehyde	50	1.10		395
Butyraldehyde	50	5.80	1 kg cm^{-2}	369
		5.00	1000 kg cm^{-2}	369
		4.20	2000 kg cm^{-2}	369
		3.50	4000 kg cm^{-2}	369
Carbon tetrabromide	50	~5000		396
		470	[a], 1,2-Dichloroethane	397
Carbon tetrachloride	55	3.86		398
	105	3.84		398
	145	6.01		398
Chlorobenzene	25	0.074	[b]	390
Deuterotrichloroethylene	40	6.4 ± 1.5	Chlorobenzene	399
	45	5.3	Chlorobenzene	399
	50	8.3	Chlorobenzene	399
	55	9.1 ± 0.5	Chlorobenzene	399
1,2-Dichloroethane	25	0.035	[b]	390
	50	0.045		394
		0.05		400
		0.051		401
2,4-Dichloropentane	50	0.05		402
N,N-Dimethylaniline	50	27		155
Dodecyl vinyl ether	50	1.56		150
Tetrahydrofuran	25	0.0025	[b]	390
	40	0.3		392
	50	0.24		395
Trichloroethylene	40	4.6 ± 0.3	[c], Chlorobenzene	399
		24.0	[d], Chlorobenzene	399
	45	8.8 ± 1.0	[c], Chlorobenzene	399
		24.0	[d], Chlorobenzene	399
	50	9.2 ± 0.2	[c], Chlorobenzene	399
		24.0	[d], Chlorobenzene	399
	55	13.3 ± 0.9	[c], Chlorobenzene	399
		24.0	[d], Chlorobenzene	399
2,4,6-Trichloroheptane	50	0.05		402

[a] Apparent value due to retardation.
[b] Retardation taken into consideration.
[c] Calculated using Mayo equation.
[d] From end-group studies.

10. Ethylene

Even though ethylene is the parent of all vinyl monomers and its free-radical polymerization has been of considerable commercial and historical importance, the scientific details of the polymerization kinetics are often neglected in treatments of vinyl polymerizations. Undoubtedly this situation has arisen because commercial polymerizations are performed under high pressures and at high temperatures and are less amenable than other polymerizations to laboratory investigation. In addition, under normal circumstances polyethylene is insoluble in virtually all convenient solvents. However, the polymer and monomer can form a single fluid phase at the high pressures and temperatures used in industrial plants [403].

Recently, Ehrlich and Mortimer [404] have presented a detailed discussion of the phase diagrams and equilibria in polymer—monomer and polymer—monomer—diluent systems and of the kinetics of the polymerization in homogeneous media. These workers have assessed available data from reactions which have been performed under homogeneous conditions. They conclude that, under homogeneous conditions, the polymerization kinetics are fairly straightforward if such effects as transfer to initiator are avoided. They also conclude that, at least for temperatures below 200 °C disproportionation, termination and transfer to monomer are unimportant.

Many older studies of ethylene polymerizations have employed conditions under which the nature and composition of phases are indeterminate and kinetic data cannot be satisfactorily analyzed. Ehrlich and Mortimer [404] have discussed experimental methods for obtaining rate data under well-defined conditions and the analysis of such data. Although values of k_p and k_t have not been obtained under conditions of high pressure and temperature, meaningful values of these parameters have been obtained at lower temperatures and pressures. Laita and Machecek [405] found $k_p = 470$ l mole^{-1} sec^{-1} and $k_t = 1.05 \times 10^9$ l mole^{-1} sec^{-1} at 83 °C and Rabel and Ueberreiter [406] found $k_p = 191$ l mole^{-1} sec^{-1} and $k_t = 4.6 \times 10^8$ l mole^{-1} sec^{-1} at −20 °C.

Since the propagating radical is a highly-reactive unstabilized alkyl radical and polymerizations are carried out at high temperatures, chain transfer is important. In a recent series of publications Mortimer [408—415] has analyzed a considerable amount of transfer data to obtain meaningful transfer constants and these are presented in Table 33. Woodbrey and Ehrlich [407] have determined transfer constants to monomer (C_M) and quote $C_M < 3 \times 10^{-5}$ at 130 °C and $(7 \pm 2) \times 10^{-5}$ at 250 °C.

An important feature of ethylene polymerizations is transfer to polymer which occurs by an intramolecular "back-biting" process which leads

TABLE 33
Transfer constants (C_s) to various transfer agents for the free-radical polymerization of ethylene

Transfer Agent	Temp. (°C)	$10^2 C_s$	Pressure (atm)	Ref.
Acetic anhydride	130	1.30 ± 0.05[a]	1360	408
Acetone	130	1.68 ± 0.05[a]	1360	408
		1.8 ± 0.1[b]	1360	409
		1.65 ± 0.10[c]	1360	409
		1.6 ± 0.3[d]	1360	409
		1.6 ± 0.01[a]	2380	410
	200	2.82 ± 0.06	1360	411
Acetonitrile	130	1.1 ± 0.1	1360	412
Benzaldehyde	130	19.7 ± 0.7	1360	413
Benzene	130	0.09 ± 0.02	1360	412
1,1-Bis(dimethylamino)ethylene	130	10.7 ± 0.2	1360	413
1-Bromo-2-chloroethane	130	3.9 ± 0.2[a]	1360	408
Bromotrichloromethane	130	100 ± 10[a]	1360	408
Butane	130	0.45 ± 0.02[a]	2380	410
n-Butanethiol	130	580 ± 40	1360	413
t-Butanethiol	130	1500 ± 100	1360	413
Butanone	130	6.0 ± 0.5[a]	1360	408
		5.5 ± 0.5[a]	2380	410
	200	7.5 ± 0.4	1360	411
Butene-1	130	4.7 ± 0.2[a]	1360	408
	200	5.7 ± 0.4	1360	411
Butene-2	130	3.8 ± 0.4	1360	408
n-Butyraldehyde	130	32.5 ± 0.9	1360	413
γ-Butyrolactone	130	5.1 ± 0.2	1360	412
n-Butyl acetate	130	0.89 ± 0.07[a]	1360	408
t-Butyl acetate	130	0.40 ± 0.05[a]	1360	408
t-Butyl alcohol	130	0.01 ± 0.05	1360	412
n-Butylamine	130	2.2 ± 0.4[a]	1360	408
n-Butylbenzoate	130	1.4 ± 0.1[a]	1360	408
n-Butylisocyanate	130	2.12 ± 0.07[a]	1360	408
n-Butylisothiocyanate	130	7.5 ± 0.1[a]	1360	408
Carbon tetrachloride	130	90 ± 10	1360	412
Chloroform	130	27 ± 3	1360	412
Cumene	130	4.6 ± 0.6	1360	412
Cyclohexane	130	0.95 ± 0.03[a]	1360	408
		1.06 ± 0.06[b]	1360	409
		0.91 ± 0.06[c]	1360	409

TABLE 33—*continued*

Transfer Agent	Temp. (°C)	$10^2 C_s$	Pressure (atm)	Ref.
		0.9 ± 0.2[d]	1360	409
		0.90 ± 0.09[a]	2380	410
	200	1.9 ± 0.1	1360	411
Cyclopentane	130	1.26 ± 0.05[a]	1360	408
		1.09 ± 0.09	2380	410
	200	2.28 ± 0.09	1360	411
Cyclopropane	130	0.00	1360	412
n-Decane	130	1.24 ± 0.09	1360	412
Deuterobenzene	130	0.05 ± 0.03	1360	412
1,2-Dibromoethane	130	12.5 ± 0.4[a]	1360	408
Di-*n*-butylamine	130	10.7 ± 0.5[a]	1360	408
Di-*n*-butylphosphine	130	360 ± 10[a]	1360	408
		420 ± 10	1360	413
1,4-Dichlorbutene-2	130	41 ± 1[a]	1360	408
1,2-Dichloroethane	130	1.1 ± 0.2	1360	412
N,N-Diethylacetamide	130	1.25 ± 0.6[a]	1360	408
Diethyl sulphide	130	2.7 ± 0.04[a]	1360	408
Diisobutylene	130	2.7± 0.2	1360	412
N,N-Diisopropylacetamide	130	0.57 ± 0.06[a]	1360	408
Dimethoxymethane	130	0.73 ± 0.03[a]	1360	408
N,N-Dimethylacetamide	130	1.82 ± 0.05[a]	1360	408
Dimethylamine	130	18 ± 2	1360	412
N,N-Dimethylformamide	130	2.6 ± 0.2	1360	412
4,4-Dimethylpentene-1	130	1.75 ± 0.08[a]	1360	408
Dimethylphosphite	130	51 ± 3	1360	412
2,2-Dimethylpropane	130	0.08 ± 0.02	1360	412
Dimethylsulphoxide	130	0.11 ± 0.03	1360	412
1,4-Dioxan	130	0.32 ± 0.3	1360	412
Ethane	130	0.06 ± 0.05	1360	412
N-Ethylacetamide	130	1.15 ± 0.06[a]	1360	408
Ethyl acetate	130	0.45 ± 0.03[a]	1360	408
	200	1.21 ± 0.04	1360	411
Ethyl alcohol	130	0.75 ± 0.03[a]	1360	408
		0.68 ± 0.06[a]	2380	410
	200	1.35 ± 0.02	1360	411
Ethylbenzene	130	5.2 ± 0.02[a]	1360	408
		4.3 ± 0.3[a]	2380	410
	200	5.0 ± 0.4[a]	1360	411

References pp. 276–285

TABLE 33—*continued*

Transfer Agent	Temp. (°C)	$10^2 C_s$	Pressure (atm)	Ref.
Ethylbenzoate	130	0.55 ± 0.06[a]	1360	408
Ethylene oxide	130	0.06 ± 0.01	1360	412
2-Ethylhexene-1	130	33 ± 2	1360	412
Formaldehyde	130	5.6 ± 0.2	1360	413
n-Heptaldehyde	130	39 ± 1	1360	314
		26 ± 3	2380	413
	200	48 ± 2	1360	413
n-Heptane	130	0.8 ± 0.2	1360	412
n-Hexadecane	130	5.8 ± 0.3	1360	414
n-Hexane	130	0.67 ± 0.06	1360	412
Hydrogen	130	1.59 ± 0.08[a]	1360	408
	200	4.0 ± 0.2	1360	411
Isobutane	130	0.72 ± 0.03[a]	1360	408
	200	1.36 ± 0.05	1360	411
Isobutylene	130	2.06 ± 0.10	1360	412, 414
Isobutyronitrile	130	10.6 ± 0.6	1360	412
Isopropyl alcohol	130	1.44 ± 0.05[a]	1360	408
		1.2 ± 0.1[a]	2380	410
	200	2.34 ± 0.05	1360	411
Methane	130	0.00 ± 0.02	1360	412
N-Methylacetamide	130	0.61 ± 0.05[a]	1360	408
Methyl acetate	130	0.31 ± 0.04[a]	1360	408
Methyl alcohol	130	0.21 ± 0.05	1360	412
Methylamine	130	0.53 ± 0.05[a]	1360	408
Methyl benzoate	130	0.53 ± 0.05[a]	1360	408
3-Methyl-2-butanone	130	8.4 ± 0.6	1360	412
2-Methylbutene-1	130	5.2 ± 0.2	1360	412
2-Methylbutene-2	130	4.5 ± 0.7	1360	412
3-Methylbutene-1	130	12 ± 2	1360	412
2-Methyl-3-buten-2-ol	130	0.31 ± 0.03	1360	414
Methyl butyrate	130	2.2 ± 0.1[a]	1360	408
Methyl chloroacetate	130	11.2 ± 0.5[a]	1360	408
Methyl cyanoacetate	130	67 ± 2[a]	1360	408
Methyl 3-cyanopropionate	130	7.9 ± 0.2[a]	1360	408
Methylcyclohexane	130	1.1 ± 0.2	1360	412
Methyl formate	130	0.42 ± 0.05[a]	1360	408

TABLE 33—*continued*

Transfer Agent	Temp. (°C)	$10^2 C_s$	Pressure (atm)	Ref.
Methylisothiocyanate	130—135	3.8	1360	415
4-Methylpentene-1	130	3.1 ± 0.3[a]	1360	408
Methylvinyl sulphide	130	5.0 ± 0.2	1360	414
Octene-1	130	0.40 ± 0.2	1360	412
Perfluoropropane	130	0.03 ± 0.03	1360	412
Phenylisothiocyanate	130—135	5.8	1360	415
Phosphine	130	2070 ± 800	1360	413
Propane	130	0.31 ± 0.02	1360	409
		0.276 ± 0.011[a]	1360	410
	200	0.652 ± 0.002	1360	411
Propionaldehyde	130	33 ± 1	1360	413
		23 ± 3	2380	413
	200	28.3 ± 4	1360	413
Propylene	130	1.22 ± 0.08[a]	1360	408
	200	2.00 ± 0.06	1360	411
Sulphurhexafluoride	130	0.00	1360	412
Tetrahydrofuran	130	2.88 ± 0.06[a]	1360	408
	200	4.01 ± 0.09	1360	411
Tetramethylsilane	130	0.00	1360	412
Toluene	130	1.54 ± 0.05[a]	1360	408
		1.8 ± 0.1[a]	2380	410
	200	2.2 ± 0.1	1360	411
Tri-*n*-butylamine	130	8.2 ± 0.4[a]	1360	408
Tri-*n*-butylphosphine	130	45 ± 2[a]	1360	408
Tridecane	130	1.35 ± 0.10	1360	412
Trimethylamine	130	3.3 ± 0.1[a]	1360	408
2,2,4-Trimethylpentane	130	0.64 ± 0.15	1360	412
Triphenylphosphine	130	0.5 ± 0.2[a]	1360	408
Vinyl methyl ether	130	0.206 ± 0.009	1360	414
p-Xylene	130	3.17 ± 0.09[a]	1360	408
		3.4 ± 2[b]	1360	409
		3.0 ± 2[c]	1360	409
		2.6 ± 5[d]	1360	409
		4.0 ± 0.03[a]	2380	410
	200	4.34 ± 0.06	1360	411

[a] In propane; [b] 12 % propane; [c] 14.4 % propane; [d] 20 % propane.

to the formation of branches, including many short-chain branches. This process has been discussed in detail by Ehrlich and Mortimer [404] (also see Chap. 2, Sect. 2.6).

TABLE 34
Absolute rate coefficients and activation energies for free-radical polymerizations of miscellaneous monomers

Monomer	Temp. (°C)	k_p (l mole^{-1} sec^{-1})	E_p (kcal mole^{-1})	$10^{-6} k_t$ (l mole^{-1} sec^{-1})	E_t (kcal mole^{-1})	$10^6 k_p/k_t$	Comments	Ref.
Acrylamide	25	18000 ± 1500		14.5 ± 2.0			[a], aq.	416
Acrylic acid	23	650 ± 85		260 ± 27			[a], aq. NaOH, pH 7.9	417
		3150 ± 140		260 ± 26			[a], aq. NaOH+ 1.5 M NaCl, pH 7.9	417
		6600 ± 925		210 ± 20			[a], aq. NaOH, pH 11	417
		2500 ± 380		225 ± 23			[a], aq. NaOH, pH 13.6	417
Butadiene	10	8.4	9.3				[b]	418
Isoprene	5	2.8	9.8				[b]	10
Methacrylamide	25	800	3.7	16.5	4.0		[a], aq.	419
Methacrylic acid	23	670 ± 80		210 ± 20			[a], aq. NaOH, pH 8.8	417
		1950 ± 160		225 ± 23			[a], aq. NaOH, pH 13.6	417
Methacrylonitrile	20					1.04	[a]	420
	25	26	11.5	21	5.0		[a]	420
	30					1.51	[a]	420
		~21		~27				421
2-Methyl-5-vinyl pyridine	25	47		3.5		13.4	[c]	313

Vinyl bromide	($k_p = 4.8 \times 10^7 \exp[-7100/RT]$; $k_t = 2.8 \times 10^{11} \exp[-3900/RT]$)[d]							422
	−50	5.8		42			[a], THF	422
	−30	22.7		92			[a], THF	422
	20	570		385				423
	50	995		705				423
N-Vinylcarbazole	($k_p = 1.2 \times 10^6 \exp[-6900/RT]$)[d]							422
	−30	0.84		0.0092			[a]	422
	−10	2.2		0.11			[a]	422
	0					17.5	[a]	422
	10	6.0		0.306			[a]	422
	30					13.7	[a]	422
Vinyl chloride	25	6200	3.7	1100	4.2		[a], THF	424
		11000		2100		71	[a]	308
	50	11000		2100			[a], THF	424
Vinylidene chloride	15	2.3	25	0.023	40		[a], hexane	425
	25	8.6		0.175			[a], hexane	425
	35	36.8		1.80			[a], hexane	425
2-Vinylpyridine	25	186	8.0	33	5.0		[a]	426
4-Vinylpyridine	25	12		3			[c]	427
						19	[c]	313

[a] Rotating-sector technique; [b] Emulsion polymerization; [c] Direct use of pre- or after-effect; [d] R in cal K^{-1} $mole^{-1}$; THF, Tetrahydrofuran solution.

TABLE 35
Monomer transfer constants (C_M) for free-radical polymerizations of miscellaneous monomers

Monomer	Temp. (°C)	$10^4 C_M$	Ref.
Allyl acetate	80	176	429
		700	193
Allyl chloride	80	1600	193
N,N-Dimethylacrylamide	50	1.5	47
Ethyl-2-chloroacrylate	60	3.0	430
Maleic anhydride	75	750	431
Methacrylonitrile	($C_M = 3.0 \exp[-5673/RT]$)[a]		420
	25	2.08	420
	60	5.81	420
	70	8.00	420
	80	10.05	420
3-Methyl-3-buten-2-one	80	4.00	432
Vinyl benzoate	60	6.0	433, 434
	80	7.0	435
	95	2.0	433
		4.0	436
Vinyl butyrate	50	26.7	350
	80	22.3	437
Vinyl decanoate	50	45.5	350
Vinyl hexanoate	80	36	437
Vinyl isobutyrate	80	46	437
Vinyl 4-methylvalerate	80	24.8	437
Vinyl laurate	50	45.5	350
1-Vinylnaphthalene	50	290	176
	60	310	176
	70	300	176
Vinyl propionate	50	48.9	350
4-Vinylpyridine	25	6.7	427
N-Vinylpyrrolidone	20	4.0	79
Vinyl salicylate	70	80	438
Vinyl stearate	50	69.8	350
N-Vinylurethane	60	0.25	439

[a] Units of R, cal K^{-1} $mole^{-1}$.

TABLE 36
Transfer constants to initiator (C_I) for free-radical polymerizations of miscellaneous monomers

Monomer	Initiator	Temp. (°C)	C_I	Ref.
Acrylamide	Bisulphite ion	75	0.17[a]	440
Acrylonitrile	2,2′-Azo-bis-isobutyronitrile	50	0[b]	319
		60	0	321
Ethylene	1,1-Dimethylazoethane	83	0.5	405
			0.51	405
Maleic anhydride	Benzoyl peroxide	75	2.63	431
Methacrylonitrile	2,2′-Azo-bis-isobutyronitrile	60	0	420
3-Methyl-3-buten-2-one	Benzoyl peroxide	80	0.0509	432
Vinyl benzoate	2,2′-Azo-bis-isobutyronitrile	80	0	435
	Benzoyl peroxide	80	0.0527	435

[a] Aqueous solution; [b] SO_2 solution.

11. Miscellaneous monomers

Although a wide variety of vinyl monomers are known to undergo free-radical polymerization and copolymerization, apart from the few monomers considered in previous sections very few definitive kinetic data are available. Indeed, so few data are available in some cases that it is by no means certain that the polymerization kinetics are simple and that due allowance has been made for complicating effects; the presence of complicating features may not yet have been recognized. At the present time it is impossible from the data available to make general comments on and comparisons of the kinetics and mechanisms of the polymerization of the less well-known monomers.

Published values of k_p and k_t (including data for vinyl chloride polymerizations) are given in Table 34. Data for butadiene and isoprene are only available from studies of emulsion polymerization since the monomers do not polymerize to high polymer in solution. Transfer constants to monomer, polymer, initiators and other transfer agents are given in Tables 35, 3, 36 and 37, respectively. Data on inhibition and retardation are presented in Table 38.

TABLE 37
Transfer constants (C_s) to various transfer agents for the free-radical polymerizations of miscellaneous monomers

Monomer	Transfer Agent	Temp. (°C)	$10^4 C_s$	Comments	Ref.
Acenaphthylene	Carbon tetrabromide	60	7.8	1 atm	445
		60	6.0	1400 atm	445
	Tetrachloroethylene	60	29	1 atm	445
			30	2800 atm	445
	Toluene	60	6.8	1 atm	445
			3.7	1400 atm	445
			2.0	2800 atm	445
Acrylamide	Isopropyl alcohol	50	19	aq. soln.	184
Allyl acetate	Benzene	80	21.0		104
	Carbon tetrachloride	100	2×10^4		137
	Phosphorous trichloride	40	10^4		178
Allyl alcohol	Piperidine	~125	880		441
	Pyrrolidone	120	1260		441
Allyl chloride	Carbon tetrachloride	100	4800		137
Allyl cyanide	Piperidine	~120	890		441
Diethyl maleate	*N*-Allylstearimide	90	22.4	a	97
Hexene-1	Cyclohexanol	125—135	390		442
	Ethyl alcohol	30—35	170		442
	Piperidine	125	1410		441
Methacrylonitrile	Carbon tetrabromide	100	900	DMF	133
		120	1000	DMF	133
	Toluene	60	0.996		179

3-Methyl-3-buten-2-one	Benzene	80	2.489	432
	Ethylbenzene	80	6.934	432
	Methylcyclohexane	80	0.500	432
	Toluene	80	3.282	432
2-Methyl-1-propene	Carbon tetrachloride	100	1.7×10^5	137
Octene-1	2-Aminopropane	120	950	441
	Butyl alcohol	117—118	270	442
	sec-Butyl alcohol	117—118	520	442
	Butylamine	125	620	441
	Cyclohexylamine	125	810	441
	Ethyl alcohol	115—118	230	442
	Hexylamine	125	700	441
	Isopropyl alcohol	30—35	630	442
		120—125	520	442
	Methyl alcohol	116—120	110	442
	N-Methylpiperidine	125	530	441
	Piperidine	120	2590	441
Propene	Carbon tetrachloride	55	1.11×10^6	428
		100	$(5—10) \times 10^4$	137
		105	86×10^4	428
		145	73×10^4	428
	Chloroform	100	10300	137
Vinyl benzoate	Benzene	60	1.5	434
		80	0.4	435
	2-Butanone	80	29.0	435
	Carbon tetrachloride	80	730	435

TABLE 37—*continued*

Monomer	Transfer Agent	Temp. (°C)	$10^4 C_s$	Comments	Ref.
	Chloroform	80	105		435
	Cyclohexane	80	2.3		435
	Isopropyl benzoate	60	1.0		434
Vinyl butyrate	Benzene	80	3.28		443
Vinyl hexanoate	Benzene	80	4.9		443
Vinylidene chloride	Phosphorous trichloride	60	14×10^4		178
Vinyl isobutyrate	Benzene	80	4.49		443
Vinyl 4-methylvalerate	Benzene	80	6.2		443
1-Vinylnaphthalene	Naphthalene	60	30—50		176
Vinyl nonanoate	Toluene	50	11.1		378
		60	13.9		378
		70	15.2		378
Vinyl octadecanoate	Toluene	50	20.7		378
		60	20.9		378
		70	21.6		378
Vinyltriethyoxysilane	Diethylphosphite	150	5000		444
N-Vinylurethan	Acetone	60	8.5		439
	Benzene	60	1.25		439

[a] Apparent value, retardation observed.
DMF, *N,N*-Dimethylformamide solution.

TABLE 38
Kinetic parameters for inhibition of free-radical polymerization of miscellaneous vinyl monomers

Monomer	Temp. (°C)	Inhibitor	k_z/k_p	k_z (l $mole^{-1}$ sec^{-1})	Comments	Ref.
Allyl acetate	80	*p*-Benzoquinone	52			193
	80	2,3,5,6-Tetrachloro-*p*-benzoquinone	160			193
	80	2,3,5,6-Tetramethyl-*p*-benzoquinone	4.1			193
	80	2,3,5-Trichloro-*p*-benzoquinone	55			193
N-*t*-Butylacrylamide	25	Iron(III) chloride	1670		Methanol	446
Methacrylamide	15	Iron(III) chloride		32 ± 8	[a], aq.	447
				(47.4 ± 1.4) x 10^3	[b], aq.	447
	18			(1.16 ± 0.19) x 10^3	[c], aq.	447
	25			53 ± 12	[a], aq.	447
				(1.73 ± 0.32) x 10^3	[c], aq.	447
				(6.1 ± 1.7) x 10^3	[b], aq.	447
	35			78 ± 20	[a], aq.	447
				(3.25 ± 0.45) x 10^3	[c], aq.	447
				(8.7 ± 2.4) x 10^3	[b], aq.	447
Methacrylonitrile	60	Copper(II) chloride	4500	~9 x 10^5	DMF	448
	60	Iron(III) chloride	3.08	620	DMF	198, 199
Vinyl chloride	40	Iron(III) chloride	980		DMF	449
	50		870		DMF	449
	60		855		DMF	449
	70		705		DMF	449

[a] Species responsible for termination $[Fe(aq.)]^{3+}$.
[b] Species responsible for termination $[Fe(aq.)\ monomer]^{3+}$.
[c] Species responsible for termination $[Fe(aq)OH]^{2+}$.
DMF, *N*,*N*-Dimethylformamide solution.

REFERENCES

1 G. M. Burnett, G. G. Cameron and S. N. Joiner, J. Chem. Soc., Faraday Trans. 1, 69 (1973) 322.
2 G. C. Eastmond and D. G. Phillips, unpublished data.
3 K. Yokota, H. Tomioka and A. Tazumi, J. Macromol. Sci., Chem., 6 (1972) 1337.
4 C. M. Starks, Free Radical Telomerization, Academic Press, New York, 1974.
5 J. Brandrup and E. H. Immergut (Eds.), Polymer Handbook, Interscience, New York, 1965.
6 G. G. Cameron and N. Grassie, Makromol. Chem., 51 (1962) 130.
7 S. E. Bresler, E. N. Kazbekov, V. N. Formichev and V. N. Shadrin, Makromol. Chem., 157 (1972) 167.
8 C. H. Bamford and M. J. S. Dewar, Proc. R. Soc. London, Ser. A, 192 (1948) 308.
9 M. S. Matheson, E. E. Auer, E. B. Bevilacqua and E. J. Hart, J. Am. Chem. Soc., 73 (1951) 1700.
10 M. Morton, M. P. Salatiello and H. Landfield, J. Polym. Sci., 8 (1952) 279.
11 G. M. Burnett, Trans. Faraday Soc., 46 (1950) 772.
12 T. G. Majury and H. W. Melville, Proc. R. Soc. London, Ser. A, 205 (1951) 496.
13 N. Grassie and H. W. Melville, Proc. R. Soc. London, Ser. A, 207 (1951) 285.
14 K. Ueberreiter and G. Sorge, Z. Phys. Chem. (Frankfurt am Main), 13 (1957) 158.
15 W. G. Barb, Proc. R. Soc. London, Ser. A, 212 (1952) 177.
16 S. Fujii, Bull. Chem. Soc. Jpn., 27 (1954) 216.
17 G. G. Cameron and J. Cameron, Polymer, 14 (1973) 107.
18 H. W. Melville and L. Valentine, Trans. Faraday Soc., 46 (1950) 210.
19 H. Miyama, Bull. Chem. Soc. Jpn., 29 (1956) 711, 715.
20 E. A. Nicholson and R. G. W. Norrish, Discuss. Faraday Soc., 22 (1956) 104.
21 M. Kinoshita and M. Imoto, Kobunshi Kagaku, 20 (1963) 231.
22 Y. Ogo, M. Yokawa and T. Imoto, Makromol. Chem., 171 (1973) 123.
23 K. P. Paoletti and F. W. Billmeyer, J. Polym. Sci., Part A, 2 (1964) 2049.
24 E. Bartholomé, H. Gerrens, R. Herbeck and H. Weitz, Z. Elektrochem., 60 (1956) 334.
25 H. Gerrens, Ber. Bunsenges. Phys. Chem., 67 (1963) 741.
26 G. Henrici-Olivé and S. Olivé, Makromol. Chem., 37 (1960) 71.
27 A. V. Tobolsky and J. Offenbach, J. Polym. Sci., 16 (1955) 311.
28 G. M. Burnett, Q. Rev., 4 (1950) 292.
29 G. Henrici-Olivé and S. Olivé, Fortschr. Hochpolym. Forsch., 2 (1961) 496.
30 N. G. Saha, U. S. Nandi and S. R. Palit, J. Chem. Soc, (1958) 12.
31 G. V. Schulz, G. Henrici and S. Olivé, Z. Elektrochem., 60 (1956) 296.
32 G. Henrici-Olivé and S. Olivé, Makromol. Chem., 53 (1962) 122.
33 G. C. Bhadhuri and U. S. Nandi, Makromol. Chem., 128 (1969) 183.
34 B. Baysal and A. V. Tobolsky, J. Polym. Sci., 8 (1952) 529.
35 D. H. Johnson and A. V. Tobolsky, J. Am. Chem. Soc., 74 (1952) 938.
36 T. Manabe, T. Utsumi and S. Okamura, J. Polym. Sci., 58 (1962) 121.
37 F. R. Mayo, R. A. Gregg and M. S. Matheson, J. Am. Chem. Soc., 73 (1951) 1691.
38 E. P. Bonsall, L. Valentine and H. W. Melville, J. Polym. Sci., 7 (1951) 39.
39 M. R. Gopalan and M. Santhappa, J. Polym. Sci., 25 (1957) 333.
40 T. E. Ferrington and A. V. Tobolsky, J. Am. Chem. Soc., 77 (1955) 4510.
41 J. W. Breitenbach and A. Schindler, Monatsh. Chem., 88 (1957) 53.

42 K. W. Doak, M. A. Deahl and I. H. Christmas, Abstracts 137th Am. Chem. Soc. Meeting, Cleveland, 1960.
43 V. V. Zaitseva, V. D. Yenal'ev and A. I. Yurzhenko, Polym. Sci., U.S.S.R., 9 (1967) 2209; Vysokomol. Soedin., Ser. A, 9 (1967) 1958.
44 J. Ulbricht, Z. Phys. Chem., 221 (1962) 346.
45 G. E. Ham, J. Polym. Sci., 21 (1956) 337.
46 R. A. Hayes, J. Polym. Sci., 13 (1954) 583.
47 A. M. North and A. M. Scallan, Polymer, 5 (1964) 447.
48 D. Lim and O. Wichterle, J. Polym. Sci., 29 (1958) 579.
49 T. G. Fox and S. Gratch, Ann. N.Y. Acad. Sci., 57 (1953) 367.
50 G. Henrici-Olivé, S. Olivé and G. V. Schulz, Makromol. Chem., 23 (1957) 207.
51 G. V. Schulz, G. Henrici and S. Olivé, J. Polym. Sci., 17 (1955) 45.
52 M. Morton and I. Piirma, J. Am. Chem. Soc., 80 (1958) 5596.
53 G. Henrici-Olivé, S. Olivé and G. V. Schulz, Z. Phys. Chem. (Frankfurt am Main), 20 (1959) 176.
54 H. Kaemmerer and F. Rocaboy, C. R. Acad. Sci., 256 (1963) 4440.
55 H. Kaemmerer and F. Rocaboy, Makromol. Chem., 72 (1964) 76.
56 M. Cantow, G. Meyerhoff and G. V. Schulz, Makromol. Chem., 49 (1961) 1.
57 J. W. Breitenbach, O. F. Olaj and A. Schindler, Monatsh. Chem., 91 (1960) 205.
58 S. Okamura and K. Katagiri, Makromol. Chem., 28 (1958) 177.
59 J. C. Bevington, G. M. Guzman and H. W. Melville, Proc. R. Soc. London, Ser. A, 221 (1954) 437.
60 G. Dixon-Lewis, Proc. R. Soc. London, Ser. A, 198 (1949) 510.
61 D. J. Stein and G. V. Schulz, Makromol. Chem., 52 (1962) 249.
62 S. Nozakura, Y. Morishima and S. Murahashi, J. Polym. Sci., Part A-1, 10 (1972) 2853.
63 R. Autrata and J. Müller, Collect. Czech. Chem. Commun., 24 (1959) 3442.
64 M. Morton and I. Piirma, J. Polym. Sci., Part A, 1 (1963) 3043.
65 H. Nakamoto, Y. Ogo and T. Imoto, Makromol. Chem., 111 (1968) 93.
66 S. Imoto, J. Ukida and T. Kominami, Kobunshi Kagaku, 14 (1957) 127, 384.
67 D. J. Stein, Makromol. Chem., 76 (1964) 170.
68 G. V. Schulz and L. Roberts-Nowakowska, Makromol. Chem., 80 (1964) 36.
69 G. V. Schulz and D. J. Stein, Makromol. Chem., 52 (1962) 1.
70 S. Imoto and T. Kominami, Kobunshi Kagaku, 15 (1958) 60.
71 S. Okamura and Y. Yamashita, Kaken Kouenshyu, 18 (1946) 46.
72 O. L. Wheeler, E. Lavin and R. N. Crozier, J. Polym. Sci., 9 (1952) 157.
73 D. Lim and M. Kolinsky, J. Polym. Sci., 53 (1961) 173.
74 W. Cooper, J. Chem. Soc., (1952) 2408.
75 A. I. Lowell and J. R. Price, J. Polym. Sci., 43 (1960) 1.
76 W. A. Pryor and T. R. Fiske, Macromolecules, 2 (1969) 62.
77 J. A. May and W. B. Smith, J. Phys. Chem., 72 (1968) 2993.
78 J. W. Breitenbach and A. Schindler, Monatsh. Chem., 83 (1952) 724.
79 J. W. Breitenbach, Makromol. Chem., 8 (1952) 147.
80 C. Walling and L. Heaton, J. Am. Chem. Soc., 87 (1965) 38.
81 C. Walling and Y. Chang, J. Am. Chem. Soc., 76 (1954) 4878.
82 V. S. Vaidhyanathan and M. Santhappa, Makromol. Chem., 16 (1955) 140.
83 W. A. Pryor and G. L. Kaplan, J. Am. Chem. Soc., 86 (1964) 4234.
84 W. A. Pryor, D. M. Huston, T. R. Fiske, T. L. Pickering and E. Ciuffarin, J. Am. Chem. Soc., 86 (1964) 4237.
85 W. A. Pryor, J. Phys. Chem., 67 (1963) 519.
86 W. A. Pryor and T. L. Pickering, J. Am. Chem. Soc., 84 (1962) 2705.
87 W. A. Pryor, A. Lee and C. E. Witt, J. Am. Chem. Soc., 86 (1964) 4229.
88 J. C. Bevington and A. Wahid, Polymer, 3 (1962) 585.

89 B. Baysal, J. Polym. Sci., 33 (1958) 381.
90 W. A. Pryor and E. P. Pultinas, J. Am. Chem. Soc., 85 (1963) 133.
91 J. T. Clarke, R. O. Howard and W. H. Stockmayer, Makromol. Chem., 44/46 (1961) 427.
92 R. B. Seymour and V. Patel, J. Macromol. Sci., Chem., 7 (1973) 961.
93 R. A. Gregg and F. R. Mayo, J. Am. Chem. Soc., 75 (1953) 3530.
94 S. L. Kapur, J. Polym. Sci., 11 (1953) 399.
95 J. A. Gannon, E. M. Fettes and A. V. Tobolsky, J. Am. Chem. Soc., 74 (1952) 1854.
96 Quoted in ref. 97, calculated from data in ref. 98.
97 E. F. Jordan, B. Artymyshyn and A. N. Wrigley, J. Polym. Sci., Part A-1, 6 (1968) 575.
98 T. Alfrey and J. G. Harrison, J. Am. Chem. Soc., 68 (1946) 299.
99 F. Hrabak and L. Jiresova, Collect. Czech. Chem. Commun., 26 (1961) 1283.
100 G. Platau, F. R. Eirich, R. B. Mesrobian and A. E. Woodward, J. Polym. Sci., 39 (1959) 357.
101 R. A. Gregg and F. R. Mayo, Discuss. Faraday Soc., 2 (1947) 328.
102 R. N. Chadha and G. S. Misra, Curr. Sci. (India), 24 (1955) 26.
103 G. S. Misra and R. N. Chadha, Makromol. Chem., 23 (1957) 134.
104 K. Nozaki, Discuss. Faraday Soc., 2 (1947) 337.
105 H. Kawakami, N. Mori, K. Kawashima and M. Sumi. Kobunshi Kagaku, 20 (1963) 408.
106 F. R. Mayo, J. Am. Chem. Soc., 65 (1943) 2324.
107 V. A. Dinaburg and A. A. Vansheidt, Zh. Obshch. Khim., 24 (1954) 840.
108 R. Pierson, A. Costanza and A. Weinstein, J. Polym. Sci., 17 (1955) 221.
109 R. Hiddema, Proefschrift Rijks Universität Gröningen, 1953.
110 R. A. Bird, G. A. Harpell and K. E. Russell, Can. J. Chem., 40 (1952) 701.
111 M. P. Godsay, G. A. Harpell and K. E. Russell, J. Polym. Sci., 57 (1962) 641.
112 A. J. Costanza, R. J. Coleman, R. M. Pierson, C. S. Marvel and C. King, J. Polym. Sci., 17 (1955) 319.
113 E. Schonfeld and P. Waltcher, J. Polym. Sci., 35 (1959) 536.
114 T. Ostu and K. Nayatani, Makromol. Chem., 27 (1955) 149.
115 J. C. Saam and D. J. Gordon, J. Polym. Sci., Part A-1, 8 (1970) 2509.
116 F. R. Mayo, J. Am. Chem. Soc., 75 (1953) 6133.
117 T. Yamamoto and T. Otsu, J. Polym. Sci., 7 (1969) 1279.
118 K. Higashiura and M. Iowa, J. Polym. Sci., Part A-1, 6 (1968) 1857.
119 C. A. Barson, R. R. Mather and J. C. Robb, Trans. Faraday Soc., 66 (1970) 2585.
120 D. A. J. Harker, R. A. M. Thomson and I. R. Walters, Trans. Faraday Soc., 67 (1971) 3057.
121 K. Katagiri, K. Uno and S. Okamura, J. Polym. Sci., 17 (1955) 142.
122 R. A. M. Thomson, I. R. Walters and J. R. King, J. Polym. Sci., Part B, 10 (1972) 63.
123 M. H. Jones, Can. J. Chem., 34 (1956) 108.
124 C. Sivertz, J. Phys. Chem., 63 (1959) 34.
125 L. A. Wall and D. W. Brown, J. Polym. Sci., 14 (1954) 513.
126 C. Walling, J. Am. Chem. Soc., 70 (1948) 2561.
127 J. Guillot and A. Guyot, C. R. Acad. Sci., Ser. C, 266 (1968) 1209.
128 A. P. Titov and I. A. Livshits, Zh. Obshch. Khim., 29 (1959) 1605.
129 M. Santhappa and V. S. Vaidhyanathan, Curr. Sci. (India), 23 (1954) 259.
130 M. Morton, J. A. Cala and I. Piirma, J. Am. Chem. Soc., 78 (1956) 5394.
131 J. P. Fischer and W. Lüders, Makromol. Chem., 155 (1972) 239.
132 T. Otsu, K. Tsuda and N. Kita, Mem. Fac. Eng., Osaka City Univ., 7 (1965) 95.
133 N. Fuhrman and R. B. Mesrobian, J. Am. Chem. Soc., 76 (1954) 3281.
134 R. A. M. Thomson and I. R. Walters, Trans. Faraday Soc., 67 (1971) 3046.

135 J. W. Breitenbach and H. Karlinger, Monatsh. Chem., 82 (1951) 245.
136 R. A. Gregg and F. R. Mayo, J. Am. Chem. Soc., 70 (1954) 2373.
137 F. M. Lewis and F. R. Mayo, J. Am. Chem. Soc., 76 (1954) 457.
138 C. Walling and J. J. Pellon, J. Am. Chem. Soc., 79 (1957) 4776.
139 M. H. George and P. F. Onyon, Trans. Faraday Soc., 59 (1963) 134.
140 M. H. George and P. F. Onyon, Trans. Faraday Soc., 59 (1963) 1390.
141 F. R. Mayo, J. Am. Chem. Soc., 70 (1948) 3689.
142 M. P. Godsay, D. H. Lohmann and K. E. Russell, Chem. Ind. (London), (1959) 1603.
143 Y. Minoura and Y. Enomoto, J. Polym. Sci., Part A-1, 6 (1968) 13.
144 G. Machell and G. N. Richards, J. Chem. Soc., (1961) 3308.
145 Y. Minoura, T. Sugimura and T. Hirahara, Kogyo Kagaku Zasshi, 70 (1967) 357.
146 T. Huff and E. Perry, J. Polym. Sci., Part A, 1 (1963) 1553.
147 A. V. Tobolsky and B. Baysal, J. Am. Chem. Soc., 75 (1959) 1757.
148 E. Perry, J. Polym. Sci., 54 (1961) S46.
149 L. A. Miller and V. Stannett, J. Polym. Sci., Part A-1, 7 (1969) 3159.
150 G. Akazome, S. Sakai and K. Maurai, Kogyo Kagaku Zasshi, 63 (1960) 592.
151 R. A. Gregg, D. M. Alderman and F. R. Mayo, J. Am. Chem. Soc., 70 (1948) 3740.
152 R. G. W. Norrish and J. P. Simons, Proc. R. Soc. London, Ser. A, 251 (1959) 4.
153 J. Ulbricht, Faserforsch. Textiltech, 11 (1960) 62.
154 M. H. George, J. Polym. Sci., Part A, 2 (1964) 3169.
155 M. Imoto, T. Otsu, T. Ota, H. Takatsugi and M. Matsuda, J. Polym. Sci., 22 (1956) 137.
156 G. S. Misra, R. C. Rastogi and V. P. Gupta, Makromol. Chem., 50 (1961) 72.
157 P. W. Allen, F. M. Merrett and J. Scanlan, Trans. Faraday Soc., 54 (1958) 715.
158 J. Scanlan, Trans. Faraday Soc., 50 (1954) 756.
159 S. N. Gupta and U. S. Nandi, J. Polym. Sci., Part A-1, 8 (1970) 1493.
160 J. C. Bevington and H. G. Troth, Trans. Faraday Soc., 59 (1963) 1348.
161 T. Otsu, Y. Kinoshita and M. Imoto, Makromol. Chem., 73 (1964) 225.
162 E. Y. C. Chang and C. C. Price, J. Am. Chem. Soc., 83 (1961) 4650.
163 I. Kar, B. M. Mandal and S. R. Palit, J. Polym. Sci., Part A-1, 7 (1969) 2829.
164 J. W. Breitenbach, Z. Elektrochem., 60 (1956) 286.
165 G. P. Scott and J. C. Wang, J. Org. Chem., 28 (1963) 1314.
166 B. R. Bhattacharya and U. S. Nandi, Makromol. Chem., 149 (1971) 231.
167 A. C. Toohey and K. E. Weale, Trans. Faraday Soc., 58 (1962) 2439.
168 W. A. Pryor and T.-L. Huang, Macromolecules, 2 (1969) 70.
169 F. R. Mayo, J. Am. Chem. Soc., 76 (1954) 5392.
170 P. Ghosh, J. Polym. Sci., Part B, 7 (1969) 293.
171 T. Huff and E. Perry, J. Am. Chem. Soc., 82 (1960) 4277.
172 Y. Minoura, N. Yasumoto and T. Ishii, Kogyo Kagaku Zasshi, 65 (1962) 1299.
173 Y. Minoura, N. Yasumoto and T. Ishii, Makromol. Chem., 71 (1964) 159.
174 E. F. Jordan, B. Artymyshyn and A. N. Wrigley, J. Polym. Sci., Part A-1, 7 (1969) 2605.
175 G. Smets and L. De Haes, Bull. Soc. Chim. Belg., 59 (1950) 13.
176 S. Loshaek, E. Broderick and P. Bernstein, J. Polym. Sci., 39 (1959) 223.
177 J. J. Pellon, J. Polym. Sci., 43 (1960) 537.
178 J. R. Little and P. F. Hartman, J. Am. Chem. Soc., 88 (1966) 96.
179 C. H. Bamford, A. D. Jenkins and R. Johnston, Trans. Faraday Soc., 55 (1959) 418.
180 C. H. Bamford and E. F. T. White, Trans. Faraday Soc., 52 (1956) 716.
181 R. N. Chadha and G. S. Misra, Makromol. Chem., 14 (1954) 97.
182 R. N. Chadha and G. S. Misra, Indian J. Phys., 28 (1954) 37.
183 R. N. Chadha and G. S. Misra, Curr. Sci. (India), 24 (1955) 26.

184 T. J. Suen, A. M. Schiller and W. N. Russell, Polymerization and Polycondensation Processes, Advances in Chemistry Series, No. 34, Am. Chem. Soc., Wash., D.C., 1962, p. 217.
185 A. C. Toohey and K. E. Weale, Trans. Faraday Soc., 58 (1962) 2446.
186 J. Curtice, H. Gilman and G. S. Hammond, J. Am. Chem. Soc., 79 (1957) 4754.
187 A. T. Coscia, R. L. Kugel and J. J. Pellon, J. Polym. Sci., 55 (1961) 303.
188 F. Tüdös, I. Kende and M. Azori, J. Polym. Sci., 53 (1961) 17; J. Polym. Sci., Part A, 1 (1963) 1353, 1369.
189 J. C. Bevington and H. G. Troth, Trans. Faraday Soc., 58 (1962) 2005.
190 J. L. Ihrig and S. P. Sood, J. Polym. Sci., Part A, 3 (1965) 1573.
191 F. Tüdös and L. Simandi, Vysokmol. Soedin., 4 (1962) 1271.
192 J. C. Bevington, N. A. Ghanem and H. W. Melville, J. Chem. Soc., (1955) 2822.
193 P. D. Bartlett, G. S. Hammond and H. Kwart, Discuss. Faraday Soc., 2 (1947) 342.
194 W. I. Bengough and T. O'Neill, Trans. Faraday Soc., 64 (1968) 1014.
195 F. Tüdös, L. Simandi and M. Azori, Vysokomol. Soedin., 4 (1962) 1431.
196 F. Tüdös and L. Simandi, Vysokomol. Soedin., 4 (1962) 1425.
197 F. Tüdös, J. Kende and M. Azori, Vysokomol. Soedin., 4 (1962) 1262.
198 C. H. Bamford, A. D. Jenkins and R. Johnston, Proc. R. Soc. London, Ser. A, 293 (1957) 214.
199 C. H. Bamford, A. D. Jenkins and R. Johnston, J. Polym. Sci., 29 (1958) 355.
200 N. N. Dass and M. H. George, J. Polym. Sci., Part A-1, 7 (1969) 269.
201 G. Henrici-Olivé and S. Olivé, Makromol. Chem., 24 (1957) 64.
202 M. Kinoshita, Kobunshi Kagaku, 20 (1963) 237.
203 D. W. E. Axford, Proc. R. Soc. London, Ser. A, 197 (1949) 374.
204 W. Kern and D. Braun, Makromol. Chem., 27 (1958) 23.
205 D. Braun, T.-O. Ahn and W. Kern, Makromol. Chem., 53 (1962) 154.
206 C. Walling and E. R. Briggs, J. Am. Chem. Soc., 68 (1946) 1141.
207 C. H. Bamford and M. J. S. Dewar, Discuss. Faraday Soc., 2 (1947) 310.
208 C. H. Bamford and M. J. S. Dewar, Proc. R. Soc. London, Ser. A, 197 (1959) 356.
209 M. H. Mackay and H. W. Melville, Trans. Faraday Soc., 46 (1950) 63.
210 G. V. Schulz and G. Harborth, Makromol. Chem., 1 (1947) 106.
211 S. E. Bresler, E. N. Kazbekov and V. N. Shadrin, Makromol. Chem., 175 (1974) 2875.
212 J. Hughes and A. M. North, Trans. Faraday Soc., 60 (1964) 960.
213 G. V. Schulz and J. P. Fischer, Makromol. Chem., 107 (1967) 253.
214 M. S. Matheson, E. E. Auer, E. B. Bevilacqua and E. J. Hart, J. Am. Chem. Soc., 71 (1949) 497.
215 W. J. Bengough and H. W. Melville, Proc. R. Soc. London, Ser. A, 225 (1954) 330.
216 G. M. Burnett, Mechanism of Polymer Reactions, Interscience, New York, 1954.
217 P. Hayden and H. W. Melville, J. Polym. Sci., 43 (1960) 201.
218 M. H. Mackay and H. W. Melville, Trans. Faraday Soc., 45 (1949) 323.
219 B. R. Chinmayanandam and H. W. Melville, Trans. Faraday Soc., 50 (1954) 73.
220 C. H. Bamford and S. Brumby, Makromol. Chem., 105 (1967) 122.
221 G. M. Burnett, G. G. Cameron and M. M. Zafar, Eur. Polym. J., 6 (1970) 823.
222 C. H. Bamford and S. Brumby, Chem. Ind. (London), (1969) 1020.
223 G. V. Schulz, G. Henrici-Olivé and S. Olivé, Z. Phys. Chem. (Frankfurt am Main), 27 (1960) 1.
224 L. Valentine, quoted in ref. 216.
225 S. Fujii, S. Tanaka and S. Sutani, J. Polym. Sci., 20 (1956) 586.
226 S. W. Benson and A. M. North, J. Am. Chem. Soc., 81 (1959) 1339.
227 A. M. North and D. Postlethwaite, Trans. Faraday Soc., 61 (1966) 2843.

228 K. Yokota, M. Kani and Y. Ishii, J. Polym. Sci., Part A-1, 6 (1968) 1325.
229 M. Yokawa, Y. Ogo and T. Imoto, Makromol. Chem., 175 (1974) 179.
230 W. I. Bengough and H. W. Melville, Proc. R. Soc. London, Ser. A, 249 (1959) 455.
231 J. L. O'Brien and F. Gornick, J. Am. Chem. Soc., 77 (1955) 4757.
232 J. C. Bevington, H. W. Melville and R. P. Taylor, J. Polym. Sci., 14 (1954) 463.
233 M. M. Zafar, Makromol. Chem., 157 (1972) 219.
234 S. R. Palit, U. S. Nandi and N. G. Saha, J. Polym. Sci., 14 (1954) 295.
235 N. G. Saha, U. S. Nandi and S. R. Palit, J. Chem. Soc., (1958) 7.
236 N. G. Saha, U. S. Nandi and S. R. Palit, J. Chem. Soc. (1956) 427.
237 G. C. Badhuri and U. S. Nandi, Makromol. Chem., 128 (1969) 176.
238 G. Ayrey and A. C. Haynes, Makromol. Chem., 175 (1974) 1463.
239 S. Basu, J. N. Sen and S. R. Palit, Proc. R. Soc. London, Ser. A, 202 (1950) 485.
240 K. Yokoto and M. Itoh, J. Polym. Sci., Part B, 6 (1968) 825.
241 R. N. Chadha, J. S. Shukla and G. S. Misra, Trans. Faraday Soc., 53 (1957) 240.
242 C. H. Bamford and S. Brumby, Makromol. Chem., 134 (1970) 159.
243 B. R. Bhattacharya and U. S. Nandi, Makromol. Chem., 149 (1971) 243.
244 C. A. Barson, A. R. Luxton and J. C. Robb, J. Chem., Soc., Faraday Trans. 1, 68 (1972) 1666.
245 S. L. Kapur, J. Sci. Ind. Res., Sect. B, 10 (1951) 186.
246 S. C. Barton, R. A. Bird and K. E. Russell, Can. J. Chem., 41 (1963) 2737.
247 T. Yamamoto and T. Otsu, J. Polym. Sci., Part A-1, 6 (1968) 49.
248 J. C. Robb and R. Senogles, Trans. Faraday Soc., 58 (1962) 708.
249 T. R. Fukuto and J. P. Kispersky, U. S. Dept. Comm. Office Tech. Serv. P.B. Rept., 147211 (1953).
250 T. Otsu, A. Shimizu and M. Imoto, J. Polym. Sci., Part B, 2 (1964) 973.
251 E. H. Riddle, Monomeric Acrylic Esters, Reinhold, New York, 1954, ref. 14, p. 64.
252 K. Noma, Y. Tajima and M. Niwa, Sci. Eng. Rev. Doshisha Univ., Kyoto, 3 (1962) 91.
253 Kh. S. Bagdasar'ian and Z. A. Sinitsina, J. Polym. Sci., 52 (1961) 31.
254 J. L. Kice, J. Am. Chem. Soc., 76 (1954) 6274.
255 B. R. Bhattacharya and U. S. Nandi, Makromol. Chem., 116 (1968) 8.
256 M. Lazar and J. Pavlinec, Chem. Zvesti, 15 (1961) 428.
257 K. K. Roy, D. Pramanick and S. R. Palit, Makromol. Chem., 153 (1972) 71.
258 J. L. O'Brien, J. R. Panchak and T. G. Fox, Abstracts 124th Am. Chem. Soc. Meeting, Chicago, 1953.
259 S. Basu, J. N. Sen and S. R. Palit, Proc. R. Soc. London, Ser. A, 214 (1952) 247.
260 C. H. Bamford and E. F. T. White, Trans. Faraday Soc., 54 (1958) 268.
261 U. S. Nandi, P. Ghosh and S. R. Palit, Nature (London), 195 (1962) 1197.
262 K. Jijie, M. Santappa and V. Mahadevan, J. Polym. Sci., Part A, 4 (1966) 377.
263 W. I. Bengough and W. H. Fairservice, Trans. Faraday Soc., 67 (1971) 414.
264 W. I. Bengough and W. H. Fairservice, Trans. Faraday Soc., 61 (1965) 1206.
265 J. L. Kice, J. Polym. Sci., 19 (1956) 123.
266 Y. Minoura and T. Sugimura, J. Polym. Sci., Part A, 4 (1966) 2721.
267 G. V. Schulz and G. Henrici, Makromol. Chem., 18/19 (1956) 437.
268 H. W. Melville and A. F. Bickel, Trans. Faraday Soc., 45 (1949) 1049.
269 W. I. Bengough and H. W. Melville, Proc. R. Soc. London, Ser. A, 249 (1959) 445.
270 M. Yokawa, Y. Ogo and T. Imoto, Makromol. Chem., 175 (1974) 2913.
271 Values quoted in ref. 12, calculated from unpublished data of Ross and Melville.
272 Values quoted in ref. 12, calculated from unpublished data of Matheson.
273 Unpublished data of Ross, quoted in ref. 274.
274 G. M. Burnett, P. Evans and H. W. Melville, Trans. Faraday Soc., 49 (1953) 1096.

275 W. I. Bengough and A. K. C. Smith, Trans. Faraday Soc., 54 (1958) 1553.
276 M. S. Matheson, E. E. Auer, E. B. Bevilacqua and E. J. Hart, J. Am. Chem. Soc., 73 (1951) 5395.
277 Z. A. Sinitsyno and K. L. Bagdasaryan, Zh. Fiz. Khim., 32 (1958) 1319.
278 V. Mahadevan and M. Santhappa, Makromol. Chem., 16 (1955) 119.
279 M. Santhappa and V. M. lyer, Curr. Sci. (India), 24 (1955) 173.
280 A. S. Nair and M. S. Muthana, Makromol. Chem., 47 (1961) 114, 128.
281 P. V. T. Raghuram and U. S. Nandi, J. Polym. Sci., Part A-1, 8 (1970) 3079.
282 P. V. T. Raghuram and U. S. Nandi, J. Polym. Sci., Part A-1, 7 (1969) 2379.
283 Y. Hachihama and H. Sumitomo, Technol. Rep. Osaka Univ., 5 (1955) 497.
284 T. L. Dawson and R. D. Ludenberg, J. Polym. Sci., Part A, 3 (1965) 1801.
285 Y. Hachihama and H. Sumitomo, Technol. Rep. Osaka Univ., 5 (1955) 491.
286 I. M. Likhterova and E. M. Lukina, Zh. Obshch. Khim., 42 (1970) 194.
287 S. R. Chatterjee, S. N. Khanna and S. R. Palit, J. Ind. Chem. Soc., 41 (1964) 622.
288 S. L. Kapur and S. D. Gadkary, J. Sci. Ind. Res., Sect. B, 17 (1958) 152.
289 D. Hardy, K. Nytrai, N. Federova and G. Kovacs, Polym. Sci., U.S.S.R., 4 (1968) 584.
290 J. N. Sen, U. S. Nandi and S. R. Palit, J. Ind. Chem. Soc., 40 (1963) 729.
291 S. Gadkary and S. Kapur, Makromol. Chem., 17 (1955) 29.
292 E. I. Kalinina, E. M. Lukina and V. P. Maslennikov, Vysokomol. Soedin., Ser. A, 11 (1969) 1039, quoted in ref. 4.
293 Z. A. Sinytsina and Kh. S. Bagdasaryan, Zh. Fiz. Khim., 32 (1958) 2663.
294 K. Jijie, M. Santhappa and V. Mahadevan, J. Polym. Sci., Part A-1, 4 (1966) 393.
295 V. A. Popov, G. P. Gladyshev and I. L. Sokolova, Polym. Sci., U.S.S.R., 14 (1972) 1598; Vysokomol. Soedin., Ser. A, 14 (1972) 1424.
296 G. Mayer and G. V. Schulz, Makromol. Chem., 173 (1973) 101.
297 K. Yokota and A. Kondo, Makromol. Chem. 171 (1973) 113.
298 M. Yokawa, Y. Ogo and T. Imoto, Makromol. Chem., 175 (1974) 2903.
299 D. H. Grant and N. Grassie, Trans. Faraday Soc., 55 (1959) 1042.
300 G. M. Burnett, P. Evans and H. W. Melville, Trans. Faraday Soc., 49 (1953) 1105.
301 M. M. Zafar and R. Mahmud, Makromol. Chem., 175 (1974) 2627.
302 S. K. Patra and D. Mangaraj, Makromol. Chem., 111 (1968) 168.
303 M. Imoto, T. Otsu, K. Tsuda and T. Ito, J. Polym. Sci., Part A, 2 (1964) 1407.
304 S. N. Khanna, S. R. Chatterjee, U. S. Nandi and S. R. Palit, Trans. Faraday Soc., 58 (1962) 1827.
305 N. Colebourne, E. Collinson, D. J. Currie and F. S. Dainton, Trans. Faraday Soc., 59 (1963) 1357.
306 W. M. Thomas and R. L. Webb, J. Polym. Sci., 25 (1957) 124.
307 F. S, Dainton and R. S. Eaton, J. Polym. Sci., 39 (1959) 313.
308 W. I. Bengough, J. Polym. Sci., 28 (1958) 475.
309 F. S. Dainton and D. G. L. James, J. Polym. Sci., 39 (1959) 299.
310 W. I. Bengough, S. A. McIntosh and R. A. M. Thomson, Nature (London), 184 (1959) 266.
311 W. I. Bengough, Proc. R. Soc. London, Ser. A, 260 (1961) 205.
312 E. F. T. White and M. J. Zissell, J. Polym. Sci., Part A, 1 (1963) 2189.
313 A. F. Revzin and Kh. S. Bagdasaryan, Zh. Fiz. Khim., 38 (1964) 1020.
314 S. K. Saha and A. K. Chaudhuri, J. Polym. Sci., Part A-1, 9 (1971) 1505.
315 A. Prevot-Bernas and J. Sebban-Danon, J. Chim. Phys., 53 (1956) 418.
316 P. F. Onyon, J. Polym. Sci., 22 (1956) 19.
317 H. Kitagawa, Kobunshi Kagaku, 20 (1963) 5.
318 M. Yoshida and K. Tanouchi, Kobunshi Kagaku, 20 (1965) 545.
319 N. Tokura, M. Matsuda and F. Yazaki, Makromol. Chem., 42 (1960) 108.
320 L. H. Peebles, J. Polym. Sci., Part A, 3 (1965) 341.

321 S. K. Das, S. R. Chatterjee and S. R. Palit, Proc. R. Soc. London, Ser. A, 227 (1955) 252.
322 C. H. Bamford, A. D. Jenkins and E. F. T. White, J. Polym. Sci., 34 (1959) 271.
323 G. Oster and Y. Mitzutani, J. Polym. Sci., 22 (1956) 173.
324 T. Toda, J. Polym. Sci., 58 (1962) 411.
325 J. Ulbricht, Faserforsch. Textiltech., 10 (1959) 166.
326 G. Vidotto, S. Brugnaro and G. Talamini, Makromol. Chem., 140 (1970) 263.
327 C. H. Bamford, A. D. Jenkins and R. Johnston, Proc. R. Soc. London, Ser. A, 241 (1957) 364.
328 T. Shimomura, F. Tsuchida and I. Shinohara, Mem. Sch. Sci. Eng., Waseda Univ., 28 (1964) 37.
329 M. Watanabe and H. Kiuchi, J. Polym. Sci., 58 (1962) 103.
330 H. Kiuchi and M. Watanabe, Kobunshi Kagaku, 21 (1964) 37.
331 F. S. Dainton and R. G. James, Trans. Faraday Soc., 63 (1967) 1512.
332 N. T. Srinivasan and M. Santhappa, Makromol. Chem., 26 (1958) 80.
333 W. M. Thomas, E. H. Gleason and J. J. Pellon, J. Polym. Sci., 17 (1955) 275.
334 C. H. Bamford, A. D. Jenkins, R. Johnston and E. F. T. White, Trans. Faraday Soc., 55 (1959) 168.
335 P. W. Allen and G. P. McSweeney, Trans. Faraday Soc., 54 (1958) 715.
336 L. H. Peebles, J. Polym. Sci., Part A, 3 (1965) 341.
337 G. E. Ham and E. L. Ringwald, J. Polym. Sci., 8 (1952) 91.
338 Z. A. Sinytsina and Kh. S. Bagdasaryan, Zh. Fiz. Khim., 34 (1960) 2736.
339 W. I. Bengough and W. H. Fairservice, Trans. Faraday Soc., 63 (1967) 382.
340 W. I. Bengough and I. C. Ross, Trans. Faraday Soc., 62 (1966) 2251.
341 W. I. Bengough, Trans. Faraday Soc., 54 (1958) 868.
342 G. M. Burnett and H. W. Melville, Nature (London), 156 (1945) 661.
343 G. M. Burnett and H. W. Melville, Proc. R. Soc., London, Ser. A, 189 (1947) 456.
344 G. M. Burnett and H. W. Melville, Proc. R. Soc. London, Ser. A, 189 (1947) 494.
345 H. Miyama, Bull. Chem. Soc. Jpn., 29 (1956) 720; 30 (1957) 10.
346 C. G. Swain and P. D. Bartlett, J. Am. Chem. Soc., 68 (1946) 2381.
347 M. S. Matheson, E. E. Auer, E. B. Bevilacqua and E. J. Hart, J. Am. Chem. Soc., 71 (1949) 497.
348 H. Kwart, H. S. Broadbent and P. D. Bartlett, J. Am. Chem. Soc., 72 (1950) 1060.
349 W. I. Bengough and H. W. Melville,Proc. R. Soc. London, Ser. A, 230 (1955) 429.
350 A. J. Buselli, M. K. Lindemann and C. E. Blades, J. Polym. Sci., 28 (1958) 485.
351 S. R. Potnis and A. M. Deshpande, Makromol. Chem., 153 (1972) 139.
352 M. Matsumoto and M. Maeda, J. Polym. Sci., 17 (1955) 438.
353 D. J. Stein and G. V. Schulz, Makromol. Chem., 38 (1960) 248.
354 L. H. Peebles, J. T. Clarke and W. H. Stockmayer, J. Am. Chem. Soc., 82 (1960) 4780.
355 R. N. Chadha and G. S. Misra, Trans. Faraday Soc., 54 (1958) 1227.
356 S. R. Potnis and A. M. Deshpande, Makromol. Chem., 125 (1969) 48.
357 R. D. Spencer, M. B. Fulton and B. H. Beggs, Abstracts 137th Am. Chem. Soc. Meeting, Cleveland, 1960.
358 M. Matsumoto, J. Ukida, G. Takayama, T. Eguchi, K. Mukumoto, K. Imai, Y. Kazusa and M. Maeda, Makromol. Chem., 32 (1959) 13.
359 S. R. Palit and S. K. Das, Proc. R. Soc. London, Ser. A, 226 (1954) 82.
360 A. A. Vansheidt and G. Khardi, Acta Chim. Acad. Sci. Hung., 20 (1959) 381.
361 S. L. Kapur and R. M. Joshi, J. Polym. Sci., 14 (1954) 489.
362 G. Takayama, Kobunshi Kagaku, 15 (1958) 89.
363 A. A. Vansheidt and G. Khardi, Acta Chim. Acad. Sci. Hung., 20 (1959) 261.
364 H. C. Haas and H. Husek, J. Polym. Sci., Part A, 2 (1964) 2297.

365 R. L. Vale and W. G. P. Robertson, J. Polym. Sci., 33 (1958) 518.
366 W. I. Bengough and R. A. M. Thomson, Trans. Faraday Soc., 56 (1960) 407.
367 M. Lazar, J. Pavlinec and Z. Manasek, Collect. Czech. Chem. Commun., 26 (1961) 1380.
368 W. H. Stockmayer, R. O. Howard and J. T. Clarke, J. Am. Chem. Soc., 75 (1953) 1756.
369 V. M. Shulin, M. G. Gonikberg, A. L. Goff and V. N. Zagorbinina, Polym. Sci. U.S.S.R., 11 (1969) 877; Vysokomol. Soedin., Ser. A, 11 (1969) 777.
370 A. J. Restaino and W. N. Reed, J. Polym. Sci., 36 (1959) 499.
371 T. Asahara and T. Makishima, Kogyo Kagaku Zasshi, 69 (1966) 2173.
372 R. D. Spencer and M. B. Fulton, U.S. Dept. Commer., Office Tech. Serv., PB Rep., 144900 (1961).
373 J. C. Bevington, M. Johnson and J. P. Sheen, Eur. Polym. J., 8 (1972) 209.
374 I. Sakurada, Y. Sakaguchi and K. Hashimoto, Kobunshi Kagaku, 19 (1962) 593.
375 A. Alexandru and M. Oprish, Polym. Sci. U.S.S.R., 3 (1962) 99.
376 H. Kawakami, N. Mori and M. Sumi, Kobunshi Kagaku, 20 (1963) 408.
377 J. C. Bevington and H. G. Troth, Trans. Faraday Soc., 59 (1963) 127.
378 C. F. Thompson, W. S. Port and L. P. Witnauer, J. Am. Chem., Soc., 81 (1959) 2552.
379 Calculated from data of ref. 91, quoted in ref. 97.
380 T. Berezhnykh-Földes and F. Tüdös, Eur. Polym. J., 2 (1966) 219.
381 F. Tüdös and T. Berezhnykh-Földes, Eur. Polym. J., 2 (1966) 229.
382 Z. A. Sinytsina and Kh. S. Bagdasaryan, Zh. Fiz. Khim., 34 (1960) 1110.
383 T. Földes-Berezhnykh, F. Tüdös and S. Szakacs, Eur. Polym. J., 8 (1972) 1247.
384 T. Földes-Berezhnykh, S. Szakacs and F. Tüdös, Eur. Polym. J., 8 (1972) 1237.
385 R. A. Bird and K. E. Russell, Can. J. Chem., 43 (1965) 2123.
386 M. Simonyi, F. Tüdös and J. Pospisil, Eur. Polym. J., 3 (1967) 101.
387 P. D. Bartlett and H. Kwart, J. Am. Chem. Soc., 72 (1950) 1051.
388 P. D. Bartlett and H. Kwart, J. Am. Chem. Soc., 74 (1952) 3969.
389 H. Lillies, R. D. Parnell and K. E. Russell, Can. J. Chem., 50 (1972) 3304.
390 W. H. Atkinson, C. H. Bamford and G. C. Eastmond, Trans. Faraday Soc., 65 (1969) 1446.
391 M. Ryska, M. Kolinsky and D. Lim, J. Polym. Sci., Part C, 16 (1967) 621.
392 H. S. Mickley, A. S. Michaels and A. L. Moore, J. Polym. Sci., 60 (1962) 121.
393 T. Uno and K. Yoshida, Kobunshi Kaguku, 15 (1958) 819.
394 G. Vidotto, A. Crosato-Arnaldi and G. Talamini, Makromol. Chem., 114 (1968) 217.
395 M. Imoto, K. Takemoto and Y. Nakai, Makromol. Chem., 48 (1961) 80.
396 J. W. Breitenbach and A. Schindler, Monatsh. Chem., 86 (1955) 437.
397 G. Vidotto, S. Bugnaro and G. Talamini, Makromol. Chem., 140 (1970) 249.
398 B. A. Englin and T. A. Onishchenko, Isv. Akad. Nauk S.S.S.R., Ser. Khim., (1960) 1906, quoted in ref. 4.
399 G. S. Park and W. H. Thomas, Eur. Polym. J., 7 (1971) 343.
400 G. V. Tkachenko, P. M. Khomikovkni and S. S. Medvedev, Zh. Fiz. Khim., 25 (1951) 823.
401 G. V. Tkachenko, G. G. Petuklov and V. A. Dodonov, Vysokomol. Soedin., 3 (1961) 1549.
402 D. Lim and M. Kolinsky, J. Polym. Sci., 53 (1961) 173.
403 G. A. M. Diepen, Rec. Trav. Chim. Pays-Bas, 84 (1965) 261. P. Ehrlich, J. Polym. Sci., Part A, 3 (1965) 131. A. Gilchrist, Faraday Soc. informal discussion, London, March 25, 1964.
404 P. Ehrlich and G. A. Mortimer, Adv. Polym. Sci., 7 (1970) 386.

405 Z. Laita, J. Polym. Sci., 38 (1959) 247. Z. Laita and Z. Machacek, J. Polym. Sci., 38 (1959) 459.
406 W. Rabel and K. Ueberreiter, Ber. Bunsenges. Phys. Chem., 67 (1963) 514, 710.
407 J. C. Woodbrey and P. Ehrlich, J. Am. Chem. Soc., 85 (1963) 1580.
408 G. A. Mortimer, J. Polym. Sci., Part A-1, 8 (1970) 1513.
409 G. A. Mortimer, J. Polym. Sci., Part A-1, 4 (1966) 1895.
410 G. A. Mortimer, J. Polym. Sci., Part A-1, 8 (1970) 1543.
411 G. A. Mortimer, J. Polym. Sci., Part A-1, 8 (1970) 1535.
412 G. A. Mortimer, J. Polym. Sci., Part A-1, 4 (1966) 881.
413 G. A. Mortimer, J. Polym. Sci., Part A-1, 10 (1972) 163.
414 Quoted in ref. 404.
415 G. A. Mortimer, J. Polym. Sci., Part B, 8 (1970) 323.
416 F. S. Dainton and M. Tordoff, Trans. Faraday Soc., 53 (1957) 499.
417 V. A. Kabanov, D. A. Topchiev and T. M. Karaputadze, J. Polym. Sci., Part C, 42 (1973) 173.
418 M. Morton, P. P. Salatiello and H. Landfield, J. Polym. Sci., 8 (1952) 215.
419 F. S. Dainton and W. D. Sisley, Trans. Faraday Soc., 59 (1963) 1369.
420 N. Grassie and E. Vance, Trans. Faraday Soc., 52 (1956) 727.
421 P. J. Flory, Principles of Polymer Chemistry, Cornell University Press, New York, 1953.
422 J. Hughes and A. M. North, Trans. Faraday Soc., 62 (1966) 1866.
423 M. Kryszewsk, Rocz. Chem., 31 (1957) 893.
424 G. M. Burnett and W. W. Wright, Proc. R. Soc. London, Ser. A, 211 (1954) 41.
425 J. D. Burnett and H. W. Melville, Trans. Faraday Soc., 46 (1950) 976.
426 W. I. Bengough and W. Henderson, Trans. Faraday Soc., 61 (1965) 141.
427 P. F. Onyon, Trans. Faraday Soc., 51 (1955) 400.
428 B. A. Englin, B. N. Osipov and R. Kh. Freidlina, Isv. Akad. Nauk S.S.S.R., Ser. Khim. (1967) 2223.
429 P. D. Bartlett and F. A. Tate, J. Am. Chem. Soc., 75 (1953) 91.
430 G. Smets, L. Convent and X. van der Borght, Makromol. Chem., 23 (1957) 162.
431 R. M. Joshi, Makromol. Chem., 53 (1962) 33.
432 A. K. Chaudhuri, Makromol. Chem., 31 (1959) 214.
433 G. F. Santee, R. H. Marchessault, H. G. Clark, J. J. Kearny and V. Stannett, Makromol. Chem., 73 (1964) 177.
434 M. Vrancken and G. Smets, Makromol. Chem., 30 (1959) 197.
435 S. Banerjee and M. S. Muthana, J. Polym. Sci., 37 (1959) 469.
436 M. Litt and V. Stannett, Makromol. Chem., 37 (1960) 19.
437 C. J. Kurian and M. S. Muthana, Makromol. Chem., 29 (1959) 1.
438 D. Hardy, V. Spiegel and K. Nytrai, Polym. Sci. U.S.S.R., 2 (1961) 528.
439 L. Ghosez and G. Smets, J. Polym. Sci., 37 (1959) 445.
440 T. J. Suen, Y. Jen and J. V. Lockwood, J. Polym. Sci., 31 (1958) 481.
441 W. H. Urry and O. O. Juveland, J. Am. Chem. Soc., 80 (1958) 3322.
442 W. H. Urry, F. W. Stacey, E. S. Huyser and O. O. Juveland, J. Am. Chem. Soc., 76 (1954) 450.
443 C. J. Kurian and M. S. Muthana, Makromol. Chem., 29 (1959) 19.
444 O. N. Dolgov and M. G. Voronkov, Zh. Obshch. Khim., 40 (1970) 1668.
445 M. N. Romani and K. E. Weale, Trans. Faraday Soc., 62 (1966) 2264.
446 E. A. S. Cavell and I. T. Gilson, J. Polym. Sci., Part A-1, 7 (1969) 1307.
447 F. S. Dainton and W. D. Sisley, Trans. Faraday Soc., 59 (1963) 1377.
448 W. I. Bengough and T. O'Neill, Trans. Faraday Soc., 64 (1968) 2415.
449 W. I. Bengough and N. M. Chawdry, J. Chem. Soc., Faraday Trans. 1, 68 (1972) 1807.

Chapter 4

Copolymerization

G. C. EASTMOND

1. Introduction

This chapter is concerned with the formation of copolymers, that is, polymers constructed from more than one distinct type of monomer unit. For convenience we can classify the binary copolymers of monomers, A, B into three major categories,

```
       ·······ABBABAAAB········                     (I)
      ······AAAAAAABBBBBB·······                    (II)
     ·····AAAAA·······AAAAAA······                  (III)
            B           B
            B           B
            B           B
            B           B
```

Copolymers of structure (I) are known as random copolymers and formed by the polymerization of a mixture of the constituent monomers. These polymers are not truly random and their detailed structures are controlled by kinetic parameters. It is the copolymers of this type with which this chapter is mainly concerned.

Materials of structure (II), consisting of long sequences of the individual components, are termed block copolymers. Although copolymers of this type can be prepared by free-radical processes, the kinetic aspects of their formation are not so important as those of type (I), since they would normally be formed by preparing a homopolymer of, say, monomer A possessing reactive end-groups which can then be used to initiate the polymerization of B in a subsequent reaction. The kinetics are thus resolved into the kinetics of two separate homopolymerization processes. Graft copolymers, of type (III) may be formed by preparing a homopolymer of A containing side-groups which can be used to initiate the polymerization of B. Or, monomer B may be polymerized in the presence of polymer A using a conventional initiator, relying on transfer processes to produce grafts of B. Alternatively, the precursor may be polymer B having polymerizable end-groups which are incorporated into the A-chains during polymerization of A. A special case of polymers with structure (III) are the branched polymers (A and B are identical) which have already been met in a previous discussion of transfer reactions, but, if so desired, they could be prepared using graft polymerization techniques. If both ends of the B-chain in (III) are attached to A-chains, then the B-chain is a cross-link between A-chains, and an indefinite continuation of this structure produces an infinite three-dimensional network. Some general kinetic features of the formation of structures (II) and (III) are discussed.

There is, of course, no necessity to restrict copolymers to materials based on only two monomers. In principle any number of monomers may be present and in structures (II) and (III) the A- and B-chains may themselves be copolymers.

The study of copolymerization reactions is important both commercially and scientifically. From the purely scientific point of view it is important for the data made available for theoretical treatments of monomer and radical reactivities (these aspects are discussed in detail in a subsequent chapter). Commercially, copolymerization is important in increasing the range of polymeric materials available and in modifying the properties of existing polymers. In addition to increasing the range of available materials, simply by using combinations of normally homopolymerizable monomers, many monomers which will not homopolymerize can be incorporated into copolymers to the extent of 50 mole %. Such monomers include 1,2-disubstituted ethylenes, sterically hindered 1,1-disubstituted ethylenes and more unlikely monomers such as carbon monoxide and sulphur dioxide. Although such monomers are capable of adding to a radical they are unable to homopolymerize, often as a result of steric and/or polar factors.

An early and major activity in the field of copolymerization was the attempt to prepare synthetic rubbers as an alternative to a reliance on natural rubber supplies. More recently rubbery components have been introduced into otherwise brittle materials to improve their impact resistance, e.g. ABS, which contains polybutadiene chains grafted onto acrylonitrile—styrene copolymer. The incorporation of small amounts of a co-monomer can be used, for example, to suppress polymer degradation or to introduce dye-sites into otherwise undyeable fibres.

2. Random copolymerization

A kinetic scheme describing the copolymerization of two monomers must contain, in addition to the reactions involved in the homopolymerizations of the individual monomers, cross-propagation and cross-termination reactions. These are the addition of one monomer to a polymer chain carrying the conjugate radical of the other monomer, and the bimolecular termination of propagating radicals derived from the different monomers. For the purposes of discussion the kinetic aspects may be conveniently divided into two sections, factors affecting copolymer composition and structure and those controlling rates of copolymerization. These will be considered in turn.

2.1 COPOLYMER COMPOSITION AND STRUCTURE

2.1.1 The composition equation for binary systems

To derive an expression describing the overall copolymer composition it is only necessary to consider the rates of the propagation reactions. Taking the simple case of the copolymerization of two monomers M_1, M_2, and assuming that the reactivity of a propagating radical is determined only by the nature of the terminal unit and not by the nature of the penultimate or other monomer units, four propagation reactions must be considered, viz.

$$PM_1\cdot + M_1 \xrightarrow{k_{11}} PM_1\cdot \tag{1a}$$

$$PM_1\cdot + M_2 \xrightarrow{k_{12}} PM_2\cdot \tag{1b}$$

$$PM_2\cdot + M_1 \xrightarrow{k_{21}} PM_1\cdot \tag{1c}$$

$$PM_2\cdot + M_2 \xrightarrow{k_{22}} PM_2\cdot \tag{1d}$$

where $PM_1\cdot$ represents a propagating chain in which the terminal unit is derived from M_1 and k_{12} is the rate coefficient for the addition of M_2 to this radical. It is necessary to make two further assumptions, (i) the chains are long, so that monomer is effectively consumed only in the propagation reactions, when we can write

$$\begin{aligned} -d[M_1]/dt &= k_{11}[PM_1\cdot][M_1] + k_{21}[PM_2\cdot][M_1] \\ -d[M_2]/dt &= k_{12}[PM_1\cdot][M_2] + k_{22}[PM_2\cdot][M_2] \end{aligned} \tag{2}$$

(ii) that a stationary state in radical concentrations exists, so that

$$d[PM_1\cdot]/dt = k_{21}[PM_2\cdot][M_1] - k_{12}[PM_1\cdot][M_2] = 0$$

or (3)

$$[PM_2\cdot]/[PM_1\cdot] = k_{12}[M_2]/k_{21}[M_1]$$

From eqn. (2), at any instant the ratio of the number of moles of the monomers entering the copolymer is

$$n = \left(\frac{-d[M_1]}{dt}\right) \Big/ \left(\frac{-d[M_2]}{dt}\right) = \frac{[M_1]}{[M_2]} \cdot \frac{k_{11}[PM_1\cdot] + k_{21}[PM_2\cdot]}{k_{12}[PM_1\cdot] + k_{22}[PM_2\cdot]}$$

Using eqn. (3) we have

$$n = \frac{[M_1]}{[M_2]} \left\{ \frac{(k_{11}/k_{12})[M_1] + [M_2]}{(k_{22}/k_{21})[M_2] + [M_1]} \right\}$$

Defining two parameters r_1 and r_2 such that

$$r_1 = k_{11}/k_{12}, \qquad r_2 = k_{22}/k_{21}$$

we obtain

$$n = \frac{[M_1]}{[M_2]} \cdot \frac{r_1[M_1] + [M_2]}{r_2[M_2] + [M_1]} \tag{4}$$

which is the well-known copolymer composition equation [1—4]. The kinetic parameters r_1 and r_2 are known as the *reactivity ratios* of the respective monomers and they define the relative rates of addition of each monomer to a particular radical. It should be noted that the principles involved in the derivation of eqn. (4) are not restricted to free-radical reactions.

The literature contains many examples of binary copolymerizations one component of which is incapable of homopolymerization. If, for example, M_2 will not homopolymerize, i.e. $k_{22} = 0$, $k_{12} \neq 0$, then $r_2 = 0$ and eqn. (4) simplifies to

$$n - 1 = r_1 [M_1]/[M_2] \tag{5}$$

Equation (4) may also be derived statistically without recourse to the stationary-state assumption (ii), as shown by Goldfinger and Kane [5]. Radical $PM_1\cdot$ may propagate by addition of either M_1 or M_2. The probability, p_{11}, that $PM_1\cdot$ propagates by addition of M_1 is

$$p_{11} = \frac{k_{11}[M_1]}{k_{11}[M_1] + k_{12}[M_2]} = \frac{r_1[M_1]}{r_1[M_1] + [M_2]}$$

and the probability that it will add M_2 is

$$p_{12} = \frac{[M_2]}{r_1[M_1] + [M_2]} = 1 - p_{11}$$

Corresponding probabilities may be written for the propagation of $PM_2\cdot$. The probability of finding a sequence of m M_1 units is, therefore,

$$p_1(m) = p_{11}^{(m-1)} p_{12} \tag{6}$$

and the average sequence length of M_1 units is

$$s_1 = \frac{p_{12}}{p_{11}} \sum_{m=1}^{\infty} m p_{11}^m = p_{12}^{-1} \tag{7}$$

Similarly the average sequence length of M_2 units is $(p_{21})^{-1}$ and the overall copolymer composition, n, is given by

$$n = \frac{s_1}{s_2} = \frac{p_{21}}{p_{12}} = \frac{[M_1]}{[M_2]} \cdot \frac{r_1[M_1] + [M_2]}{r_2[M_2] + [M_1]}$$

identical with eqn. (4).

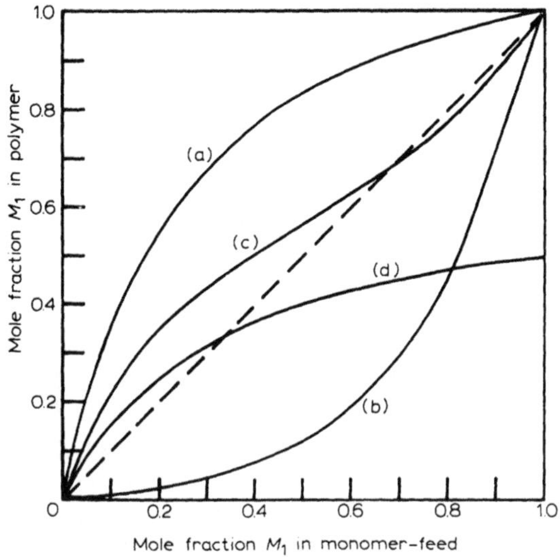

Fig. 1. Variation in instantaneous copolymer composition with monomer-feed composition. (a) $r_1 = 1/r_2$; $r_1 = 5$, $r_2 = 0.2$, (b) $r_1 < 1 < r_2$; $r_1 = 0.5$, $r_2 = 10$, (c) r_1, $r_2 < 1$; $r_1 = 0.7$, $r_2 = 0.3$, (d) $r_1 = 0$, $r_2 = 0.5$. (- - -) Azeotropic line corresponding to $r_1 = r_2 = 1$.

Inspection of eqn. (4) shows that in general the composition of the copolymer is not the same as that of the monomer feed. Indeed the copolymer composition can only equal the monomer feed composition at all values of the latter in the comparatively rare event that $r_1 = r_2 = 1$, i.e. all the rate coefficients for propagation are equal. Under these conditions the curve relating polymer and monomer compositions corresponds to the diagonal straight line in Fig. 1, often referred to as the azeotropic line.

It was originally suggested by Wall [6] that the relative reactivities of a series of monomers would be independent of radical structure, i.e.

$$k_{11}/k_{12} = k_{21}/k_{22} \quad \text{or} \quad r_1 = 1/r_2 \tag{8}$$

Under these conditions eqn. (4) reduces to

$$n = r_1[\mathrm{M}_1]/[\mathrm{M}_2] \tag{9}$$

so that, unless $r_1 = 1$, the copolymer composition never corresponds to the monomer feed composition (Fig. 1). Although Wall's suggestion is an oversimplification, monomer feed and copolymer compositions can never be identical if $r_1 < 1 < r_2$, or vice versa (Fig. 1), except in the trivial cases when $[\mathrm{M}_1]$, $[\mathrm{M}_2] \rightarrow 0$. Under such circumstances the copolymer composition curve lies wholly below or above the azeotropic line. These conditions correspond to the rate coefficients for one monomer being greater than for the other monomer, irrespective of radical structure (e.g. $k_{12} > k_{11}$, $k_{22} > k_{21}$), although the relative reactivities of the monomers (r_1, r_2) may depend on radical structure.

The alternative situation is when both r_1 and r_2 are greater than or less than unity, when the copolymer composition does correspond to the monomer feed composition at a specific value of the latter, i.e. at the azeotropic composition (Fig. 1). Simply, this arises because the copolymer composition curve lies on opposite sides of the azeotropic line at low values of $[M_1]$ and $[M_2]$, the slope of the composition curve approaching $1/r_2$ as $[M_1] \rightarrow 0$ and $1/r_1$ as $[M_2] \rightarrow 0$.

2.1.2 Structure of binary copolymers

Although the copolymers prepared by the polymerization of monomer mixtures are often referred to as "random copolymers" this term is a misnomer since it implies that the constituent monomer units are distributed at random along the polymer chain. This only occurs in the "ideal copolymerization" envisaged by Wall [6], where $r_1 r_2 = 1$, i.e. when eqn. (8) holds and the probabilities of M_1 adding to radicals $PM_1\cdot$ and $PM_2\cdot$ are identical.

The two situations, r_1, $r_2 > 1$ and r_1, $r_2 < 1$, referred to above (Sect. 2.1.1), have different implications regarding the detailed structure of the copolymer molecules. If r_1, $r_2 > 1$ the rate coefficients for homopolymerization are greater than those for cross-propagation. Consequently, once a radical is formed it will tend to add a long sequence of that type of monomer unit. An examination of reactivity ratios (Appendix) shows that while the above situation does not normally occur in free-radical polymerization the situation r_1, $r_2 < 1$ is quite common. Under these conditions the rate coefficients for cross-propagation are greater than those for homopropagation, so that a given radical will tend to add the alternative monomer.

We may illustrate the difference in these two situations by comparing the average sequence lengths of M_1 and M_2 units in the copolymers formed by polymerization of equimolar mixtures of M_1 and M_2 in specific examples. First, we consider the cases (i) $r_1 = 5$, $r_2 = 3$ and (ii) $r_1 = 0.5$, $r_2 = 0.3$. In these systems the ratio of the rate coefficients for cross-propagation (k_{12}/k_{21}) are identical. The average sequence lengths of M_1 and M_2 units (eqn. (7)) are given by p_{12}^{-1} and p_{21}^{-1}, respectively, and have the values

Case (i) $s_1 = 6$, $s_2 = 4$

Case (ii) $s_1 = 1.5$, $s_2 = 1.3$

Secondly, we consider the cases (iii) $r_1 = 10$, $r_2 = 10$ and (iv) $r_1 = 0.1$, $r_2 = 0.1$. In this example both copolymers have the same overall composition and the average sequence lengths are given by

Case (iii) $s_1 = s_2 = 11$

Case (iv) $s_1 = s_2 = 1.1$

TABLE 1

Probabilities of finding M_1 sequences of length m in two copolymers of the same overall composition ($n = 1$)

m	1	2	3	4	5	6	7
$r_1 = r_2 = 0.1$	0.909	0.083	0.008				
$r_1 = 1/r_2 = 10$	0.500	0.250	0.125	0.063	0.031	0.016	0.008

In both examples it is seen that if $r_1, r_2 > 1$ there is a tendency towards block copolymer formation, while if $r_1, r_2 < 1$ the copolymer approaches an alternating structure. Although this comparison is somewhat artificial, since the condition $r_1, r_2 > 1$ does not occur in free-radical copolymerization, it provides an illustration of the principles involved and the differences which can occur between copolymers prepared by free-radical and ionic processes; in the latter there are a number of instances where both reactivity ratios exceed unity.

To compare the different structures which do occur in free-radical copolymerization we may compare the sequence distributions (calculated from eqn. (6)) in two different copolymers of the same overall composition ($n = 1$). In Table 1 we compare the probabilities of finding sequences of length m in two copolymers, the "alternating" copolymer of case (iv) above and a random copolymer ($r_1 = 1/r_2$). These data illustrate how a tendency for cross-propagation generates a high proportion of very short sequences compared with the truly random copolymer. Sequence distributions and related quantities for a number of systems have been tabulated recently by Tosi [7].

A limiting case is $r_1 = r_2 = 0$, when the copolymer formed at all monomer feed compositions has the regular alternating structure

$$\cdots\cdots\cdot M_1M_2M_1M_2M_1M_2\cdot\cdots\cdots$$

This situation arises if the monomers can add to the conjugate radical of the other monomer but are unable to homopolymerize, and is approximated to if the rate coefficients for homopropagation are extremely small compared to those for cross-propagation.

For cases where M_1 will not homopolymerize the copolymer will not normally contain more than 50 mole % M_1 (Fig. 1). In the limit as $[M_2] \rightarrow 0$ a regular alternating polymer is formed and, if r_2 is small, a regular alternating structure is approximated to for low values of $[M_2]$. For example, in the copolymerization of maleic anhydride (M_1) and styrene (M_2), for which values of $r_1 = 0$, $r_2 = 0.02$ have been quoted [8], the sequence lengths of maleic anhydride are always unity and the average sequence lengths of styrene units are < 1.01 if the mole fraction of styrene in the monomer feed is less than 0.33.

A detailed discussion of molecular-weight distributions in copolymers is outside the scope of this chapter, we may, however, note a consequence of this distribution in regard to molecular structure and composition. All polymers contain a distribution of molecular sizes, and for polymers prepared by free-radical copolymerization with disproportionation, or if the molecular weights are controlled by transfer, the overall polymers contain large numbers (but a small weight fraction) of short chain molecules. The copolymer composition eqn. (4) describes the composition of the whole polymer, and for long chains the composition of the individual molecules will approximate to this. On the other hand, the composition of a short chain, comprising only a few sequences, may deviate considerably from the overall composition. The distribution of compositions arising from variations in molecular weights has been treated by Stockmayer [9].

2.1.3 Variation of composition and structure with conversion

Equation (4) describes the instantaneous relationship between copolymer and monomer-feed compositions. In any practical copolymerization experiment the relative rates of consumption of the two monomers will normally differ from the relative concentrations in the initial feed. Hence, if a reaction is taken to finite conversion the composition of the unpolymerized monomer mixture and of the copolymer will vary with time. To describe the results of a copolymerization experiment, therefore, it is necessary to consider the integrated form of eqn. (4) [1], viz.

$$\log \frac{[\mathrm{M}_1]}{[\mathrm{M}_1]_0} = \frac{r_1}{1-r_1} \log \frac{[\mathrm{M}_1]_0[\mathrm{M}_2]}{[\mathrm{M}_2]_0[\mathrm{M}_1]} - \frac{1-r_1 r_2}{(1-r_1)(1-r_2)} \log \frac{(r_2-1)[\mathrm{M}_2]/[\mathrm{M}_1]-r_1+1}{(r_2-1)[\mathrm{M}_2]_0/[\mathrm{M}_1]_0-r_1+1} \tag{10}$$

where $[\mathrm{M}_1]_0$, $[\mathrm{M}_2]_0$ are the initial and $[\mathrm{M}_1]$, $[\mathrm{M}_2]$ the final concentrations of M_1, M_2, respectively. Equation (10) can be usefully transformed into

$$r_1 = \frac{\log \dfrac{[\mathrm{M}_1]_0}{[\mathrm{M}_1]} - \dfrac{1}{p} \log \left(1-p\dfrac{[\mathrm{M}_2]}{[\mathrm{M}_1]}\right) \Big/ \left(1-p\dfrac{[\mathrm{M}_2]_0}{[\mathrm{M}_1]_0}\right)}{\log \dfrac{[\mathrm{M}_2]_0}{[\mathrm{M}_2]} + \log \left(1-p\dfrac{[\mathrm{M}_2]}{[\mathrm{M}_1]}\right) \Big/ \left(1-p\dfrac{[\mathrm{M}_2]_0}{[\mathrm{M}_1]_0}\right)} \tag{11}$$

where $p = (1-r_2)/(1-r_1)$.

Equation (10) may be used directly to calculate $[\mathrm{M}_2]$ if $[\mathrm{M}_1]$, $[\mathrm{M}_1]_0$, $[\mathrm{M}_2]_0$, r_1 and r_2 are known, and hence the mean value of n may be computed. As we shall see later, when transformed into eqn. (11), it can be used to calculate values of the kinetic parameters r_1, r_2 for reactions taken to finite conversion.

To predict mean values of n for finite conversions Skeist's method of graphic integration may be used [10]. The basis of this method is to express eqn. (4) in the form

$$\frac{d[M_1]}{d([M_1] + [M_2])} = F_1 = \frac{r_1 f_1^2 + f_1 f_2}{r_1 f_1^2 + 2f_1 f_2 + r_2 f_2^2}$$

where f_1 and f_2 are, respectively, the mole fractions of M_1 and M_2 in the monomer feed and F_1 is the mole fraction of M_1 in the polymer. Putting $[M] = ([M_1] + [M_2])$, when a total of d[M] moles of monomer have polymerized the polymer will contain $F_1\,d[M]$ moles of M_1 and the concentration of M_1 in the monomer feed will be $([M] - d[M])(f_1 - df_1)$. Hence

$$f_1[M] - ([M] - d[M])(f_1 - df_1) = F_1 d[M]$$

or

$$d[M]/[M] = df_1/(F_1 - f_1)$$

and

$$\ln([M]/[M]_0) = \int_{f_1^0}^{f_1} \frac{df_1}{F_1 - f_1} \qquad (12)$$

where $f_1^0 = [M_1]_0/[M]_0$. Corresponding values of f_1 and F_1 are calculated from eqn. (4) for various values of f_1. Then, by graphical or numerical integration of eqn. (12) ln $([M]/[M]_0)$ is obtained and hence, the fraction of total monomer which must react to change the composition of the residual monomers to any desired value may be calculated. The mean value of n at that point is given by

$$n = \frac{f_1^0[M]_0 - f_1[M]}{(1 + f_1^0)[M]_0 - (1 - f_1)[M]}$$

The application of this equation is not restricted to cases where eqn. (4) holds, and in all cases corresponding values of f_1 and F_1 may be obtained by direct experiment.

Recently, Meyer and Lowry [11] have described an analytical solution of Skeist's equation.

Skeist's method can also be used to determine the distribution curves describing the variation in copolymer composition which results from polymerization to finite conversion. Spinner et al. [2] have shown that, when eqn. (4) holds, the relative abundances, y, of copolymers of composition F_1, defined by

$$y = \frac{1}{[M]_0} \frac{d[M]}{dF_1} \qquad (13)$$

can be obtained as an explicit function of r_1, r_2, f_1 and f_1^0. When both

References pp. 329—332

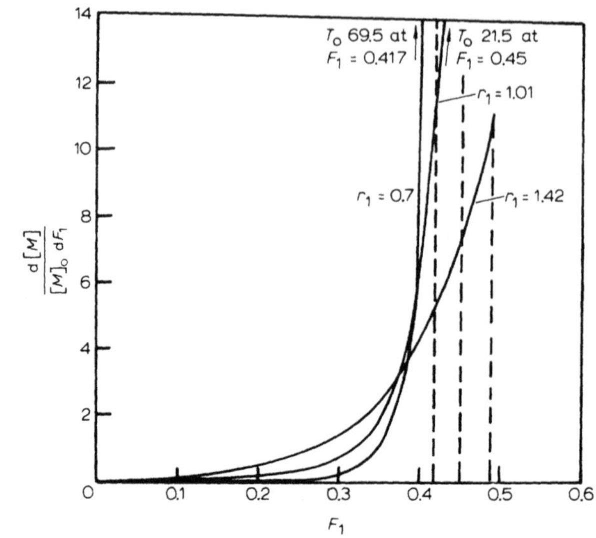

Fig. 2. Representative copolymer composition distribution functions in cases where both monomer reactivity ratios are >0.5. Dashed lines indicate copolymer compositions in the initial stage of reaction for various experimental conditions indicated. In each case r_2 = 0.7, f_1^0 = 0.4. (After Spinner et al. [12], courtesy Am. Chem. Soc.)

reactivity ratios are greater than 0.5 the relative abundances of copolymer compositions at all conversions varies monotonically with copolymer composition (Fig. 2), while, if at least one reactivity ratio falls below 0.5, the copolymer composition distribution curves show two peaks (Fig. 3)

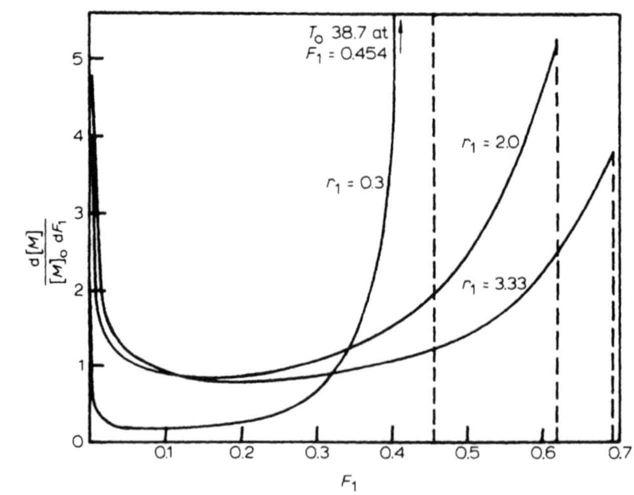

Fig. 3. Representative copolymer composition distribution functions in cases where, say, $r_2 < 0.5$. Dashed lines indicate copolymer compositions in the initial stage of reaction for various experimental conditions indicated. In each case r_2 = 0.3, f_1^0 = 0.4. (After Spinner et al. [12], courtesy Am. Chem. Soc.)

and the final copolymer may contain appreciable amounts of material of widely differing composition, with little material of intermediate composition [11—13]. This result can, of course, have important industrial implications, when it is economically desirable to take reactions to high conversion and yet avoid significant heterogeneity in the composition of the product.

Variations in sequence distributions with conversion, reflecting variations in composition, have been detected in vinyl chloride—vinylidene chloride systems by NMR spectroscopy [14]. Inagaki et al. [15] have developed a thin-layer chromatographic technique for the fractionation of copolymers according to composition, rather than by molecular weight. These workers have examined styrene—methyl acrylate copolymers prepared under conditions of high conversion and have demonstrated good agreement between the observed composition distribution and the two-peak distribution predicted theoretically.

When preparing large amounts of polymer with a specific composition, as an alternative to stopping reaction at low conversion and recovering residual monomer, it is possible to maintain a constant copolymer composition by continuously replenishing the supply of the more rapidly consumed monomer [16—18]. The possibility of using temperature control as a means of maintaining a desired copolymer composition has also been discussed [19].

2.1.4 Composition and structure in multicomponent systems

Multicomponent copolymerizations can be treated by a direct extension of the principles used in binary copolymerizations. Thus, in a terpolymer system there are nine propagation reactions to be considered [20], viz.

$$\begin{array}{lcl}
PM_1\cdot + M_1 & \xrightarrow{k_{11}} & PM_1M_1\cdot \\
PM_1\cdot + M_2 & \xrightarrow{k_{12}} & PM_1M_2\cdot \\
PM_1\cdot + M_3 & \xrightarrow{k_{13}} & PM_1M_3\cdot \\
PM_2\cdot + M_1 & \xrightarrow{k_{21}} & PM_2M_1\cdot \\
PM_2\cdot + M_2 & \xrightarrow{k_{22}} & PM_2M_2\cdot \\
PM_2\cdot + M_3 & \xrightarrow{k_{23}} & PM_2M_3\cdot \\
PM_3\cdot + M_1 & \xrightarrow{k_{31}} & PM_3M_1\cdot \\
PM_3\cdot + M_2 & \xrightarrow{k_{32}} & PM_3M_2\cdot \\
PM_3\cdot + M_3 & \xrightarrow{k_{33}} & PM_3M_3\cdot
\end{array} \qquad (14)$$

The three equations describing the rates of disappearance of the individual monomers are

$$
\begin{aligned}
-\mathrm{d}[\mathrm{M}_1]/\mathrm{d}t &= k_{11}[\mathrm{PM}_1\cdot][\mathrm{M}_1] + k_{21}[\mathrm{PM}_2\cdot][\mathrm{M}_1] + k_{31}[\mathrm{PM}_3\cdot][\mathrm{M}_1] \\
-\mathrm{d}[\mathrm{M}_2]/\mathrm{d}t &= k_{12}[\mathrm{PM}_1\cdot][\mathrm{M}_2] + k_{22}[\mathrm{PM}_2\cdot][\mathrm{M}_2] + k_{32}[\mathrm{PM}_3\cdot][\mathrm{M}_2] \\
-\mathrm{d}[\mathrm{M}_3]/\mathrm{d}t &= k_{13}[\mathrm{PM}_1\cdot][\mathrm{M}_3] + k_{23}[\mathrm{PM}_2\cdot][\mathrm{M}_3] + k_{33}[\mathrm{PM}_3\cdot][\mathrm{M}_3]
\end{aligned}
\tag{15}
$$

and application of the stationary-state assumption for each radical type provides three equations relating their rates of appearance and disappearance, viz.

$$
\begin{aligned}
&k_{21}[\mathrm{PM}_2\cdot][\mathrm{M}_1] + k_{31}[\mathrm{PM}_3\cdot][\mathrm{M}_1] - k_{12}[\mathrm{PM}_1\cdot][\mathrm{M}_2] \\
&\quad - k_{13}[\mathrm{PM}_1\cdot][\mathrm{M}_3] = 0 \\
&k_{12}[\mathrm{PM}_1\cdot][\mathrm{M}_2] + k_{32}[\mathrm{PM}_3\cdot][\mathrm{M}_2] - k_{21}[\mathrm{PM}_2\cdot][\mathrm{M}_1] \\
&\quad - k_{23}[\mathrm{PM}_2\cdot][\mathrm{M}_3] = 0 \\
&k_{13}[\mathrm{PM}_1\cdot][\mathrm{M}_3] + k_{23}[\mathrm{PM}_2\cdot][\mathrm{M}_3] - k_{31}[\mathrm{PM}_3\cdot][\mathrm{M}_1] \\
&\quad - k_{32}[\mathrm{PM}_3\cdot][\mathrm{M}_2] = 0
\end{aligned}
\tag{16}
$$

Defining six reactivity ratios

$$
\begin{aligned}
&r_{12} = k_{11}/k_{12}, \quad r_{21} = k_{22}/k_{21}, \quad r_{31} = k_{33}/k_{31} \\
&r_{13} = k_{11}/k_{13}, \quad r_{23} = k_{22}/k_{23}, \quad r_{32} = k_{33}/k_{32}
\end{aligned}
$$

we can obtain relation (17) describing the relative rates of incorporation of M_1, M_2 and M_3 into the terpolymer, viz.

$$
\begin{aligned}
&\mathrm{d}[\mathrm{M}_1]:\mathrm{d}[\mathrm{M}_2]:\mathrm{d}[\mathrm{M}_3]:: \\
&[\mathrm{M}_1]\left(\frac{[\mathrm{M}_1]}{r_{21}r_{31}} + \frac{[\mathrm{M}_2]}{r_{21}r_{32}} + \frac{[\mathrm{M}_3]}{r_{31}r_{23}}\right)\left([\mathrm{M}_1] + \frac{[\mathrm{M}_2]}{r_{12}} + \frac{[\mathrm{M}_3]}{r_{13}}\right) \\
&:[\mathrm{M}_2]\left(\frac{[\mathrm{M}_1]}{r_{12}r_{31}} + \frac{[\mathrm{M}_2]}{r_{12}r_{32}} + \frac{[\mathrm{M}_3]}{r_{32}r_{13}}\right)\left([\mathrm{M}_2] + \frac{[\mathrm{M}_1]}{r_{21}} + \frac{[\mathrm{M}_3]}{r_{23}}\right) \\
&:[\mathrm{M}_3]\left(\frac{[\mathrm{M}_1]}{r_{13}r_{21}} + \frac{[\mathrm{M}_2]}{r_{23}r_{12}} + \frac{[\mathrm{M}_3]}{r_{13}r_{23}}\right)\left([\mathrm{M}_3] + \frac{[\mathrm{M}_1]}{r_{31}} + \frac{[\mathrm{M}_2]}{r_{32}}\right)
\end{aligned}
\tag{17}
$$

Relation (17) shows that the composition of the terpolymer can be predicted from a knowledge of the reactivity ratios of the binary copolymerizations of pairs of the constituent monomers. Walling and Briggs [21] have extended this treatment to systems containing l monomers and have shown that the copolymer composition can be predicted from a knowledge of $l(l-1)/2$ monomer pairs. The equations are not presented here since they are complex, involve no new principles and are not widely used.

Price [22] has adopted the statistical approach in deriving the composition equation for terpolymerizations. Sequence distributions in multicomponent systems have been considered by Alfrey and Goldfinger [3]. As with binary copolymerizations it is possible to have azeotropic copolymers in multicomponent systems, and Walling and Briggs [21] have demonstrated that, except in trivial cases, in any system only one azeotropic composition exists, irrespective of the number of monomers involved. Other workers [23—26] have described methods for determining the conditions for azeotrope formation. Slocombe [23] has described a graphical method of determining azeotropic compositions, and according to Tarasov et al. [24] the condition is that P, Q and R, in the relations

$$P = \left(1 - \frac{1}{r_{13}}\right)\left(\frac{1}{r_{32}} - 1\right) - \left(\frac{1}{r_{12}} - \frac{1}{r_{13}}\right)\left(\frac{1}{r_{31}} - 1\right)$$

$$Q = \left(\frac{1}{r_{21}} - \frac{1}{r_{23}}\right)\left(\frac{1}{r_{12}} - \frac{1}{r_{13}}\right) - \left(1 - \frac{1}{r_{23}}\right)\left(1 - \frac{1}{r_{13}}\right)$$

$$R = \left(\frac{1}{r_{31}} - 1\right)\left(1 - \frac{1}{r_{23}}\right) - \left(\frac{1}{r_{32}} - 1\right)\left(\frac{1}{r_{21}} - \frac{1}{r_{23}}\right)$$

have the same sign and are not zero, when the feed composition for the azeotrope is

$$[M_1]:[M_2]:[M_3]::R\left(\frac{Q}{r_{13}r_{32}} + \frac{P}{r_{12}r_{23}} + \frac{R}{r_{12}r_{13}}\right)$$

$$:P\left(\frac{P}{r_{21}r_{23}} + \frac{Q}{r_{23}r_{31}} + \frac{R}{r_{21}r_{13}}\right)$$

$$:Q\left(\frac{Q}{r_{31}r_{32}} + \frac{P}{r_{21}r_{32}} + \frac{R}{r_{12}r_{31}}\right)$$

As for binary copolymerizations, to describe the variation in composition with conversion it is necessary to integrate the composition equation. Equation (17) and the corresponding equations for multicomponent systems are not readily integrated, but Walling and Briggs [21] have described an approximate form and Skeist [10] and Seiner [27] have described numerical methods for obtaining diagrams relating instantaneous compositions to conversion.

Whereas considerable simplification of eqn. (4) occurs if one reactivity ratio is zero, under similar circumstances eqn. (17) breaks down [28], and the appropriate relation is determined by the detailed mechanism of the system involved. If, for example, M_3 will not homopolymerize it is necessary to carry out at least one terpolymerization experiment to

References pp. 329—332

determine the parameter k_{13}/k_{23}, which appears in the composition equation under these conditions. When both M_2 and M_3 will not homopolymerize the ratio k_{12}/k_{32} also appears in the composition equation, and must be determined experimentally. Walling et al. [29] have considered systems of the type maleic anhydride (M_1), and two substituted α-methyl styrenes (M_2, M_3) where none of the monomers homopolymerize and M_2, M_3 do not cross-propagate. Under such circumstances

$$\frac{d[M_2]}{d[M_3]} = \frac{k_{12}[M_2]}{k_{13}[M_3]}$$

and the ratio k_{12}/k_{13}, the relative rates of addition of M_2 and M_3 to the conjugate radical of M_1, can be obtained. In this way the relative reactivities of a number of substituted α-methyl styrenes towards the maleic anhydride radical have been determined, as shown in Table 2, where the relative reactivities are the reactivities of the substituted monomers relative to α-methyl styrene.

TABLE 2
Relative reactivities of substituted α-methyl styrenes with the maleic anhydride radical [29]

Substituent	Reactivity
p-$N(CH_3)_2$	300
p-OCH_3	18.5 ± 4
p-CH_3	1.72 ± 0.12
None	1.00
m-Br	0.96 ± 0.14
p-CN	0.96 ± 0.57
p-Cl	0.79 ± 0.02
p-Br	0.73 ± 0.15
p-F	0.72 ± 0.10

Thus, from a study of terpolymerizations it is sometimes possible to obtain the relative rates of addition of one monomer to two different radicals (e.g. k_{13}/k_{23}) and relative rates of addition of two monomers to one radical (e.g. k_{12}/k_{13}), the radicals not being conjugate radicals of the monomers concerned. Such ratios of rate coefficients are not normally obtainable by direct experiment in other ways and the values of these ratios may be compared with the values calculated from schemes devised to correlate radical reactivities (Chap. 5).

In general, terpolymer compositions calculated from eqn. (17) agree well with experimental results. Table 3 shows the agreement achieved in terpolymerizations of styrene (S), methyl acrylate (M), vinyl chloride (V) and acrylonitrile (A). Ham [31, 32] has suggested that the sensitivity of terpolymer compositions to the values of reactivity ratios may provide a check of doubtful values of the latter quantities, and may also be used to provide evidence for penultimate effects (Sect. 2.1.5).

It has been suggested by Valverossi and Sartori [25, 33] that the rather complex composition equations for multicomponent systems may be simplified by modifying the stationary-state expressions. For a terpolymer system they re-write the stationary-state expressions (16) as

$$\begin{aligned} &k_{21}[\mathrm{PM_2}\cdot][\mathrm{M_1}] - k_{12}[\mathrm{PM_1}\cdot][\mathrm{M_2}] = 0 \\ &k_{32}[\mathrm{PM_3}\cdot][\mathrm{M_2}] - k_{23}[\mathrm{PM_2}\cdot][\mathrm{M_3}] = 0 \\ &k_{13}[\mathrm{PM_1}\cdot][\mathrm{M_3}] - k_{31}[\mathrm{PM_3}\cdot][\mathrm{M_1}] = 0 \end{aligned} \tag{18}$$

If these relations are now used in conjunction with the monomer consumption relations (15) a simplified terpolymer composition equation

$$\begin{aligned} \mathrm{d}[\mathrm{M_1}]:\mathrm{d}[\mathrm{M_2}]:\mathrm{d}[\mathrm{M_3}]::[\mathrm{M_1}]&\left([\mathrm{M_1}] + \frac{[\mathrm{M_2}]}{r_{12}} + \frac{[\mathrm{M_3}]}{r_{13}}\right) \\ :[\mathrm{M_2}]\frac{r_{21}}{r_{12}}&\left([\mathrm{M_2}] + \frac{[\mathrm{M_1}]}{r_{21}} + \frac{[\mathrm{M_3}]}{r_{23}}\right) \\ :[\mathrm{M_3}]\frac{r_{31}}{r_{13}}&\left([\mathrm{M_3}] + \frac{[\mathrm{M_1}]}{r_{31}} + \frac{[\mathrm{M_2}]}{r_{32}}\right) \end{aligned} \tag{19}$$

TABLE 3
Terpolymerizations of styrene, methyl acrylate, vinyl chloride and acrylonitrile at 50 °C [30]

Monomers	Weight % in feed	Weight % in polymer	
		Experiment	Calculated
S	20	46.5	45.2
M	60	52.0	53.7
V	20	1.5	1.1
S	60	76.1	75.7
M	20	22.8	23.6
V	20	1.1	0.7
S	32.0	67.1	66.1
A	48.8	32.5	33.4
V	19.2	0.4	0.5
S	30.2	70.4	71.0
A	15.4	26.2	26.8
V	54.4	3.4	2.2

is obtained. By multiplying the modified stationary-state expressions (18) together the relation

$$k_{12}k_{23}k_{31} = k_{21}k_{13}k_{32}$$

is obtained, and by dividing both sides by $k_{11}k_{22}k_{33}$

$$r_{12}r_{23}r_{31} = r_{21}r_{13}r_{32} \tag{20}$$

The implications of these simplifications are that the rate of conversion of $PM_1\cdot$ radicals to $PM_2\cdot$, by simple addition of M_2, is the same as the rate of conversion of $PM_2\cdot$ directly to $PM_1\cdot$, and similarly for other radical pairs. Thus, the rate of formation of M_1—M_2 bonds is equal to the rate of formation of M_2—M_1 bonds (this equality must exist in binary systems, but its existence in multicomponent systems is not obvious). This situation can be represented schematically as

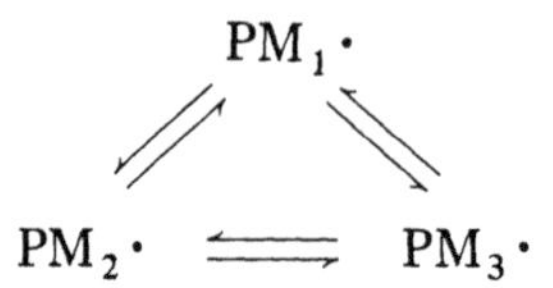

(21)

where the two reactions on each side of the triangle are in equilibrium. These ideas are the same as those embodied in the scheme proposed by Ham [34, 35], who suggested that the probability of forming any sequence, e.g. $M_1M_2M_3M_1$, is equal to that of forming its inverse, $M_1M_3M_2M_1$. This view was expressed in the form

$$P_{12}P_{23}P_{31} = P_{13}P_{32}P_{21} = \mathscr{P} \tag{22}$$

where P_{12} is the probability that a radical $PM_1\cdot$ adds a unit of M_2. In these terms the terpolymer composition equation becomes

$$\frac{d[M_1]}{d[M_2]} = \frac{P_{21}}{P_{12}} = \frac{P_{21}P_{31} + P_{32}P_{21} + P_{23}P_{31}}{P_{12}P_{32} + P_{31}P_{12} + P_{13}P_{31}}$$

$$\frac{d[M_1]}{d[M_3]} = \frac{P_{31}}{P_{13}} = \frac{P_{21}P_{31} + P_{23}P_{31} + P_{32}P_{21}}{P_{13}P_{23} + P_{12}P_{23} + P_{21}P_{13}}$$

These techniques have also been used to simplify the relations for multicomponent systems [25, 33, 35]. As in the more complex expressions, if one monomer will not homopolymerize the copolymer composition is indeterminate unless at least one experiment is carried out to determine the rate coefficients for addition of that monomer to conjugate radicals of the other monomers. Simplified expressions for azeotropic terpolymer formation have also been described [25].

There would appear to be no good grounds for adopting these simplifying assumptions, except that they appear to work in that they adequately predict copolymer compositions in many ternary systems. Indeed Ham [31] has suggested that deviations from the equality (22) could be taken to identify errors in reactivity ratios determined from binary copolymerizations. It was proposed [35] that, for equimolar concentrations of monomers the products in eqn. (22) could be equated to $\mathscr{P}$ which takes on values of 0.037 for systems of three conjugated monomers (acrylic and methacrylic acids and esters, acrylonitrile, styrenes, etc.) and 0.006 when one or two monomers have low conjugation (vinyl esters, vinyl chloride, etc.) Equation (22) can be expressed, assuming equimolar monomer concentrations, in terms of reactivity ratios as

$$\left(\frac{r_{13}}{r_{13}+r_{12}r_{13}+r_{12}}\right)\left(\frac{r_{21}}{r_{21}+r_{23}r_{21}+r_{23}}\right)\left(\frac{r_{32}}{r_{32}+r_{32}r_{31}+r_{31}}\right)=$$

$$\left(\frac{r_{12}}{r_{12}+r_{13}r_{12}+r_{13}}\right)\left(\frac{r_{31}}{r_{31}+r_{31}r_{32}+r_{32}}\right)\left(\frac{r_{23}}{r_{23}+r_{23}r_{21}+r_{21}}\right)=\mathscr{P}$$

(from which eqn. (20) may be derived). From this relation and eqn. (20), and accepting the proposed values of $\mathscr{P}$, if the reactivity ratios for the binary copolymerizations M_1—M_2 and M_2—M_3 are known, the reactivity ratios for the copolymerization M_1—M_3 may be computed.

The adoption of this simplified approach has received both criticism and support from various workers [36—39]. Mayo [36] suggested that the equality (22) should be recast in the form

$$\frac{P_{12}P_{23}P_{31}}{P_{13}P_{32}P_{21}}=H$$

and demonstrated that considerable deviations from $H = 1$ occur in practice; the significance of the deviations has been the subject of some debate [37, 39]. Mayo also pointed out that the Alfrey—Price "*Q*—*e* scheme" [40] and the Bamford—Jenkins "Patterns scheme" of radical reactivity [41] (when a simplifying assumption which does not hold generally is used) (Chap. 5) also predict $H = 1$. But, as Jenkins [38] has pointed out, if the simplifying assumption in the patterns scheme is removed H is not restricted to unity and he demonstrated that observed values of H are closer to values calculated from the patterns scheme in full than they are to unity.

O'Driscoll and Izu [42] have considered the situation where the simplified stationary-state equations (18) do not apply, so that, while stationary concentrations of the individual radical species are maintained, there is a net flux in the cycle

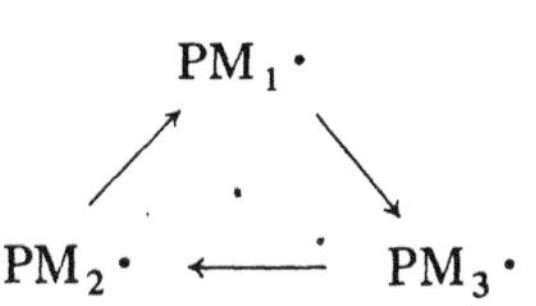

(23)

If such a flux exists there is a net tendency for $PM_1\cdot$ radicals to add M_3, and so on around the cycle. The consequence of this is that in the direction of chain propagation there would be more $M_1M_3M_2$ sequences than the reverse, and the polymer chain would have a direction. These authors were unable to obtain a distinction between the models (21) and (23) from calculated values of the flux, probably because of the lack of precision in values of reactivity ratios.

Although reactivities are considered in detail in the next chapter, we may note here that in the early days of copolymerization studies Mayo and Walling [13] demonstrated that monomers could be arranged in a general order of reactivity. Furthermore, a reactive monomer normally leads to an unreactive conjugate radical, and vice versa. While this general pattern of behaviour holds true, there are a number of exceptions and these manifest themselves in apparently unexpected reactivity ratios for binary copolymerizations. It might be anticipated that such effects would also be apparent in ternary systems. The inclusion of the third monomer, however, appears to mask these deviations and prevent distinction between the various models. Eventually it may be possible to distinguish between schemes (21), (23) from an analysis of sequence distributions. For example, for any particular system, scheme (21) would predict a certain proportion of $M_1M_3M_1$ sequences, while scheme (23) would predict a smaller proportion of these and a larger proportion of $M_1M_3M_2$ sequences.

Additional relationships for multicomponent systems have been compiled in a recent review of multicomponent copolymerization theory [25].

2.1.5 Effects of non-terminal monomer units

In the whole of the preceding discussion it has been assumed that the reactivity of a propagating radical is determined solely by the nature of the terminal monomer from which that radical was derived. The possibility that the penultimate monomer unit may influence the reactivity of the radical (i.e. the rate coefficients of the addition of, say, M_2 to $PM_1M_1\cdot$ and $PM_2M_1\cdot$ are *not* equal) was envisaged by Merz et al. [43]. To derive the copolymer composition equation in such circumstances, for the binary copolymerization of M_1 M_2, it is necessary to consider the eight propagation reactions

$$PM_1M_1\cdot + M_1 \xrightarrow{k_{111}} PM_1M_1M_1\cdot$$

$$PM_1M_1\cdot + M_2 \xrightarrow{k_{112}} PM_1M_1M_2\cdot \qquad (24)$$

$$PM_2M_1\cdot + M_1 \xrightarrow{k_{211}} PM_2M_1M_1\cdot$$

$$\begin{aligned}
&PM_2M_1\cdot + M_2 \xrightarrow{k_{212}} PM_2M_1M_2\cdot \\
&PM_2M_2\cdot + M_1 \xrightarrow{k_{221}} PM_2M_2M_1\cdot \\
&PM_2M_2\cdot + M_2 \xrightarrow{k_{222}} PM_2M_2M_2\cdot \\
&PM_1M_2\cdot + M_1 \xrightarrow{k_{121}} PM_1M_2M_1\cdot \\
&PM_1M_2\cdot + M_2 \xrightarrow{k_{122}} PM_1M_2M_2\cdot
\end{aligned} \tag{24 (contd.)}$$

The composition equation for scheme (24) is

$$n = \frac{d[M_1]}{d[M_2]} = \left\{1 + r_1' \frac{[M_1]}{[M_2]} \cdot \frac{r_1[M_1] + [M_2]}{r_1'[M_1] + [M_2]}\right\} \left\{1 + r_2' \frac{[M_2]}{[M_1]} \cdot \frac{r_2[M_2] + [M_1]}{r_2'[M_2] + [M_1]}\right\}^{-1} \tag{25}$$

where

$$r_1 = k_{111}/k_{112}, \quad r_1' = k_{211}/k_{212}$$

$$r_2 = k_{222}/k_{221}, \quad r_2' = k_{122}/k_{121}$$

Merz et al. [43] have also provided relations describing sequence distributions in copolymerizations involving penultimate effects. Price [22] has used a statistical approach to copolymerizations of this type and has extended the treatment to cases in which the unit preceding the penultimate can influence the reactivity of the terminal unit of the growing chain.

If one monomer, say M_2, in a binary copolymerization will not homopolymerize (i.e. $r_2 = r_2' = 0$) the composition equation (25) reduces to

$$n - 1 = r_1' \frac{[M_1]}{[M_2]} \cdot \frac{r_1[M_1] + [M_2]}{r_1'[M_1] + [M_2]} \tag{26}$$

in which, unlike eqn. (5), $(n-1)$ is a non-linear function of $[M_1]/[M_2]$ unless $r_1' = r_1$ (Fig. 4). Barb [44] has analysed certain experimental data in terms of eqn. (26) and has demonstrated that the inclusion of penultimate effects provides a better description of copolymerizations of styrene with maleic anhydride, citraconic anhydride and fumaronitrile. For example, in the copolymerization of styrene (M_1) and maleic anhydride (M_2) at 60 °C, values of $r_1 = 0.017$, $r_1' = 0.063$ and $r_2 = r_2' = 0$ were estimated, in comparison with previously reported values [8] of $r_1 = 0.02$, $r_2 = 0$. Barb [44] suggested that penultimate effects should be most noticeable in copolymerizations where one monomer is CHX=CHX and X is a strong dipole, e.g. fumaronitrile.

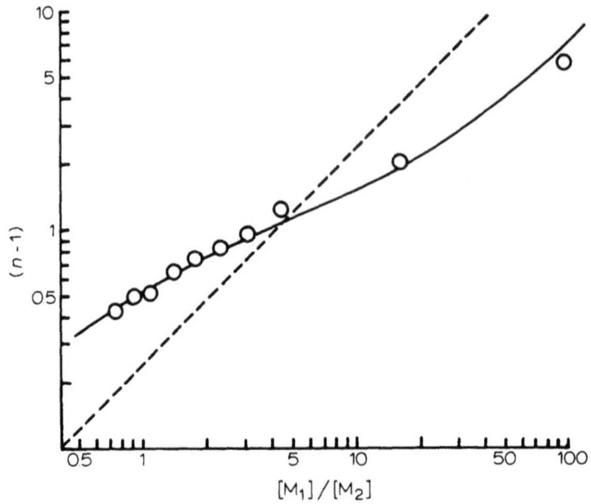

Fig. 4. Composition of styrene (M_1)/fumaronitrile (M_2) copolymers prepared at 40 °C in dioxan solution. Solid line is theoretical curve calculated from eqn. (26) with $r_1 = 0.07$, $r_1' = 1.0$. Dashed line indicates type of dependence expected in the absence of penultimate effects, according to eqn. (5). (After Barb [44], courtesy, J. Polym. Sci.)

Ham [45] has extended the statistical approach to copolymer compositions to include possible effects of more remote units and has applied this approach to a number of systems. In this way it was demonstrated that, in the copolymerization of styrene and fumaronitrile, the inclusion of effects of fumaronitrile units up to four units from the chain-end can provide a better description of this system than the inclusion of penultimate effects alone. From the reactivity ratios estimated on this basis the fumaronitrile content in the polymer cannot exceed 40 mole %, in agreement with experiment. In most cases non-homopolymerizable monomers can be incorporated into copolymers to the extent of 50 mole %, but a model of an alternating styrene (M_1)—fumaronitrile (M_2) copolymer suggests serious steric and dipole repulsions, while a model of the regular structure

$$\cdots\cdots M_1M_2M_1M_2M_1M_1M_2M_1M_2M_1M_1M_2\cdots\cdots$$

(which contains 40 mole % M_2) relieves these effects [45].

An examination of the data in the literature shows that the vast majority of binary copolymerizations are adequately described by eqn. (4), and there would appear to be no reason for invoking penultimate effects generally. This situation, however, may arise because the experimental data are not sufficiently precise to detect penultimate effects, in which case the quoted values of r will be some function of r and r' which provide an adequate description of copolymer composition over the entire composition range. This approach was adopted by Berger and Kuntz

[46], who re-examined certain experimental data and maintained that, even in cases where penultimate effects had been reported, the experimental data were not sufficiently precise to allow distinction between the models. These workers suggested that a determination of sequence distributions would provide a more sensitive method of distinguishing between the two models. Although a full analysis of sequence distributions would be a formidable task, sufficient data may become available as the analysis of high-resolution NMR spectra obtained at high frequencies is developed. Guillot et al. [47] have calculated sequence distributions in acrylonitrile—methyl methacrylate copolymers, on the basis of their reactivity ratio data [48] which indicate a penultimate effect, and have used the distributions to suggest assignments of bands in the 60 MHz NMR spectrum of the copolymer.

The direct extension of the Alfrey—Goldfinger treatment of terpolymerizations to include penultimate effects is complex. Ham [49] has extended the simplified equations for terpolymerizations to include penultimate effects and has used the resulting relations to demonstrate the existence of such effects in some cases.

2.1.6 Copolymerization with depropagation

We have already met the concept of "ceiling-temperature" in Chap. 1 and this phenomenon occurs in copolymerization. Thus, near the ceiling-temperature of one component, in addition to propagation, depropagation reactions are present, and these affect copolymer compositions. Whereas depropagation can occur at all stages of growth in the homopolymerization of a monomer, near its ceiling-temperature, the same is not necessarily true in copolymerization.

Consider the case of the copolymerization of M_1 and M_2 well below the ceiling-temperature of M_1, but close to that of M_2. A radical of the type $PM_2M_2\cdot$ would, under these conditions, have a tendency to depropagate. However, if a unit of M_1 adds to this radical rapidly to produce $PM_2M_2M_1\cdot$ depropagation may be prevented and the sequence of M_2 units will be permanently built into the copolymer. The tendency for depropagation often arises as a result of steric effects, e.g. if M_2 is α-methyl styrene there is a considerable overcrowding in the species $PM_2M_2M_2\cdot$. This strain may be relieved if the penultimate unit is $M_1\cdot$, i.e. $PM_2M_1M_2\cdot$, and such a radical may have little or no tendency to depropagate. Terminal sequences of two or more M_2 units may be required before depropagation is important. It is also conceivable that a radical $PM_2M_2M_1\cdot$ may be induced to undergo depropagation if steric effects from a sequence of M_2 units can exert sufficient influence on the terminal M_2—M_1 bond.

Lowry [50] has derived copolymer composition equations for various combinations of the above effects using a combination of the stationary-state and probability approaches. The resulting equations are complex and are not reproduced here. Unlike simple coplymerizations the copolymer composition in these systems depends on the total monomer concentration, and not just on the ratios of the concentrations of the component monomers. As would be anticipated, the composition curves are markedly temperature dependent in the vicinity of the ceiling-temperature of the appropriate monomer. When the mole fraction of M_2 in the monomer feed approaches unity and at temperatures above the ceiling-temperature of M_2, the mole fraction of M_2 in the copolymer may approach one-half or two-thirds, depending on the nature of the component reactions.

Howell et al. [51] have adopted the probability approach and have extended the treatment to systems in which all propagation reactions are reversible, penultimate effects are present in propagation and the rates of depropagation depend on the nature of the terminal triad, and have obtained a very general copolymer composition equation. These workers have also given expressions for probabilities of sequence distributions in the copolymers. They have also investigated the conditions for equilibrium between monomer and polymer and have shown that, in principle, the ceiling-temperature for copolymerization may be greater than or less than those for individual homopolymerizations [52].

Of the various cases considered by Lowry [50], his case (II) is likely to be the most widely applicable. In this example radicals possessing a terminal M_1 unit never depropagate and radicals having a terminal M_2 unit only tend to depropagate if that terminal M_2 unit is attached to a sequence of two or more M_2 units. These conditions are summarized in the reactions

$$PM_1\cdot + M_1 \longrightarrow PM_1\cdot$$

$$PM_1\cdot + M_2 \longrightarrow PM_2\cdot$$

$$PM_1M_2\cdot + M_1 \longrightarrow PM_1\cdot$$

$$PM_1M_2\cdot + M_2 \longrightarrow PM_1M_2M_2\cdot$$

$$PM_2M_2\cdot + M_2 \rightleftharpoons PM_2M_2M_2\cdot$$

O'Driscoll and Gasparro [53] have demonstrated that this reaction scheme provides an adequate description of the free-radical copolymerizations of styrene and acrylonitrile with α-methyl styrene and methyl methacrylate in the vicinities of the ceiling-temperatures of the latter monomers. Ivin and Spensley [54] also studied the copolymerization of styrene and methyl methacrylate at 132 °C and suggested that terminal sequences of at least three methyl methacrylate units are required to cause depropagation.

The copolymerization of α-methyl styrene and methyl methacrylate has been investigated at temperatures between the ceiling-temperatures of the respective homopolymerizations. Application of the simple copolymer composition equation to polymers prepared at about and above 100 °C gave an anomalous negative value for the reactivity ratio for α-methyl styrene [55, 56]. Izu and O'Driscoll [56] have shown that it is possible to predict copolymer compositions without apparent negative rate coefficients if all propagation reactions are considered to be reversible.

2.1.7 Determination of reactivity ratios in binary copolymerizations

Of fundamental importance in studies of monomer and radical reactivities is an accurate knowledge of the relative rates of reactions of monomers and radicals, i.e. of the reactivity ratios. This can only be achieved by quantitative isolation and accurate analysis of the copolymer and correct treatment of the data.

To use eqn. (4) to determine values of r in a binary copolymerization it is, in principle, only necessary to carry out two copolymerization experiments with different values of $[M_1]/[M_2]$, determine the corresponding values of n and solve the simultaneous equations. However, this method can lead to considerable inaccuracies and should be avoided. A somewhat more satisfactory approach is to carry out a number of copolymerizations over as wide a range of monomer feed compositions as possible, when values of r may be determined by fitting theoretical equations to plots of copolymer and monomer feed compositions, as in Fig. 1. The final choice of the values of r_1 and r_2 is subjective, especially under conditions where the composition curve is not very sensitive, to variations in r_1 or r_2. If one monomer does not homopolymerize more reliable values of r for the other monomer may be obtained by use of the simplified composition eqn. (5), when the effects of non-terminal units may become apparent (cf. Sect. 2.1.5, eqn. (26)).

A still more reliable approach is to use the graphical method of "intersecting lines" due to Mayo and Lewis [1], who re-wrote the instantaneous composition eqn. (4) in the form

$$r_2 = \frac{[M_1]}{[M_2]}\left\{n\left(1 + \frac{[M_1]}{[M_2]}r_1\right) - 1\right\}$$

Substitution of values of n and $[M_1]/[M_2]$ from one experiment into this equation provides a linear relation between possible values of r_1 and r_2. When this is done for a series of experiments a series of such relations is obtained. If the experimental data are sufficiently accurate the plots of r_1 vs. r_2 should intersect at a common point, giving the correct values of r_1, r_2. More usually there is an area of the graph in which the lines intersect and "best values" of r_1 and r_2 must be chosen. The area of intersection provides an estimate of the errors involved.

Another variation on eqn. (4) was proposed by Fineman and Ross [57], who arranged eqn. (4) in the form

$$\left(\frac{n-1}{n}\right)\frac{[\mathrm{M}_1]}{[\mathrm{M}_2]} = \left(\frac{[\mathrm{M}_1]}{[\mathrm{M}_2]}\right)^2 \frac{r_1}{n} - r_2$$

so that a plot of $\{(n-1)/n\}\,[\mathrm{M}_1]/[\mathrm{M}_2]$ vs. $([\mathrm{M}_1]/[\mathrm{M}_2])^2$ gives r_1 and r_2 directly.

The various methods described above are based on the use of the instantaneous copolymer composition equation in its various forms, and their use requires that the copolymerization reactions are taken to only low conversion, hence, careful isolation of small amounts of polymer is required. It is possible to determine r_1 and r_2 from experiments taken to higher conversion by adopting the approach of Lewis and Mayo [1] with the use of eqn. (11). This involves calculating r_1 for selected values of p and then using the values r_1 to calculate r_2. The intersection of r_1 vs. r_2 plots for the various values of p gives the correct values of r_1 and r_2.

A second method described by Meyer [58], which uses data obtained at finite conversions, is based on the use of Skeist's eqn. (12), which on integration [11] yields the relation

$$\frac{[\mathrm{M}]}{[\mathrm{M}]_0} = \left(\frac{f_1}{f_1^0}\right)^{\alpha}\left(\frac{f_2}{f_2^0}\right)^{\beta}\left(\frac{f_1^0-\delta}{f_1-\delta}\right)^{\gamma} \tag{27}$$

where

$$\alpha = r_2/(1-r_2), \quad \beta = r_1/(1-r_1),$$

$$\gamma = (1-r_1r_2)/[(1-r_1)(1-r_2)], \tag{28}$$

$$\delta = (1-r_2)/(2-r_1-r_2)$$

The reader is referred to the original article for details of this method but in principle it depends on selecting combinations of r_1 and r_2 which satisfy relations (27) and (28) and then using the method of intersecting lines to determine the best values.

The precautions required in using these various techniques and methods of estimating the uncertainties in values of r have been discussed by Alfrey et al. [40]. More recently Behnken [59] and Tidwell and Mortimer [60] have discussed the subjective nature of the above methods and have described non-linear methods of data analysis which minimize the errors between the calculated and experimental copolymer compositions. Both discussions give accounts of estimations of errors and the design of experiments to give the most suitable results for evaluating reactivity ratios.

2.1.8 *Effects of reaction conditions on reactivity ratios*

a. Precipitation of polymer

Many polymers and copolymers are insoluble in their monomers and in common organic solvents, consequently the polymers precipitate as they are formed. This leads to considerable complications in the kinetics of homopolymerizations of monomers such as acrylonitrile and vinyl chloride. It has been demonstrated [61] that radicals are occluded in the precipitating polymer during the polymerization of acrylonitrile and that when the polymer particles are suitably swollen these radicals are able to propagate [62]. Under certain conditions the precipitated polymer forms an effective barrier around the radical and prevents access of monomer while ferric chloride is able to diffuse into the polymer particle and deactivate the radicals [63]. It is therefore possible to envisage the situation in a copolymerization reaction where one monomer is able to diffuse into a polymer particle more readily than the other monomer. The swollen polymer particles would then contain an excess concentration of one monomer and polymerization in these regions would preferentially incorporate that monomer, leading to erroneous results.

In practice many copolymerizations are carried out in precipitating media and from the values quoted in the literature there appears to be little evidence that this effect has any marked influence on the results. Mayo and Walling [13], however, have reported that the copolymerization of vinyl chloride and vinyl acetate yields inconsistent results when phase separation occurs and that the inconsistencies disappear when a suitable solvent is added. Hinsch and Curro [64] observed that the copolymerization of acrylic acid and a fluorinated heptyl ester in swollen ethylene–propylene rubber appeared to produce anomalous reactivity ratios, but when the true monomer concentrations in the rubber were used the anomaly disappeared, and the simple composition eqn. (4), with normal reactivity ratios, applied. Any apparent variations in reactivity ratios arising from such effects are, of course, physical in origin and lead to an incorrect assessment of reactivities.

b. Copolymerization in emulsion

Commercially many polymerizations and copolymerizations are carried out in aqueous emulsions, when, ideally, polymerization occurs only in the swollen polymer particles. It is, therefore, important to know if the normal reactivity ratios apply under such circumstances. In practice it is found that, if the monomers are water insoluble and the monomer-feed composition in the polymer particles is equal to the overall feed composition, the reactivity ratios in emulsion and in solution are essentially the

TABLE 4
Compositions of copolymers of styrene and itaconic acid prepared in dioxan solution and in emulsion at 70 °C [69]

Method of polymerization	Mole % itaconic acid in	
	Monomer Feed	Copolymer
Solution	8.17	21.4
	16.68	28.0
	34.80	42.5
	54.56	49.0
Emulsion	16.67	2.6
	34.80	0.3
	54.56	2.0

same [65, 66]. However, if one monomer is appreciably water soluble its relative concentration in the polymer particle is reduced and the copolymer composition varies accordingly. Thus, in the copolymerization of styrene and acrylonitrile the copolymers prepared in emulsion contain slightly less acrylonitrile than the corresponding copolymer prepared in solution [67] as a result of the solubility of acrylonitrile in water and the reduction of its concentration at the locus of reaction [66—68]. A more striking example is provided by the copolymerization of styrene and itaconic acid, when copolymers prepared in emulsion contain very little itaconic acid compared with those prepared in solution (Table 4). This is a direct result of the high solubility of itaconic acid in water [69]. Obviously care should be taken in using data from emulsion systems for predicting compositions or assessing reactivities.

c. *Temperature*

From the rather scattered data in the literature it is apparent that reactivity ratios are normally insensitive to temperature changes. This is to be expected since we may write the reactivity ratio r_1 as

$$r_1 = \frac{k_{11}}{k_{12}} = \frac{A_{11}\exp(-E_{11}/RT)}{A_{12}\exp(-E_{12}/RT)} = \frac{A_{11}}{A_{12}}\exp(E_{12}-E_{11})/RT$$

Thus, the temperature variation of r_1 is determined by the difference in the activation energies for the addition of the two monomers to the $PM_1\cdot$ radical and this difference is usually small [70], in most cases probably less than 2 kcal mole^{-1}. O'Driscoll [71] has derived expressions for expected variations of r with temperature and has shown that all values of r tend to unity at high temperatures. However, significant changes in r over practicable temperature ranges are small unless r is high (>10) or small (<0.1).

d. *Pressure*

The variation in rate of reaction with pressure is dependent on $\Delta V^{\dagger}$, the difference in partial molar volumes of the transition state and of the reactants. Since the reactivity ratios are the ratios of the rate coefficients for two similar reactions, the variation in r with pressure depends on the difference in the values of $\Delta V^{\dagger}$ for these reactions. Thus, we may anticipate that reactivity ratios will be fairly insensitive to variations in pressure. Although little work of this type has been reported this expectation appears to be borne out in practice. Lamb and Weale [72] found that, while the rates of copolymerization of styrene and methyl methacrylate increased significantly on increasing the pressure up to 400 atm, the reactivity ratios did not change. This indicates that $\Delta V^{\dagger}$ for all four propagation reactions is the same.

Burkhart and Zutty [73] investigated the copolymerizations of acrylonitrile with styrene and with methyl methacrylate in the pressure range 1—1000 atm, and in both cases it was found that the reactivity ratios increased with increasing pressure (Table 5).

TABLE 5
Effect of pressure on monomer reactivity ratios in copolymerizations of acrylonitrile (M_1) [72]

Pressure (atm)	M_2	r_1	r_2
1	Styrene	0.07	0.37
10^3	Styrene	0.14	0.55
1	Methyl methacrylate	0.12	1.34
10^3	Methyl methacrylate	0.45	2.01

The consistent increase in all reactivity ratios in Table 5 indicates that in these systems the value of $\Delta V^{\dagger}$ for homopropagation is greater than that for cross-propagation. Furthermore, comparing the data with those of Lamb and Weale [72], the same proportional increase in r_2 for both cases in Table 5 indicates that the values of $\Delta V^{\dagger}$ for the addition of acrylonitrile to styrene and methyl methacrylate radicals are identical, while the larger proportional increase in r_1 with methyl methacrylate suggests that $\Delta V^{\dagger}$ is larger for the addition of styrene to acrylonitrile than that for the addition of methyl methacrylate.

e. *Solvent effects*

A few specific copolymerization systems, such as the copolymerizations of acidic with basic monomers, exhibit marked variations in reactivity

References pp. 329—332

ratios as the solvent medium is modified. One example is the copolymerization of methacrylic acid (M_1) with diethyl-aminoethyl methacrylate (M_2) in aqueous solutions of different pH [74]. At pH 1.2, when the acid monomer is unionized, $r_1 = 0.98$ and $r_2 = 0.90$ while at pH 7.2, when both monomers are ionized, $r_1 = 0.08$ and $r_2 = 0.65$. These effects arise since both monomers can exist in two forms in solution and the lower reactivity ratios quoted above reflect a greater tendency for alternation at pH 7.2 due to the interaction of oppositely charged species.

Apart from systems of the type mentioned above, an examination of published reactivity ratios indicates that variations in reactivity ratios with changes in the solvent medium are only minor and often within experimental error. It has sometimes been concluded that reactivity ratios are solvent independent. However, there is an increasing amount of evidence that many of the small variations in reactivity ratios are real.

Since polar factors are important in influencing the tendency for alternation in many systems (Chap.5,), it might be expected that the dielectric constant of the reaction medium would influence the values of r. In general, variations in dielectric constant of the medium do not appear to have a marked effect. For example, reactivity ratios in copolymerizations of styrene with methyl methacrylate appear to be similar in benzene and in acetonitrile solution [75, 76]. Ito and Otsu [77] have correlated the small variations in reactivity ratios observed for this monomer pair with changes in the C=C stretching frequencies of the monomers in the different solvents. However, in copolymerization of α-methyl styrene (M_1) with maleic anhydride (M_2), where the range of polar factors is greater, reactivity ratios are reported to change [76] from $r_1 = 0.005$ and $r_2 = 0.27$ in benzene solution to $r_1 = 0.038$ and $r_2 = 0.008$ in acetonitrile solution, but it has been argued in this case that the variation is not due to polar factors [76].

Ryabov et al. [78] observed that in copolymerizations of methacrylic acid with methyl methacrylate and with styrene, reactivity ratios were modified when dimethylformamide was used as solvent. Aso et al. [79 (a, b)] found reactivity ratios for copolymerizations of styrene (M_1) with acrylonitrile (M_2) to be virtually identical with benzene, acetonitrile and dimethylformamide as solvents ($r_1 = 0.36 \pm 0.04$, $r_2 = 0.14 \pm 0.04$) but very different for carbon disulphide solutions ($r_1 = 0.04$, $r_2 = 0.32$). It was suggested [79(b)] that this latter variation arose from a strong interaction between the propagating radical and carbon disulphide.

Consequently, although solvent effects in free-radical copolymerizations are generally minor it must be acknowledged that real effects exist. More detailed investigation of these effects using accurate values of reactivity ratios should provide interesting information on interactions of radicals and monomers with solvents and on the reactivities of the various species involved in reaction.

f. Complex formation

While the formation of alternating copolymers from styrene and maleic anhydride can be reconciled with low values of reactivity ratios for the two monomers, it has been proposed that alternation could arise from the direct participation of a 1 : 1 monomer—monomer complex between styrene (donor) and maleic anhydride (acceptor) in the propagation reaction [80]. Although evidence may be quoted in favour of the participation of such complexes in copolymerizations of certain monomer pairs the mechanisms of the reactions are the subject of continuing debate; a recent review has been presented by Iwatsuki and Yamashita [81]. The role of such complexes may be expected to vary with the nature of any solvent present in the system and the simultaneous formation of monomer—solvent complexes.

A new area of research in the field of copolymerization stems from the observation of Hirooka et al. [82] that alternating copolymers can be formed from propene and acrylonitrile in the presence of ethylaluminium dichloride. Subsequently a large number of papers have been published in which it has been reported that alternating copolymers are formed from combinations of donor and acceptor monomers in the presence of Lewis acids [83]. Although a number of these polymerizations proceed in the absence of added initiator, they are often accelerated by addition of free-radical sources, but the kinetics do not always conform to those of simple free-radical reactions. The detailed mechanisms of these reactions remain obscure although it is certain that monomer—Lewis acid complexes play a major role in the reactions. While a number of workers assume the presence of radical-type intermediates in the reactions the nature of the reactive species remains uncertain and it is probable that no single mechanism is universally applicable. Additional evidence is required before the kinetic aspects of this area can be discussed in detail.

2.2 RATES OF COPOLYMERIZATION

2.2.1 The rate equation

While a study of copolymer compositions provides a wealth of data on reactivities it gives no information on the overall rates of reaction. In order to obtain an expression describing rates of copolymerization it is necessary to include, in addition to the various propagation reactions considered in previous sections, one initiation reaction for each monomer and all possible bimolecular termination reactions. A typical reaction scheme for a catalysed binary copolymerization consists of the reactions

Initiation

$$\text{Initiator} \longrightarrow \text{R}\cdot$$
$$\text{R}\cdot + \text{M}_1 \longrightarrow \text{PM}_1\cdot$$
$$\text{R}\cdot + \text{M}_2 \longrightarrow \text{PM}_2\cdot$$

Propagation

$$\text{PM}_1\cdot + \text{M}_1 \longrightarrow \text{PM}_1\cdot$$
$$\text{PM}_1\cdot + \text{M}_2 \longrightarrow \text{PM}_2\cdot$$
$$\text{PM}_2\cdot + \text{M}_1 \longrightarrow \text{PM}_1\cdot \qquad (29)$$
$$\text{PM}_2\cdot + \text{M}_2 \longrightarrow \text{PM}_2\cdot$$

Termination

$$2\text{PM}_1\cdot \xrightarrow{k_{t11}} \text{Polymer}$$
$$2\text{PM}_2\cdot \xrightarrow{k_{t22}} \text{Polymer}$$
$$\text{PM}_1\cdot + \text{PM}_2\cdot \xrightarrow{k_{t12}} \text{Polymer}$$

where the rate coefficients for propagation are the same as those defined in Sect. 2.1.1. Transfer reactions may be included in the above scheme, but, as in simple homopolymerizations, if they are not accompanied by retardation, they have no influence on the rate of reaction and may be neglected. If it is assumed that kinetic chains are long, so that monomer is effectively consumed only in propagation, all initiation reactions may be combined into a total rate of chain starting $\mathscr{I}$. Making the usual assumptions that a stationary state exists and that radical reactivities are independent of chain length (p. 3), and are also unaffected by the nature of non-terminal units, the overall rate of polymerization is

$$-\frac{\mathrm{d}([\text{M}_1] + [\text{M}_2])}{\mathrm{d}t} =$$
$$\frac{\mathscr{I}^{1/2}(k_{11}k_{21}[\text{M}_1]^2 + 2k_{12}k_{21}[\text{M}_1][\text{M}_2] + k_{22}k_{12}[\text{M}_2]^2)}{(k_{t11}k_{21}^2[\text{M}_1]^2 + 2k_{t12}k_{12}k_{21}[\text{M}_1][\text{M}_2] + k_{t22}k_{12}^2[\text{M}_2]^2)^{1/2}}$$

which, by substituting

$$\delta_1 = k_{t11}^{1/2}/k_p, \quad \delta_2 = k_{t22}^{1/2}/k_{22} \qquad (30)$$
$$\phi = k_{t12}/(k_{t11}k_{t22})^{1/2}$$

becomes

$$\frac{-\mathrm{d}([\mathrm{M}_1] + [\mathrm{M}_2])}{\mathrm{d}t} =$$

$$\frac{\mathscr{I}^{1/2}(r_1[\mathrm{M}_1]^2 + 2[\mathrm{M}_1][\mathrm{M}_2] + r_2[\mathrm{M}_2]^2)}{(r_1^2\delta_1^2[\mathrm{M}_1]^2 + 2\phi r_1 r_2 \delta_1 \delta_2[\mathrm{M}_1][\mathrm{M}_2] + r_2^2\delta_2^2[\mathrm{M}_2]^2)^{1/2}} \tag{31}$$

The rate of consumption of M_1 is given by

$$\frac{-\mathrm{d}[\mathrm{M}_1]}{\mathrm{d}t} = \frac{\mathscr{I}^{1/2}(r_1[\mathrm{M}_1]^2 + [\mathrm{M}_1][\mathrm{M}_2])}{(r_1^2\delta_1^2[\mathrm{M}_1]^2 + 2\phi r_1 r_2 \delta_1 \delta_2[\mathrm{M}_1][\mathrm{M}_2] + r_2^2\delta_2^2[\mathrm{M}_2]^2)^{1/2}} \tag{32}$$

and a corresponding equation describes the rate of consumption of M_2.

Equation (31) may be extended to ternary systems [84]. Expressions for degrees of polymerization, in the absence of transfer, are obtained in exactly the same way as for homopolymerizations. The influence of transfer to solvent on degrees of polymerization is readily allowed for by an expression of the Mayo form [65, 85]

$$\frac{1}{\bar{P}} = \frac{1}{\bar{P}_0} + C\frac{[\mathrm{S}]}{[\mathrm{M}]}$$

where $\bar{P}$ and $\bar{P}_0$ are the degrees of polymerization in the presence and absence of transfer, respectively, [M] is the total monomer concentration, [S] is solvent concentration and C is a composite transfer constant given by

$$C = \frac{r_1 C_1[\mathrm{M}_1] + r_2 C_2[\mathrm{M}_2]}{r_1[\mathrm{M}_1]^2 + 2[\mathrm{M}_1][\mathrm{M}_2] + r_2[\mathrm{M}_2]^2} \tag{33}$$

where C_1 and C_2 and the transfer constants for S in the homopolymerizations of M_1 and M_2, respectively. Transfer to monomer is not so readily allowed for, since it is not possible to separate the rate coefficients of the cross-transfer processes [86]

$$\mathrm{PM_1\cdot + M_2 \longrightarrow Polymer + M_2\cdot}$$

$$\mathrm{PM_2\cdot + M_1 \longrightarrow Polymer + M_1\cdot}$$

Experimental studies of rates of copolymerization in terms of eqn. (31) are somewhat inconvenient, since some suitable method of determining the total rate of consumption of monomer (or the rate of consumption of one monomer if eqn. (32) is used) must be devised. The conventional dilatometric technique may be used but the conversion factor relating contraction to monomer consumption is a function of the copolymer composition. Use of eqn. (31) can potentially provide information on values of $\mathscr{I}$ and ϕ. For polymerizations initiated by azo-bis-isobutyronitrile the total rate of initiation appears to be the same as that expected from rates of initiation of homopolymerizations [87] but deviations from this appear to occur with benzoyl peroxide as initiator [88, 89]. Potentially the most interesting information available from eqn. (31) is on the nature of the cross-termination reaction (i.e. the significance of ϕ eqn. (30)).

2.2.2 Termination processes

Kinetic studies have demonstrated that for the copolymerization of similar monomers (e.g. styrene and *p*-methoxy styrene) the value of ϕ is unity [90]. On the other hand, for many copolymerizations of dissimilar monomers (e.g. styrene with an acrylate, methacrylate or fumarate) values of ϕ greatly in excess of unity have been observed [87, 89, 91–93], i.e., cross-termination is favoured in comparison with termination of like radicals. This phenomenon has often been explained in terms of polar factors in the transition state such as those responsible for the alternating tendency of the propagation reactions in such systems (Chap. 5). In some cases ϕ is found to vary with copolymer composition. For example, with styrene and butyl acrylate it has been observed [89] that, as the butyl acrylate content in the polymer decreases from 90 % to 20 %, ϕ increases from about 2 to more than 100. Such variations in ϕ have been attributed by Barb [44] to the influence of penultimate effects in the termination reaction.

Before discussing the nature of the termination reaction further, we may briefly mention the qualitative influence of high values of ϕ on the overall rate of copolymerization. The rate of growth of the copolymer chain is determined by the rate coefficients of the various propagation reactions. Some copolymerizations are expected to have low rates of chain growth, while others, as a result of two rapid cross-propagation reactions may be expected to have high rates of chain growth. High values of ϕ will counteract the tendency to produce long chains in the latter systems and may result in low observed rates of copolymerization.

The interpretation of ϕ values in terms of chemical effects, as outlined above, is only valid if the rate of termination is uninfluenced by physical effects. As we saw in an earlier chapter (p. 67) the termination reaction in homopolymerizations is often diffusion controlled. Atherton and North [94] have demonstrated that such effects are also present in copolymerizations. Under these conditions the conventional significance of the ϕ factor is invalid [95], since the rate of termination is influenced by the physics of the polymer chain, such as its flexibility, and by the viscosity of the solvent. Atherton and North [94] produced an alternative derivation of the copolymer rate equation in which they used a single rate coefficient $k_{t(12)}$ to describe the termination reaction, when they obtained the relation

$$\frac{-\mathrm{d}([\mathrm{M}_1] + [\mathrm{M}_2])}{\mathrm{d}t} = \frac{\mathcal{I}^{1/2}(r_1[\mathrm{M}_1]^2 + 2[\mathrm{M}_1][\mathrm{M}_2] + r_2[\mathrm{M}_2]^2)}{k_{t(12)}^{1/2}(r_1[\mathrm{M}_1]/k_{11} + r_2[\mathrm{M}_2]/k_{22})}$$

$k_{t(12)}$ would be expected to be determined by the composition of the copolymer, since this will affect the flexibility of the growing chain, and they suggested that $k_{t(12)}$ may be written as a linear combination of the rate coefficients of homopolymerizations, viz.

$$k_{t(12)} = k_{t(11)}F_1 + k_{t(22)}F_2 \tag{34}$$

This approach has been applied with some success, substituting mole fractions of the respective monomers in the copolymer for F_1 and F_2. In general $k_{t(12)}$ may not vary with copolymer composition in the simple manner suggested in eqn. (34), particularly if the flexibility of the copolymer chain does not vary monotonically with composition. Thus, high values of ϕ do not necessarily signify high rates of cross-termination due to chemical effects, and it is notable that high values of ϕ are not found in gas-phase reactions [96] where diffusion control is absent.

In view of the development of the concept of diffusion-controlled termination, it is necessary to reconsider the mechanism of the cross-termination reaction. Apparently, the only information available is the old data of Bevington et al. [97] on the copolymerization of methyl methacrylate and styrene. These workers calculated the fractions of terminations between like and unlike radicals from rates of polymerization and initiation. They concluded that, in all cases studied, at least 85 % of all termination reactions were cross-termination processes, i.e. consistent with a high value of ϕ. They further concluded, from tracer studies, that the cross-termination reaction occurs mainly by radical combination. Indeed, this is the result we might expect in view of studies of termination reactions in homopolymerizations (Chap. 1), since the only radical possessing an α-methyl group is the methyl methacrylate radical and the styryl radical is probably too unreactive to abstract a hydrogen atom from it. Also the transition state for combination of unlike radicals would not be seriously strained, as in the termination of two methacrylate radicals. However, we have recalculated the data of Bevington et al. [97], assuming a single value of k_t. This produces a considerable change in the proportions of the various radical—radical reactions (calculated from the relative populations of the two radical species determined from composition data), considerably increasing the proportions of like-radical terminations. The net result of these calculations was that the fraction of disproportion in the cross-termination reaction was apparently numerically significant but negative. This impossible result possibly implies that even when termination is diffusion controlled a single value of k_t may not provide an adequate description of the termination process, but that some preference for cross-termination still exists.

Russo and Munari [98] have attempted to combine both physical and chemical aspects into a common scheme. They adopt the approach that it is only the flexibility, and hence composition, of the chain-end which is important physically, rather than the overall chain composition and flexibility. They proposed the existence of ten termination reactions to take account of both chemical and physical effects of the penultimate unit. It was then assumed that the rate coefficients for the cross-termination reactions were given by the geometric means of appropriate homo-termination reactions. In this way they obtained an expression for the rate of polymerization and, by suitable choice of parameters, they were able to fit the earlier kinetic data on the copolymerizations of styrene with methyl methacrylate and with *n*-butyl acrylate to their equation.

During copolymerization studies of styrene and α-methyl styrene at termperatures close to the ceiling-temperature of the latter monomer, it was noted that α-methyl styrene had a retarding effect on the copolymerization [53]. O'Driscoll and Dickson [99] have studied the kinetic implications in copolymerizations carried out near the ceiling-temperature of one monomer for Lowry's case (II) [50] (see p. 308). They concluded that during such a copolymerization large numbers of short radicals of the monomer, subject to depropagation, are formed and that these short radicals have a high rate coefficient for termination. Thus, they dominate the kinetics by terminating with long-chain radicals, reducing both the rate and degree of polymerization.

2.2.3 *The integrated rate equation*

Integration of eqn. (31) to obtain a relation between conversion and time has been performed by de Butts [100], but the result is unwieldy. A more convenient integration has been performed by O'Driscoll and Knorr [101]. This was achieved by rearranging the equations for consumption of the individual monomers (eqn. (2)) into the form

$$\frac{-\mathrm{d}\ln[\mathrm{M}_1]}{\mathrm{d}t} = [\mathrm{R}\cdot]\left\{(k_{11} - k_{21})\frac{[\mathrm{PM}_1\cdot]}{[\mathrm{R}\cdot]} + k_{21}\right\}$$

$$\frac{-\mathrm{d}\ln[\mathrm{M}_2]}{\mathrm{d}t} = [\mathrm{R}\cdot]\left\{k_{22} + (k_{12} - k_{21})\frac{[\mathrm{PM}_1\cdot]}{[\mathrm{R}\cdot]}\right\}$$

where $[\mathrm{R}\cdot] = [\mathrm{PM}_1\cdot] + [\mathrm{PM}_2\cdot]$. Manipulation of these equations yielded the relation

$$\ln\left\{\left(\frac{f_1}{f_1^0}\right)^a\left(\frac{1-f_1}{1-f_1^0}\right)^b\left(\frac{f_1^0-\delta}{f_1-\delta}\right)^c\right\} = -(k_{21} - Xk_{22})\int_0^t [\mathrm{R}\cdot]\,\mathrm{d}t \qquad (35)$$

where f_1^0 and f_1 are the initial and instantaneous mole fractions of M_1, in the feed, respectively, and

$$X = \frac{k_{11} - k_{21}}{k_{12} - k_{22}}$$

$$a = \alpha(1 - X) + 1 \qquad (36)$$

$$b = \beta(1 - X) - X$$

$$c = \gamma(1 - X)$$

(α, β, γ and δ are defined in eqn. (28), p. 310).

In many cases the rate of termination is insensitive to composition over wide ranges so that it is possible to introduce a single overall rate coefficient, k_t, to describe this process. If termination is purely chemically controlled k_t is given by

$$k_t = k_{t(11)} \frac{[PM_1\cdot]^2}{[R\cdot]^2} + 2k_{t(12)} \frac{[PM_1\cdot][PM_2\cdot]}{[R\cdot]^2} + k_{t(12)} \frac{[PM_2\cdot]^2}{[R\cdot]^2}$$

and if termination is diffusion controlled it may be given by an expression such as eqn. (34). In either case it is possible to introduce the normal stationary-state assumptions and write

$$\frac{-d[R\cdot]}{dt} = k_t[R\cdot]^2$$

(O'Driscoll and Knorr [101] use the alternative definition of k_t, $-d[R\cdot]/dt = 2k_t[R\cdot]^2$) when eqn. (35) becomes

$$\ln\left\{\left(\frac{f_1}{f_1^0}\right)^a \left(\frac{1-f_1}{1-f_1^0}\right)^b \left(\frac{f_1^0-\delta}{f_1-\delta}\right)^c\right\} = -(k_{21} - Xk_{22}) \int_0^t \left(\frac{\mathscr{I}}{k_t}\right)^{1/2} dt \qquad (37)$$

Equation (37) describes f_1 as a function of time if the variation of $\mathscr{I}$ with time is known, and can be used to relate conversion, m, to time through the relation [101]

$$m = 1 - \left(\frac{f_1}{f_1^0}\right)^\alpha \left(\frac{1-f_1}{1-f_1^0}\right)^\beta \left(\frac{f_1^0-\delta}{f_1-\delta}\right)^\gamma \qquad (38)$$

with the parameters defined in eqns. (28) and (36). Equations (37) and (38) apply except under conditions of azeotropic copolymerization, or when $r_1, r_2 = 1$.

A test of eqns. (37) and (38) was carried out by introducing a first-order decay of initiator and appropriate rate coefficients from published data for the copolymerization of methyl methacrylate with vinyl acetate, and comparing calculated and experimental variations of conversion and rate with time [101]. Qualitatively good agreement was achieved in that both sets of data showed a gradual decrease in rate in the early stages of reaction and a marked increase in rate in the later stages, when the supply of methyl methacrylate is diminished. The lack of quantitative agreement is seen in the time at which the increase in rate occurs. O'Driscoll and Knorr [101] point out that the agreement can be improved by adjusting the kinetic parameters, particularly the value of k_t and the efficiency of initiation. They suggest that such studies may be used to determine values of k_t, which is sometimes composition dependent [94, 95], and hence aid the understanding of the physical processes involved in chain termination.

3. Block and graft copolymers and networks

In this section we are concerned with the formation of polymers with structures (II) and (III), defined earlier (p. 287). The polymer literature contains numerous examples of the formation of such materials [102—104] and in many of these systems free-radical processes are involved. The main emphasis in this work has been on the preparation of new materials and an examination of their properties, rather than on kinetic studies of their formation. Indeed many of the processes do not lend themselves to kinetic analysis. For example, various mechanochemical techniques have been employed to prepare block copolymers. If a polymer is swollen with a monomer and the mixture masticated in an inert atmosphere main chain scission of the polymer occurs, producing shorter polymer molecules possessing terminal free radicals. The terminal free radicals then initiate polymerization of the monomer producing block copolymers. Similar effects may be achieved by subjecting polymer in a monomer containing solution to ultrasonic irradiation when cavitation results in main chain scission and the formation of polymer molecules with radical end-groups.

In view of the variety and complexity of the systems used to prepare materials under discussion in the remaining sections, we shall restrict our attention to some of the underlying principles rather than attempt a series of kinetic analyses.

3.1 BLOCK COPOLYMERS

It is possible to prepare block copolymers by free-radical processes which are more amenable to kinetic analysis than those outlined above. For example, a polyfunctional initiator, such as a dihydroperoxide, may be used in the preparation of the prepolymer when a large proportion of the polymers so obtained will contain terminal groups (e.g. hydroperoxide groups) capable of initiation [105—110]. Similar prepolymer molecules can be prepared by using peroxy radicals to initiate the polymerization of the first monomers [111, 112]. Block copolymer formation is then achieved by initiating the polymerization of a second monomer by radicals derived from the terminal groups. There is nothing conceptually new in the polymerization kinetics in these systems, but the detailed mechanisms control the structures of the products obtained.

If the radical HOOṘ is used to initiate the polymerization of monomer A in the preparation of the prepolymer, then the prepolymer will consist of molecules with structures HOORAA·····AAROOH and HOORAA······AA, the proportions of which will depend on the nature of the termination reaction of the propagating radical of A, transfer processes being neglected. Initiation of the second monomer B can result in the formation of a variety of structures. Simple thermal initiation by HOORAA····AA produces two initiating radicals, viz.

$$\text{AA}\cdots\cdots\text{AAROOH} \rightarrow \text{AA}\cdots\cdots\text{AARO}\cdot + \cdot\text{OH}$$

Initiation by the former radical gives rise to a block copolymer radical AA·····AAROBB····B·, while the ·OH radical gives a homopolymer radical of B. Depending on the nature of the termination reaction of B radicals the final reaction mixture will contain a mixture of AA······AABB····BB, AA·····AABB·····BBAA······AA block copolymers and homopolymers in various proportions. Formation of homopolymer of B can be minimized if the initiating group on the prepolymer is a hydroperoxide by use of redox initiation when the formation of HO· radicals is avoided (Chap. 1).

A wider range of products are formed if the prepolymer is HOORAA·····AAROOH, since initiation in the second stage can occur at both ends of the prepolymer molecule. In the early stages of polymerization the block copolymers formed will be mainly AA·····AB·····BB and/or AA·····AABB·····BBAA·····AA through initiation at one end of the prepolymer only. As reaction proceeds the probability of initiation by the second peroxy group on the prepolymer increases. If all B radicals terminated by disproportionation the final products would be BB·····BBAA·····AABB·····BB block copolymers. While if all B radicals terminated by combination long sequences of ·····BA·····AB·····BA·····AB····· would tend to form; the lengths of the sequences would be limited by transfer processes. For termination of B radicals by both disproportionation and combination a whole range of block copolymer structures would be formed, the distribution of which would be determined by the conversion of peroxy groups and the ratio of the termination rate coefficients, k_{tc}/k_{td}. In principle, distributions of the populations of the different structures for various situations may be computed. The individual blocks will also have a distribution of lengths controlled by normal polymerization kinetics.

Essentially similar results may be achieved by introducing photosensitive groups at the ends of the polymer chains [113, 114]. This is usually done through a transfer process, e.g. with $CClBr_3$, in the preparation of the prepolymer which leads to the incorporation of one reactive group per prepolymer molecule. Block copolymer formation then involves photolysis of the end groups in the presence of the second monomer. Polyhalogenated end-groups can also be used in the preparation of block copolymers, without formation of homopolymer of the second monomer, by interaction with metal carbonyls in the presence of a vinyl monomer [115].

As a final general method of preparing block copolymers we may mention the use of prepolymers containing terminal amine groups [116, 117]. Such groups act as powerful transfer agents in the polymerization of the second monomer and both the rate of formation of block copolymers and the lengths of the blocks of the second monomer are determined, in part, by the transfer constant to the terminal group.

3.2 GRAFT COPOLYMERS AND NETWORKS

Essentially the same chemical processes as used in the preparation of block copolymers may be used in graft copolymerization, except that in this case the reactive groups are incorporated into side groups of the prepolymer instead of on the end of the prepolymer molecule.

Transfer processes are often used in graft copolymerizations, and an outline of a typical reaction scheme is

$$\text{Initiator} \longrightarrow \text{R}\cdot \tag{39a}$$

$$\text{R}\cdot + \sim\!\text{AAA}\!\sim \longrightarrow \text{RH} + \sim\!\text{A}\dot{\text{A}}\text{A}\!\sim \tag{39b}$$

$$\text{R}\cdot + n\text{M} \longrightarrow \text{RM}_n\cdot \tag{39c}$$

$$\text{RM}_n\cdot + \sim\!\text{AAA}\!\sim \longrightarrow \text{RM}_n\text{H} + \sim\!\text{A}\dot{\text{A}}\text{A}\!\sim \tag{39d}$$

$$\sim\!\text{A}\dot{\text{A}}\text{A}\!\sim + n\text{M} \longrightarrow \begin{array}{c}\sim\!\text{AAA}\!\sim \\ | \\ \text{M}_n\cdot\end{array} \tag{39e}$$

Free radicals are generated from low-molecular-weight initiator in eqn. (39a) in the presence of prepolymer and monomer. The primary radicals may undergo transfer directly with the prepolymer (39b) to form a macroradical, or may initiate the polymerization of the monomer (39c) to produce a propagating radical which may then undergo transfer with the prepolymer to generate a macroradical (39d). Initiation by the macroradical results in the formation of a graft (39e). Various termination reactions should be included to complete the scheme.

The rate of graft formation in a scheme such as eqn. (39) depends on the normal kinetic parameters, including the transfer coefficients for the prepolymer with both primary and propagating radicals. Some of the possible variations have been illustrated by Smets et al. [118] who attempted to graft methyl methacrylate chains onto polystyrene. When azo-bis-isobutryonitrile was used as initiator no grafting was achieved since neither the isobutyronitrile radical nor the propagating methyl methacrylate radical is sufficiently reactive to abstract a hydrogen from polystyrene. However, when benzoyl peroxide was used as initiator grafting was achieved, the more reactive phenyl and benzoyloxy radicals having high transfer coefficients to polystyrene. Transfer to prepolymer is usually slow unless a reactive propagating radical, such as the vinyl acetate radical, is used.

Higher rates of graft formation can be achieved if side groups having high transfer coefficients are introduced into the prepolymer. If high concentrations of such groups are introduced, so that efficient use is made of the primary radicals in producing grafts, there will also be a higher probability of a propagating graft (i.e. a propagating radical attached to prepolymer — an attached radical) undergoing transfer to prepolymer. This process could occur either intra- or intermolecularly, and will occur randomly if the reactions are carried out under kinetically good conditions, avoiding physical effects which we shall refer to subsequently. The net result of this process is to increase the number of grafts but to decrease their average length, and the consequences of various reaction schemes are calculable in principle if the appropriate kinetic parameters are known.

Many polymers have been carefully oxidized to introduce hydroperoxide side-groups which may subsequently be employed to initiate graft formation. As in block copolymer formation, the detailed nature of the products depends on the mode of initiation, as indicated in scheme (40), viz.

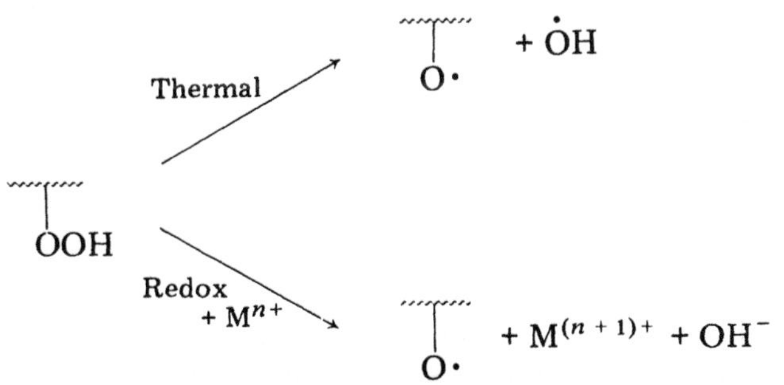

(40a)

(40b)

Initiation by the attached radical O· produces graft copolymer, while, in (40a), the unattached ȮH radical will give rise to the formation of homopolymer. The formation of unattached radicals may be minimized by employing redox. initiation (40b), the proportions of attached and unattached radicals therefore depends on the conditions of initiation

(Chap. 1). Combination termination of propagating radicals attached to prepolymer gives rise to cross-links, and in (40a) this mode of termination would incorporate some chains initiated by $\dot{O}H$ into the macromolecular system as grafts. Both disproportionation termination, and termination by metal ions in (40b) (see Chap. 2) result in graft formation only.

Complications arising from both the formation of unattached radicals and termination by metal ions have been avoided by Bamford et al. [115, 119—121], who have employed a prepolymer containing $-CCl_3$ side-groups (polyvinyltrichloracetate) and effected initiation with molybdenum and manganese carbonyls, when all radicals are attached radicals (Chap. 1). The specificity of the initiation system allowed kinetic studies of cross-linking to be carried out, and these were used to determine the nature of the bimolecular termination reaction in the free-radical polymerization of a number of monomers (see Chap. 1). The detailed mechanism of radical formation, which involves abstraction of a chlorine atom from the organic halide by the metal derivative, was described in Chap. 1. When the halide component of the initiating system is a polymer (such as polyvinyltrichloracetate) a macroradical is formed which initiates polymerization as in eqn. (41a), viz.

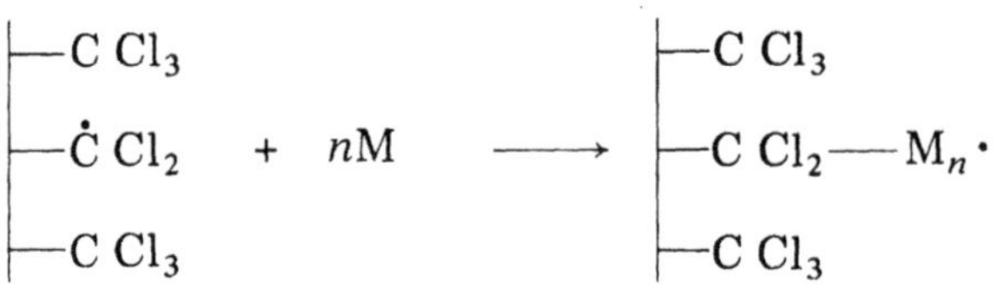

(41a)

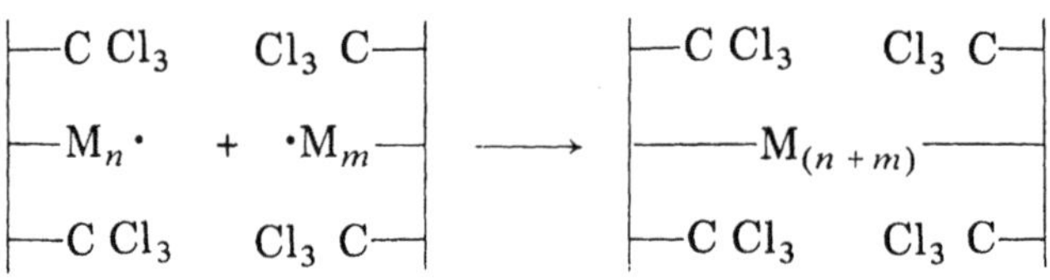

(41b)

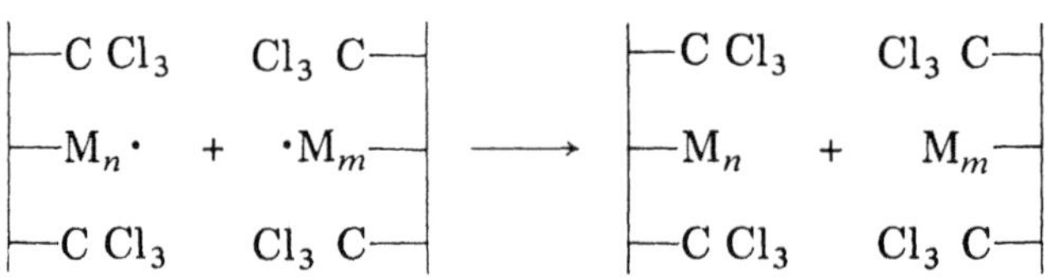

(41c)

Obviously combination termination produces cross-links as in (41b), while disproportionation termination produces branches (eqn. (41c)). If any proportion of the termination reaction occurs by combination, continuation of this process, utilizing further halogenated groups on the prepolymer, eventually leads to the formation of an infinite network, at which point the reaction mixture gels.

Gelation theory, developed by Flory, is considered in detail elsewhere [124], but we may state the critical condition for network formation; this is "the probability that any cross-linkable unit forms part of a cross-link is $1/\bar{P}_w$, where $\bar{P}_w$ is the weight-average degree of polymerization of cross-linkable units in the prepolymer". Thus, if the base-molar concentration of cross-linkable units in the reaction mixture is c, the rate of initiation $\mathscr{I}$ and all termination is by combination the critical condition for network formation is achieved in time t_g, the gel-time, given by [119]

$$t_g = c/\bar{P}_w \mathscr{I} \tag{42}$$

If only a fraction of the termination occurs by combination then gel formation is delayed and the gel-time is

$$t_g = \left(\frac{k_{tc}}{k_{tc} + k_{td}}\right)^{-1} \frac{c}{\bar{P}_w \mathscr{I}} \tag{43}$$

The branch/cross-link ratio can be determined from a knowledge of the termination reaction. Equations (42) and (43) show that, whatever the mode of termination, the gel-time increases linearly as the concentration of prepolymer increases and as the rate of initiation decreases. It has been demonstrated [119] that these conditions hold if the rate of initiation is low and the concentration of prepolymer is reasonably high. For relations (42) and (43) to be obeyed cross-linking must be random. At high rates of initiation and low prepolymer concentrations the probability of forming intramolecular cross-links, i.e. loops such as

increases, and the probability of introducing additional cross-links between prepolymer units already cross-linked increases, thus reducing the rate of incorporation of further prepolymer molecules into the developing network structure. Such non-random cross-linking represents a wastage in the formation of intermolecular cross-links, and the observed gel-time increases. This behaviour has also been observed in radiation induced cross-linking at low polymer concentrations [122], as shown in Fig. 5, where at constant dose-rate, the radiation dose required for gel formation is ideally proportional to the polymer concentration. Other possible causes of non-random cross-linking have also been discussed [119].

Transfer processes produce complications in these systems, but they can usually be allowed for mathematically [120]. Transfer of a growing graft to a prepolymer molecule results in the formation of a dead graft and regenerates a radical attached to prepolymer. Simple transfer to prepolymer, therefore, affects the cross-link/branch ratio but has no influence on the gel-time. On the other hand, transfer of a propagating attached radical to monomer or solvent produces an unattached radical which can never partake in cross-link formation, but can only give rise to branches or soluble polymer, the proportions of which depend on the

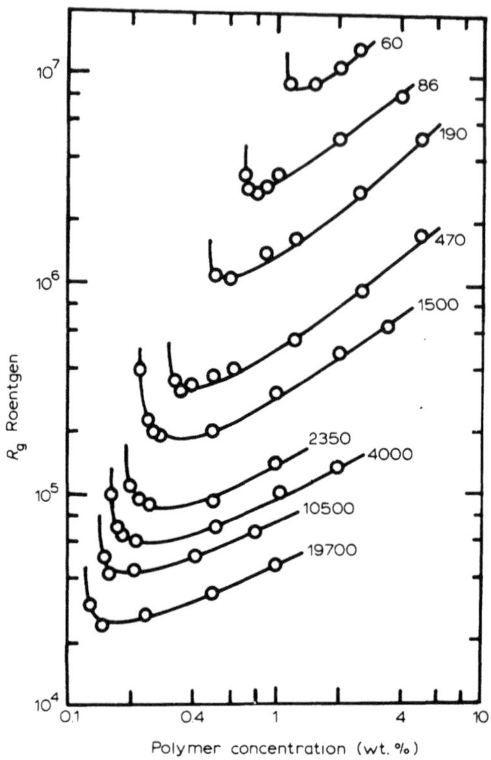

Fig. 5. Total radiation doses (R_g) required (at constant dose rate) to produce gelation during the irradiation cross-linking of poly(vinyl alcohol) in aqueous solution. Numbers in parentheses refer to initial degree of polymerization. (After Sakurada and Ikada [122], courtesy of Inst. Chem. Res. Kyoto Univ.)

mechanism of termination. Transfer of this type increases the gel-time. If t_g^0 is the gel-time in the absence of transfer, and the quantity which should be used in eqns. (42) and (43), and t_g is the observed gel time, t_g^0 and t_g are related by the equations

$$t_g^0 = t_g/(1+B)^2$$

$$B = \frac{k_p}{(k_{tc}+k_{td})^{1/2}} \cdot \frac{C_M[M] + C_S[S]}{\mathscr{I}^{1/2}} \tag{44}$$

where C_M and C_S are the transfer constants to monomer and solvent, respectively.

If a large proportionation of the termination reaction occurs by disproportionation, gel-times may be long and $\mathscr{I}$ may vary due to initiator consumption. Assuming first-order decay of initiator this effect may be allowed for [120] (assuming negligible transfer) by use of relation (45)

$$t_g^0 = (1/k)\{1 - \exp(-kt_g)\} \tag{45}$$

where t_g^0 is again the corrected gel-time and k is the first-order rate coefficient for initiator decay.

Under conditions where both initiator consumption and transfer are significant the corrections to the observed gel-time may be combined in the relations

$$t_g^0 = \frac{1}{k}\left\{1 - y^2 - 4B_0(1-y) + 6B_0^2 \ln\left(\frac{1+B_0}{y+B_0}\right) - \frac{2B_0^3(1-y)}{(1+B_0)(y+B_0)}\right\} \tag{46}$$

$$y = \exp(-kt_g/2)$$

where B_0 is B in eqn. (44) with $\mathscr{I}$ replaced by $\mathscr{I}_0$, the initial rate of initiation. Further generalization of this treatment to include transfer from unattached radicals has not yet been achieved.

In many graft copolymerization systems, e.g. initiation by hydroperoxides, and polyvinyltrichloracetate using a nickel-containing catalyst to generate free radicals [123], a fraction of the primary radicals are not attached to the prepolymer chain. The net result of this is to reduce the rate of cross-link formation and to increase the gel-time. If f is the ratio of the rate of formation of attached radicals to the rate of formation of unattached radicals the observed gel-times require further correction, which is achieved [123] by multiplying the relations for t_g^0 above by $f/(1+f)^2$. In the polyvinyltrichloracetate/nickel-containing systems good agreement has been achieved between values of f determined by tracer and gelation studies [123] (see Chap. 1).

The lengths of cross-links and branches in systems of the type described above are determined, at least up to the gel-point, by normal kinetic relations. Thus, it is possible, from a knowledge of the appropriate kinetic parameters to prepare cross-linked materials with known cross-link/branch ratios, and various densities and lengths of cross-links.

REFERENCES

1 F. R. Mayo and F. M. Lewis, J. Am. Chem. Soc., 66 (1944) 1594.
2 F. T. Wall, J. Am. Chem. Soc., 66 (1944) 2050.
3 T. Alfrey and G. Goldfinger, J. Chem. Phys., 12 (1944) 205.
4 R. Simha and H. Branson, J. Chem. Phys., 12 (1944) 253.
5 G. Goldfinger and T. Kane, J. Polym. Sci., 3 (1948) 462.
6 F. T. Wall, J. Am. Chem. Soc., 63 (1941) 1862.
7 C. Tosi, Adv. Polym. Sci., 5 (1967—8) 451.
8 C. H. Bamford and W. G. Barb, Discuss. Faraday Soc., 14 (1953) 208.
9 W. H. Stockmayer, J. Chem. Phys., 12 (1944) 253.
10 I. Skeist, J. Am. Chem. Soc., 68 (1946) 1781.
11 E. Meyer and G. G. Lowry, J. Polym. Sci., Part A, 32 (1965) 2483.
12 I. H. Spinner, B. C. Y. Lee and W. F. Graydon, J. Am. Chem. Soc., 77 (1955) 2198.
13 F. R. Mayo and C. Walling, Chem. Rev., 46 (1950) 204.

14 Y. Yamashita, K. Ito, H. Ishii, S. Hoshino and M. Kai, Macromolecules, 1 (1968) 529.
15 H. Inagaki, H. Matsuda and F. Kamiyama, Macromolecules, 1 (1968) 520.
16 R. H. Hoffman, S. Schneider and G. Rosen, Ind. Eng. Chem., 56 (1964) 51.
17 L. F. Beste and H. K. Hall, J. Macromol. Chem., 1 (1966) 122.
18 J. T. O'Toole, J. Macromol. Sci., Chem., 2 (1968) 183.
19 W. H. Roy and C. E. Gall, Macromolecules, 2 (1969) 425.
20 T. Alfrey and G. Goldfinger, J. Chem. Phys., 12 (1944) 322.
21 C. Walling and E. R. Briggs, J. Am. Chem. Soc., 67 (1945) 1744.
22 F. C. Price, J. Chem. Phys., 36 (1962) 209.
23 R. J. Slocombe, J. Polym. Sci., 26 (1957) 9.
24 A. I. Tarasov, V. A. Tskhai and S. S. Spasski, Vysokomol. Soedin., 2 (1960) 1601; 3 (1961) 14.
25 A. Valvassori and G. Sartori, Adv. Polym. Sci., 5 (1967—8) 28.
26 R. K. S. Chan and V. Meyer, J. Macromol. Sci., Chem., 1 (1967) 1089.
27 J. A. Seiner, J. Polym. Sci., Part A, 3 (1965) 2401.
28 T. Alfrey and G. Goldfinger, J. Chem. Phys., 14 (1946) 115.
29 C. Walling, D. Seymour and K. B. Wolfstirn, J. Am. Chem. Soc., 70 (1948) 1544.
30 E. C. Chapin, G. E. Ham and R. G. Fordyce, J. Am. Chem. Soc., 70 (1948) 538.
31 G. E. Ham, in G. E. Ham (Ed.), High Polymers, Vol. XVIII, Copolymerization, Interscience, New York, 1964.
32 G. E. Ham, J. Macromol. Sci. Chem., 2 (1968) 877.
33 A. Valvassori and G. Sartori, Rend. Ist. Lomb. Sci. Lett. A, 96 (1962) 107.
34 G. E. Ham. J. Polym. Sci., Part A, 2 (1964) 2735.
35 G. E. Ham. J. Polym. Sci., Part A, 2 (1964) 4169.
36 F. R. Mayo, J. Polym. Sci., Part A, 2 (1964) 4207.
37 K. F. O'Driscoll, J. Polym. Sci., Part B, 3 (1965) 305.
38 A. D. Jenkins, Eur. Polym. J., 1 (1965) 177.
39 P. W. Tidwell and G. A. Mortimer, J. Polym. Sci., Part B, 4 (1966) 527.
40 T. Alfrey and C. C. Price, J. Polym. Sci., 2 (1947) 101. T. Alfrey, J. J. Bohrer and H. F. Mark, Copolymerization, Interscience, New York, 1952, Chap. 3.
41 C. H. Bamford, A. D. Jenkins and R. Johnston, Trans. Faraday Soc., 15 (1959) 418. C. H. Bamford and A. D. Jenkins, J. Polym. Sci., 53 (1961) 149. C. H. Bamford and A. D. Jenkins, Trans. Faraday Soc., 59 (1963) 530.
42 K. F. O'Driscoll and M. Izu, J. Macromol. Sci. Chem., 4 (1970) 311.
43 E. Merz, T. Alfrey and G. Goldfinger, J. Polym. Sci., 1 (1946) 75.
44 W. G. Barb, J. Polym. Sci., 11 (1953) 117.
45 G. E. Ham, J. Polym. Sci., 45 (1960) 169, 177, 183; J. Polym. Sci., Part A, 2 (1964) 3633.
46 M. Berger and I. Kuntz, J. Polym. Sci., Part A, 2 (1964) 1687.
47 J. Guillot, A. Guyot and T. Q. Pham, J. Macromol. Sci. Chem., 2 (1968) 1303.
48 A. Guyot and J. Guillot, J. Macromol. Sci. Chem., 2 (1968) 889.
49 G. E. Ham, J. Polym. Sci., Part A, 2 (1964) 4191.
50 G. G. Lowry, J. Polym. Sci., 42 (1960) 463.
51 J. A. Howell, M. Izu and K. F. O'Driscoll, J. Polym. Sci., Part A-1, 8 (1970) 699.
52 M. Izu and K. F. O'Driscoll, Polym. J., 1 (1970) 27.
53 K. F. O'Driscoll and F. P. Gasparro, J. Macromol. Sci. Chem., 1 (1967) 643.
54 K. J. Ivin and R. H. Spensley, J. Macromol. Sci. Chem., 1 (1967) 653.
55 M. Izu and K. F. O'Driscoll, J. Polym. Sci., Part A-1, 8 (1970) 1687.
56 M. Izu and K. F. O'Driscoll, J. Polym. Sci., Part A-1, 8 (1970) 1675.
57 M. Fineman and S. D. Ross, J. Polym. Sci., 5 (1950) 260.
58 V. Meyer, J. Polym. Sci., Part A-1, 4 (1966) 2819.
59 D. W. Behnken, J. Polym. Sci., Part A, 2 (1964) 645.

60 P. W. Tidwell and G. A. Mortimer, J. Polym. Sci., Part A, 3 (1965) 369; J. Macromol. Sci., Rev. Macromol. Chem., 4 (1970) 281.
61 C. H. Bamford, D. J. E. Ingram, A. D. Jenkins and M. C. R. Symons, Nature (London) 175 (1955) 894.
62 C. H. Bamford and A. D. Jenkins, J. Polym. Sci., 20 (1956) 405.
63 C. H. Bamford, A. D. Jenkins and R. Johnston, Proc. R. Soc. London, Ser. A, 239 (1957) 214.
64 J. E. Hinsch and J. G. Curro, Macromolecules, 1 (1968) 405.
65 F. A. Bovey, I. M. Kolthoff, A. I. Medalia and E. J. Meehan, Emulsion Polymerization, High Polymers, Vol. 9, Interscience, New York, 1955.
66 W. V. Smith, J. Am. Chem. Soc., 70 (1948) 2177.
67 R. G. Fordyce and E. C. Chapin, J. Am. Chem. Soc., 69 (1947) 581.
68 R. G. Fordyce and G. E. Ham, J. Polym. Sci., 3 (1948) 891.
69 R. G. Fordyce and G. E. Ham, J. Am. Chem. Soc., 69 (1947) 695.
70 T. Kagiya and Y. Sumida, Polym. J., 1 (1970) 137.
71 K. F. O'Driscoll, J. Macromol. Sci., Chem., 3 (1969) 307.
72 J. A. Lamb and K. E. Weale, Proc. Symp. Phys. Chem. High Press., London, (1962) 229.
73 R. D. Burkhart and N. L. Zutty, J. Polym. Sci., 57 (1962) 793.
74 T. Alfrey, C. G. Overberger and S. H. Pinner, J. Am. Chem. Soc., 75 (1953) 4221.
75 F. M. Lewis, C. Walling, W. Cummings, E. Briggs and F. R. Mayo J. Am. Chem. Soc., 70 (1948) 1519.
76 C. C. Price and J. G. Walsh, J. Polym. Sci., 6 (1951) 239.
77 I. Ito and T. Otsu, J. Macromol. Sci., Chem., 3 (1969) 197.
78 A. V. Ryabov, Yu. D. Semichikova and N. N. Slavnitskaya, Dokl. Akad. Nauk S.S.S.R., 145 (1962) 822.
79 (a) C. Aso, T. Kunitake, T. Nakayas and S. Ando, private communication.
(b) C. Aso, T. Kunitake and S. Ando, J. Macromol. Sci., Chem., 5 (1971) 2019.
80 P. D. Bartlett and K. Nozaki, J. Am. Chem. Soc., 68 (1946) 1495.
81 S. Iwatsuki and Y. Yamashita, Radical Alternating Copolymerization, in M. Imoto and S. Onogi (Eds.), Progress in Polymer Science, Japan, Vol. 2, Kodansha, Tokyo, 1971.
82 M. Hirooka, H. Yabuuchi, S. Morita, S. Kawasumi and K. Nakaguchi, J. Polym. Sci., Part B, 5 (1967) 47.
83 M. Hirooka, H. Yabuuchi, J. Iseki and Y. Nakai, J. Polym. Sci., Part A-1, 6 (1968) 1381. J. Furukawa, Y. Iseda, H. Haga, N, Kataoka, Y. Yoshimoto, T. Imamura, Y. Yoshida, A. Miyagi, K. Tanaka and K. Sakamoto, J. Polym. Sci., Part B, 7 (1969) 561. M. Taniguchi, A. Kawasaki and J. Furukawa, Bull. Chem. Soc. Jpn., 42 (1969) 2728. J. Furukawa and I. Iseda, J. Polym. Sci., Part B, 7 (1969) 47. J. Furukawa, E. Kobayashi, Y. Iseda and Y. Arai, Polym. J., 1 (1970) 442. N. G. Gaylord and A. Takahashi, J. Polym. Sci., Part B, 6 (1968) 743, 749; J. Polym. Sci., Part B, 7 (1969) 145; Adv. Chem. Ser., 91 (1969) 94. N. G. Gaylord and H. Antropiusova, J. Polym. Sci., Part B, 7 (1969) 1451; Macromolecules, 2 (1969) 442. A Takahashi and N. G. Gaylord, J. Macromol. Sci., Chem., 4 (1970) 127. N. G. Gaylord and S. Maiti, Makromol. Chem., 142 (1971) 101.
84 C. H. Bamford and M. J. S. Dewar, J. Chem. Phys., 17 (1949) 1188.
85 T. Alfrey and V. Hardy, J. Polym. Sci., 3 (1948) 500.
86 G. M. Burnett, Mechanisms of Polymer Reactions, Interscience, New York, 1954.
87 J. H. Bradbury and H. W. Melville, Proc. R. Soc. London, Ser. A, 222 (1954) 456.
88 H. W. Melville and L. Valentine, Proc. R. Soc. London, Ser. A, 200 (1950) 337, 358.
89 E. J. Arlman and H. W. Melville, Proc. R. Soc. London, Ser. A, 203 (1950) 301.

90 E. P. Bonsall, L. Valentine and H. W. Melville, J. Polym. Sci., 7 (1951) 39.
91 C. Walling, J. Am. Chem. Soc., 71 (1949) 1930.
92 H. W. Melville and L. Valentine, Proc. R. Soc. London, Ser. A, 227 (1954) 10.
93 C. Walling and E. A. McElkill, J. Am. Chem. Soc., 73 (1951) 2819.
94 J. N. Atherton and A. M. North, Trans. Faraday Soc., 58 (1962) 2049.
95 A. M. North, Polymer, 4 (1963) 134.
96 A. F. Trotman-Dickenson, Ann. Rep. Progr. Chem., Chem. Soc., London, 55 (1958) 36.
97 J. C. Bevington, H. W. Melville and R. P. Taylor, J. Polym. Sci., 14 (1954) 463.
98 S. Russo and S. Munari, J. Macromol. Sci., Chem., 2 (1968) 1321.
99 K. F. O'Driscoll and J. R. Dickson, J. Macromol. Sci., Chem., 2 (1968) 449.
100 E. H. de Butts, J. Am. Chem. Soc., 72 (1950) 411.
101 K. F. O'Driscoll and R. Knorr, Macromolecules, 1 (1968) 367.
102 R. J. Ceresa, Block and Graft Copolymers, Butterworths, London, 1962.
103 A. S. Hoffman and R. Bacskai, in G. E. Ham (Ed.), High Polymers, Vol. XVIII, Copolymerization, Interscience, New York, 1964, p. 335.
104 H. Battaerd and G. W. Tregear, Graft Copolymers, Polymer Reviews, No. 16, 1967.
105 A. E. Woodward and G. Smets, J. Polym. Sci., 17 (1955) 51.
106 G. Smets and R. Hart, Fortschr. Hochpolym. Forsch., 2 (1960) 173.
107 R. J. Orr and H. L. Williams, J. Am. Chem. Soc., 78 (1956) 3273; 79 (1957) 3137.
108 P. E. M. Allen, J. M. Downer, G. W. Hastings, H. W. Melville, P. Molyneux and J. R. Unwin, Nature (London), 177 (1956) 910.
109 J. R. Unwin, J. Polym. Sci., 27 (1958) 580.
110 P. Molyneux, Makromol. Chem., 37 (1960) 165; 43 (1961) 31.
111 W. Kern, M. A. Achan, G. Schroeder and R. Schulz, Z. Elektrochem., 60 (1956) 309.
112 J. Brepoels and G. Smets, J. Polym. Sci., 56 (1962) 359.
113 A. S. Dunn, B. D. Stead and H. W. Melville, Trans. Faraday Soc., 50 (1954) 274.
114 T. Otsu, J. Polym. Sci., 26 (1957) 236.
115 C. H. Bamford and G. C. Eastmond, British Pat. 1,027,148 (1966).
116 C. H. Bamford and E. F. T. White, Trans. Faraday Soc., 52 (1956) 716.
117 C. H. Bamford, A. D. Jenkins and E. F. T. White, J. Polym. Sci., 34 (1959) 271.
118 G. Smets, J. Roovers and W. van Humbleek, J. Appl. Polym. Sci., 5 (1961) 149.
119 C. H. Bamford, R. W. Dyson and G. C. Eastmond, J. Polym. Sci., Part C, 16 (1967) 2425.
120 C. H. Bamford, R. W. Dyson, G. C. Eastmond and D. Whittle, Polymer, 10 (1969) 759.
121 C. H. Bamford, R. W. Dyson and G. C. Eastmond, Polymer, 10 (1969) 885.
122 I. Sakurada and Y. Ikada, Bull. Inst. Chem. Res., Kyoto Univ., 42 (1964) 22.
123 C. H. Bamford, G. C. Eastmond and D. Whittle, Polymer, 10 (1969) 771.
124 P. J. Flory, Principles of Polymer Chemistry, Cornell University Press, Ithaca, New York, 1953.

Appendix. Reactivity ratios in free-radical copolymerizations

G. C. EASTMOND and E. G. SMITH

COPOLYMERIZATION REACTIVITY RATIOS

This appendix contains values of reactivity ratios for free-radical copolymerizations in homogeneous solution. In compiling these data considerable reliance has been placed on the accuracy of data in and references to the less accessible literature quoted in earlier compilations [1—3]. In the earlier compilations data relating to the less accessible literature were often those cited in Chemical Abstracts and locations in Chemical Abstracts were quoted; these locations have been omitted in the present compilation. For more recent data (1965—1972) the more popular journals have been scanned. Reactivity ratios quoted are those calculated on the assumption that the simple copolymerization mechanism operates; where relevant, references are given to studies of penultimate effects. In the present compilation each set of data is presented once, usually under the more common monomer of the pair.

Where practicable, heterogeneous systems have been excluded (except for a few cases retained for comparative purposes). However, it is not always clear from the original literature whether or not reaction mixtures remained homogeneous during the course of the polymerization. Although it is probable that many of the copolymers prepared in the studies quoted were insoluble in the reaction mixtures used, in the absence of specific information the data have been retained since the copolymers may have remained soluble at the low conversions which should be employed in determining reactivity ratios.

It is not possible to quote definitive values for the reactivity ratios for any copolymerization under a specific set of reaction conditions, as will be appreciated from the scatter of data obtained by different workers for the same system. In an attempt to provide some background information as to the reliability of the published data the more readily available papers have been examined and the composition range of monomer mixtures employed and the conversions to which the reactions were taken have been noted. Reactivity ratios quoted in some early studies were based on the results of a single experiment; such data have been excluded from this compilation. Most of the data quoted are based on data from at least three experiments; an indication in the Table that a wide range of monomer feed compositions have been used may hide the fact that the ratios quoted are based on a minimal amount of information. An increasing number of workers are reporting solvent effects on reactivity ratios and, where possible, the solvents employed have been noted, together with other relevant experimental details.

M_1/M_2	r_1		r_2		T(°C)	Remarks	Ref.
Acrolein/							
Sodium *p*-styrene sulphonate	0.33	±0.15	0.32	±0.05	52	f, m aqueous pH 3	4
	0.23	±0.12	0.05	±0.03	54	f, m aqueous pH 5	4
	0.26	±0.03	0.025	±0.025	54	f, m aqueous pH 7	4
Acrylamide/							
Acrolein	0.76	±0.02	2.0	±0.05	20	f, m aqueous	5
	0.19	±0.02	1.65	±0.1	50		6
	0.21	±0.02	1.69	±0.1	50	a, m DMF	7
cis-3-Bromoacrylic acid	12	±0.5	−0.14	±0.03	20	f, m aqueous DMF	8
trans-3-Bromoacrylic acid	3.0	±0.2	−0.21	±0.08	20	f, m aqueous DMF	8
cis-Cinnamic acid	6.7	±0.6	−0.22	±0.07	20	f, m aqueous DMF	8
trans-Cinnamic acid	3.0	±0.2	−0.26	±0.03	20	f, m aqueous DMF	8
cis-Crotonic acid	4.0	±0.3	−0.21	±0.2	20	f, m aqueous DMF	8
trans-Crotonic acid	6.5	±0.4	−0.09	±0.2	20	f, m aqueous DMF	8
1,2-Dimethyl-5-vinylpyridinium methyl sulphate	0.19		2.7		48	f, n aqueous	9
cis-3-Ethylacrylic acid	7.1	±0.5	−0.11	±0.03	20	f, m aqueous DMF	8
trans-3-Ethylacrylic acid	11	±0.6	−0.14	±0.03	20	f, m aqueous DMF	8
Methacrylamide	0.74	±0.11	1.1	±0.20	25	a, l aqueous	10
N-Methylolacrylamide	2.9	±0.4	0.9	±0.2	—		11
2-Methyl-5-vinyl pyridine	0.56	±0.09	0.01	±0.09	60		12
Sodium acrylate (90 %)	1.10	± 0.05	0.35	±0.03	60	a, n	13
Vinyl pyrrolidone	0.66		0.17		60	a, m aqueous	14
	10.6		0.11		60	a, m aqueous (80 % glycerol)	14
p-Vinyl phenyl boronic acid	0.0		1.0		60	f, n aqueous butyl alcohol	15
Acrylic acid/							
Acrolein	1.15	±0.2	0.50	±0.30	54	g, l aqueous pH3	4
Acrylamide	1.43	±0.03	0.60	±0.02	25	a, m	13
	0.36		1.38		60	g, p acetone, benzene	16
	1.73	±0.21	0.48	±0.06	60	f, m aqueous pH 2.17	17

	0.56	±0.09	0.56	±0.09	60	f, m aqueous pH 3.77	17
	0.45	±0.03	0.67	±0.04	60	f, l aqueous pH 4.25	17
	0.42	±0.02	0.95	±0.03	60	f, l aqueous pH 4.73	17
	0.35	±0.03	1.32	±0.12	60	f, l aqueous pH 6.25	17
N-Allyl acetamide	1.3	±0.3	0.07	±0.06	80	ethanol	18
N-Allyl benzamide	4.7	±0.2	0.05	±0.03	80	ethanol	18
N-Allyl chloroacetamide	3.6	±0.3	0.04	±0.04	80	ethanol	18
N-Allyl formamide	2.6	±0.2	0.06	±0.03	80	ethanol	18
N-Allyl isovaleramide	4.3	±0.7	0.05	±0.05	80	ethanol	18
n-Butyl acrylate	0.58		1.07		50	f, n ethanol	19
n-Butyl methacrylate	0.29		3.67		50	f, n ethanol	19
2-Chloroallyl acetate	1.0		0.0		100		20
N,N-Dimethylacrylamide	0.4	±0.05	0.5	±0.1	75	i, m dioxan	22
Ethyl ethacrylate	0.9	±0.1	0.45	±0.05	80	f, m benzene	23
Sodium styrene sulphonate	0.10	±0.02	1.0	±0.2	70	f, m aqueous	24
Vinyl *N,N*-diethylcarbamate	5.55		0.09		80	f, m	25
Vinyl hydroquinone dibenzoate	0.44	±0.13	0.95	±0.002	60	i, o toluene	26
Vinylidene cyanide	0.26	±0.06	0.29	±0.08	50	a, m, s (see ref. 588)	27
N-Vinylpyrrolidone	1.3	±0.2	0.15	±0.1	75		28
Acrylonitrile/							
α-Acetoxyacrylonitrile	0.09	±0.17	7.4	±0.40	60		1
1-Acetoxy-1,3-butadiene	0.0		0.7		70	g, m	29
α-Acetoxystyrene	0.08	±0.01	0.4	±0.05	75	g, l	30
Acrolein	0.77	±0.1	1.09	±0.05	20	f, m aqueous	5
	0.52	±0.02	1.60	±0.04	50	a, m DMF	7
Acrolein dibutylether acetal	7.20		0.01		60	g, m DMF	31
Acrolein diethyl acetal	12.00		0.01		—		32
Acrolein diethyl ether acetal	7.56		0.01		60	g, l DMF	31
Acrylic acid	0.35		1.15		50		33
	0.13	±0.02	6.0	±2.0	80		1
Allyl acetate	8.2		0.0		—		34
Allyl alcohol	5.5		0.1		—		34
	3.96	±0.53	0.11	±0.10	25	f, l, t aqueous	35
	0.03		1.99		—		36

Copolymerization reactivity ratios—*continued*

M_1/M_2	r_1	r_2	T(°C)	Remarks	Ref.
Allyl chloride	3.0 ±0.2	0.05 ±0.01	60	f, l	30
N-Allylstearimide	3.61 ±0.087	0.118 ±0.084	80		37
Allyl triethoxysilane	1.7	0	50	h, m	38
Benzyl acrylate	1.490	0.631	25		39
	0.294 ±0.08	0.636 ±0.07	60	a	40
Benzyl methacrylate	1.055	1.15	25		39
	0.206 ±0.02	0.810 ±0.02	60	a	40
1-Bicyclobutane carbonitrile	1.10	0.50	50	k, r γ-butyrolactone	41
Bicyclo(2.2.1)-hepta-2,5-diene	0.67	0.08	50	d, m	42
	0.65 ±0.02	0.47 ±0.08	60	g, l	43
Butadiene	0.03	0.18	5	u	44
	0.02	0.3	40	g	45
	0.0 ±0.04	0.35 ±0.08	50		46, 47
	0.04 ±0.01	0.04 ±0.02	50		48
	0.25	0.33	60		21
	—	—	—	s DMF	49
1-Butene : SO_2	1.0	0.8	−78		50
2-Butenyl triethoxy silane	10.0	0	50	j, m	38
Butyl acrylate	1.2 ±0.1	0.89 ±0.08	60	e, m	51
	1.52 ±0.03	0.75 ±0.18	60		52
	1.003 ±0.012	1.005 ±0.005	60	a, l	53, 54
n-Butyl acrylate	1.0 ±0.1	0.9 ±0.1			55
Butyl methacrylate	0.31	1.08	60	f, m	56
α-*n*-Butylstyrene	—	—	60	s	57
cis-β-*n*-Butylstyrene	3.0	0	60	s	57
trans-β-*n*-Butylstyrene	—	—	60	s	57
α-*s*-Butylstyrene	—	—	60	s	57
cis-β-*s*-Butylstyrene	—	0	60	s	57
trans-β-*s*-Butylstyrene	6.15	0	60	s	57
α-*t*-Butylstyrene	—	—	60	s	57
trans-β-*t*-Butylstyrene	18.8	0	60	s	57

t-Butyl vinyl ether	0.14	±0.04	0.0032	±0.0002	60		1
Butyl vinyl sulphide	0.097		0.046		60	g, m	58
t-Butyl vinyl sulphide	0.13		0.01		60	g, p	58
2-Chloro-1,3-butadiene	0.005	±0.011	6.07	±0.53	50		47
N-p-β-Chloromethylamino-sulphonylbenzene acrylamide	0.20	±0.03	1.00	±0.05	70	h, r DMF	59
N-p-β-Chloromethylamino-sulphonylbenzene methacrylamide	0.05	±0.03	1.03	±0.03	70	f, r DMF	59
α-Chlorovinyl triethoxy silane	0.7		0		50	h, p	38
Crotonic acid	21	±10	0		60	v	60
1,3-Cyclohexadiene	0.026	±0.014	0.20	±0.01	60	g, l	43
Cyclohexa-1,3-diene	0.05	±0.02	0.11	±0.02	65		61
Δ^3-Cyclohexenyl triethoxy silane	12.0		0		50	j, l	38
4-Cyclopentene-1,3-dione	3.67		0.21		60	a, b benzene	62
2,5-Dichlorostrene	0.26	±0.02	0.09	±0.02	38.5	f, m	63
	0.22	±0.05	0.07	±0.05	67.5	f, m	63
	0.25	±0.11	0.07	±0.06	86.5	f, m	63
Diethylene glycol monovinyl ether	0.59	±0.02	0.0021	±0.0004	60		1
2,4-Dimethoxy-6-(*β*-itaconylhydra-zino)-*s*-triazine	0.22	±0.2	0.7	±0.5	60	<10 % M_1, m	64
3,3-Dimethyl-1,4-pentadiene	0		3.31		50	i, m, x DMF	65
N-N-Divinylaniline	0.42		0.003		60	g, n DMF	66
Divinyl benzene	0.364		1.94		50	i, m, x DMF	65
Divinyl ether	0.938		0.024		50	a, DMF	67
Dodecyl acrylate	3.2	±0.5	1.3	±0.1	60	i, m	51
n-Dodecyl vinyl ether	0.82	±0.05	0.0	±0.2	50		68
Ethyl acrylate	1.17	±0.1	0.67	±0.02	50		69
	0.44		0.95		80		1
Ethylene	7.0		0		20		70
Ethyleneglycol dimethacrylate	0.22	±0.01	1.24	±0.10	—		71
Ethyl-2-ethoxy-acrylate	10.5	±1.5	0.02	±0.02	80	f, n	72
α-Ethyl styrene	—		—		60	s	57
cis-β-Ethylstyrene	2.5		0		60	s	57
trans-β-Ethyl styrene	—		—		60	s	57

Copolymerization reactivity ratios—*continued*

M_1/M_2	r_1		r_2		T(°C)	Remarks	Ref.
1-Ethyl-2-vinyl-acetylene	0.17	±0.01	0.63	±0.4	60	f, m	73
Ethyl vinyl ether	0.7	±0.2	0.03	±0.02	80	f, r	72, 74
Ethyl vinyl oxalate	2.0		0.2		60	c, m	75
5-Ethyl-2-vinyl pyridine	0.02	±0.02	0.43	±0.05	60	a, m	76
Ethyl vinyl sulphoxide	3.0		0.2		60	g, l	77
2-Fluorobutadiene	0.07	±0.03	0.50	±0.10	50	h, m	78
Glycidyl acrylate	1.01		1.02		60	a, r tetrahydrofuran	79
Glycidyl bicyclo(2.2.1)-hept-5-en-2-carboxylate	1.4		0		100	d, r DMF	80
Glycidyl methacrylate	0.14	±0.001	1.32	±0.03	60		81
Glycidyl 2-methyl-bicyclo-(2.2.1)-hept-5-en-2-carboxylate	1.5		0		100	h, r DMF	80
Glycidyl 1-methyl-3-cyclohexen-1-carboxylate	18.4		0.02		100	f, r DMF	80
Hydronopyl acrylate	1.1	±0.1	0.9	±0.1	60	b, p benzene	82
6-Hydroxy-2-(*p*-chlorophenyl)-3-pyridazinone	15		0		60	DMSO	83
6-Hydroxy-2-phenyl-3-pyridazinone	14		0		60	DMSO	83
6-Hydroxy-2-(*p*-tolyl)-3-pyridazinone	12		0		60	DMSO	83
Isobutene	1.8	±0.2	0.02	±0.02	50		48
	1.02		0		60		84
Isobutyl methacrylate	0.21		1.04		60	g, n	56
Isobutyl vinyl ether	1.0		0.2		60	g, m	85
Isobutyl vinyl sulphide	0.068		0.050		60	g, m	58
Isoprene	0.03	±0.03	0.45	±0.05	50		46
Isopropenyl acetylene	0.33	±0.01	0.47	±0.01	60	f, n	248
β-Isopropenyl naphthalene	0.05	±0.01	0.23	±0.02	99	k, r	86
α-Isopropenyl styrene	—		—		60	s	57
cis-β-Isopropylstyrene	15.0		0		60	s	57

trans-β-Isopropylstyrene	3.5		0		60	s	57
Itaconic acid	0.25		1.57		50		87
Methacrolein	0.15	±0.04	1.7	±0.3	30		88
	0.06		2.0		70		29
Methacrylic anhydride	1.94		0.21		60		89
Methacrylonitrile	0.32		2.68		60		84
Methacryloxymethylpentamethyl-disiloxane	0.19	±0.04	1.44	±0.15	50	k, r	90
Methacryloyl-*D*-glucose	0.25		0		50	c, r DMF	92
α-Methoxystyrene	0.06	±0.005	0		60	a, m	93
Methyl acrylate	0.86	±0.05	1.25	±0.15	30		94
	0.70	±0.20	1.22	±0.20	20		95
	1.5	±0.1	0.84	±0.05	50	d, l benzene	82
	1.02	±0.02	0.70	±0.02	50	DMSO	96
	1.26	±0.1	0.67	±0.1	60		97
	1.4	±0.1	0.95	±0.05	60	f, m	51
	0.84		0.83		65		84
	0.50	±0.47	0.71	±0.012	80	a, r	98
Methyl bicyclo-(2,2,1)-2-heptene-5-carboxylate	1.5	±0.5	0.2	±0.1	60	j, m DMF	99
Methyl α-bromoacrylate	0.08		4.2		60	g, m DMF	100
Methyl-1-chloroacrylate	2.0		0.15		60		101
	0.15	±15 %	2.0	±15 %	70	d, l	102
Methyl cinnamate	6	±2	0		60		60
N-Methylcitraconimide	0.527		0.604		60		103
1-Methyl cyclohexa-1,3-diene	0.18	±0.05	0.18	±0.05	65		61
Methyl isopropenyl ketone	0.36	±0.08	0.70	±0.14	80		1
N-Methylolacrylamide	0.7	±0.1	1.2	±0.1	—		104
Methyl α-phenylacrylate	0.075		6.7		60	f, m	105
α-Methylstyrene	0.08	±0.03	0.25	±0.06	—	liq. SO_2	106
	—		—		60	s	57
	0.06	±0.02	0.1	±0.02	75	a, l, s (see refs. 526, 589)	107
cis-β-Methylstyrene	1.6		0		60	s	57

Copolymerization reactivity ratios—*continued*

M_1/M_2	r_1		r_2		T(°C)	Remarks	Ref.
trans-β-Methylstyrene	—		—		60	s	57
o-Methylstyrene	0.06	±0.05	0.33	±0.1	—		97
m-Methylstyrene	0.07	±0.04	0.43	±0.1	—		97
p-Methylstyrene	0.05	±0.02	0.33	±0.1	—		97
Methyl vinyl ketone	0.61	±0.04	1.78	±0.22	60	f, n	108
2-Methyl-5-vinyl pyridine	0.16	±0.003	0.27	±0.04	60	a, l	109
2-Methyl-5-vinylpyridinium HCl	0.15		2.3		—		110
N-Methyl-*N*-vinyl-*p*-toluene sulphonamide	0.42		0		60	i, n	111, 112
2-Nitrobutyl acrylate	0.90		1.58		70		113
N-*n*-Octadecylacrylamide	1.10	±0.035	1.44	±0.019	60		37
n-Octadecylacrylate	1.61	±0.34	0.40	±0.12	60	n benzene	114
Octadecyl acrylate	1.74	±0.04	0.68	±0.18	60		52
	4.1	±0.8	1.2	±0.1	60	a, m	51
n-Octadecyl vinyl ether	0.85	±0.05	0.0	±0.2	50		68
Octyl acrylate	1.93	±0.08	0.83	±0.23	60		115
n-Octyl vinyl ether	0.81	±0.05	0.0	±0.2	50		68
1,4-Pentadiene	1.116		<0.01		50		116
	1.13		0		50	a DMF	67
Phenyl acetylene	0.30	±0.02	0.69	±0.04	—		117
Potassium itaconate	0.43		0.10		50		87
Potassium styrene sulphonate	0.07	±0.02	1.5	±0.4	70	a, m aqueous	24
Propenyl triethoxy silane	20.0		0		50	i, m	38
α-*n*-Propylstyrene	—		—		60	s	57
cis-β-*n*-Propylstyrene	3.0		0		60	s	57
trans-β-*n*-Propylstyrene	—		—		60	s	57
trans-1,2-bis-(2-Pyridyl)-ethylene	0.95	±0.05	0.02	±0.05	50	h, m DMF	118
Sodium acrylate	0.21		0.77		50	aqueous	33
Sodium allylsulphonate	4.94	±0.06	0.07	±0.06	30		119
	1.00	±0.01	0.38	±0.02	45	a, m DMSO pH7	120
	1.25	±0.01	0.28	±0.02	45	a, m DMSO + H_2O (94:6)	120

	1.85	±0.01	0.43	±0.01	45	a, m DMSO pH 1.5	120
	0.69	±0.05	0.18	±0.05	60		119
Sodium styrene sulphonate	0.05	±0.02	1.5	±0.2	40	a, m aqueous	24
p-Sodium styrenesulphonate	0.05	±0.01	1.40	±0.04	45	a, m aqueous pH 7.0	121
	0.15	±0.02	0.55	±0.03	45	a, m DMSO	122
	0.10		1.20		45	aqueous pH 3	122
	0.05		1.40		45	aqueous pH 7	122
Sodium *p*-vinyl benzoate	0.09	±0.08	0.26	±0.06	50	i, o aqueous	123
p-Styrenesulphonic acid	0.10	±0.02	1.20	±0.10	45	a, n aqueous pH 3.0	121
trans-2,3,4,5-Tetrachlorohexa-1,3,5-triene	0.20		4.05		70		124
Tetrahydrofurfuryl acrylate	0.689		0.665		70		125
Tetrahydrofuryl acrylate	1.577		0.91		25		39
	0.602	±0.10	0.773	±0.07	60	a	40
Tetrahydrofuryl methacrylate	0.477		1.01		25		39
	0.154	±0.03	0.710	±0.01	60	a	40
3,3,3-Trichloropropene	12.2	±1.2	0.100	±0.015	60		60
6-Triethoxy silyl bicyclo-(2,2,1)-2-heptene	0.7		0		50	j, m	38
Trimethyl aconitate	5.50	±0.50	—0.10	±0.10	60	h, l DMF	126
Trimethylamine methacrylamide	0.37	±0.04	0.10	±0.01	70	f, m	127
(Trimethylsiloxyl)-vinyldimethyl-silane	8.0	±15 %	0.1	±15 %	50		128
bis(Trimethysiloxy)-vinylmethyl-silane	8.0	±15 %	0.1	±15 %	50	i, n acetone	128
tris-(Trimethylsiloxy)-vinyl-silane	8.0	±15 %	0.1	±15 %	50	i, n acetone	128
Vinyl acetate	3.88		0.009		25		129
	4.2		0.05		50	a, m DMF	130, 131
	4.05	±0.3	0.061	±0.013	60	f, o acetonitrile	132
	6	±2	0.02	±0.02	60	a, l	107
	6.0	±15 %	0.07	±15 %	70	a, l	102
Vinyl acetylene	0.13	±0.01	0.60	±0.02	60	f, m	73
Vinyl benzoate	5.0	±0.05	0.05	±0.05	75	g, l	30
p-Vinyl benzoic acid	0.076	±0.01	1.8	±0.05	50	j, n DMF	123

References pp. 404—418

Copolymerization reactivity ratios—*continued*

M_1/M_2	r_1		r_2		T(°C)	Remarks	Ref.
Vinyl benzyl chloride (60 % *meta*, 40% *para*)	0.06		0.67		60		12
Vinyl caproate	4.0	±0.3	0.04	±0.03	60		133
Vinyl caprylate	4.0	±0.3	0.04	±0.03	60		133
N-Vinyl carbazole	0.28	±0.02	0.04	±0.02	30	k, l tetrahydrofuran	134
Vinyl chloride	—		—		—	s	135
	3.28	±0.06	0.02	±0.02	60	f, q	108
	2.8	±0.5	0.04	±0.02	40	a, r	136
	3.7		0.074		50	a, m acetone	137
	3.6	±0.2	0.052	±0.009	50	a, m	138
Vinyl chloroacetate	0.34		0.09		80		1
2-Vinyl-4-chloromethyl-1,3-dioxolane	6.25		0.01		60	i, m DMF	139
Vinyl dichloroacetate	0.25		0.18		80		1
2-Vinyl-1,3-dioxane	4.80		0.15		60	DMF	140
Vinyl esters of fatty acids in C_6—C_{18} range	4.0	±0.3	0.04	±0.03	60		133
Vinyl ethyl diethoxysilane	9.0		0		50	i, n	38
Vinyl-2-ethylhexanoate	12	±2	0.01	±0.01	30	g, l	30
Vinyl ferrocene	0.16	±0.05	0.15	±0.05	70	a, m benzene	141
Vinyl formate	3.0	±0.05	0.04	±0.005	60	f, l	30
Vinyl fluoride	~0.001		24	±2	30	a, m	142
N-Vinyl glycidylurethan	0.066	±0.004	1.5	±0.5	60	a, n tetrahydrofuran	143
2-Vinyl-4-hydroxymethyl-1,3-dioxalane	4.25		0.05		60	DMF	140
Vinyl isocyanate	0.19	±0.03	0.16	±0.06	60		144
Vinyl isothiocyanate	0.36		1.4		—		1
Vinyl 9(10)-ketostearate	2.96	±0.12	0.0	±0.091	60	i, l	145
Vinyl 12-ketostearate	3.11	±0.05	0.0	±0.02	60	i, l	145
Vinyl laurate	4.0	±0.3	0.04	±0.03	60		133
Vinyl mesitylene	0.98	±0.2	0.16	±0.02	138		146

Vinyl methyl diethoxysilane	6.0		0		50	i, m	38
2-Vinyl-4-methyl-1,3-dioxolane	5.28		0.01		60	i, m DMF	139
Vinyl myristate	4.0	±0.3	0.04	±0.03	60		133
Vinyl nonanoate	3.57	±0.16	0.059	±0.095	60	d, l benzene, DMF	147
Vinyl octadecanoate	4.0	±0.03	0.04	±0.03	60		133
	4.20	±0.02	0.064	±0.005	60	e, m benzene, DMF	147
	4.3		0.03		70	k, m	148
Vinyl palmitate	4.0	±0.3	0.04	±0.03	60		133
2-Vinyl-4-phenoxymethyl-1,3-dioxolane	6.45		0.01		60	i, m DMF	139
Vinyl phenyl diethoxysilane	8.3		0		50	j, m	38
p-Vinylphenyldiethylcarbinol	0.04		0.38		60	k, m dioxan	149
p-Vinylphenyldimethylcarbinol	0.05		0.41		60	k, m dioxan	149
p-Vinylphenylmethylethylcarbinol	0.06		0.39		60	k, m dioxan	149
Vinyl pinonate	3.40	±0.04	0.143	±0.046	60	c, m DMF	147
2-Vinylpyridine	0.113	±0.002	0.47	±0.03	60	a, m	76
	0.086		2.13		—		150
	0.05	±0.01	21.9	±5.52	60		151, 152
4-Vinylpyridine	0.113	±0.005	0.14	±0.09	60	a, m acetic acid, γ-butyrolactone	76
N-Vinylpyridinium fluoroborate	1.06		0.20		60	k, r	153
N-Vinylsuccinimide	0.54		0.16		60	a, m	111
Vinylsulphonic acid	1.5		0.15		—		154
Vinyl triethoxysilane	5	±1	0		60	f, m DMF	155
	4.5		0		50	i, n	38
Vinyl tri-isopropoxysilane	6.5		0		50	j, m	38
Vinyl trimethoxysilane	6.0		0		50	i, l	38
Vinyl trimethylsilane	3.9	±3	0.07	±0.03	60	h, l DMF	155
Vinyl undecenoate	1.82	±0.04	0.0	±0.10	60	c, l DMF	147
Allyl acetate/							
2-Chloroethyl acrylate	0		5.5	±1.0	60		108
Allyl chloride/							
2-Hydroxyethyl acrylate	0.016	±0.030	8.85	±1.32	60		156
2-Hydroxy ethyl methacrylate	0.033	±0.048	35.6	±1.7	60		156
Isobutyl methacrylate	0.047	±0.056	45.5	±10.2	60		156
n-Propyl acrylate	0.091	±0.013	7.56	±0.10	60		156
n-Propyl methacrylate	0.067	±0.027	52.0	±3.2	60		156

Copolymerization reactivity ratios—*continued*

M_1/M_2	r_1		r_2		T(°C)	Remarks	Ref.
Vinyl benzoate	0.88	±0.470	0.46	±0.16	60		156
Vinyl butanoate	1.15	±0.59	0.31	±0.35	60		156
N-Vinyl carbazole	0		Very large		70		157
Vinyl 2-ethylhexanoate	1.52	±0.130	0.71	±0.18	60		156
Vinyl formate	0.78	±0.100	0.57	±0.08	60		156
Vinyl pentanoate	1.62	±0.310	0.59	±0.20	60		156
Vinyl propionate	0.68	±0.110	0.62	±0.09	60		156
Vinyl valerate	1.62		0.59		60		156
Benzyl acrylate/							
Allyl chloride	9.90	±0.14	0.058	±0.003	60		156
Benzyl methacrylate/							
Allyl chloride	58.7	±4.1	0.016	±0.018	60		156
3-Chloromethyl acrylate	0.96		1.04		60	a, m	158
Cyclohexyl methacrylate	0.93		1.15		60	l benzene	159
Phenyl methacrylate	0.65		1.42		60	l benzene	159
Butadiene/							
2-Chlorobenzalacetophenone	0.78		−0.02		59	c, m	160
2-Chloro-1,3-butadiene	0.059	±0.014	3.41	±0.07	50		47
2,5-Dichlorostyrene	0.65	±0.1	0.2	±0.04	70	a, n	161
Diethyl fumarate	2.13		0.25		—		162
Diethyl maleate	8.08		0.11		—		162
Dimethyl dithiolfumarate	0.0106	±0.0175	−0.0014	±0.027	50	i, l benzene	163
Dinonyl fumarate	2.02		0.32		—		162
Dinonyl maleate	5.36		0.12		—		162
Methyl-2-chlorocinnamate	1.07	±0.12	<0.03		60	f, m	160
	1.20	±0.12	<0.02		80	f, m	160
Methyl-4-chlorocinnamate	2.73	±0.3	<0.05		80	g, m	160
Methyl thiolacrylate	0.35	±0.01	0.20	±0.05	70	g, l	164
Trichloroethylene	12.32	±0.31	−0.007	±0.022	80	a, m	165

Triethyl aconitate			0.02	±0.01	50	h, p	166
			0.00	±0.01	60	f, m	126
Trimethyl aconitate	0.40	±0.03	0.00	±0.03	60	f, m	126
Vinyl formate	5.0	±0.01	0.2	±0.05	120		167
Vinyl nonanoate	26.3	±10.0	0.02	±0.02	60	f, l benzene	147
Vinyl octadecanoate	34.5	±6.0	0.034	±0.034	60	f, l benzene	147
Vinyl pinonate	37.8	±6.5	0.015	±0.015	60	f, l benzene	147
Vinyl undecenoate	37.9	±4.0	0.015	±0.015	60	f, l benzene	147
n-Butyl acrylate/							
Allyl chloride	5.83	±1.32	0.10	±0.170	60		156
Butyl acrylate/							
Di(2-cyanoethyl)fumarate	2.65	±0.35	0.03		90	k, r	168
N-Methylolacrylamide	0.87	±0.05	0.16	±0.17	—		169
Methyl vinyl ketone	0.65	±0.07	1.6	±0.1	50	a, p benzene	170
4-Vinyl pyridine	0.46	±0.09	5.51	±0.09	60	a, l	171
n-Butyl methacrylate/							
Allyl chloride	46.3	±2.6	0.058	±0.024	60		156
3-Chloromethylacrylate	0.85		1.10		60	a, m	158
t-Butyl methacrylate/							
3-Chloromethyl acrylate	0.88		1.14		60	a, m	158
Butyl methacrylate/							
Glycidyl methacrylate	0.79		0.94		—	h, m	172
n-Butyl vinyl ether/							
Poly(1,3-butanediol fumarate)	0		1.8	±0.5	60		173
N-Vinyl succinimide	0		15		60	a, n	111
2-Chloro-1,3-butadiene/							
Fluoroprene	3.70		0.22		50	f, r benzene	174, 175
Hexachlorobutadiene	5.47		0.10		50	f, r benzene	174
trans-2,3,4,5-Tetrachloro-1,3,5-hexatriene	3.6	±0.17	0.2	±0.07	70		176
Vinyl butanoate	90		0.02		65		177
Vinyl formate	30		0.01		65		177

Copolymerization reactivity ratios—*continued*

M_1/M_2	r_1		r_2		T(°C)	Remarks	Ref.
Vinylidene cyanide	0.016		0.0048		40	a, o benzene	27
Vinyl isopropyl ether	0.164	±0.043	11.45	±0.54	65		178
Vinyl propionate	70		0.05		65		177
2-Vinyl pyridine	5.19	±0.03	0.06	±0.01	60	k, r	151, 152
2-Vinylquinoline	2.10	±0.13	0.38	±0.03	60	k, r	151, 152
2-Chloroethyl acrylate/							
Dodecyl vinyl ether	1.97	±0.08	0.013	±0.010	60	c, m	179
Methallyl acetate	4	±1	0		60	k, r	108
p-Chlorophenyl vinyl sulphide/							
p-Bromophenyl vinyl sulphide	0.98		1.00		60	f, m	180
o-Chlorostyrene/							
Anethole	22	±8	0	±0.01	70		105
Indene	3.5	±15 %	0		70	a, n	139
p-Chlorostrene/							
Bicyclo(2.2 1)hepta-2,5-diene	85		0.01		60		181
p-Cyanostyrene	0.34		1.4		30	k, r dimethylacetamide	182
Dimethyl itaconate	0.69		0.15		—		183
p-Methoxystyrene	0.70		0.48		30	k, r	182
	0.86	±0.08	0.58	±0.03	60		184
m-Nitrostyrene	0.25	±0.05	1.3	±0.1	75	a, l benzene	173
p-Nitrostyrene	0.70	±0.08	0.91	±0.37	60		184
Vinyl mesitylene	10		0.06		130		146
p-Vinylphenyl dimethyl carbinol	1.24		0.53		60	k, m dioxan	149
p-Vinylphenyl methyl ethyl carbinol	1.43		0.33		60	k, m dioxan	149
Chlorotrifluoroethylene/							
Propene	0.016	±0.005	0.022	±0.005	−40	a, l, w	185
	0.017	±0.001	0.03	±0.02	−40	a, l, w	185
	−0.03	±0.09	0.03	±0.1	0	a, l, w	185

	0.016	±0.001	0.017	±0.005	0	a, l, w	185
	0.06	±0.05	0.06	±0.02	40	f, l, w	185
	0.028	±0.006	0.028	±0.015	40	f, l, w	185
bis-(Trimethylsiloxy)-vinyl methylsilane	0.05		0.20		60	g, m benzene	128
tris-(Trimethylsiloxy)-vinyl silane	0.05		0.20		60	g, n benzene	128
Decyl methacrylate/							
N-Vinyl oxazolidone	12.8	±0.5	0.015	±0.05	50	c, m	186
2,5-Dichlorostyrene/							
2,5-Dimethylstyrene	1.55		0.27		70	d, l	29
Methyl vinyl ketone	2.0		0.5		70		187
N-Vinyl carbazole	8	±0.5	0.016	±0.002	70		188
Diethylaminoethyl methacrylate/							
Sodium methacrylate	0.65	±0.03	0.08	±0.015	70		190
Diethyl fumarate/							
3,3,3-Trichloropropene	1.10	±0.10	1.46	±0.35	60		60
2,5-Dimethylstyrene/							
3-Methyl-α,β,β-trifluorostyrene	0.20		0.50		60		191
4-Methyl-α,β,β-trifluorostyrene	0.44		0.7		60		191
α,β,β-Trifluorostyrene	0.32		0.67		60		191
Divinyl ether/							
Tetracyanoethylene	0.2		0.1		50	d, m	192
Dodecyl methacrylate/							
Diallyl butylphosphonate	18.7	±0.11	0.091	±0.02	80	b, n benzene	193
Diallyl phenylphosphine oxide	18.1	±0.4	−0.011	±0.013	80	g, o, s benzene	194
	11.1	±0.4	−0.013	±0.025	120	g, p, s benzene	194
	5.6	±0.4	−0.089	±0.022	140	g, o, s benzene	194
Diallyl phenylphosphonate	19.5	±1.8	0.072	±0.22	80	f, n benzene	193
N,N-Diallylurea	27.4	±0.8	0.014	±0.05	100	a	195
N,N'-Diallylurea	14.5	±0.7	−0.12	±0.04	100	d	195
Diethyl allylphosphonate	52.5	±1.8	0.066	±0.06	80	a, n benzene	193
Monoallylurea	29.9	±1.2	−0.22	±0.06	100	a	195

Copolymerization reactivity ratios—*continued*

M_1/M_2	r_1		r_2		T(°C)	Remarks	Ref.
Ethyl acrylate/							
Allyl chloride	2.3		0.48	±0.05	70		157
	7.73	±0.35	0.084	±0.100	60		156
Bicyclo(2.2.1)hepta-2,5-diene	3.05		0.01		50	a, m	42
Di(2-cyanoethyl)fumarate	8.3	±0.5	0.03	±0.01	90	k, r	168
Ethyl 1-acetoxy acrylate	1.0	±0.1	1.0	±0.05	60	a, n	196
Methylstyrene, mixt. of *o*- and *p*- isomers (ca. 1:2)	0.15	+0.23 −0.13	0.48	±0.06	80		12
N-Methylolacrylamide	1.4	±0.2	1.4	±0.2		ethanol	104
Sodium acrylate	5.7		1.5		50		197
2-Sulphoethyl methacrylate	0.30	±0.05	3.2	±0.6	50—60	a, n 1,2-dimethoxyethane	198
2-Vinylpyridine	0.19	±0.06	0.23	±0.05	75		28
2-Vinylthianthrene	0.07		0.314		60	j, r	199
Ethylene/							
Butene-1	3.25	±0.38	—		130—220	j, r 25,000 psi	200
	3.6		0.16		—	10 cm Hg press.	201
	3.2		<0.64		130—220	1020—1700 atm	202
trans-Butene-2	7.5		—		130—220	1020—1700 atm	202
n-Butyl acrylate	0.03		11.9	±2.5	70	a, l toluene 1000 atm	203
n-Butyl isothiocyanate	5.5		0		130—135	m 1360 atm	204
Carbon monoxide	0.0424		0		20	i, s (ref. 592) 170—680 atm	205
	0.045		0		20	680 atm	206
	0.16	±0.02	0		70	850—1000 atm	207
	0.147		0		120—130	1000 atm	208
	0.500		0		135	136 atm	209
Chlorotrifluoroethylene	0.070	±0.0004	0.001	±0.017	−78	l, w	210
	0.070	±0.0015	0.001	±0.002	−78	l, w	210
	0.116	±0.001	0.005	±0.014	−40	l, w	210
	0.116	±0.001	0.004	±0.001	−40	l, w	210
	0.172	±0.0001	0.0096	±0.0003	0	l, w	210
	0.173	±0.001	0.009	±0.001	0	l, w	210

	0.252	±0.0002	0.0290	±0.0013	60	l, w	210
	0.250	±0.0025	0.0285	±0.001	60	l, w	210
Isobutene	2.1		<0.49		130—220	1020—1700 atm	202
Methyl isothiocyanate	7.0		0		130—135	m 1360 atm	204
Octene-1	3.2		—		130—135	1020—1700 atm	202
	3.66	±0.15	—		130—220	j, r 25,000 psi	200
Phenyl isothiocyanate	6.7		0		130—135	m, 1360 atm	204
Propene	3.2		<0.62		130—135	1020—1700 atm	202
	3.43	±0.15	—		130—220	j, r 25000 psi	200
Vinyl fluoride	0.3	±0.03	1.7	±0.1	30	a, m	142
	4.39	±0.77	0.16	±0.05	160	a, l toluene 1000 atm	211
Ethyl methacrylate/							
Allyl chloride	57.9	±15.0	0.082	±0.083	60		156
2-Sulphoethyl methacrylate	1.0	±0.1	2.0	±0.4	50—60	a, n 1,2-dimethoxyethane	198
Fluoroprene/							
Hexafluorobutadiene	2.93		0.24		40	f, r benzene	174
Fumaronitrile/							
2-Chloroethyl acrylate	0.11	±0.03	10.2	±0.3	60	f, m benzene	179
Dodecyl vinyl ether	0.19	±0.003	0.004	±0.003	60	d, l benzene	179
Glycidyl methacrylate/							
2-Ethyl hexyl acrylate	0.98		1.08		138	e, m	212
2-Hydroxy ethyl methacrylate/							
2-Sulphoethyl methacrylate (sodium salt)	1.6	±0.1	0.7	±0.1	50—60	g, p aqueous	198
Isobutene/							
Tetrafluoroethylene	0		0		—78		213
Isobutyl methacrylate/							
3-Chloromethyl acrylate	0.88		1.07		60	a, m	158
Isoprene/							
2-Chloro-1,3-butadiene	0.133	±0.025	3.65	±0.11	50	f, r benzene	174
Dimethylbutadiene	1.18		0.84		—18	f, m	214
2-Fluorobutadiene	0.19	±0.10	2.05	±0.19	50	f, m	214

Copolymerization reactivity ratios—*continued*

M_1/M_2	r_1		r_2		T(°C)	Remarks	Ref.
Hexafluorobutadiene	1.19	±0.12	0.78	±0.05	40	f, r benzene	174
trans-2,3,4,5-Tetrachloro-1,3,5-hexatriene	0.58	±0.02	1.58	±0.1	70		176
2-Vinylpyridine	0.59	±0.05	0.47	±0.07	60	k, r	151
2-Vinyl quinoline	1.88	±0.02	0.53	±0.01	60	k, r	151
N-Isopropyl acrylamide/							
S-Vinyl-*O*-*t*-butyl thiocarbonate	1.17		0.3		70	a, n benzene	215
Maleic anhydride/							
Dodecyl vinyl ether	0.046	±0.052	−0.046	±0.054	50	benzene	216
Isopropenyl acetate	0.002		0.032	±0.005	75	a, l	217
Isostilbene	0.08	±0.08	0.07	± 0.07	60	p chloroform	218
Stilbene	0.03	±0.03	0.03	±0.03	60	p chloroform	218
Trimethyl stannyl methacrylate	0.0		0.20		50	g, r DMF	219
Vinyl *N*,*N*-diethylcarbamate	0		0.035		80	f, m	25
Vinylidene cyanide	~0		45		50	a, n	27
β-Vinyloxyethyl(3,5-di-*t*-butyl-4-hydroxy)benzoate	0		0		70	f, q dioxan	220
Methacrylamide/							
1,2-Dimethyl-5-vinylpyridinium methyl sulphate	0.14		1.2		48	f, o aqueous	9
Methacrylic acid/							
N-Allyl acetamide	16.9	±0.7	0.04	±0.05	80	ethanol	18
N-Allyl benzamide	13.3	±0.4	0.03	±0.02	80	ethanol	18
N-Allyl chloroacetamide	12.2	±0.4	0.05	±0.05	80	ethanol	18
N-Allyl formamide	4.6	±0.2	0.025	±0.03	80	ethanol	18
N-Allyl isovaleramide	8.1	±0.3	0.05	±0.03	80	ethanol	18
Bis(*β*-chloroethyl)vinyl phosphonate	1.7	±0.5	0.1	±0.075	—		221
n-Butyl acrylate	1.31		0.35		50	f, n ethanol	19
n-Butyl methacrylate	0.75		1.20		50	f, n ethanol	19

Butyl methacrylate	0.53		1.1		—	c, m	172
2-Chloroallyl alcohol	4.5		0.0		100		20
2-Chloroallyl chloride	4.0		0.0		100		20
o-Chlorostyrene	0.7	±15 %	0.12	±15 %	70	a, m	102
Diethylaminoethyl methacrylate	0.98	±0.016	0.90	±0.23	70	a, n aqueous pH 1.2	190
	0.08	±0.015	0.65	±0.03	70	a, n aqueous pH 7.2	190
1,2-Dimethyl-5-vinylpyridinium methyl sulphate	0.58		0.60		50	a, m aqueous	9
Glycidyl methacrylate	0.85		1.18		—	c, m	172
Methacrylamide	2.0		0.22		50		222
	2.0		0.3		70	a, n	223
	2.50		0.30		70	a, n	223
Methacrylonitrile	1.63	±0.08	0.59	±0.08	80	f, l	224
	1.64	±0.05	0.62	±0.05	65	i, l	225
α-Phenyl vinyl phosphinic acid	3.50	±0.2	0.36	±0.12	80	g, p	226
Vinyl bromide	0	±0.1	90	±15	50		227
Vinyl hydroquinone dibenzoate	1.91	±0.23	0.91	±0.25	60	i, o toluene	26
2-Vinylpyridine	0.44	±0.02	1.38	±0.04	—		228
Methacrylonitrile/							
Acrolein	1.20	±0.08	0.72	±0.06	50	g, l	6
	1.20	±0.08	0.72	±0.06	50	d, m dioxan	7
Acrylic anhydride	0.4		0.9		35	f, m cyclohexanone	229
n-Amyl methacrylate	0.55		0.51		80	g, l	224
n-Butyl methacrylate	0.51		0.69		80		224
t-Butyl methacrylate	0.37		0.70		80		224
o-Chlorostyrene	0.86		0.78		80		1
p-Chlorostyrene	0.73		0.45		80		1
Diphenyl ethylene	—		—		—	s	230
Ethyl methacrylate	0.46		0.83		80	e, l	224
n-Hexyl methacrylate	0.75		0.56		80	j, l	224
Isopropyl methacrylate	0.43		0.92		80	g, l	224
Methacrolein	0.40	±0.04	1.78	±0.06	50	a, m DMF	7
Methacrylic anhydride	0.27		1.6		36.6	a, m cyclohexanone	229
α-Methoxy styrene	—		—		—	s	230

Copolymerization reactivity ratios—*continued*

M_1/M_2	r_1		r_2		T(°C)	Remarks	Ref.
Methyl α-phenylacrylate	—		—		—	s	230
α-Methyl styrene	0.38		0.54		60	g, m toluene	231
	0.35	±0.02	0.12	±0.02	80	a, l	107
	0.21		0.15		80		1
cis-β-Methylstyrene	27		0		60	f, m	232
trans-β-Methylstyrene	4.4		0		60	f, m	232
Octadecyl methacrylate	0.90		1.13		80	i, l	224
n-Octyl methacrylate	0.75		0.58		80		224
n-Propyl methacrylate	0.29		0.79		80		224
Vinyl cinnamate	4.6	±0.4	0.15	±0.06	70	a, r cyclohexanone	223
p-Methoxystyrene/							
p-Cyanostyrene	0.093		0.85		30	k, r dimethylacetamide	182
p-Bromostyrene	0.43		1.1		30	k, r	182
p-Methoxyphenyl vinyl sulphide/							
p-Bromophenyl vinyl sulphide	0.58		0.93		60	f, m	180
p-Chlorophenyl vinyl sulphide	0.59		0.98		60	f, m	180
p-Methylphenyl vinyl sulphide	0.96		1.61		60	f, m	180
Phenyl vinyl sulphide	0.66		1.09		60	f, m	180
Methyl acrylate/							
Acrolein	10.0	±0.2	0.2	±0.05	20	c, m aqueous	5
	7.7	±0.2	0.0	±0.05	20		6
Acrylamide	0.05	±0.05	1.30	±0.05	60		12
Allyl chloride	9.100	±0.930	−0.020	±0.100	36		234
	8.450	±0.370	0.054	±0.039	60		234
	22	±4	—		60	j, m	235
	6.460	±0.570	0.404	±0.073	90		234
	5.450	±0.070	0.071	±0.011	120		234
Allyl ethene sulphonate	10.7	±0.2	0.07	±0.02	60	a, m benzene	236
2-Bromovinyl ethyl ether	31.5	±2.5	0		80	g, o	72
Butadiene	0.05	±0.03	1.0	±0.15	5	d, l	164

Butyl vinyl sulphonate	5.0	±1.5	0.11	±0.03	70	a, n	237
2-Chloroethyl acrylate	0.9	±0.1	0.9	±0.1	60	k, r	108
Deuteromethyl methacrylate	0.50	±0.08	2.3	±0.5	130	g, p	238
2,5-Dichlorostyrene	0.15	±0.03	3.4	±1.4	70	a, l	239
	0.25	±0.04	4.27	±0.28	60	f, m	240
1,1-Dichloro-2-vinylcyclopropane	3.2		0.33		60	g, n	241
Diethylphosphonomethyl acrylate	0.88	±0.09	0.29	±0.05	75	h, m	242
5-Ethyl-2-vinylpyridine	0.179	±0.006	1.16	±0.08	60	a, l	243
Ferrocenyl methyl acrylate	4.46	±0.2	0.14	±0.02	70	p benzene	244
Ferrocenyl methyl methacrylate	0.82	±0.07	0.08	±0.03	70	m benzene	244
Maleic anhydride	2.8	±0.05	0.02		75	a, m	217
Methacrylamide	2.0		0.22	±0.0	65		245
Methacrylic anhydride	0.16	±0.02	4.75	±0.15	60	k, r benzene	246
α-Methoxy styrene	0.17	±0.01	0		60	m	93
N-Methylolacrylamide	1.3	±0.2	1.9	±0.7	—		169
Methyl α-phenylacrylate	~0.055		~1.0		60	g, m	105
Methylstyrene, mixture of *o*- and *p*-isomers (1:2)	0.15	+0.25 −0.14	0.56	±0.06	80		12
cis-β-Methylstyrene	6.0		0		60	f, m	232
trans-β-Methylstyrene	2.7		0		60	g, m	232
2-Methyl-5-vinyl pyridine	0.172	±0.007	0.88	±0.10	60	a, l	109
Methyl vinyl sulphide	0.35	±0.04	0.05	±0.03	60	a, n	247
Pentachlorophenyl vinyl sulphide	0.89		0.25		80	a, m	248
Phenylacetylene	0.65	±0.02	0.47	±0.1	—		249
Phenyl vinyl sulphide	0.40		0.05	±0.02	60	f, m	250
m-Styrenesulphonyl fluoride	0.14		1.50		75	a, n benzene	251
p-Styrenesulphonyl fluoride	0.20		4.0		75	a, n benzene	251
trans-2,3,4,5-Tetrachlorohexa-1,3,5-triene	0.27	±0.04	3.17	±0.27	—		252
Tetravinyl silane	8.87	±0.07	0.133	±0.006	—		253
Vinyl acetate	6.3	±0.4	0.031	±0.006	60	j, l	254
	9	±2.5	0.1	±0.1	60	f, m	132
N-Vinylcarbazole	0.43	±0.02	0.11	±0.02	30	k, l tetrahydrofuran	134
	0.50		0.050		75	g, m benzene	255

Copolymerization reactivity ratios—*continued*

M_1/M_2	r_1		r_2		T(°C)	Remarks	Ref.
Vinyl chloride	4		0.06		45		256, 257
	4.4	±0.5	0.12	±0.01	50	b, m benzene	82
	9.0		0.083		50	d, l	137
Vinyl *N,N*-diethylcarbamate	4.45		0.01		80	f, m	25
Vinyl ferrocene	0.63	±0.03	0.82	±0.05	70	q benzene	141
Vinylhydroquinone dibenzoate	0.46		0.75		78	k, m benzene	258
Vinylidene chloride	1		1		70		188
Vinyl isocyanate	1.38	±0.27	0.14	±1.0	60		144
2-Vinylmercaptobenzothiazole	0.1		1.0		60	f, q	259
Vinyl octadecanoate	5.8		0.03		70	k, m	148
2-Vinyl phenanthrene	0.1	±0.03	2.0	±0.2	60	a, m	260
3-Vinyl phenanthrene	0.8	±0.05	1.75	±0.25	60	a, m	260
2-Vinyl pyridine	0.20	±0.09	2.03	±0.49	60	a, l	240
	0.168	±0.003	1.58	±0.05	60	a, l	243
4-Vinyl pyridine	0.22	±0.01	1.7	±0.2	60	a, l	243
N-Vinyl pyridinium fluoroborate	1.5		0.2		60	k, r acetic acid γ-butyrolactone	153
N-Vinyl pyrrolidone	0.27	±0.16	0.041	±0.024	60		1
N-Vinyl succinimide	1.2		0.4		60	a, n	111
Vinyl undecenoate	3.69	±0.12	0.031	±0.12	60	c, l benzene	147
N-Methylcitraconimide/							
2-Vinylnaphthalene	0.256		0.187		70		103
Methyl methacrylate/							
p-Acetyl amino-styrene	0.30	±0.05	0.50	±0.05	65	a, m DMF	261
Acrylamide	3.3		0.2		60	f, r DMSO	262
	3.2		1.00		60	f, r DMSO + 10 % (v/v) H_2O	262
	3.5		0.4		60	f, r DMSO + 2.5 % (v/v) acetic acid	262
	3		0.15		60	f, r DMSO + 10 % (v/v) H_2O + 2.5 % (v/v) acetic acid	262
	0.30		1.0		60	f, r chloroform	262
	0.25		3.5		60	f, r chloroform + 1 % (v/v) H_2O	262

	2.55	±0.40	2.45	±0.35	70	a, m dioxan	263
	2.53	±0.22	0.82	±0.07	70	a, m dioxan/ethanol	263
	2.60	±0.21	0.44	±0.06	70	a, m ethanol	263
1-Acrylamido-1-deoxy-*D*-glucitol	4.22		0.206		50	f, m aq. methanol	264
Acrylic acid	2.89	±0.05	0.24	±0.03	50	b DMF	265
	2.64		0.12		50	b DMSO	265
	2.46	±0.04	0.25	±0.02	50	b tetrahydrofuran	265
	2.32	±0.08	0.30	±0.04	50	b dioxan	265
	2.30	±0.04	0.31	±0.04	50	b methyl ethyl ketone	265
	2.17	±0.04	0.33	±0.05	50	b acetonitrile	265
	1.86	±0.04	0.24	±0.05	50	b	265
	1.5	±0.1	0.40	±0.15	50	b benzene	265
	1.13		0.29		65	c, l benzene	266
Acrylonitrile	—		—		—	s	267
	1.01	±0.07	0.09	±0.05	—	liq. SO_2	106
	1.45		0.15		−80	a, m toluene (n-Bu_3B)	268
	1.34		0.12		−12	a, m ethylene dichloride (n-Bu_3B)	268
	1.34		0.12		12	a, m (n-Bu_3B)	268
	1.65	±0.40	0.15	±0.03	20		95
	1.34		0.12		26	a, m (n-Bu_2BCl)	268
	1.05		0.15		26	a, m (i-Bu_3Al)	268
	0.85		0.09		26	a, m (n-Bu_2Zn)	268
	1.20		0.03		26	a, m (Me_2Cd)	268
	1.351	±0.133	0.100	±0.070	40		234
	1.25	±0.03	—		50		269
	1.20	±0.14	0.15	±0.07	60	f, n, t	270
	1.46		0.37		70	a, l 100 atm, toluene	271
	2.01		0.45		70	a, l 1000 atm, toluene	271
	1.224	±0.100	0.150	±0.080	80		234
	1.34		0.12		90	1 atm	272
	1.46		0.37		90	100 atm	272
	2.01		0.45		90	1000 atm	272
	1.186	±0.120	0.160	±0.100	100		234
	1.14	±0.02	0.46	±0.08	100	g, m DMSO	269
	1.15	±0.06	—		120		269

Copolymerization reactivity ratios—*continued*

M_1/M_2	r_1		r_2		$T(^\circ C)$	Remarks	Ref.
	1.09	±0.06	—		140		269
	1.03	±0.06	—		160		269
	1.01	±0.04	—		180		269
Acrylyl chloride	0.48		1.51		45		273
Allyl allyl sulphonate	5.3	±0.7	0.37	±0.09	—	a	274
Allyl chloride	48.1	±5.00	0.048	±0.038	60		156
	56		—		60	j, l	235
Allyl glycidyl ether	40.7		0.035		70		275
Allyl methacrylate	0.87	±0.025	0.69	±0.02	65		276
m-Aminophenyl vinyl ether	0.07	±0.02	0.75	±0.05	60		277
N-(*p*-Anisyl)-methacrylamide	0.14		0.57		70		278
Benzyl methacrylate	0.93		1.05		60	benzene	159
1-Bicyclobutane carbonitrile	1.70		0.44		50	k, r γ-butyrolactone	41
Bicyclo(2,2,1)hepta-2,5-diene	10.0		0		50	f, m	42
2,5-Bis(trifluoromethyl)styrene	0.57	±0.07	1.35	±0.05	60	d, m	279
N-Bornyl maleimide	2.02		0.16		50	f, o tetrahydrofuran	280
p-Bromophenyl vinyl sulphide	0.85		0.20		60	f, m	180
m-Bromostyrene	0.48	±0.02	1.17	±0.25	60	k, r	184
p-Bromostyrene	0.395	±0.02	1.10	±0.25	60	k, r	184
Butadiene	0.25	±0.03	0.75	±0.05	90	k, r	108
	0.32		0.70		—		281
Butylacrylate	1.81	±0.1	0.37	±0.1	60	a, l	282
n-Butyl acrylate	1.74		0.20		60	f, l	283
N-Butyl maleimide	1.33	±0.03	0.12	±0.02	50	k, r	284
n-Butyl methacrylate	0.79		1.27		60	f, l	283
Δ^3-Carene	3.58		0.22		—		285
β-Chloroethyl itaconate	2.63	±0.2	0.10	±0.05	50		286
β-Chloroethyl methacrylate	0.88		1.13		60	g, m	158
N-(*p*-Chlorophenyl)-methacryl-amide	0.61		0.24		70		278
p-Chlorophenyl vinyl sulphide	0.89		0.08		60	f, m	180

m-Chlorostyrene	0.47	±0.075	0.91	±0.11	60	k, r	184
o-Chlorostyrene	0.50	±0.03	1.37	±0.1	60	k, r	287
p-Chlorostyrene	0.415	±0.02	0.89	±0.05	60	k, r	184
Chlorotrifluoroethylene	75.0		0.005		60	f, n	288
4-Chloro-1-vinyl naphthalene	0.7	±0.2	0.7	±0.2	60	g, m	260
6-Chloro-2-vinyl naphthalene	0.45	±0.05	1.6	±0.2	60	a, n	260
α-Cyanoacrylate	0.10		0.13		60	a, l acetic acid	289
p-Cyanostyrene	0.22	±0.02	1.41	±0.13	60	k, r	184
Cyclohexyl methacrylate	0.86		1.15		60	l benzene	159
4-Cyclopenten-1,3-dione	7.4		0.083		65	a, l benzene	62
1-Deoxy-1-methacrylamido-*D*-glucitol	4.20		0.036		50	e, m aq. methanol	264
Deuteromethyl acrylate	2.2	±0.4	0.42	±0.06	130	g, o	238
Diallyl melamine	27.8	±3.2	0		60	i, o DMF, w	290
	28.9	±2.3	0		60	i, o DMF, w	290
Diallyl phenyl phosphate	9.78		0.225		70		291
Diallyl phenyl phosphonate	22.96		0.135		70		292
N,*N*-Di-*n*-butyl acrylamide	1.85	±0.05	0.42	±0.04	50		293
Di-*n*-butyl itaconate	0.8		0.4		—		183
Di-β-chloroethyl itaconate	2.63	±0.2	0.10	±0.05	50		286
2,5-Dichlorostyrene	0.44		2.25		68	a, n	294
1,1-Dichloro-2-vinylcyclopropane	11.0		0.07		60	g, n	241
Diethyl fumarate	40.1	±0.9	0.03	±0.02	60	g, l	295
	2.10	±0.4	0.05	±0.1	60	k, r dioxan	296
Diethyl maleate	354	±57	0		60	g, l	295
Diethylphosphonoethyl acrylate	2.02	±0.05	0.15	±0.03	65	h, m	242
Diethylphosphonomethyl acrylate	1.88	±0.03	0.27	±0.01	75	h, n	242
Diethylphosphonomethyl methacrylate	1.43	±0.05	0.50	±0.03	68	h, m	242
Diethynyl benzene	0.60	±0.1	—		—		249
	0.60	±0.15	—		—		297
1,1-Dihydroperfluorobutyl acrylate	1.4		0.25		50	a, l	298
N-(1,1-Dihydroperfluorobutyl)-*N*-ethyl acrylamide	0.77		0.89		66	f, m	299

Copolymerization reactivity ratios—*continued*

M_1/M_2	r_1		r_2		T(°C)	Remarks	Ref.
Diisobutyl vinyl phosphine oxide	30	±10	0		60	i, n	300
1,2:3,4-Di-*O*-isopropylidene-6-*O*-methacryloyl-α-*D*-galactopyranose	0.93	±0.03	1.30	±0.19	50	j, m	301
2,5-Dimethoxystyrene	0.25	±0.01	0.72	±0.04	70	a, r	302
2,6-Dimethoxystyrene	0.15		0.80		60		303
N,N-Dimethylacrylamide	1.8	±0.18	0.45	±0.08	50	f, l	304
	2.04	±0.11	0.51	±0.07	70	g, m dioxan	305
	2.30	±0.24	0.42	±0.10	70	g, m ethanol	305
p-Dimethylaminostyrene	0.205	±0.02	0.11	±0.02	60	f, m	184
1,1-Dimethyl-1-(2,3-dihydroxypropyl)amine-methacrylamide	1.60		0.04		75	a, m acetonitrile	306
1,1-Dimethyl-1-(2-hydroxypropyl)-amine-methacrylamide	1.99		0.0		75	a, m acetonitrile	306
Dimethyl itaconate	1.2	±1.3	0.3		—		183
Dimethyl methacrylaminodiacetate	16.67		0.013		70		278
N-(1,1-Dimethyl-3-oxobutyl)-acrylamide	1.68	±0.06	0.57	±0.03	60		307
1,2-Dimethyl-5-vinylpyridinium methyl sulphate	0.12		1.8		50	c, m aqueous methanol	9
2,4-Dimethyl-6-vinyl-*s*-triazine	0.37		1.75		60	g, m benzene	308
Dipentene	2.35		0.30		—		285
Diphenyl-4-vinylphenylphosphine	0.32	±0.02	0.91	±0.12	60	k, r benzene	309
Diphenyl-4-vinylphenylphosphine oxide	0.38	±0.02	1.46	±0.10	60	k, r benzene	98
Diphenyl-4-vinylphenylphosphine sulphide	0.29	±0.01	1.22	±0.34	60	k, r benzene	309
Diphenyl vinylphosphine	5.1	±0.2	0		60	j, n	300
Diphenyl vinylphosphine oxide	11	±4	0		60	i, n benzene	300
Diphenyl vinylphosphine sulphide	13	±3	0		60	i, n benzene	300

N,*N*-Divinylaniline	2.0		0.01		60	d, n benzene	66
m-Divinyl benzene	0.41		0.61		70	a, m	587
p-Divinyl benzene	~0.26		~1.3		70	a, l	587
Divinyl ether	10.0		0.006		60	f, n	66
Divinyl sulphide	0.85	±0.05	0.13	±0.05	60	f, n	310
Divinyl sulphone	8.5		0.045		60	f, m benzene	66
Divinyl tartrate	0.97		0.04		60		311
Ethyl 1-acetoxy acrylate	1.65	±0.07	0.65	±0.05	60	a, m	196
Ethyl acrylate	2.03	±0.12	0.24	±0.12	60	g, l	282
	1.83		0.47		60	l	283
N-Ethyl,*N*-1,1-dihydroperfluoro-butylacrylamide	0.77		0.89		66	g, m	299
N,*N*-Ethylene-*N'*-vinylurea	1.70	±0.10	0.08	±0.04	60	f, m tetrahydrofuran	143
N-Ethyl methacrylamide	1.75		0.11		70		278
Ethyl methacrylate	0.92		1.08		60	f, l benzene	159
	1.09	±0.1	0.98	±0.1	60	g, l	282
Ethyl methacrylylaminoacetate	1.09		0.9		70		278
Ethyl vinyl oxalate	6.0		0.1		60	c, m	75
5-Ethyl-2-vinylpyridine	0.395	±0.003	0.69	±0.03	60	a, l	243
5-Ethyl-2-vinylpyridine-*N*-oxide	0.11	±0.01	5.5	±0.5	60	g, m	312
	0.12	±0.02	4.7	±0.6	60	i, m ethanol	312
Ethyl vinyl sulphide	2.7	±1.5	0.3	±0.1	60		313
	0.93		0.07		60	a, m	58
Ethyl vinyl sulphoxide	10.0		0.01		60	f, m	77
N-Ethyl-*N'*-vinylurea	1.8		0.015		75		314
Ferrocenyl methyl acrylate	2.9	±0.5	0.08	±0.03	70	m benzene	244
Ferrocenyl methyl methacrylate	3.27		0.12		70	o benzene	244
2-Fluorobutadiene	0.64	±0.08	1.54	±0.08	50	f, n	78
Fumaronitrile	3.5	±0.5	0.01	±0.01	79	a, m chloroform	315
Furfuryl methacrylate	0.75		1.19		60	l	283
Glycidyl methacrylate	0.80		1.05		60	k, r	316
	0.76	±0.07	0.88	±0.10	60	k, m	317
	0.75		0.94		80		318
Glycidyl vinylsulphonate	14.4	±2.4	0.01	±0.13	70		319
o-Hydroxystyrene	0.33	±0.08	0.21	±0.06	60	a, m tetrahydrofuran	320

Copolymerization reactivity ratios—*continued*

M_1/M_2	r_1		r_2		T(°C)	Remarks	Ref.
p-Iodostyrene	0.36	±0.03	0.95	±0.20	60	k, r	184
Isobutyl acrylate	1.71		0.20		60	f, l	283
Isobutyl methacrylate	0.91		1.09		60	f, l	283
Isobutyl vinyl sulphide	0.94		0.04		60	g, n	58
Isopropenyl acetate	30		0.017		75	a, m benzene	321
N-Isopropenylglycidyl urethane	1.70		0.12		60	h, m tetrahydrofuran	143
Isopropenyl isocyanate	3.10	±0.29	0.14	±0.10	—		322
β-Isopropenylnaphthalene	0.45	±0.03	0.00		99		86
Isopropenyl-2-oxazoline	0.89		1.12		60	a benzene	323
Isopropylstyrene	0.44		0.39		70	a, l	324
Isopropyl vinyl sulphide	0.85		0.07		70	d, n	58
Itaconic acid	1.14		0		90		325
	1.23		0		—		326
Maleic anhydride	4.63	±1.14	−0.18	±0.28	30	g, m ethyl acetate	327
	6.7	±0.2	0.02		75	g, m	217
	3.5		0.03		60	d, r	328
Maleimide	2.50		0.17		75	a, n cyclohexane	329
Methacrylamide	1.5	±0.02	0.47	±0.04	65		245
	1.65	±0.05	0.49	±0.02	70		245
	1.55	±0.22	1.27	±0.19	70	f, m dioxan	330
	1.68	±0.07	0.43	±0.04	70	f, m ethanol	330
	1.65	±0.05	0.49	±0.02	70	a, m 2-ethoxyethanol	331
Methacrylic acid	0.55	±0.02	1.55	±0.06	45		332
	0.98	±0.04	0.68	±0.05	45	DMF	332
	0.345	±0.05	0.956	±0.02	65	a, l benzene	266
Methacrylic anhydride	0.22	±0.02	1.70	±0.07	60	g, r benzene	246
Methacrylonitrile	0.67	±0.01	0.65	±0.06	60	f, n	108
	0.68		0.80		80		1
	0.74		0.70		80	f, l	224
Methacryloxymethyl-pentamethyl disiloxane	0.93	±0.1	1.13	±0.1	50	k, r	90

p-Methoxystyrene	0.29	±0.03	0.32	±0.05	60	k, r	184
	0.29		0.32		60	a, m	333
α-Methoxystyrene	2.5	±0.03	0		60	a, m	93
p-Methoxyphenyl vinyl sulphide	0.77		0.04		60	f, m	180
Methyl acrylate	1.69		0.34		60	l benzene	334
	2.05		0.35		60	d, l, w	335
	1.99		0.33		60	d, l, w	335
	2.04		0.35		60	d, l, w	335
	1.8	±0.4	0.35	±0.1	65	a, l	263
	0.3	±0.005	1.5	±0.01	65		245
	1.91		0.504		130	d, r	238
Methyl α-bromoacrylate	0.23		2.08		60	l benzene	334
Methyl α-chloroacrylate	0.15		2.00		60	l benzene	334
	0.3	±15 %	1.2	±15 %	70		102
N-Methylcitraconimide	3.24		0.15		70		103
Methyl ethacrylate	2.03		0.1		30—60		336
Methylglycidyl itaconate	1.73	±0.05	0.20	±0.02	60	a, m	203
N-Methylmethacrylamide	1.54		0.24		70		278
	1.54	±0.05	0.24	±0.02	70	a, m 2-ethoxyethanol	331
N-Methylolmethacrylamide methyl ether	1.66	±0.02	0.27	±0.03	70	a, m 2-ethoxyethanol	331
Methyl α-phenylacrylate	0.21		0		60	g, m	105
p-Methylphenyl vinyl sulphide	0.81		0.07		60	f, m	180
α-Methylstyrene	0.40		0.15		60	k, l	337
	0.50	±0.03	0.14	±0.01	60	k, r	287
	0.89	±0.03	—0.01	±0.01	99		86
m-Methylstyrene	0.53	±0.025	0.49	±0.02	60	f, n	184
p-Methylstyrene	0.405	±0.025	0.44	±0.02	60	k, r	184
2-Methyl-*N*-vinylimidazole	3.48		0.003		—		338
2-Methyl-4-vinylpyridine	0.46	±0.02	0.61	±0.08	60	a, l	109
Methyl vinyl sulphone	14	±2	0		60	v	60
N-Methyl-*N*-vinyl-*p*-toluenesulphonamide	4.68		0		60	j, n	111, 112
2-Nitrobutyl acrylate	1.60		0.22		70		113
m-Nitrostyrene	0.35	±0.05	0.85	±0.2	75	a, m benzene	339

Copolymerization reactivity ratios—*continued*

M_1/M_2	r_1		r_2		T(°C)	Remarks	Ref.
Nonyl methacrylate	1.10		0.86		60	l benzene	159
N-n-Octadecylacrylamide	3.85	±0.17	0.44	±0.04	60	n benzene	114
n-Octadecylacrylate	2.36	±0.04	0.48	±0.12	60	n benzene	114
N-n-Octylacrylamide	3.50	±0.06	0.24	±0.06	50		293
N-tert-Octylacrylamide	3.6	±0.04	0.24	±0.04	50		293
Oleylacrylate	2.23	±0.01	0.30	±0.03	60	m benzene	114
Pentachlorostyrene	4.0	±0.4	0.35	±0.05	70	d, m	340
Phenylacetylene	0.85	±0.1	0.05		—		249
	0.81		—		—		341
N-Phenyl methacrylamide	0.54		0.46		70		278
Phenyl methacrylate	0.53		1.67		30	l benzene	159
	0.56		1.72		60	l benzene	159
	0.56		1.30		110	l benzene	159
Phenyl vinyl ether	3.5		0.01		—		342
2-Phenyl-*N*-vinylimidazole	3.50		0.006		—		338
2-Phenyl-5-(4′-vinyl)phenyl-tetrazole	0.73	±0.03	0.22	±0.10	60	g, m, w benzene (C analysis)	343
	0.44	±0.03	0.16	±0.10	60	g, m, w benzene (C analysis)	343
	0.50	±0.03	0.26	±0.10	60	g, m, w benzene (C analysis)	343
	0.86	±0.20	0.55	±0.12	60	g, m, w benzene (N analysis)	343
	0.82	±0.20	0.45	±0.12	60	g, m, w benzene (N analysis)	343
	0.86	±0.20	0.68	±0.12	60	g, m, w benzene (N analysis)	343
α-Phenyl vinyl phosphinic acid	3.30	±0.2	0.06	±0.04	80	g, p	226
Phenyl vinyl sulphide	0.66		1.09		60	f, m	180
α-Pinene	6.60		0.07		—		285
Poly(ethylene glycol fumarate)	17.5	±7.5	0.35	±0.35	60	b, c	344
Propyl acrylate	1.61	±0.1	0.29	±0.1	60	g, l	282
Styrene tricarbonyl chromium	0.75		0		70	e, p benzene	345
Tetrachloroethylene	240	±20	—		60		235
trans-2,3,4,5-Tetrachloro-1,3,5-hexatriene	0.465	±0.05	1.82	±0.075	—		252
	0.09	±0.105	0.315	±0.14	70		176

N-p-Tolylmethacrylamide	0.39		0.67		70		278
Triallyl cyanurate	50.2	±7.5	0		60	g, n, w	290
	46.3	±5.3	0		60	g, n, w	290
Triallyl isocyanurate	48.9	±7.3	0		60	i, n, w	290
	45.0	±5.1	0		60	i, n, w	290
Tri-*n*-butylvinyltin	27.9		0.03		50	a, m benzene	346
3-Trifluoromethylstyrene	0.60	±0.10	0.98	±0.15	60	a, m	279
Trimethylvinyl germane	19.98		0.05		50	f, n benzene	347
Trimethylvinyl tin	25.1		0.03		50	a, m benzene	346
N-Vinyl acetanilide	8.90		0.005		—		348
Vinyl acetate	28.6		0.035		30		349
	20	±3	0.015	±0.015	60	f, n	132
	22.21	±0.89	0.072	±0.026	60		350
	20	±2	0.03	±0.02	60		351
	26.0	±2.5	0.03	±0.01	60	l benzene	352
Vinyl benzoate	20.3	±2.5	0.07	±0.01	60	l benzene	352
	16	±1	0.02	±0.01	60		351
Vinyl bromide	20	±2	0.05	±0.015	28	d, m	353
	25	±2	0.05	±0.01	0	d, n	353
S-Vinyl-*O*-*t*-butylthiocarbonate	1.40		0.17		70	a, l	215
Vinyl butyrate	16	±1	0.01	±0.01	60		351
	25.0	±2.5	0.03	±0.01	60	l benzene	352
N-Vinyl carbazole	2.7	±0.1	0.07	±0.01	30	k, l tetrahydrofuran	134
	2.0		0.04		75	a, m benzene	255
	2.0	±0.3	0.20	±0.03	70	a, l	354
Vinyl chloride	10		0.1		68	d, o	294
	15		0.02		45		257
Vinyl chloroacetate	19	±2	0.015	±0.01	60		351
Vinyl ethyl sulphide	2.7	±1.5	0.3	±0.1	60	k, r	355
1-Vinyl-3-ethyl urea	1.8		0.015		70		356
Vinyl ferrocene	1.22	±0.37	0.52	±0.27	70	p benzene	141
Vinyl formate	23	±2	0.01	±0.01	60		351
	28.6	±2.5	0.05	±0.01	60	l benzene	352
N-Vinylglycidylurethan	2.75		0.10		60		143
Vinylhydroquinone dibenzoate	0.34		0.41		78		258

Copolymerization reactivity ratios—*continued*

M_1/M_2	r_1		r_2		T(°C)	Remarks	Ref.
Vinylidene carbonate	70		0.005		70	a, o acetone	357
Vinylidene chloride	2.53	±0.01	0.24	±0.03	60	f, q	270
Vinylidene cyanide	0.046		0.031		50	a, m	27
	0.046		0.031		—		358
Vinyl isocyanate	5.57	±0.33	0.16	±0.08	60		144
Vinyl isothiocyanate	0.85	±0.03	0.6	±0.02	60		359
Vinyl mesitylene	1.6	±0.3	0.05	±0.01	130		146
	1.4		0.08		85		146
2-Vinyl naphthalene	0.4	±0.05	1.0	±0.15	60	g, m	260
N-Vinyloxazolidinone	6.00		0.03		75	i, n benzene	360
	9.6	±0.2	0.035		50	d, l	186
Vinyl phenylacetate	22.5	±2	0.005	±0.003	60		351
	26.4	±2.5	0.03	±0.01	60	l benzene	352
Vinyl phosphonic dichloride	6.6		0.084		25	a, l	361
Vinyl propionate	24.0	±2.5	0.03	±0.01	60	l benzene	352
2-Vinylpyridine	0.20	±0.03	0.84	±0.13	25		362
	0.439	±0.002	0.77	±0.02	60	a, l	243
	0.395	±0.025	0.86	±0.06	60	k, r	287
	0.33	±15 %	0.70	±15 %	70	a, m	102
4-Vinylpyridine	0.574	±0.004	0.79	±0.05	60	g, l	243
2-Vinylpyridine-*N*-oxide	0.13	±0.03	3.9	±0.8	60	g, m ethanol	312
N-Vinylpyridinium fluoroborate	4.75		0.008		60	k, r acetic acid, γ-butyrolactone	153
N-Vinylpyrrolidinone	4.7	±0.5	0.005	±0.05	50	f, l	186
N-Vinyl succinimide	9.5		0.064		60	f, m	594
2-Vinylthianthrene	2.99		0.07		60		199
p-Methyl phenyl vinyl sulphide/							
p-Bromophenyl vinyl sulphide	0.58		0.73		60	f, m	180
p-Chlorophenyl vinyl sulphide	0.59		0.78		60	f, m	180
α-Methylstyrene/							
β-Cyanoethyl methacrylate	0.12	±0.02	0.35	±0.02	70	o	363
Cyanomethyl methacrylate	0.12	±0.01	0.36	±0.02	70	o	363

2,5-Dichlorostyrene	0.14		3		70		189
2-Fluorobutadiene	0.38	±0.11	1.71	±0.19	50	f, m	78
Fumaronitrile	0.022	±0.005	0.00		60	g, l dioxan	364
Maleic anhydride	0.005	±0.005	0.27	±0.03	60	a, m, benzene	365
	0.038	±0.003	0.08	±0.03	60	a, m acetonitrile	365
Methyl isopropenyl ketone	0.03		1.7		—		1
Vinylhydroquinone dibenzoate	0.11		0.30		78		258
p-Methylstyrene/							
N,N-Divinylaniline	11.8		0.05		60		66
Methyl vinyl ketone/							
α-(2-Cyanoethyl)acrylonitrile	5.05	±0.95	1.24	±0.76	—		366
4,6-Diamino-2-vinyl-*s*-triazine	0.26	±0.04	1.2	±0.15	60	a, m	367
Phenyl vinyl ether	4.4		0.01		—		342
2-Methyl-5-vinyl pyridine/							
Triethyleneglycol dimethacrylate	0.32	±0.065	0.59	±0.06	80	a, n	368
Pentachlorophenyl vinyl sulphide/							
Vinyl sulphide styrene	0.24		3.9		80		248
Phenyl methacrylate/							
3-Chloromethyl acrylate	0.98		0.84		60	a, m	158
Phenyl vinyl sulphide/							
p-Bromophenyl vinyl sulphide	0.86		0.96		60	f, m	180
p-Chlorophenyl vinyl sulphide	0.86		0.91		60	f, m	180
Poly(1,3-butylene glycol fumarate)/							
Allyl diethylphosphinic acid	9.25	±3.00	0.12	±0.008	—		369
Allyl diethylphosphonate	10.00	±2.00	0.075	±0.075	80		370
Allyl ethyl hydrogen phosphite	5.50	±2.50	0.035	±0.035	80		370
n-Amyl vinyl ether	0		0		60		173
n-Butyl vinyl ether	1.8	±0.5	0		60		173
Diethyl allylphosphonate	9.22	±3.00	0.12	±0.008	—		369
Isoamyl vinyl ether	3.8	±0.7	0		60		173
Isobutyl vinyl ether	2.0	±0.7	0		60		173
n-Propyl vinyl ether	1.6	±0.5	0		60		173

References pp. 404—418

Copolymerization reactivity ratios—*continued*

M_1/M_2	r_1		r_2		T(°C)	Remarks	Ref.
Poly(ethylene glycol fumarate)/							
Allyl diethyl phosphinate	1.73	±0.03	0.15	±0.06	—		370
Diethyl allyl phosphinic acid	2.07	±1.12	0.09	±0.95	—		370
Poly(ethyleneglycol fumarate phosphinate)/							
Allyl diethylphosphinic acid	2.07	±1.12	0.09	±0.05	—		370
Sodium acrylate/							
Sodium styrene sulphonate	0.34	±0.23	2.3	±1.2	40	f, m aqueous	24
Styrene/							
Acenaphthylene	3.81		0.33		90	g, r	371
α-Acetoxyacrylonitrile	0.19	±0.056	0.20	±0.052	60		1
1-Acetylbicyclobutane	0.59		0.46		50	k, r methyl ethyl ketone	41
Acrylamide	1.27	±0.17	1.38	±0.09	70	a, m dioxan	263
	1.13	±0.05	0.59	±0.10	70	dioxan/ethanol	263
	1.44	±0.22	0.30	±0.09	70	ethanol	263
	1.13		0.59		70	m dioxan/ethanol	372
1-Acrylamido-1-deoxy-*D*-glucitol	2.72		0.056		50	f, m aq. methanol	264
Acrylic acid	0.25		0.07		50	a, l	373
	0.70		0.15		50	a, l methyl ethyl ketone	373
	0.75		0.13		50	a, l dioxan	373
	0.90		0.14		50	a, l tetrahydrofuran	373
	1.03		0.15		50	a, l DMF	373
	0.25	±0.02	0.15	±0.01	60	a, l	30
	0.29	±0.02	0.075	±0.002	65	a, l benzene	266
	0.22		0.35		70		374
	0.25	±0.05	0.45	±0.1	80		1
Acrylic anhydride	0.17		0.1		35	d, m cyclohexanone	229
Acrylonitrile	0.36		0.03		20	a, r	375
	0.33		0.05		26	a, m	268
	0.32		0.02		26	a, m	268

	0.35		0.02		26	a, m	268
	0.09		0.17		40	d	45
	0.38	±0.03	0.05	±0.02	41.5	f, m	63
	0.37	±0.03	0.07	±0.006	50	a, n	138
	0.37	±0.02	0.05	±0.02	50		46
	0.41	±0.08	0.04	±0.04	60	f, p heterogeneous	270
	0.45	±0.03	0.02	±0.02	65	f, m	63
	0.43		0.13		70	a, l 100 atm, toluene	271
	0.55		0.14		70	a, l 1000 atm, toluene	271
	0.47	±0.03	0.02	±0.02	86.5	f, m	63
	0.37		0.07		90	1 atm	272
	0.43		0.13		90	100 atm	272
	0.55		0.14		90	1000 atm	272
	0.39	±0.02	0.06	±0.01	99		86
	0.36	±0.04	0.14	±0.04	—	benzene, acetonitrile, *N,N*-dimethylformamide	91
	0.04		0.32		—	carbon disulphide	91
	—		—		—	s	267, 376
Allyl allyl sulphonate	13.1		0.01	±0.01		a	274
Allyl chloride	30		—		60	j, m	235
	36.8	±0.60	0.029	±0.006	60		156
	31.5	±4	0.016	±0.016	70	a, l	377
Allyl ethene sulphonate	1.6	±0.1	0.23	±0.03	60	a, m benzene	236
n-Amyl methacrylate	0.55	±0.02	0.40	±0.05	60	g, l	158, 378
Atroponitrile	0.02		0.7		80	i, l	379
Benzylidene malononitrile	0.51		0		70	g, m	380
	0.125		0		80	i, m, s	379
N-Benzylidene-4-methacryloxy-aniline	0.24	±0.01	0.25	±0.05	60	j, l	381
N-Benzoyloxyisomaleimide	0.22	±0.08	0.09	±0.06	70	a, n dioxan	382
Benzoyl vinyl sulphide	4.7		0.27		60	g, n	58
Benzyl methacrylate	0.45	±0.01	0.51	±0.01	60	g, l	158, 378
N-[1-(Benzyl-2-oxopropyl)] acrylamide	2.06	±0.05	0.48	±0.05	60	a, m	383

Copolymerization reactivity ratios—*continued*

M_1/M_2	r_1		r_2		T(°C)	Remarks	Ref.
N-[1-(Benzyl-2-oxopropyl)] methacrylamide	1.30	±0.1	0.57	±0.1	60	a, m	383
1-Bicyclobutane carbonitrile	1.35		0.16		50	f, l DMSO	41
	1.43		0.12		50	k, r γ-butyrolactone	41
1,3-Bicyclobutane dicarbonitrile	1.09		0.01		50	k, r DMSO	41
	0.93		~0		50	k, r γ-butyrolactone	41
Bis(β-chloroethyl)vinyl phosphonate	2.2	±0.4	0.2	±0.2	—		221
Bis(2-chloro-1-methylpropyl) fumarate	0.25	±0.02	−0.05	±0.01	—	a, m	384
Bis(1-methylpropyl)fumarate	0.50	±0.03	0	±0.01	—	a, m	384
2,5-Bis(trifluoromethyl)styrene	0.45	±0.05	1.15	±0.08	60	a, m	279
1,4-Bis(4-vinylphenyl)butane	1.13		0.78		61	f, m benzene	385
1,2-Bis(4-vinylphenyl)ethane	1.05		0.87		61	f, m benzene	385
Bis(4-vinylphenyl)ether	0.94		1.06		61	f, m benzene	385
Bis(4-vinylphenyl)methane	1.01		0.93		61	f, m benzene	385
1,3-Bis(4-vinylphenyl)propane	1.11		0.89		61	f, m benzene	385
Bis(4-vinylphenyl)sulphide	0.60		1.50		61	f, m benzene	385
N-Bornylmaleimide	0.13		0.05		50	f, o tetrahydrofuran	280
Bornyl methacrylate	0.49		0.44		60	f, n	386
p-(β-Bromoethyl)styrene	0.85	±0.01	0.79	±0.01	50	a, m	387
p-Bromophenyl vinyl sulphide	3.42		0.31		60	f, m	180
m-Bromostyrene	0.55	±0.03	1.05	±0.21	60	k, r	184
p-Bromostyrene	0.60		1.1		30	k, r	182
	0.695	±0.02	0.99	±0.07	60	f, o	184
	0.71	±0.02	1.05	±0.05	60	e, m	388
2-Bromovinyl ethyl ether	37.5	±2.0	0.02	±0.02	80	d, n	72
Butadiene	0.68		0.50		25		389
	0.23	±0.07	1.48	±0.08	50		46
	0.58	±0.15	1.35	±0.12	50		47
	0.78	±0.01	1.39	±0.03	60	f, p	108

Butyl acrylate	0.48	±0.04	0.15	±0.04	25			390
	0.76		0.15		60			30
	0.68		0.19		60		benzene (0.60 mole fraction)	391
	0.64		0.19		60		benzene (0.90 mole fraction)	391
	1.03		0.34		60		benzene (0.975 mole fraction) (benzene said to enter into polymer)	391
n-Butyl cinnamate	0.87		0.10		80			1
n-Butylmaleimide	0.025	±0.025	0.06	±0.02	50	k, r	benzene	284
n-Butyl methacrylate	0.97		0.67		30	e, r		392
	0.63		0.64		50			393
	0.56	±0.03	0.40	±0.03	60	a, l		158, 378
	0.54		0.64		70			393
t-Butyl methacrylate	0.56	±0.03	0.60	±0.04	60	i, l		158, 378
α-*t*-Butyl styrene	>30		—		110			394
n-Butyl vinyl sulphide	4.0		0.20		60	a, m		58
t-Butyl vinyl sulphide	4.7		0.20		60	g, n		58
n-Butyl vinyl sulphonate	2.5	±1.0	0.13	±0.03	90	a, n		237
α-Chloroacrolein	0.78	±0.022	0.18	±0.026	60	a, m		395
α-Chloroacrolein diacetate	4.6	±0.21	0.14	±0.018	60	a, l		395
α-Chloroacrolein dimethyl acetal	7.75	±0.076	0.05	±0.036	65	a, m		395
α-Chloroacrylonitrile	0.06		0.13		60	a, l		396
2-Chloroallyl acetate	4.10		0		50			20
2-Chloroallyl alcohol	12.5		0		40			20
2-Chloroallyl chloride	5.0	±15 %	0.06	±15 %	70			354
	2.5		0		40			20
3-Chloro-1-bicyclobutane carbonitrile	1.77		0.05		50	k, r	γ-butyrolactone	41
2-Chloro-1,3-butadiene	0.00		6.3	±0.1	50			47
	0.05	±0.02	7	±2	70	a, m		161
β-Chloro-α,β-difluoro-4-methyl-styrene	2.157		0.294		60			191
β-Chloro-α,β-difluorostyrene	2.099		0.402		60			191
2-Chloroethyl acrylate	0.59	±0.03	0.08	±0.01	60			60
	0.54	±0.01	0.10	±0.01	60	f, n		108

Copolymerization reactivity ratios—*continued*

M_1/M_2	r_1		r_2		T(°C)	Remarks	Ref.
N-*p*-β-Chloroethylamino sulphonyl benzene acrylamide	0.69	±0.06	0.38	±0.06	70	f, r tetrahydrofuran	59
N-*p*-β-Chloroethylamino sulphonyl benzene methacrylamide	1.02	±0.08	0.33	±0.07	70	f, r DMF	59
2-Chloroethyl itaconate	0		0.5	±0.05	50		286
2-Chloroethyl methacrylate	0.33	±0.02	0.46	±0.04	60	g, l	158
α-Chloro-β-fluorostyrene	2.1		0.55		60		397
4-(Chloromethyl dimethylsilyl) styrene	0.69	±0.03	0.86	±0.03	80	a, m toluene	398
ω-Chloromethyl-ω′-(4-vinyl-phenyl)tetramethyldisiloxane	1.01	±0.08	1.08	±0.16	80	j, m toluene	398
ω-Chloromethyl-ω′-(4-vinyl-phenyl)hexamethyltrisiloxane	1.02	±0.04	0.97	±0.10	80	j, m toluene	398
ω-Chloromethyl-ω′-(4-vinyl-phenyl)octamethyltetrasiloxane	0.98	±0.04	1.08	±0.08	80	j, m toluene	398
p-Chlorophenylvinyl sulphide	3.57		0.32		60	f, m	180
m-Chlorostyrene	0.64	±0.05	1.09	±0.23	60	k, r	184
o-Chlorostyrene	0.56	±0.03	1.6	±0.07	60	k, r	287
p-Chlorostyrene	0.66		1.1		30	k, r	182
	0.74	±0.025	1.025	±0.05	60	f, o	399
	0.816	±0.03	1.042	±0.05	131	f, o	399
Chlorotrifluoroethylene	7.0		0.001		60	f, n	288
4-Chloro-1-vinyl-naphthalene	0.85	±0.1	0.8	±0.1	60	f, n	260
6-Chloro-2-vinyl-naphthalene	0.4	±0.1	1.5	±0.2	60	a, m benzene	260
Cinnamic acid	1.85	±0.03	0		60	l	400
Cinnamonitrile	2.3		0		70	g, m	380
	2.2		0		80	a, m, s	379
	0.9	±0.2	0.1	±0.1	90	j, n	401
3-Cinnamoyl pyridine	0.85	±0.05	0.09	±0.10	50	h, m	402
Citraconic anhydride	0.15	±0.02	0.01	±0.01	60	a, l	30
Crotonic acid	20		0		60		60

γ-Crotonolactone	8.5	±5.6	0		60	f, m	99
Crotonitrile	20.0	±0.5	0		60	e, l, s	403
	24.0		0		60	d, l, s benzene	404
β-Cyanoacrolein	0.09(>60 % M_1)		0		60	a, m, s tetrahydrofuran	117
α-(2-Cyanoethyl)acrylonitrile	0.85	±0.06	0.25	±0.15	—		366
β-Cyanoethyl methacrylate	0.45	±0.01	0.32	±0.01	70	o	405
Cyanomethyl methacrylate	0.45	±0.01	0.36	±0.02	70	o	405
p-Cyanostyrene	0.19		1.2		30	k, r dimethylacetamide	182
	0.28	±0.025	1.16	±0.13	60	k, r	184
Cyclohexyl methacrylate	0.52	±0.07	0.45	±0.09	60	a, l	158, 378
4-Cyclopentene-1,3-dione	0.02		0.4		50	a, l benzene	406
1-Deoxy-1-methacrylamido-*D*-glucitol	2.09		0.005		50	e, m aq. methanol	264
Diallyl melamine	103.0	±11.8	0		60	i, n DMF, w	290
	102.3	±15.1	0		60	i, n DMF, w	290
Diallyl phenyl phosphate	16.84		0.145		70		291
Diallyl phenyl phosphonate	28.97		0.027		70		292
Di-*n*-amyl itaconate	0.34	±0.02	0.52	±0.02	60	a, m	407
Dibenzyl maleate	0.55	±0.2	0.02	±0.01	70		408
O,O'-Dibenzoylvinyl hydroquinone	0.22		0.43		60	k, r benzene	409
Dibromostyrene (mixed isomers)	0.22	±0.05	1.40	±0.15	30	g, n	410
N,N-Di-*n*-butylacrylamide	1.65	±0.03	0.32	±0.02	50		293
Di-*n*-butyl itaconate	0.40	±0.05	0.38	±0.02	60	a, m	407
Di-*n*-butyl vinyl phosphonate	5.4		0.3		60		411
cis-Dichloroethylene	56.1		0		50		412
	75	±25	0		68	a, n	413
	47.5		0		90		412
	72.5		0		90		412
trans-Dichloroethylene	44.4		0		50		412
	44.9		0		50		412
	~10		0		68	a, n	413
	29.8		0		90		413
	25.9		0		90		413
Di-β-chloroethyl itaconate	0.50	±0.05	0		50		286

Copolymerization reactivity ratios—*continued*

M_1/M_2	r_1		r_2		T(°C)	Remarks	Ref.
2,2-Dichloro-4-isobutyl pseudoxazolone-5	0.10		0.0		60	a, m	414
2,2-Dichloro-4-isopropyl pseudoxazolone-5	0.10		0.0		60	a, m	414
2,2-Dichloro-4-methyl pseudoxazolone-5	0.10		0.0		60	a, m	414
2,5-Dichlorostyrene	0.18	±0.07	0.25	±0.09	41.5	f, m	63
	0.32	±0.06	0.08	±0.05	65	f, m	63
	0.2		0.8		70	d, n	415
	0.29		2.2		70		416
	0.30		1.8		70		416
	0.23		2.2		70		416
	0.31		1.9		70		416
	0.38		1.77		70	k, r	417
	0.40	±0.03	0.05	±0.03	86.5	f, m	63
1,1-Dichloro-2-vinylcyclopropane	17.4		0.05		60	f, m	241
Di(2-cyanoethyl)fumarate	2.65	±0.35	0.03		90	k, r	168
Diethyl chloromaleate	2.5		0		70	g, l	415
Diethyl fumarate	0.31	±0.02	0.05	±0.02	—	a, m	384
	0.35	±0.02	0.005	±0.005	60	g, m benzene	418
	0.30	±0.02	0.070	±0.007	60	f, q	399
	0.400	±0.02	0.0905	±0.007	131	f, n	399
Diethyl itaconate	0.30	±0.02	0.33	±0.02	60	d, m	407
Diethyl maleate	5	±1.5	0.0	±0.1	70	g, l	415
	6.52	±0.50	0.005	±0.01	60	f, o	399
	5.48	±0.56	0		131	j, q, v	399
Diethyl methylene malonate	0.03	±0.08	0.08		60	a, m	396
N-Diethyl-*O*-vinylcarbamate	32	±5	0.03	±0.01	66		419
N-Diethyl-*S*-vinylmonothiol-carbamate	4.4	±0.6	0.14	±0.03	66		419
Diethyl vinyl phosphonate	8.87	±0.14	0.06	±0.02	—		420

	4.1		0		60		411
	3.25		0		116	i, p	421
Diethynyl benzene	1.15	±0.1	0.87	±0.1	—		249
	1.18	±0.1	0.74	±0.1	—		297
α-Difluoromethylstyrene	0.54		1.313		60		191
2,5-Difluorostyrene	0.044		2.195		60		191
2,4-Difluorostyrene	1.05		0.75		60		397
α,β-Difluorostyrene	2.42		0.04		60		397
β,β-Difluorostyrene	10.4		0		60		397
Diglycidyl 1,2-dimethyl-3-cyclohexen-1,5-dicarboxylate	17.2		0		100	f, r	80
·1,1-Dihydroperfluorobutylacrylate	0.33		0.07		50	a, l	298
Diisobutyl vinyl phosphine oxide	17	±5	0		60	j, n	300
Diisobutyl vinyl phosphonate	4.4		0.5		60		411
Diisopropyl vinyl phosphonate	2.39		0		60		411
2,5-Dimethoxystyrene	0.77	±0.07	1.13		70	a, r	302
2,6-Dimethoxystyrene	0.98		0.55		60		303
N,N-Dimethylacrylamide	1.23	±0.43	0.23	±0.13	50	f, l	304
	1.28	±0.15	0.44	±0.08	70	f, m dioxan	305
	1.33	±0.08	0.42	±0.05	70	f, m ethanol	305
p-Dimethylaminostyrene	0.89		0.62		—	k, r	422
	1.015	±0.06	0.84	±0.05	60	k, r	184
2-*N,N*-Dimethylamino-4-vinyl pyridine	0.35	±0.02	1.4	±0.1	60	g, m	367
Dimethyl dithiolfumarate	0.098	±0.013	0.0163	±0.013	50	g, l benzene	163
Dimethyl fumarate	0.21	±0.02	0.025	±0.015	60	f, p	218
Dimethyl itaconate	0.48		0.14		—		183
	0.32	±0.02	0.25	±0.02	60	f, m	407
Dimethyl maleate	8.5	±0.2	0.03	±0.01	60	f, r	218
N-(1,1-Dimethyl-3-oxobutyl)-acrylamide	1.77	±0.08	0.49	±0.06	60		307
Dimethyl stannyl methacrylate	0.28		0.83		60	a, r DMF	219
Dimethyl vinyl phosphonate	4.61		0.4		60		411
2,4-Dimethyl-6-vinyl-*s*-triazine	0.12		0.92		60	f, n benzene	308

Copolymerization reactivity ratios—*continued*

M_1/M_2	r_1		r_2		T(°C)	Remarks	Ref.
Di-*n*-octyl itaconate	0.35	±0.03	0.60	±0.02	60	c, m	407
2,4-Diphenyl butene-1	1.6		—		110		394
1,1-Diphenyl ethylene	0.34	±0.04	0		110	a, n	423
Diphenyl 4-vinylphenylphosphine	0.52	±0.05	1.43	±0.25	60	k, r benzene	309
Diphenyl 4-vinylphenylphosphine oxide	0.42	±0.02	1.40	±0.15	60	k, r benzene	309
Diphenyl 4-vinylphenylphosphine sulphide	0.43	±0.05	1.49	±0.33	60	k, r benzene	309
Diphenyl vinylphosphine	7	±1	0		60	j, n	300
Diphenyl vinyl phosphine oxide	5	±1	0		60	j, n benzene	300
Diphenyl vinyl phosphine sulphide	2.1	±0.3	0		60	j, n benzene	300
Diphenyl vinyl phosphonate	2.03		0		60		411
Di-*n*-propyl itaconate	0.25	±0.03	0.41	±0.04	60	a, m	407
Di-*n*-propyl vinyl phosphonate	4.24		0.9		60		411
N,N-Divinylaniline	13.0		0.45		60	a, n	66
m-Divinylbenzene	0.65		0.60		60	f, l	424
	0.65		0.60		61	f, m benzene	385
	1.11		1.00		75	o	425
	0.605		0.88		80	f <3.7 % conv.	426
	1.27		1.08		80	f 2.7—7.5 % conv.	426
o-Divinylbenzene	0.92		1.0		80	h, r	427
p-Divinylbenzene	0.77		1.46		70	g, l	428
	0.20		1.00		75	o	425
	0.77		2.08		80	f <3.7 % conv.	426
	0.14		0.5		80	h, r	427
Divinyl ether	40		0.02		60	f, m	66
Divinyl sulphide	1.90	±0.1	0.41	±0.05	60	f, l	310
Divinyl sulphone	1.3		0.01		60	f, n benzene	66
n-Dodecyl methacrylate	0.57	±0.02	0.45	±0.04	60	g, l	158, 378
n-Dodecyl vinyl ether	56	±5	0.0	±0.05	70		429
	27	±2	0.0	±0.05	100		429

α-Eleostearic acid	0.46		4		—		430
Ethyl 1-acetoxyacrylate	0.57	±0.05	0.20	±0.05	60	a, m	196
Ethyl acrylate	0.79		0.19		50		393
	0.80		0.20		70		393
	0.80		0.48		80		1
Ethyl benzylidene cyanoacetate	0.51		0		70	a, m	380
	0.3		0		80	a, l, s	379
Ethyl α-bromoacrylate	0.06		0.50		60	a, m	396
Ethyl α-chloroacrylate	0.08		0.30		60	a, l	396
Ethyl cinnamate	2.7	±0.3	0.05		75		233
Ethyleneglycol dimethacrylate	0.35		0.65		60	a, l	424
	0.40		0.65		61		385
N,*N*-Ethylene-*N*′-isopropenylurea	15.7	±0.8	0.01	±0.03	60		143
N,*N*-Ethylene-*N*′ vinylurea	14.5	±0.9	0.05	±0.04	60		143
Ethyl 2-ethoxyacrylate	23.5	±1	~0		80	f, m	72
2-Ethylhexyl acrylate	0.91		0.29		138	f, m	212
Ethyl methacrylate	0.67		0.26		50		393
	0.53	±0.03	0.41	±0.03	60	f, l	158, 378
	0.65		0.29		70		393
S-Ethyl-*N*-methyl-*N*-vinyl monothiolcarbamate	13	±3	0.025	±0.01	66		419
α-Ethyl styrene	1.6		—		110		394
Ethyl vinyl ether	80	±40	0		80	f, n	72
Ethyl vinyl oxalate	8.0		0.1		60	c, m	75
5-Ethyl-2-vinyl pyridine	0.79	±0.03	1.2	±0.2	60	a, l	243
5-Ethyl-2-vinyl pyridine-*N*-oxide	0.10	±0.01	2.6	±0.3	60	g, m ethanol	312
Ethyl vinyl sulphide	6.0	±1.5	0.25	±0.1	—		313
	4.4		0.12		60	a, n	58
Ethyl vinyl sulphoxide	7.0		0.02		60	a, l	77
	6.0		0.1		60	f, m ethanol	77
	2.8		0.2		60	a, m phenol	77
	3.2		0.2		60	f, l *p*-chlorophenol	77
	3.5		0.1		60	d, l *p*-cresol	77
N-Ethyl-*N*′-vinylurea	20.0		0.020		75		356
Ferrocenyl methyl acrylate	2.3	±0.3	0.02	±0.01	70	m benzene	244

Copolymerization reactivity ratios—*continued*

M_1/M_2	r_1		r_2		T(°C)	Remarks	Ref.
Ferrocenyl methyl methacrylate	3.7	±0.2	0.03	±0.02	70	m benzene	244
2-Fluorobutadiene	0.16	±0.08	1.61	±0.24	5	f, p	78
	0.50	±0.10	1.55	±0.10	50	f, l	214
β-Fluorostyrene	5.95		0.01		60		397
p-Fluoro-α-trifluoromethylstyrene	0.56		0		60		397
4-Fluoro-4′-vinylbiphenyl	0.50	±0.15	0.88	±0.07	70		432
Fumaronitrile	0.3	±0.3	0		50—70		433
	0.09	±0.005	0.00		79	i, m, s (see refs. 590, 591)	364
	0.23	±0.01	0.01	±0.01	60	i, l chloroform, benzene	315
	0.19	±0.03	0		60	f, p	218
	0.30	±0.02	0.00	±0.02	49.65	d, l chloroform	434
Fumaryl chloride	0.04		0.00		27	a, l	29
Glycidyl acrylate	0.60		0.17		60	a, r	79
	0.73		0.25		138	f, m	212
Glycidyl methacrylate	0.45	±0.001	0.55	±0.002	60		435
	0.53	±0.07	0.44	±0.001	60		81
	0.34	±0.05	0.63	±0.1	65		437
	0.450	±0.001	0.552	±0.002	60	m	436
	0.450	±0.002	0.551	±0.005	60	m dioxan	436, 438
	0.446	±0.005	0.543	±0.003	60	m toluene	436, 438
	0.450	±0.004	0.538	±0.002	60	m cyclohexanone	436, 438
	0.441	±0.007	0.544	±0.004	60	m cyclohexanone	436, 438
	0.451	±0.007	0.550	±0.006	60	m cyclohexanone	436, 438
	0.502	±0.005	0.601	±0.008	120	g, m cyclohexanone	436, 438
	0.50		0.60		120		436
Glycidyl-1-methyl-3-cyclohexen-1-carboxylate	21.0		0		100		80
Glycidyl vinylsulphonate	1.19	±0.11	0.03	±0.08	70		319
Hexachlorobutadiene	Very large		0		70	a, l	102
n-Hexyl methacrylate	0.60	±0.02	0.45	±0.06	60	g, l	158, 378

Hydronopyl acrylate	0.66	±0.1	0.29	±0.1	50	d, n benzene	82
6-Hydroxy-2-(*p*-chlorophenyl)-3-pyridazinone	10		0		60	DMSO	439
6-Hydroxy-2-phenyl-3-pyridazinone	13		0		60	DMSO	439
o-Hydroxystyrene	0.72	±0.1	1.32	±0.25	60	a, m tetrahydrofuran	320
6-Hydroxy-2-(*p*-tolyl)-3-pyridazinone	16		0		60	DMSO	439
p-Iodostyrene	0.45	±0.02	1.03	±0.05	50	a, l	440
	0.62	±0.05	1.09	±0.30	60	k, r	184
Isobutyl methacrylate	0.55	±0.02	0.40	±0.05	60	g, l	158, 378
Isobutyl styrene	1.9		—		110		394
Isobutyl vinyl sulphide	4.0		0.10		60	g, m	58
Isobornyl methacrylate	0.70		0.32		60		386
Isoprene	0.4		2.0		—		441
	0.44		1.98		—		441
	0.48		1.30		−18	f, m	442, 214
	1.38	±0.54	2.05	±0.45	50		47
	0.54		1.92		80	a, m	443
	0.50		1.92		100	a, m	443
2-Isopropenyl-4,5-dimethoxyazole	0.15	±0.05	2.20	±0.1	70	g, m	444
2-Isopropenyl-4,4-dimethyl-2-oxazolin-5-one	0.31	±0.06	1.00	±0.09	60		445
Isopropenylisocyanate	7.0		0.14		60		446
	8.12		0.07	±0.00	—		322
2-Isopropenyl-4-methyl-4-ethyl-2-oxazolin-5-one	0.27		0.70		60		445
2-Isopropenyl-4-methyl-4-(7-methyl octyl)-2-oxazolin-5-one	0.24		0.69		60		445
2-Isopropenyl-2-oxazolin-5-one	0.12		0.82		60		445
2-Isopropenyl-2-oxazoline	0.49		0.55		60	a benzene	323
	0.10		1.50		60	a benzene/Zn Cl_2	323
	0.36		2.76		60	a benzene/Sn Cl_4	323
	0.62		2.60		60	a benzene/Hg Cl_2	323
	0.32		0.60		60	a benzene/BF_3	323

Copolymerization reactivity ratios—*continued*

M_1/M_2	r_1		r_2		T(°C)	Remarks	Ref.
p-Isopropylstyrene	11	±0.01	0.54	±0.01	50	g, m	387
	1.22		0.89		70	f, l, w	324
	1.37		0.99		70	f, l, w	324
N-[1-(1-Isopropyl-2-oxopropyl)] acrylamide	1.87	±0.1	0.25	±0.1	60	a, m	383
N-[1-(1-Isopropyl-2-oxopropyl)] methacrylamide	1.07	±0.05	0.25	±0.05	60	a, m	383
Isopropyl vinyl sulphide	4.3		0.20		60	g, n	58
Itaconic acid	0.34		0		—		326
Itaconic anhydride	0.10		0.60		65	i, m benzene; i, n tetrahydrofuran	447
Linoleic acid	16		0.29		150		448
Maleic anhydride	0.023		0		40	s 1 kg cm^{-2} acetone	449
	0.022		0		40	s 2000 kg cm^{-2} acetone	449
	0.023		0		40	s 4000 kg cm^{-2} acetone	449
	19		0		50		450
	0		0.015		50		451
	0.97	±0.002	0	±0.002	50	c, n	452
	0.02		0		60	g, l, s (see ref. 348)	453
	0.04		0		80		453
	0.042	±0.008	0		80	s (see ref. 348)	454
Maleimide	0.1		0.1		75	a, m cyclohexanone	329
Maleonitrile	0.19	±0.01	0		60	f, n	218
Methacrylamide	1.46	±0.09	1.29	±0.08	70	f, m dioxan	330
	1.44	±0.15	0.54	±0.08	70	f, m ethanol	330
Methacrylic acid	0.2	±0.02	0.66	±0.08	45		332
	0.53	±0.02	0.45	±0.06	45	DMF	332
	0.15	±0.01	0.7	±0.05	60	a, l	30
	0.15		0.37		65	a, l benzene	266
Methacrylic anhydride	0.12		0.26		36.6	a, m cyclohexanone	229
	0.10	±0.01	0.33	±0.03	60	k, r benzene	246
Methacrylonitrile	0.30	±0.10	0.16	±0.06	60	f, p	108

	0.32		0.39		60	d, o, w toluene	455
	0.43		0.44		60	d, o, w toluene	455
	0.38	±0.05	0.26	±0.05	80	d, l	224
	0.43		0.28		80		1
	0.25	±0.02	0.25	±0.02	80	a, l	107
	0.39		0.32		90	d, o, w toluene	455
	0.43		0.39		90	d, o, w toluene	455
	0.42		0.15		120	d, o, w toluene	455
	0.33		0.28		120	d, o, w toluene	455
α-Methacrylophenone	0.12—0.35		0.18—0.5		75	a, l	456
4-Methacryloxybenzylidene aniline	0.25	±0.03	2.4	±0.5	60	j, n	381
4-Methacryloxybenzylidine-4′-chloroaniline	0.18	±0.03	5.4	±0.6	60	j, n	381
Methacryloxymethylpentamethyl-disiloxane	0.77	±0.02	0.58	±0.02	50	k, r	90
N-Methacryloyl-ϵ-caprolactam	1.0		0		70—80	f, q	457
	1.86	±0.10	0		90	g, n	458
Methacrylyl triethyl germanium	1.05	±0.02	0.93	±0.08	60		459
Methallyl acetate	71	±10	0		60		460
p-Methoxyphenyl vinyl sulphide	3.98		0.18		60	f, m	180
p-Methoxystyrene	1.05		0.79		—	k, r	422
	0.70	±0.10	1.5	±0.22	25	f, l	362
	1.0		0.85		30	k, r	182
	1.16	±0.09	0.82	±0.07	60	k, r	184
	1.13		0.93		60	a, m	462
α-Methoxystyrene	2.7	±0.6	0		60	a, m	93
Methyl acrylate	0.75	±0.11	0.18	±0.03	25	f, l	362
	1.0	±0.1	0.16	±0.02	60	a, m	463
	0.7	±0.1	0.15	±0.05	60	a, m benzene	82
	0.75	±0.03	0.18	±0.02	60	f, n	399
	0.75	±0.1	0.20	±0.05	70	f, l toluene	415
	0.825	±0.0	0.238	±0.02	131	f, n	399
Methyl α-benzyl acrylate	0.59	±0.05	0.17	±0.03	65	a, n	464
Methyl-1-bicyclobutane carboxylate	1.81		0.14		50	k, r methyl ethyl ketone	41

Copolymerization reactivity ratios—*continued*

M_1/M_2	r_1		r_2		T(°C)	Remarks	Ref.
Methyl α-*n*-butyl acrylate	0.80	±0.05	0.20	±0.05	65	a, n	464
Methyl α-*sec*-butyl acrylate	2.25	±0.05	0		65	a, n	464
Methyl cinnamate	1.9	±0.2	0		60	v	60
2-Methyl-5-cinnamoyl-pyridine	0.92	±0.08	−0.15	±0.2	50	h, m	402
N-Methylcitraconimide	0.153		0.24		70		103
Methyl crotononitrile	26.0		0.01		60	a, l, s benzene	404
Methyl α-cyclohexylacrylate	1.65	±0.05	0		65	a, n	464
4-Methyl-2,4-diphenylpentene-1	1.9		—		110		394
Methyl ethacrylate	0.82	±0.08	0.21	±0.02	65	a, n	464
	0.90	±0.07	0.20	±0.02	70	a, n	464
Methyl α-isobutylacrylate	0.96	±0.05	0.20	±0.03	65	a, n	464
Methyl isopropenyl ketone	0.32		0.66		80		1
	0.44	±0.10	0.29	±0.06	80		1
Methyl α-isopropylacrylate	1.85	±0.07	0		65	a, n	464
	1.94	±0.03	0		70	a, n	464
Methyl methacrylate	0.44	±0.02	0.29	±0.02	−60	a, l toluene	465
	0.44	±0.02	0.30	±0.01	−30	k, l benzene	465
	0.63	±0.09	0.50	±0.07	25	f, l	362
	0.485		0.422		30		466
	0.50	±0.02	0.44	±0.02	35	f, n benzene	467
	0.22	±0.05	0.64	±0.08	40	g, l	468
	0.3	±0.07	0.56	±0.05	40	g, l tetrahydrofuran, dioxan	468
	0.37	±0.06	0.51	±0.04	40	g, l quinoline	468
	0.45	±0.04	0.47	±0.03	40	g, l pyridine	468
	0.63	±0.07	0.47	±0.05	40	g, l DMSO	468
	0.52		0.45		50	k, r 1 atm/C_6H_6	469
	0.58		0.48		50	k, r 100 atm/C_6H_6	469
	0.72		0.58		50	k, r 1000 atm/C_6H_6	469
	0.49		0.46		50	k, r 1 atm/$ClCH_2CH_2Cl$	469
	0.53		0.52		50	k, r 100 atm/$ClCH_2CH_2Cl$	469
	0.63		0.66		50	k, r 1000 atm/$ClCH_2CH_2Cl$	469

	0.44		0.50		50	k, r 1 atm/pyridine	469
	0.49		0.57		50	k, r 100 atm/pyridine	469
	0.56		0.80		50	k, r 1000 atm/pyridine	469
	0.45		0.47		50	k, r, t 1 atm/EtOH	469
	0.49		0.52		50	k, r, t 100 atm/EtOH	469
	0.59		0.62		50	k, r, t 1000 atm/EtOH	469
	0.52	±0.026	0.46	±0.026	60	m	436, 438
	0.57	±0.032	0.46	±0.032	60	d, l benzene	470
	0.48	±0.045	0.49	±0.045	60	e, l benzonitrile	470
	0.44	±0.054	0.39	±0.054	60	a, l benzyl alcohol	470
	0.35	±0.024	0.35	±0.024	60	f, m phenol	470
	0.51	±0.02	0.46	±0.02	60	a, m benzene	465
	0.52	±0.02	0.46	±0.02	60	k, r	158
	0.54	±0.02	0.50	±0.06	60	bulk, toluene	471
	0.45	±0.10	0.50	±0.05	60	a, m	472
	0.48		0.46		60	f, m	473
	0.44	±0.08	0.5	±0.04	60	a, m acetonitrile	365
	0.54	±0.04	0.42	±0.1	60	a, m benzene	365
	0.53		0.49		60	k, r, w	474
	0.52		0.49		60	k, r, w	474
	0.52	±0.026	0.46	±0.026	60	f, m bulk, benzene acetonitrile, methanol	399
	0.50	±0.02	0.50	±0.02	60	a, o	475
	0.55		0.58		90	Dioxan	476
	0.54	±0.03	0.49	±0.03	99		86
	0.590	±0.026	0.536	±0.026	131	f, n	399
	0.60		0.55		132	f, m	473
N-[1-(1-Methyl-2-oxopropyl)] acrylamide	1.83	±0.02	0.05	±0.02	60	a, m	383
N-[1-(1-Methyl-2-oxopropyl)] methacrylamide	0.91	±0.02	0.01	±0.02	60	g, m	383
Methyl α-phenylacrylate	0.055		1.0		60	a, m	105
	0.03	±0.02	0.4	±0.2	65	a, m	464
p-Methyl phenyl vinyl sulphide	3.65		0.25		60	f, m	180
Methyl α-*n*-propylacrylate	0.86	±0.08	0.22	±0.04	65	a, n	464

Copolymerization reactivity ratios—*continued*

M_1/M_2	r_1		r_2		T(°C)	Remarks	Ref.
p-Methylstyrene	0.83		0.96		63	d, l	477
α-Methylstyrene	1.18	±0.04	0.36		—	k, r	478
	2.3		0.38		—	a, q	479
	1.0		0.15		60	a, r	480
	1.09		0.3		60	a, r	480
	1.13		0.4		60	a, r	480
	1.2		0.8		60	a, r	480
2-Methyl-α,β,β-trifluorostyrene	1.177		0.403		60		191
3-Methyl-α,β,β-trifluorostyrene	0.40		0.73		60		191
4-Methyl-α,β,β-trifluorostyrene	0.40		0.63		60		191
2-Methyl-*N*-vinylimidazole	8.97		0.069		—		338
Methyl vinyl ketone	0.29	±0.04	0.35	±0.02	60	f, o	108
2-Methyl-5-vinylpyridine	0.6	±0.1	0.68	±0.1	70		481
	0.812	±0.005	0.91	±0.02	60	a, l	109
	0.88	±0.2	1.19	±0.12	60		482
Methyl vinyl sulphide	5.1	±1.0	0.12	±0.05	60	a, m	247
	5.0		0.15		60	g, n	58
Methyl vinyl sulphone	2.0	±0.5	0.01	±0.01	60	a, l	247
	2.4	±0.10	0.0	±0.2	60		60
	1.40		0.01		60	a, l	247
Methyl vinyl sulphoxide	4.2	±0.2	0.01	±0.01	60	i, l	483
N-Methyl-*N*-vinyl-*p*-toluene sulphonamide	12.3		0		60	h, m	111, 112
Monoethyl fumarate	0.18	±0.1	0.25	±0.10	60	f, o	431
Monoethyl maleate	0.13	±0.01	0.035	±0.01	60	f, p	218
Mono(1-methylpropyl)fumarate	0.34	±0.03	0	±0.01	—	a, m	384
Neopentyl styrene	~2		—		110		394
2-Nitrobutyl acrylate	0.58		0.115		70		113
m-Nitrostyrene	0.45	±0.05	0.85	±0.1	75	a, m benzene	339
p-Nitrostyrene	0.19	±0.02	1.15	±0.20	60	f, n	184
β-Nitrostyrene	0.4		0		80	<0.036 (mole fraction M_2), n	484

N-*n*-Octadecyl acrylamide	1.41	±0.10	0.20	±0.05	80	m benzene	114
n-Octadecyl acrylate	0.75	±0.12	0.34	±0.06	60	n	114
	0.44	±0.07	0.18	±0.45	60	m *t*-butanol	114
	0.79	±0.09	0.31	±0.31	60	m benzene	114
N-*n*-Octylacrylamide	2.70	±0.10	0.20	±0.05	50		293
N-*tert*-Octylacrylamide	2.8	±0.10	0.25	±0.06	50		293
n-Octyl methacrylate	0.67	±0.03	0.55	±0.07	60	a, l	158, 378
n-Octyl vinyl ether	65	±5	0	±0.04	70		429
α-Oleostearic acid	0.46		4		—		485
N-[1-(2-Oxopropyl)] methacryl- · amide	1.15	±0.1	0.13	±0.1	60	a, m	383
Pentachlorophenyl vinyl sulphide	3.9		0.24		80	a, n	248
Pentachlorostyrene	1.31	±0.2	0.10	±0.02	70	a, m	340
Phenyl acetylene	0.36	±0.04	0.84	±0.1	—		249
	0.34	±0.03	0.36	±0.02	—		341
Phenyl methacrylate	0.30	±0.03	0.60	±0.05	60	g, l	158, 378
Phenyl vinyl ether	1.7		0.01		—		342
2-Phenyl-5-(4′-vinyl) phenyl- tetrazole	0.89	±0.06	1.26	±0.01	60	g, m, w benzene (C analysis)	343
	1.00	±0.06	1.40	±0.01	60	g, m, w benzene (C analysis)	343
	0.90	±0.06	1.30	±0.01	60	g, m, w benzene (C analysis)	343
	0.98	±0.02	0.67	±0.01	60	g, m, w benzene (N analysis)	343
	0.92	±0.02	0.59	±0.01	60	g, m, w benzene (N analysis)	343
	0.95	±0.02	0.66	±0.01	60	g, m, w benzene (N analysis)	343
Phenyl vinyl sulphide	3.88		0.36		60	f, m	180
	4.5		0.15		60	f, n	250
Phenyl vinyl sulphone	3.3		0.01	±0.01	60	f, m	250
2-Phthalimidomethyl-1,3-butadiene	0.15	±0.02	2.0	±0.13	60	j, m DMF	486
Poly(1,3-butylene glycol)-fumarate	3.0	±0.4	0.03	±0.03	60		487
Poly(dimethylsiloxyethylene fumarate)	0.03	±0.03	0.6	±0.3	—		488
p-Potassium styrenesulphonate	—0.06		0.54		90	e, n	287
	0.02		0.93		110	e, n	287
Potassium *p*-vinylbenzene- sulphonate	0		0.56		90	f, n	287
	0.02		0.93		110	f, n	287

Copolymerization reactivity ratios—*continued*

M_1/M_2	r_1		r_2		T(°C)	Remarks	Ref.
n-Propenyl methyl ketone	13.7		0.01		60	a, l, s benzene	404
n-Propyl methacrylate	0.57	±0.01	0.38	±0.04	60	a, l	158, 378
3-Pyridalacetophenone	0.50	±0.10	0.00	±0.25	50	h, m	402
trans-1,2-bis-(2-Pyridyl)-ethylene	1.85	±0.1	0.17	±0.1	50	h, m benzene	118
Stilbene	11.2	±1.2	0		60	j, n, v	489
m-Styrenesulphonylfluoride	0.8		1.25		75	f, n benzene	251
p-Styrenesulphonylfluoride	0.25		1.30		75	g, n butanone	251
Styrene tricarbonyl chromium	1.39		0		70	e, p benzene	345
2-Sulphoethyl methacrylate	0.37	±0.03	0.6	±0.2	50—60	d, m 1,2-dimethoxyethane	198
p-Sulphonamido styrene	0.24		1.07		90	f, n DMF	490
Tetrachloroethylene	165		0		50		412
	208		0		50		412
	185	±20	0		60	e, p, v	491
	200	±20	—		60	j, m	235
	66.4		0		90		412
	129		0		90		412
	187		0		90		412
trans-2,3,4,5-Tetrachloro-1,3,5-hexatriene	0.21		0.84		70		124
Tetrahydrofurfuryl acrylate	0.501		0.485		70		125
Triallyl cyanurate	80.7	±9.3	0		60	i, o, w	290
	90.6	±13.2	0		60	i, o, w	290
Triallyl isocyanurate	87.6	±10.0	0		60	i, o, w	290
	90.1	±13.1	0		60	i, o, w	290
2,4,5-Tribromostyrene	0.05	±0.15	0.60	±0.35	30	g, n	410
Tri-*n*-butylvinyltin	16.0		0.005		50	a, m benzene	410
1,1,2-Trichloro-1,3-butadiene	0.07	±0.03	1.18	±0.08	60		492
Trichloroethylene	16	±2	0.0		60	f, n	491
	16.5		0		50		412
	17.1		0		50		412
	12.1		0		90		412

	12.7		0		90		412
3,3,3-Trichloropropene	6.9	±0.2	0.0	±0.02	60		60
Triethyl aconitate	1.10	±0.10	—0.10	±0.10	60	f, l	126
m-Trifluoromethylstyrene	0.62		0.75		60		397
3-Trifluoromethylstyrene	0.70	±0.05	1.05	±0.05	60	a, m	279
	0.603		0.806		60		191
α,β,β-Trifluorostyrene	0.81		0.50		60		191
	3.5		0.15		60		397
p-Trimethoxysilylstyrene	0.71	±0.02	1.4	±0.1	70.3	d, m	493
Trimethyl aconitate	1.10	±0.01	0.00	±0.01	60	d, l	126
(Trimethylsiloxy)vinyl dimethyl silane	60	±15 %	0.1	±15 %	50	f, m benzene	128
bis-(Trimethylsiloxy)vinyl methyl silane	60	±15 %	0.1	±15 %	50	f, m benzene	128
p-Trimethylsilylstyrene	1.0	±0.2	1.0	±0.2	70.3	d, n	493
tris-(Trimethylsiloxy)vinylsilane	60	±15 %	0.1	±15 %	50	i, m benzene	128
Trimethyl stannyl methacrylate	1.57		0.53		60	a, r DMF	219
Trimethyl vinyl germane	24.4		0.009		50	f, n benzene	347
Trimethyl vinyl tin	44.8		0.001		50	a, m benzene	346
2,4,4-Triphenyl pentene-1	1.6		—		110		394
p-Triphenylstyryltin	0.93		2.55		108	h, n	494
1,2,4-Trivinylbenzene	1.80		1.12		70	g, l	495
N-Vinyl acetanilide	13.0	±0.66	0.054	±0.017	70	a, r	496
	22.3		0.009		—		497
Vinyl acetate	70	±20	0.025	±0.02	60		351
	55	±10	0.01	±0.01	60	e, o	132
1-Vinylanthracene	0.81		0.57		65	g, m	498
9-Vinylanthracene	2.12		0.25		100	g, m	498
Vinyl benzoate	40	±20	0.02	±0.02	60		351
	38		0.05		80	h, n	499
p-Vinyl benzophenone	0.18	±0.08	3.0	±0.3	60	i, n	500
Vinyl benzyl chloride (60 % *meta*, 40 % *para*)	0.72		1.08		60		12
endo-Vinyl bicyclo(2.2.1)-heptane-2-carboxylate	30		0.01		60	k, r	501

References pp. 404—418

Copolymerization reactivity ratios—*continued*

M_1/M_2	r_1	r_2	$T(°C)$	Remarks	Ref.
exo-Vinyl bicyclo(2.2.1)-heptane-2-carboxylate	30	0.01	60	k, r	501
4-Vinylbiphenyl	0.98 ±0.04	0.92 ±0.08	70		432
Vinyl-*O*,*O*′-bis(1-ethoxyethyl) hydroquinone	0.90	0.50	60	k, r benzene	409
Vinyl bromide	20—25	<0.05	0	c, l	205
	18 ±2	0.06 ±0.015	280	c, l	205
2-Vinyl-5-*t*-butylhydroquinone dimethyl ether	0.60	1.20	60	k, r benzene	409
S-Vinyl-*O*-*t*-butylthiocarbonate	3.0	0.2	70	a, m	215
Vinyl butyrate	40 ±20	0.02 ±0.02	60		351
N-Vinyl carbazole	5.5 ±15 %	0.012 ±15 %	70	a, l	354
	5.7	0.035	75	a, l benzene	255
Vinyl chloride	35	0.067	50	d, l acetone	137
	17 ±3	0.02	60	f, n	491
	12.4	0.045	—		502
	—	—	—	s	135
Vinyl chloroacetate	1.0	0.05	60		503
	30 ±20	0.025 ±0.02	60		351
	45	0.03	80		1
Vinyl cinnamate	1.25 ±0.1	0.25 ±0.1	70	a, r, x benzene	233
N-Vinyl-5-decyl-2-oxazolidone	1.43	0.059	60	g, m	504
Vinyl dichloroacetate	20	0.28	80		1
2-Vinyl-3,6-dimethylhydroquinone dimethyl ether	3.5	0	60	k, r benzene	409
2-Vinyl-4,5-dimethyloxazole	0.18 ±0.05	1.30 ±0.1	70	g, m	444
N-Vinyl-3,5-dimethyl pyrazole	8.2 ±0.1	0.08 ±0.02	60—70	a, m	505
4-Vinyl diphenyl	0.98 ±0.04	0.92 ±0.08	75—85	i, o toluene	506
p-Vinyl-1,1-diphenylethylene	0.4	4.4	80	i, m	507
Vinyl ethyl sulphide	6.0 ±1.5	0.25 ±0.1	60	k, r	355
1-Vinyl-3-ethyl urea	20	0.020	70		356

Vinyl ferrocene	2.5	±0.5	0.08	±0.04	70	a, m, w benzene	141
	2.91	±0.07	0.097	±0.043	70	a, m, w benzene	141
4-Vinyl-4′-fluorodiphenyl	0.50	±0.15	0.88	±0.07	75—85	i, p toluene	506
Vinyl formate	60	±20	0.02	±0.02	60		351
N-Vinylglycidylurethan	10.4	±0.02	0.02	±0.02	60	f, m tetrahydrofuran	143
Vinylhydroquinone dibenzoate	0.22		0.43		78	a, m toluene	258
Vinylhydroquinone dimethyl ether	0.77		1.13		60	k, r benzene	409
Vinylhydroquinone di(2-pentyl) ether	0.96		0.20		60	k, r benzene	409
Vinylidene chloride	1.80	±0.27	0.15	±0.02	25	f, l	362
	2.1	±0.2	0.145	±0.009	50	a, n	138
	1.85	±0.05	0.085	±0.010	60	f, m	491
	2.0	±0.1	0.14	±0.05	60	f, p	270
	0		0.046		—		508
Vinylidene cyanide	0.005		0.001		45	a, o benzene	27
	0		0.0459		—		508
Vinyl isobutyl ether	50		0.01		—		479
Vinyl isocyanate	8.13	±0.04	0.08	±0.04	60		144
	6.9	±0.5	0.1	±0.05	60		359
Vinyl isothiocyanate	0.8		0.5		—		1
	0.65	±0.02	0.37	±0.02	60		359
Vinyl methyl adipate	49.5	±0.7	0.02	±0.01	80	g, r	509
Vinyl methyl ether	100		0.01		—		479
2-Vinyl-5-methylhydroquinone dimethyl ether	0.88		1.30		60	k, r benzene	409
2-Vinyl-6-methylhydroquinone dimethyl ether	0.88		1.30		60	k, r benzene	409
1-Vinyl naphthalene	0.67	±0.03	1.35	±0.15	60	a, n	510
2-Vinyl naphthalene	0.5	±0.1	1.4	±0.1	60	a, n	260
Vinyl nonanoate	49.5	±15	0.01	±0.01	60	d, l	147
Vinyl octadecanoate	68	±30	0.01	±0.01	60	d, l	147
N-Vinyl oxazolidinone	30	±0.5	0.05	±0.05	50	e, m	186
β-Vinyloxyethyl(3,5-di-*t*-butyl-4-hydroxy)benzoate	30		0		70	g, m toluene	220

Copolymerization reactivity ratios—*continued*

M_1/M_2	r_1		r_2		T(°C)	Remarks	Ref.
Vinyl perfluorobutyrate	8.73		0.017		—		1
9-Vinylphenanthrene	0.58		2.36		60	g, n	498
m-Vinylphenol	0.9		1.21		60	a, m	333
4-Vinylphenyl diphenyl arsine	0.46	±0.03	1.18	±0.04	65	f, n toluene	511
4-Vinylphenyl diphenyl bismuthine	0.63	±0.05	1.35	±0.10	65	f, n toluene	511
4-Vinylphenyl diphenyl phosphine	0.46	±0.07	1.11	±0.12	65	f, n toluene	511
4-Vinylphenyl diphenyl stibine	0.6		1.3		65	e, n toluene	511
1-*p*-Vinylphenyl disiloxane	1.04	±0.06	1.2	±0.4	80	j, n	359
4-Vinylphenyl heptamethyl-trisiloxane	0.90	±0.03	1.2	±0.2	80	j, o toluene	512
p-Vinylphenylmethyl ethyl carbinol	0.79		1.25		60	k, m dioxan	149
4-Vinylphenyl nonamethyltetra-siloxane	1.15	±0.05	1.1	±0.3	80	j, o toluene	512
4-Vinylphenyl pentamethyl-disiloxane	1.04	±0.06	1.2	±0.4	80	j, n toluene	512
2-[(4-Vinyl)phenyl]-5-phenyl-1,3-oxazole	0.68	±0.04	4.5	±0.4	70	a, r toluene	363
4-Vinylphenyl undecamethyl-pentasiloxane	1.11	±0.01	1.2	±0.1	80	j, o toluene	512
N-Vinyl phthalimide	8.3	±0.3	0.075	±0.03	65—85	a, m	513
	8.3	±0.3	0.075	±0.03	85		514
Vinyl pinonate	65	±17	0.01	±0.01	60	d, l benzene	147
2-Vinylpyridine	0.50	±0.07	1.27	±0.19	25	f, l	362
	0.57	±0.03	1.33	±0.04	50		515
	0.56	±0.02	0.9	±0.2	60	a, l	243
	0.55	±0.025	1.135	±0.08	60	k, r	287
	0.55	±0.03	1.81	±0.05	60	k, r	151
4-Vinylpyridine	0.62	±0.02	0.52	±0.06	80		516
	0.54	±0.03	0.7	±0.1	60	g, l	243
2-Vinylpyridine-*N*-oxide	0.11	±0.01	2.1	±0.6	60	a, m ethanol	312
2-Vinylpyridinium acetate	0.16	±0.02	0.36	±0.04	22		515

References pp. 404—418

4-Vinylpyrimidine	0.17	±0.02	1.2	±0.1	60	g, m	367
N-Vinyl pyrrolidone	15.7	±0.5	0.045	±0.05	50	f, m benzene	186
	9.0		0.11		80		1
2-Vinylquinoline	0.49	±0.14	2.09	±0.55	60	k, r	151
N-Vinylsuccinimide	7.0		0.09		60	a, n	111
	10.5		0.07		60	f, m	594
2-Vinylthianthrene	0.477		0.133		60	j, r	199
2-Vinyl thiazole	0.14		3.32		80	a, n	517
4-Vinyl thiazole	0.66		0.82		80	a, n	517
2-Vinylthiophene	0.35	±0.025	3.10	±0.45	60	k, r	287
Vinyl trimethoxysilane	22	±5	0		60	f, m benzene	155
2-Vinyl-3,5,6-trimethylhydro- quinone dimethyl ether	3.0		0		60	k, r benzene	409
Vinyl trimethylsilane	26	±8	0		60	f, m benzene	155
p-Vinyl triphenyl carbinol	0.47	±0.10	1.60	±0.15	50	g, m benzene	500
Vinyl undecenoate	29	±9	0.02	±0.02	60	d, l benzene	147
Tetrafluoroethylene/							
3,3,3-Trifluoropropene	0.15		2.5		21	g, m	518
	0.12		5.0		21	g, n, >5000 atm	518
Vinyl acetate/							
N-Acetoxyethyl maleimide	0.25		0.85		70		519
1-Acrylamido-1-deoxy-*D*-glucitol	0.18		2.41		50	f, m water	264
Acrylic acid	0.1		2		70		188
	0.01	±0.003	10	±1	70	g, l benzene	13
Allyl acetate	1.0		0.7		60		520
Allyl allyl sulphonate	0.5	±0.15	1.54	±0.08	—	a	274
Allyl chloride	1.30		—		60		235
	0.34	±0.26	0.75	±0.46	60		156
	0.7		0.67		68	a, m	294
Allyl ethene sulphonate	0.38	±0.08	3.6	±0.2	60	a, m benzene	236
Allylidene diacetate	0.48	±0.3	1.34	±0.05	60	a, m	521
Allyl laurate	0.71		0.8		60		522
N-Allylstearamide	0.740	±0.087	0.532	±0.012	70		37
Bicyclo(2.2.1)hepta-2,5-diene	0.82		1.28		60		181

Copolymerization reactivity ratios—*continued*

M_1/M_2	r_1		r_2		$T(°C)$	Remarks	Ref.
p-Bromophenylvinylsulphide	0.85		0.20		60	f, m	180
p-Bromovinylbenzoate	0.60		1.70		30	benzene	523
n-Butyl methacrylate	0.023		28.8		60	h, m	56
	0.127	±0.015	62.1	±3.8	60		350
n-Butyl vinyl ether	0.127	±0.015	62.1	±3.8	60		350
	0.71	±0.1	0.2	±0.05	—		524
Butyl vinyl sulphonate	0.04	±0.01	0.20	±0.05	70	a, n	237
Carbon monoxide	0.24	±0.05	0.33	±0.05	60		525
2-Chloro-1,3-butadiene	0.01		50		65		177
p-Chlorophenylvinylsulphide	0.89		0.08		60	f, m	180
Chlorotrifluoroethylene	0.6		0.01		60	f, m methylethylketone	288
p-Chlorovinylbenzoate	0.62		1.85		30	Benzene	523
Crotonic acid	0.33	±0.05	0.01	±0.01	68	i, l	30
	0.33		0		—	a, r	526
	0.3		0.01		70		527
Crotonamide	0.01	±0.01	2.0	±0.05	110		528
p-Cyano vinyl benzoate	0.61		1.63		30	Benzene	523
1,Deoxy-1-methacrylamido-*D*-glucitol	0.16		0.56		50	h, m water	264
Diallyl melamine	1.41	±0.06	0.19	±0.004	60	f, n DMF, w	529
	1.44	±0.28	0.20	±0.03	60	f, n DMF, w	529
Diallyl phthalate	0.72		2.0		—		530
Di-*n*-butyl itaconate	0.02		6.3		—		183
1,3-Dichloro-2-butene	0.0		4.8	±0.9	50		531
1,1-Dichloro-2-2-difluoroethylene	0.6	±15 %	0	±15 %	70	a, n	102
cis-1,2-Dichloroethylene	5—7		~0		—50	m	532
	3—4		~0		30	p	532
	6.3	±0.2	0.018	±0.003	60	f, q	218
	2.8		0		68	g, n	413
trans-1,2-Dichloroethylene	0.95		0.06		—45	m	532
	0.95		0.05		30	p	532

	0.99	±0.02	0.086	±0.01	60	f, q	218
	0.85		0		68	j, n	413
2,5-Dichlorostyrene	<0.04		—		70		416
Di(2-cyanoethyl)fumarate	0.02	±0.02	0.02	±0.01	90	k, r	168
Diethylfumarate	0.031	±0.005	0.32	±0.02	60	k, m benzene	418
	0.011	±0.001	0.444	±0.003	60	f, n	218
Diethylmaleate	0.17	±0.01	0.043	±0.005	60	f, n	218
N-Diethyl-*O*-vinyl carbamate	1.8	±0.4	0.25	±0.08	66		419
N-Diethyl-*S*-vinyl monothiol-carbamate	0.16	±0.08	1.5	±0.3	66		419
Di-*n*-hexylfumarate	0.031	±0.005	0.27	±0.02	60	k, m benzene	418
Dimethylmaleate	0.12		0.028		60		533
Divinyl acetal	1.005	±0.08	0.99	±0.08	60		534
N,*N*-Divinylaniline	0.1		2.0		60	d, l	66
Divinyl *n*-butyral	1.005	±0.015	1.06	±0.01	60		534
Divinyl formal	1.012	±0.107	1.005	±0.105	60		534
Divinyl isoamylal	1.04	±0.08	0.987	±0.057	60		534
Divinyl isobutyral	1.002	±0.047	0.985	±0.05	60		534
n-Dodecyl vinyl ether	3.67	±0.45	0	±0.23	50		535
Ethyl 1-acetoxyacrylate	0.08	±0.03	5.4	±0.5	60	g, n	196
Ethylene	1.14		0.16		60	c, r 100 atm	536
	3.70		0.70		60	c, r 1200 atm	536
	1.08	±0.19	1.07	±0.06	90	a, l 1000 atm toluene	211, 272
Ethyl *n*-hexyl fumarate	0.034	±0.005	0.26	±0.02	60	k, m benzene	418
dl-Ethyl 2-methyl-2-ethyl-1-butenoate	0.3	±0.2	3.2	±0.50	60		537
d-Ethyl 2-methyl-2-ethyl-1-butenoate	0.1	±0.1	2.2	±0.4	60		537
S-Ethyl-*N*-methyl-*N*-vinyl monothiol carbamate	0.6	±0.1	1.3	±0.2	66		419
Ethyl vinyl oxalate	0.3		3.0		60	c, m	75
N-Ethyl-*N'*-vinylurea	0.45		0.63		75		356
Ferrocenyl methyl acrylate	0.46		1.44		70	p benzene	244
Ferrocenyl methyl methacrylate	0.02	±0.10	1.52	±0.30	70	m benzene	244
Fumaryl chloride	0.14		0.00		70	a, l r_1 appears to vary with comp. of monomer mixture	29

Copolymerization reactivity ratios—*continued*

M_1/M_2	r_1	r_2	T(°C)	Remarks	Ref.
Heptyl methacrylate	0.271 ±0.039	60.4 ±0.4	60		350
Hexadecyl methacrylate	0.135 ±0.055	68.3 ±3.2	60		350
Hydroxy methylcrotonamide	0.01 ±0.01	0.045 ±0.1	110		528
Isobutyl methacrylate	0.025	29.8	60	k, m	56
Isopropenyl acetate	1.0	1.0	75	a, m benzene	321
Maleic anhydride	0.072 ±0.04	0.01	—		538
	0.055 ±0.015	0.003	75	g, l	217
Methacrylic acid	0.01	20	70		188
Methacrylonitrile	0.01 ±0.01	12 ±2	70	a, l	107
Methacryloxymethyl-pentamethyldisiloxane	0.16 ±0.16	24 ±5	50		144
Methallylchloride	0.13	0	73—90		539
p-Methoxyphenylvinylsulphide	0.77	0.04	60	f, m	180
p-Methoxyvinylbenzoate	0.78	1.45	30	benzene	523
Methyl bicyclo(2,2,1)-2-heptene-5-carboxylate	1.5 ±0.24	0.45 ±0.07	60	f, l	99
Methylolcrotonamide	0.01 ±0.01	0.045 ±0.1	110		528
p-Methylphenyl vinylsulphide	0.81	0.07	60	f, m	180
p-Methyl vinylbenzoate	0.73	1.73	30	benzene	523
Methyl vinyl ketone	0.05	7.00	70	f, l	29
Methyl vinyl sulphone	0.29	0.35	60	c, n	247
Monomethyl maleate	0.04681	0.4747	56		595, 596
	0.522	0.035	56		597
	0.09 ±0.005	0.0 ±0.007	60		533
	0.0345	0.522	65		595
	0.13 ±0.03	0.015 ±0.01	75		533
	0.1168	0.99959	78		595
Methyl vinyl sulphone	0.0 ±0.01	0.4 ±0.08	60		60
N-*n*-Octadecylacrylamide	0.027 ±0.009	6.11 ±0.045	60		37
n-Octadecyl vinyl ether	4.50 ±0.58	0 ±0.35	50		535

n-Octyl vinyl ether	3.47	±0.51	0	±0.31	50		535
Phenyl vinyl sulphide	0.85		0.08		60	f, m	180
Phenyl vinyl sulphone	0.28		0.35		60	f, l	250
Propyl methacrylate	0.186	±0.038	73.3	±7.4	60		350
Sodium acrylate	0.01		2		60	a, l ethanol	13
Tetrachloroethylene	6.8	±0.5	0		60	j, p	491
	5		0		68	a, o	294
trans-2,3,4,5-Tetrachloro-1,3,5-hexatriene	0.013		32		70		124
Triallyl cyanurate	0.71	±0.02	0.62	±0.05	60	g, n, w	290
	0.77	±0.15	0.52	±0.07	60	g, n, w	290
Triallyl isocyanurate	0.91	±0.03	0.75	±0.06	60	g, n, w	290
	0.95	±0.18	0.70	±0.09	60	g, n, w	290
Trichloroethylene	0.66	±0.04	0.01	±0.01	60	i, n	132
	0.67		0		68	g, o	413
3,3,3-Trichloropropane	0.19	±0.04	0.19	±0.03	60		60
Trimethylsiloxyvinyl-dimethylsilane	0.99	±15 %	0.01	±15 %	70	i, m benzene	128
bis-(Trimethylsiloxy)-vinylmethylsilane	0.99	±15 %	0.01	±15 %	70	i, m benzene	128
tris-(Trimethylsiloxy)-vinylsilane	0.99	±15 %	0.01	±15 %	70	g, l benzene	128
N-Vinyl acetanilide	0.15	±0.015	1.6	±0.13	70	g, r	496
Vinyl benzoate	0.66		1.74		30	benzene	523
	0.70	±0.07	1.13	±0.07	60	a, l benzene	352
	0.35	±0.09	0.99	±0.13	60	f, l benzene	598
	0.7		1.5		80	h, n	499
Vinyl bromide	0.35	±0.09	4.5	±1.2	60	f, n	132
Vinyl butyl ether	0.71	±0.1	0.2	±0.05	—		599
S-Vinyl-*O*-*t*-butylthiocarbonate	0.04		11.0		70	a, n	215
Vinyl butyrate	1.00	±0.07	0.97	±0.07	60	a, l benzene	352
N-Vinylcarbazole	0.13	±0.03	3.9	±0.2	30	k, l tetrahydrofuran	134
	0.126		6.05		65		600
	0.126	±0.32	2.68	±0.10	65		601, 602
	0.152	±0.018	3.02	±0.24	100		601, 602

Copolymerization reactivity ratios—*continued*

M_1/M_2	r_1		r_2		T(°C)	Remarks	Ref.
Vinyl chloride	0.6	±0.2	1.8	±0.6	40	h, l	603, 479
	0.65	±0.04	1.35	±0.05	40		115
	0.23	±0.02	1.68	±0.08	60	f, o chloroform	132
	0.30	±0.08	2.45	±0.15	60	a, r	604
	0.3		2.1		68	a, n	294
	—		—		—	s	540
Vinyl chloroacetate	0.73		1.20		60	k, r	541
Vinyl cinnamate	0.04		1.20	±0.10	70	c, r, x benzene	233
Vinylene carbonate	3.71		0.0579		55	a, l	542
	7.3	±0.7	0.13	±0.1	60	f, m	99
	3.0		0.27		70	a, o acetone	357
	4.00		0.015		70	a, m	543
1-Vinyl-3-ethylurea	0.45		0.63		70		356
Vinyl fluoride	0.16	±0.01	2.9	±0.2	30	a, m	142
Vinyl formate	0.94	±0.07	0.95	±0.07	60	a, l benzene	352
N-Vinylglycidylurethan	0.26	±0.04	3.2	±0.3	60	h, m tetrahydrofuran	143
Vinylidene chloride	—		—		—	t	540
	0.0	±0.03	3.6	±0.5	60	f, n	491
	0.1		6		68	a, m	294
Vinylidene cyanide	0.0054		0.11		45	a, m benzene	27
Vinyl isocaproate	0.67		1.14		80	j, l	544
Vinyl 4-ketostearate	1.04	±0.2	1.18	±0.22	60	i, m	145
Vinyl 9(10)-ketostearate	1.03	±0.05	1.49	±0.25	60	i, m	145
Vinyl 12-ketostearate	1.07	±0.02	1.26	±0.08	60	i, m	145
Vinyl laurate	1.4		0.7		60		522
Vinyl methyl adipate	0.57	±0.04	0.66	±0.04	80	g, r	509
Vinyl octadecanoate	0.90		0.73		70	k, m	148
	0.97		1.00		50		545
N-Vinyl oxazolidinone	0.52	±0.08	1.90	±0.10	50	e, m	186
	0.60		1.50		75	a, m acetone	360
Vinyl palmitate	1.15	±0.13	0.78	±1.0	70—73	a, q benzene, w	546

	0.84	±0.10	0.66	±0.07	70—73	a, q benzene, w	546
Vinyl phenylacetate	0.96	±0.07	0.92	±0.07	60	a, l benzene	352
N-Vinyl phthalimide	0.18		2.08		65		600
	0.07		2.4		—		547
Vinyl pivalate	0.75	±0.04	0.43	±0.05	60		548
Vinyl propionate	0.98	±0.07	0.98	±0.07	60	a, l benzene	352
2-Vinylpyridine	0.3		10		—		549
	0		30	±15	70	a, m	102
N-Vinylpyrrolidone	0.38		0.44		—		550
	0.205	±0.015	3.30	±0.15	50	h, l	186
	0.38		0.44		70	a, o acetone	357
	0.237	±0.037	2.28	±0.19	76	f, l benzene	551
N-Vinyl succinimide	0.175		5.1		50—65	j, r	552
	0.18		6.1		60	i, n	11
	0.175		5.1		65		553
	0.07		2.4		65		600
Vinyl trifluoroacetate	0.6		0.32		60	a, m	554
N-Vinyl urethane	0.33		0.33		65	d, o	555
Vinyl benzoate/							
Vinyl 4-bromobenzoate	0.71		1.18		60		523
Vinyl 4-chlorobenzoate	0.85		0.84		60		523
Vinyl 4-cyanobenzoate	0.67		0.89		60		523
Vinyl 4-methoxybenzoate	1.28		0.92		60		523
Vinyl 4-bromobenzoate/							
Vinyl 4-cyanobenzoate	0.96		1.11		60		523
N-Vinyl-*O*-*t*-butyl carbamate/							
S-Vinyl-*O*-*t*-butyl thiocarbonate	0.13		5.10		70	d, m benzene	215
N-Vinyl caprolactam/							
n-Butyl vinyl ether	1.39	±0.01	0		60	a, m	556
Isopropyl vinyl ether	1.88	±0.02	0		60	a, m	556
Vinyl phenyl ether	2.53	±0.03	0.39	±0.003	60	a, p	556
N-Vinyl carbazole/							
Cyanoacetylene	0.075	±0.005	0.030	±0.005	30	k, l tetrahydrofuran	134

Copolymerization reactivity ratios—*continued*

M_1/M_2	r_1		r_2		T(°C)	Remarks	Ref.
Vinyl butanoate	1.28	±0.06	0.059	±0.020	100		601, 602
Vinyl formate	4.22	±0.16	0.196	±0.004	100		601, 602
Vinyl propionate	1.68	±0.140	0.076	±0.018	100		601, 602
Vinyl thioacetate	0.622		0.5		65	f	600
Vinyl chloride/							
Abityl acrylate	0.13	±0.01	4.2	±0.2	60	b, n benzene	82
Acenaphthylene	0.001		64		—		557
Acrylic acid	0.025		6.4		50	f, n dioxan	558
	0.107		6.8		60	i, m	559
Acrylyl chloride	0.017		2.65		50	a, m ethylene dichloride	558
Allyl acetate	1.16		0		40		539
Allyl triethoxy silane	2.0		0		50	j, m	38
Allyl-11-iodoundecanoate	0.42		1.64		60	j, n benzene	560
Bicyclo(2.2.1)hepta-2,5-diene	0.74		0.35		50	a, m	42
Bornyl methacrylate	0.06		12.5		60		386
Butadiene	0.031		—		40	Chlorobenzene	561
	0.035		8.80		50		605
	0.035		8.8		50		562
2-Butenyl triethoxy silane	0.4		0		50	i, n	38
Butyl acrylate	0.07		4.4		45		256
t-Butylethylene	5	±15 %	0.0		70	m	102
Butyl methacrylate	0.05		13.5		45		257
Butyl vinyl sulphonate	0.35	±0.05	0.30	±0.05	70	a, n	237
α-Chloroacrolein dimethyl acetal	0.288	±0.007	0.04	±0.063	65	i, m	395
2-Chloroallyl acetate	0.70		0		100		20
1-Chloro-1-propene	1.13		0.24		50		562
2-Chloro-1-propene	0.75		0.58		50		562
Chlorotrifluoroethylene	2.53		0.01		60		563
α-Chlorovinyltriethoxysilane	0.2		0		50	j, n	38
Dibutyl maleate	1.4		0.0		40		564

Diisobutyl maleate	0.65	±0.05	0.1		40		431
Diethyl fumarate	0.12	±0.01	0.47	±0.05	60	f, o	218
Di-2-ethylhexyl maleate	0.42		0		68	f, o	294
Diethyl itaconate	0.06	±0.01	5.65	±0.25	50		565
Diethyl maleate	0.77	±0.03	0.009	±0.003	60	f, p	218
	0.9	±15 %	0	±15 %	70	g, m	102
Diisopropyl itaconate	0.06	±0.01	6.0	±0.5	50		565
Dimethyl itaconate	0.053	±0.01	5.0	±0.2	50		286
Dioctyl itaconate	0.06	±0.02	7.0	±1.0	50		565
n-Dodecyl vinyl ether	1.93	±0.15	0.15	±0.2	50		566
Ethylene	1.85	±0.2	0.20	±0.02	70	300 atm	567
	1.85		0.16		70	d, r 300 atm	536, 568
	1.85		0.20		70	d, r 1500 atm	536, 568
	3.60	±0.30	0.24	±0.07	90	g, l 1000 atm toluene	211
Glycidyl methacrylate	0.04		8.84		—		557
Hydronopyl acrylate	0.14	±0.01	4.3	±0.2	60	e, o benzene	82
Isobornyl methacrylate	0.12		10.0		60		386
Isobutylene	1.3		0.03		0		569
	2.05	±0.3	0.08	±0.1	60	f, o	108
Isopropenyl acetate	2.2		0.25		65	a, n	278
Maleic anhydride	0.296	±0.07	0.008		75	d, l	217
Methacrylic acid	0.034		23.8		60	g, o	559
Methallyl chloride	0.31		0		45		539
2-Methylpentyl vinyl brassylate	1.06		0.234		60	a, m, w benzene	570
	1.10		0.239		60	a, m, w benzene	570
Methyl vinyl ketone	0.10		8.3		70		187
n-Octadecyl vinyl ether	2.10	±0.20	—0.1	±0.2	50		566
Octyl acrylate	0.12		4.8		45		256
Octyl methacrylate	0.04		14.0		45		257
n-Octyl vinyl ether	1.90	±0.20	0.1	±0.1	50		566
Pentachlorostyrene	0.43		5.3		—		502
2-Phenyl-5-(4′-vinyl) phenyl tetrazole	0.24	±0.02	2.24	±0.25	60	g, m, w benzene (C analysis)	343
	0.22	±0.02	1.89	±0.25	60	g, m, w benzene (C analysis)	343
	0.25	±0.02	2.54	±0.25	60	g, m, w benzene (C analysis)	343

Copolymerization reactivity ratios—*continued*

M_1/M_2	r_1		r_2		$T(^\circ C)$	Remarks	Ref.
	0.26	±0.03	1.49	±0.15	60	g, m, w benzene (N analysis)	343
	0.25	±0.03	1.19	±0.15	60	g, m, w benzene (N analysis)	343
	0.29	±0.03	1.64	±0.15	60	g, m, w benzene (N analysis)	343
Propenyl triethoxysilane	8.0		0		50	j, n	38
6-Triethoxysilyl bicyclo[2,2,1]-2-heptene	1.6		0		50		38
trans-Trimethyl aconitate	0.15	±0.10	0.00	±0.50	60	j, l	126
bis-(Trimethylsiloxy)vinylmethyl silane	0.90	±15 %	0.50	±15 %	50	j, o acetone	128
tris-(Trimethylsiloxy)-vinyl silane	0.90		0.50		50	j, o acetone	128
Vinyl benzoate	0.72		0.28		45		257
Vinyl butanoate	1.35	±0.05	0.65	±0.04	40		571
Vinyl caprate	4.7		0.2		40		115
Vinyl caproate	1.35	±0.05	0.65	±0.04	40		115
N-Vinylcarbazole	0.17		4.77		—		557
Vinylene carbonate	5.2		0.09		80	a, o acetone	357
Vinyl ethyl diethoxy silane	1.0		0		50	j, n	38
Vinyl fluoride	0.05	±0.005	11.0	±1	30	a, m	142
Vinylidene chloride	—		—		—	s	540
	0.2	±0.2	1.8	±0.5	45	a, p	571
	0.5		7.5		47		572
	0.2		4.5		50		562
	0.23		3.15		55	a, l	573
	0.14		Large		68	g, l	294
Vinylidene cyanide	0.017	±0.01	0.54	±0.2	50	a, n benzene	27
	0.0093		0.72		—		574
Vinyl isobutyl ether	2.0		0.02		—		479
	2.0	±0.2	0.02	±0.01	50		30
Vinyl 4-ketostearate	0.874	±0.044	0.320	±0.076	60	i, m	145
Vinyl 12-ketostearate	0.963	±0.01	0.248	±0.01	60	i, m	145

Vinyl levulinate	1.40	±0.004	0.419	±0.002	60	f, m benzene	147
Vinyl methyl diethoxy silane	1.0		0		50	h, n	38
Vinyl nonanoate	1.16	±0.06	0.282	±0.35	60	j, m	147
Vinyl octadecanoate	0.745	±0.026	0.290	±0.025	60	f, m	147
N-Vinyl oxazolidinone	0.84	±0.02	0.35	±0.02	50	f, m	186
	0.60		1.50		75	a, n cyclohexanone	360
Vinyl phenyl diethoxy silane	0.7		0		50	i, n	38
Vinyl pinonate	1.458	±0.04	0.446	±0.028	60	j, l benzene	147
Vinyl propionate	1.35	±0.05	0.65	±0.04	40		115
N-Vinyl pyrrolidone	0.53		0.38		—		550
Vinyl tri-*t*-butoxy silane	0.84		0		55	g, m	575
Vinyl triethoxy silane	0.9		0		50	i, o	38
Vinyl trimethoxy silane	0.8		0		50	h, n	38
Vinyl triisopropoxy silane	0.8		0		50	i, o	38
Vinyl undecenoate	1.06	±0.05	0.358	±0.065	60	f, m benzene	147
Vinyl chloroacetate/							
Vinyl butyl ether	1.1	±0.02	0.5	±0.02	—		599
Vinyl 4-chlorobenzoate/							
Vinyl 4-cyanobenzoate	0.78		1.02		60		523
Vinyl cinnamate/							
N-Vinylpyrrolidone	1.20		0.01		70	a, r benzene	233
Vinylene carbonate/							
Methyl vinyl sulphide	0.50	±0.04	10.6	±1.2	60	f, m	99
Vinyl thioacetate	0.04	±0.01	12.9	±1.1	60		576
Vinyl fluoride/							
Chlorotrifluoroethylene	0.18	±0.02	0.006	±0.02	30	a, m	142
Cyclohexafluorobutane	3	±0.6	0		30	a, m	142
Hexafluoropropene	1.01		0		30	a, m	142
cis-1,2,3,3,3-Pentafluoropropene	0.9	±0.05	0		30	a, m	142
Tetrafluoroethylene	0.27	±0.03	0.05	±0.02	30	a, m	142
Vinylidene fluoride	5.5	±0.5	0.17	±0.03	30	a, m	142
Vinylidene chloride/							
Acrylamide	0.15	±0.08	4.89	±0.08	60		12

Copolymerization reactivity ratios—*continued*

M_1/M_2	r_1		r_2		$T(°C)$	Remarks	Ref.
Acrylic acid	0.46		1.26		50	f, m dioxan	558
Acrylyl chloride	0.50		1.12		50	d, m ethylene dichloride	558
Allyl chloride	3.8		0.26		68	d, m	294
N-Allylstearamide	5.23	±0.067	0	±0.136	80		37
Bicyclo(2.2.1)hepta-2,5-diene	1.41		0.08		50	a, m	42
Butylacrylate	0.84	±0.2	0.46	±0.13	70		577
	0.88	±0.10	0.83	±0.02	—		52
n-Butyl crotonate	33		0		70		578
n-Butyl methacrylate	0.35		0.22		68	a, m	294
n-Butyl vinylsulphonate	7.5	±0.6	0.065	±0.007	80	a, n	237
1-Chloro-1-bromoethylene	0.83	±0.08	2.38	±0.06	—		579
Chlorotrifluoroethylene	17.14		0.02		60		245
Coumarin	Very Large		0		70	h, m	102
Crotonaldehyde	17		0		70		578
Crotonic acid	35	±5	0.065	±0.005	60	g, l dioxan	30
Diallyl phthalate	5.0		0.2		70		189
Diethyl fumarate	12.2	±2	0.046	±0.015	60	f, n	491
Diethyl maleate	40	±8	0.0	±0.04	60		60
Dimethallyl oxalate	4.8	±0.2	0.16	±0.01	40		580
Ethyl methacrylate	0.35		0.22		68	a, n	294
4-Hydroabietyl alcohol	2.6		0		70		578
Indene	0.4	±15 %	0.33	±15 %	70	a, n	102
Isopropenyl isocyanate	0.85	±0.04	0.31	±0.02	—		322
Isopropenyl methylketone	0.15		4.5		—		479
Maleimide	0.71		0.48		75	a, m cyclohexanone	329
Methacrylic acid	0.15	±0.015	3.0	±0.45	70	a, l	102
Methyl isopropenyl ketone	0.15	±0.02	4.5	±0.1	60	a, l	30
Methyl vinyl ketone	0.55		1.8		70		187
N-*n*-Octadecylacrylamide	0.438	±0.008	1.37	±0.008	60		37
Octadecyl acrylate	0.91	±0.05	1.01	±0.01	60		52
Octyl acrylate	0.87	±0.02	0.70	±0.01	60		52

Phenyl acetylene	0.1		1.4		60		60
n-Propyl crotonate	20		0		70		578
2-Sulphoethyl methacrylate	0.22	±0.03	3.6	±0.5	50—60	d, m 1,2-dimethoxyethane	198
trans-Trimethyl aconitate	54	±5	0.01	±0.1	60	a, q	581
Vinyl benzoate	7.0	±1	0.1	±0.02	50	a, l	30, 479
N-Vinyl carbazole	0.020		3.7		75	a, m benzene	255
Vinyl cyclohexene	1.8		0		70		578
2-Vinyl furan	11.7	±0.07	0.15	±0.014	—		582
Vinylidene bromide	1.04	±0.10	1.90	±0.11	—		579
Vinylidene cyanide	0.012		0.049		22	a, n benzene	27
Vinyl isocyanate	1.46	±0.28	0.33	±0.18	60		144
Vinyl 4-ketostearate	3.8	±0.6	0	±0.17	60	i, l	145
Vinyl 9(10)-ketostearate	3.7	±0.3	0.01	±0.09	60	i, l	145
Vinyl 12-ketostearate	4.0	±0.5	0	±0.16	60	i, l	145
Vinyl nonanoate	4.08	±0.20	0.0	±0.01	60	f, l benzene	147
Vinyl octadecanoate	3.80	±0.05	0.075	±0.025	60	c, l benzene	147
N-Vinyl-2-oxazolidinone	1.35		0.08		75	a, n cyclohexanone	360
Vinyl pinonate	3.00	±0.18	0.03	±0.028	60	d, l benzene	147
N-Vinyl succinimide	1.44		0.32		60	d, m	594
Vinyl undecenoate	2.58	±0.09	0.054	±0.30	60	f, m benzene	147
Vinylidene cyanide/							
2-Chloro-1-propene	0.20		0		40	a, l	27
cis-Dichloroethylene	30		~0		40	a, m	27
trans-Dichloroethylene	30		~0		40	a, l	27
2,5-Dichlorostyrene	0.0092		0.031		40	a, n, s (see ref. 588) benzene	27
Isobutylene	0.182		0.00176		—		583
Methyl 1-chloroacrylate	0.091	±0.05	0.41	±0.14	50	a, m	27
Vinyl benzoate	0.10		0.008		43	a, n benzene	27
Vinyl chloroacetate	0.13	±0.05	<0.004		40	a, m	27
Vinyl 4-methoxybenzoate/							
Vinyl 4-bromobenzoate	0.82		1.19		60		523
Vinyl 4-chlorobenzoate	0.73		0.92		60		523
Vinyl 4-cyanobenzoate	0.93		0.75		60		523

Copolymerization reactivity ratios—*continued*

M_1/M_2	r_1		r_2		$T(°C)$	Remarks	Ref.
Vinyl 4-methylbenzoate/							
Vinyl 4-bromobenzoate	0.63		1.04		60		523
Vinyl 4-Chlorobenzoate	0.74		0.89		60		523
Vinyl 4-cyanobenzoate	0.65		0.86		60		523
Vinyl octadecanoate/							
Diisopropyl maleate	0		0.0075		70		148
N-Vinyl phthalimide/							
Vinyl thioacetate	0.23		3.2		65	f	600
2-Vinylpyridine/							
Acrolein	0		4.0		50	a, m DMF	7
2,5-Dichlorostyrene	1.1		0.9		70		188
Isopropenyl acetylene	1.65	±0.05	0.55	±0.01	60	f, l	73
Phenylacetylene	4.0	±0.07	0.2	±0.05	60	a, l	584
	4.0	±0.2	0.32	±0.15	—		249
1-Vinyl-2-ethyl acetylene	1.5	±0.5	0.6	±1	60	f, l	73
4-Vinylpyridine/							
Divinyl ether	32		~0		60	a, m	192
N,N,N-Triethyl-*N*-[2-(methacryloxy)-ethyl]-ammonium iodide	0.30	±0.02	0.61	±0.09	60	e, n	576
Vinylhydroquinone dibenzoate	0.48		0.40		78	k, m chloroform	258
N-Vinylpyrrolidone/							
n-Butyl vinyl ether	2.97	±0.01	0		60	a, q	556
Dioctyl fumarate	0.030	±0.033	0.041	±0.007	76	g, l heptane/ethyl alcohol	551
Isopropyl vinyl ether	1.68		0		60	a, n	556
tris-(Trimethylsiloxy) vinylsilane	4.0	±15 %	0.1	±15 %	125	j, n xylene	128
S-Vinyl-*o*-*t*-butyl thiocarbonate	0.12		3.94		65	d, n	215
Vinyl cyclohexyl ether	3.84		0		60	a, q	585

Vinyl cyclopentadienyl manganese tricarbonyl	0.09		0.14		70	p, w benzene	586
	0.085		0.053		70	p, w benzene	586
Vinylene carbonate	0.7		0.4		60	a, m acetone	118
Vinyl ferrocene	0.40		0.66		70	p, w benzene	586
	0.33		0.67		70	p, w benzene	586
Vinylhydroquinone dibenzoate	0.98		0.26		60		357
Vinyl phenyl ether	4.43	±0.001	0.22	±0.001	60	a, q	585
N-Vinyl urethane	2		0.42		65	d, n	555
N-Vinyl succinimide/ Vinyl thioacetate	0.095		3.1		65	f	600

Approximate range of monomer feed compositions used, quoted in terms of mole fractions of M_1 and based on total monomers employed (i.e. neglecting solvent concentrations).

a, 0—1.00; b, 0—0.25; c, 0—0.50; d, 0—0.75; e, 0.25—0.50; f, 0.25—0.75; g, 0.25—1.00; h, 0.50—0.75; i, 0.50—1.00; j, 0.75—1.00; k, not quoted.

Approximate maximum conversions employed, expressed as percentage of total monomers in feed.

l, 5 %; m, 10 %; n, 20 %; o, 35 %; p, 50 %; q, >50 %; r, not quoted.

Other symbols

s, Penultimate group effects.
t, Heterogeneous.
u, Emulsion polymerization with data corrected for partition of monomers between phases.
v, r_2 value assumed.
w, Same experimental data, different method of calculation.
x, Cyclopolymerization effects noted.

DMF, *NN'*-Dimethylformamide solution.
DMSO, Dimethylsulphoxide solution.

References pp. 404—418

REFERENCES

1 L. J. Young, J. Polym. Sci., 54 (1961) 411.
2 H. Mark, B. Immergut, E. H. Immergut, L. J. Young and K. I. Benyon, in J. Brandrup and E. H. Immergut (Eds.), Polymer Handbook, Interscience, New York, 1965.
3 L. J. Young, in J. Brandrup and E. H. Immergut (Eds.), Polymer Handbook, Interscience, New York, 1965.
4 G. F. D'Alelio and T. F. Huemmer, J. Polym. Sci., Part A-1, 5 (1967) 77.
5 R. C. Schulz, H. Cherdron and W. Kern, Makromol. Chem., 28 (1958) 197.
6 W. Kern, quoted in ref. 2.
7 R. C. Schulz, E. Kaiser and W. Kern, Makromol. Chem., 58 (1962) 160.
8 K. Matsuo and S. Machida, J. Polym. Sci., Part A-1, 10 (1972) 187.
9 W. P. Shyluk, J. Polym. Sci., Part A, 2 (1964) 2191.
10 F. S. Dainton and W. D. Sisley, Trans. Faraday Soc., 59 (1963) 1385.
11 H. Kamogawa and T. Sekiya, Kogyo Kagaku Zasshi, 63 (1960) 1631.
12 American Cyanamid Company, quoted in ref. 2.
13 J. Bourdais, Bull. Soc. Chim. Fr., (1955) 485.
14 A. M. Chatterjee and C. M. Burns, Can. J. Chem., 49 (1971) 3249.
15 J. Pellon, L. H. Schwind, M. J. Guinard and W. M. Thomas, J. Polym. Sci., 55 (1961) 161.
16 G. Smets and A. M. Hesbain, J. Polym. Sci., 40 (1959) 217.
17 W. R. Cabaness, T. Y.-C. Lin and C. Parkanyi, J. Polym. Sci., Part A-1, 9 (1971) 2155.
18 F. F. Shcherbina, I. P. Federova and A. P. Ivanova, Vysokomol. Soedin., 9 (1967) 2615; Polym. Sci. U.S.S.R., 9 (1967) 2959.
19 T. R. Paxton, J. Polym. Sci., Part B, 1 (1963) 73.
20 W. O. Kenyon and J. H. van Campen, U.S. Pat. 2,419,221 (1947), calc. by F. R. Mayo and C. Walling, ref. 21.
21 F. R. Mayo and C. Walling, Chem. Rev., 46 (1950) 191.
22 G. Saini, G. Polla-Mattiot and M. Meirone, J. Polym. Sci., 50 (1961) S13.
23 G. Smets and W. van Humbeeck, J. Polym. Sci., Part A, 1 (1963) 1227.
24 C. E. Grabiel and D. L. Decker, J. Polym. Sci., 59 (1962) 425.
25 G. Hägele, H. Fröhlich, D. Bischoff and K. Hamann, Makromol. Chem., 75 (1964) 98.
26 V. F. Kazanskaya, O. M. Klimova and B. M. Khlebnikov, Vysokomol. Soedin., 6 (1964) 1799; Polym. Sci. U.S.S.R., 6 (1964) 1993.
27 H. Gilbert, F. F. Miller, S. J. Averill, E. J. Carlson, V. L. Folt, H. J. Heller, F. D. Stewart, R. F. Schmidt and H. L. Trumbull, J. Am. Chem. Soc., 78 (1956) 1669.
28 G. van Paesschen and G. Smets, Bull. Soc. Chim. Belg., 64 (1955) 173.
29 H. C. Haas and M. S. Simon, J. Polym. Sci., 9 (1952) 309.
30 E. C. Chapin, G. E. Ham and C. L. Mills, J. Polym. Sci., 4 (1949) 597; erratum: J. Polym. Sci., 55 (1961) S6.
31 N. Yamashita, T. Seita, M. Yoshihara and T. Maeshima, J. Polym. Sci., Part B, 9 (1971) 641.
32 T. Ota, T. Otsu and M. Imoto, Kogyo Kagaku Zasshi, 71 (1968) 736.
33 H. Ito and S. Suzuki, Kogyo Kagaku Zasshi, 58 (1955) 627.
34 Y. Tsuda, Kogyo Kagaku Zasshi, 62 (1959) 1112.
35 G. Oster and Y. Mizutani, J. Polym. Sci., 22 (1956) 173.
36 Z. Csuros, M. Gara and I. Gyukovics, Acta. Chim. Acad. Sci. Hung., 29 (1961) 207.
37 E. F. Jordan and A. N. Wrigley, J. Appl. Polym. Sci., 8 (1964) 527.

38 B. R. Thompson, J. Polym. Sci., 19 (1956) 373.
39 A. S. Bank, S. D. Savranskaya and M. A. Askarov, Khim. Fiz.-Khim. Prir. Sint. Polim., Akad. Nauk Uz. S.S.R., Inst. Khim. Polimerov, No. 2 (1964) 110.
40 M. A. Askarov and A. S. Bank, Vysokomol. Soedin., 8 (1966) 1012; Polym. Sci. U.S.S.R., 8 (1966) 1110.
41 T. D. Swartz and H. K. Hall, J. Am. Chem. Soc., 93 (1971) 137.
42 N. L. Zutty, J. Polym. Sci., Part A, 1 (1963) 2231.
43 E. S. Ferdinandi, W. P. Garby and D. G. L. James, Can. J. Chem., 42 (1964) 2568.
44 W. H. Embree, J. M. Mitchell and H. L. Williams, Can. J. Chem., 29 (1951) 253.
45 I. Sakurada, T. Okada, S. Hatakeyama and F. Kimura, J. Polym. Sci., Part C, 1 (1963) 1233.
46 K. R. Henery-Logan and R. V. V. Nicholls, quoted in ref. 593.
47 K. R. Henery-Logan and R. V. V. Nicholls, quoted in ref. 21.
48 N. Ashikari, Bull. Chem. Soc. Jpn., 32 (1959) 1060.
49 J. Vialle, J. Guillot and A. Guyot, J. Macromol. Sci., Chem., 5 (1971) 1031.
50 H. Hayashi, I. Ito, T. Saegusa and J. Furukawa, Kogyo Kagaku Zasshi, 65 (1962) 1634.
51 T. Tamikado and Y. Iwakura, J. Polym. Sci., 36 (1956) 529.
52 E. F. Jordan, K. M. Doughty and W. S. Port, J. Appl. Polym. Sci., 4 (1960) 203.
53 J. Müller, Chem. Listy, 48 (1954) 1593.
54 J. Müller, Collect. Czech. Chem. Commun., 20 (1955) 241.
55 H. Tatemichi and S. Suzuki, Kogyo Kagaku Zasshi, 63 (1960) 1843.
56 A. S. Nair and M. S. Kuthana, Makromol. Chem., 47 (1961) 138.
57 Y. Inaki, S. Hirose, K. Yasufuku, S. Nozakura and S. Murahashi, Polym. J., 2 (1971) 481.
58 T. Otsu and H. Inoue, J. Macromol. Sci., Chem., 4 (1970) 35.
59 N. Nakabayashi and Y. Iwakura, Makromol. Chem., 81 (1965) 180.
60 K. W. Doak, quoted in ref. 21.
61 T. Yamaguchi, T. Ono, T. Yanoshita and R. Shishida, Kobunshi, 22 (1965) 835.
62 F. L. Hamb and A. Winston, J. Polym. Sci., Part A, 2 (1964) 4475.
63 G. Goldfinger and M. Steidlitz, J. Polym. Sci., 3 (1948) 786.
64 G. F. D'Alelio and L. X. Mallavarapu, Makromol. Chem., 37 (1960) 25.
65 G. B. Butler and R. B. Kasat, J. Polym. Sci., Part A, 3 (1965) 4205.
66 E. Y. C. Chang and C. C. Price, J. Am. Chem. Soc., 83 (1961) 4650.
67 J. M. Barton, G. B. Butler and E. C. Chapin, J. Polym. Sci., Part A-1, 3 (1965) 501.
68 G. Akazome, S. Sakai and K. Murai, Kobunshi, 17 (1960) 452.
69 J. Brandrup, Faserforsch. Textiltech., 12 (1961) 135.
70 V. F. Gromov, P. M. Khomikovskii and A. D. Abkin, Vysokomol. Soedin., 3 (1961) 1015.
71 C. Aso and K. Sadakata, Kogyo Kagaku Zasshi, 62 (1959) 1610.
72 C. C. Price and T. C. Schwan, J. Polym. Sci., 16 (1955) 577.
73 C. C. Price and T. F. McKeon, J. Polym. Sci., 41 (1959) 445.
74 T. C. Schwan and C. C. Price, Proc. Indiana Acad. Sci., 63 (1953) 103.
75 N. Kawabata, T. Tsuruta and J. Furukawa, Makromol. Chem., 48 (1961) 106.
76 Y. Iwakura, T. Tamikado, M. Yamaguchi and K. Takei, J. Polym. Sci., 39 (1959) 203.
77 H. Inoue, I. Umeda and T. Otsu, Makromol. Chem., 147 (1971) 271.
78 R. J. Orr and H. L. Williams, Can. J. Chem., 33 (1955) 1328.
79 Y. Iwakura, T. Kurosaki and N. Nakabayashi, Makromol. Chem., 46 (1961) 570.
80 Y. Iwakura, N. Nakabayashi and M. H. Lee, Makromol. Chem., 78 (1964) 157.
81 Y. Iwakura and K. Matsuzaki, Chem. High Polym. (Tokyo), 17 (1960) 187.
82 C. S. Marvel and R. Schwen, J. Am. Chem. Soc., 79 (1957) 6003.

83 Y. Matsubara, M. Yoshihara and T. Maeshima, Kogyo Kagaku Zasshi, 74 (1971) 2163.
84 S. Okamura and T. Yamashita, J. Soc. Textile Cellulose Ind. Jpn., 9 (1953) 446.
85 H. Inoue and T. Otsu, Makromol. Chem., 148 (1971) 251.
86 K. W. Doak, M. A. Deahl and I. H. Christmas, 137th Am. Chem. Soc. Meeting, Cleveland, Ohio, Abstr. Papers, 1, No. 1 (1960) 151.
87 S. Nagai, Bull. Chem. Soc. Jpn., 36 (1963) 1459.
88 A. R. Kol'k, A. A. Konkin and Z. A. Rogovin, Khim. Volokna, (1963) 12.
89 C. Aso, Kogyo Kagaku Zasshi, 63 (1960) 363.
90 R. L. Merker and M. J. Scott, J. Polym. Sci., 25 (1957) 115.
91 C. Aso, T. Kunitake, T. Nakayas and S. Ando, private communication.
92 S. Kimura and M. Imoto, Makromol. Chem., 50 (1961) 155.
93 H. Lüssi, Makromol. Chem., 103 (1967) 68.
94 S. Yuguchi and M. Watanabe, Chem. High Polym. (Tokyo), 18 (1961) 613.
95 S. Yuguchi and M. Watanabe, Kobunshi, 15 (1958) 129.
96 Z. Izumi and H. Kitagawa, J. Polym. Sci., Part A-1, 5 (1967) 1967.
97 American Cyanamid Company, The Chemistry of Acrylonitrile, 2nd edn., 1960, p. 34.
98 R. M. Joshi and S. L. Kapur, J. Polym. Sci., 14 (1954) 508.
99 J. M. Judge and C. C. Price, J. Polym. Sci., 41 (1959) 435.
100 J. Szafko and E. Turska, Makromol. Chem., 156 (1972) 311.
101 T. Yamamoto, Kogyo Kagaku Zasshi, 62 (1959) 476.
102 T. Alfrey, J. Bohrer, H. Haas and C. Lewis, J. Polym. Sci., 5 (1950) 719.
103 G. N. Larina, Z. V. Borisova and T. V. Sheremeteva, Vysokomol. Soedin., 3 (1961) 1664.
104 H. Kamogawa and T. Sekiya, Kogyo Kagaku Zasshi, 62 (1959) 1117.
105 H. Lüssi, Makromol. Chem., 103 (1967) 62.
106 M. Matsuda, M. Iino and N. Tokura, Makromol. Chem., 65 (1963) 232.
107 R. G. Fordyce, E. C. Chapin and G. E. Ham, J. Am. Chem. Soc., 70 (1948) 2489.
108 F. M. Lewis, C. Walling, W. Cummings, E. R. Briggs and W. J. Wenisch, J. Am. Chem. Soc., 70 (1948) 1527.
109 T. Tamikado, Makromol. Chem., 38 (1960) 85.
110 V. S. Dyurnbaum and V. S. Klimenkov, Khim. Volokna, (1963) 8.
111 J. Furukawa, T. Tsuruta, N. Yamamoto and H. Fukutani, J. Polym. Sci., 37 (1959) 215.
112 J. Furukawa, T. Tsuruta and N. Yameda, Kogyo Kagaku Zasshi, 61 (1958) 734.
113 A. Tanaka, K. Sasaki, Y. Hozumi and O. Hashimoto, J. Appl. Polym. Sci., 8 (1964) 1787.
114 E. F. Jordan, R. Bennett, A. C. Shuman and A. N. Wrigley, J. Polym. Sci., Part A-1, 8 (1970) 3113.
115 T. Kimura and K. Yoshida, Kagaku To Kogyo (Osaka), 32 (1958) 223, 341.
116 J. M. Barton, G. B. Butler and E. C. Chapin, Polym. Prepr. Am. Chem. Soc., Div. Polym. Chem., 5 (1964) 216.
117 H. Sumitomo and K. Azuma, J. Polym. Sci., Part B, 4 (1966) 883.
118 C. S. Marvel, A. T. Tweedie and J. P. Economy, J. Org. Chem., 21 (1956) 1420.
119 K. Myamichi, A. Suzuki, S. Harada and M. Katayama, Kobunshi, 21 (1964) 79.
120 Z. Izumi, H. Kiuchi and M. Watanabe, J. Polym. Sci., Part A, 3 (1965) 2965.
121 Z. Izumi, H. Kiuchi and M. Watanabe, J. Polym. Sci., Part A, 1 (1963) 705.
122 Z. Izumi, H. Kiuchi, M. Watanabe and H. Uchiyama, J. Polym. Sci., Part A, 3 (1965) 2721.
123 E. Osawa, K. Wang and O. Kurihara, Makromol. Chem., 83 (1965) 100.
124 A. N. Akopyan and C. Ye. Krbekyan, Vysokomol. Soedin., 5 (1963) 201; Polym. Sci., U.S.S.R., 4 (1963) 595.
125 K. Noma and M. Niwa, Kobunshi, 21 (1964) 17.

126 C. S. Marvel, J. W. Johnson, J. P. Economy, G. P. Scott, W. K. Taft and B. G. Labbe, J. Polym. Sci., 20 (1956) 437.
127 B. M. Culbertson and R. C. Slagel, J. Polym. Sci., Part A-1, 6 (1968) 363.
128 R. M. Pike and D. L. Bailey, J. Polym. Sci., 22 (1956) 55.
129 M. Taniyama and G. Oster, Bull. Chem. Soc. Jpn., 30 (1957) 856.
130 I. S. Dorokhina, A. D. Abkin and V. S. Klimenkov, Khim. Volokna, (1962) 49.
131 I. S. Dorokhina, A. D. Abkin and V. S. Klimenkov, Vysokomol. Soedin., 5 (1963) 385; Polym. Sci., U.S.S.R., 4 (1964) 1044.
132 F. R. Mayo, C. Walling, F. M. Lewis and W. F. Hulse, J. Am. Chem. Soc., 70 (1948) 1523.
133 T. Asahara and K. Mitsuhashi, Yukagaku, 6 (1957) 331.
134 A. M. North and K. E. Whitelock, Polymer, 9 (1968) 590.
135 J. Guillot, J. Vialle and A. Guyot, J. Macromol. Sci., Chem., 5 (1971) 735.
136 E. W. Rugeley, T. A. Field and G. H. Fremon, Ind. Eng. Chem., 40 (1948) 1724.
137 E. C. Chapin, G. E. Ham and R. G. Fordyce, J. Am. Chem. Soc., 70 (1948) 538.
138 B. R. Thompson and R. H. Raines, J. Polym. Sci., 41 (1959) 265.
139 N. Yamashita, S. Kunitani, S. Komiyama, M. Yoshihara and T. Maeshima, J. Polym. Sci., Part B, 10 (1972) 675.
140 N. Yamashita, T. Seita, M. Yoshihara and T. Maeshima, Kogyo Kagaku Zasshi, 74 (1971) 2157.
141 J. C. Lai, T. Rounsfeld and C. U. Pittman, J. Polym. Sci., Part A-1, 9 (1971) 651.
142 D. Sianesi and G. Caporiccio, J. Polym. Sci., Part A-1, 6 (1968) 335.
143 Y. Iwakura, N. Nakabayashi and H. Suzuki, Makromol. Chem., 78 (1964) 168.
144 Y. Iwakura, M. Sato, T. Tamikado and T. Mizoguchi, Kobunshi, 13 (1956) 390.
145 C. S. Marvel, T. K. Dykstra and F. C. Magne, J. Polym. Sci., 62 (1962) 369.
146 A. De Pauw, and G. Smets, Bull. Soc. Chim. Belg., 59 (1950) 629.
147 C. S. Marvel and W. G. DePierri, J. Polym. Sci., 27 (1958) 39.
148 L. P. Witnauer, N. Watkins and W. S. Port, J. Polym. Sci., 20 (1956) 213.
149 K. Anda and S. Iwai, J. Polym. Sci., Part A-1, 7 (1969) 2414.
150 S. Yuguchi and M. Watanabe, Chem. High Polym. (Tokyo), 18 (1961) 386; Makromol. Chem., 49 (1961) 243.
151 M. M. Koton, J. Polym. Sci., 30 (1958) 331.
152 M. M. Koton, Inst. Khim. Akad. Nauk Latv., S.S.R. Riga (1957) 119.
153 I. N. Duling and C. C. Price, J. Am. Chem. Soc., 84 (1962) 578.
154 S. Kunichika and T. Katagiri, Kogyo Kagaku Zasshi, 64 (1961) 929.
155 C. E. Scott and C. C. Price, J. Am. Chem. Soc., 81 (1959) 2670.
156 R. M. Joshi and S. L. Kapur, J. Sci. Ind. Res., Sect. B, 16 (1957) 441.
157 T. Alfrey and S. L. Kapur, quoted in ref. 189.
158 T. Otsu, T. Ito and M. Imoto, J. Polym. Sci., Part C, 16 (1965) 2126.
159 J. C. Bevington and B. W. Malpass, Eur. Polym. J., 1 (1965) 19.
160 G. P. Scott, J. Org. Chem., 20 (1955) 736.
161 T. Alfrey, A. I. Goldberg and W. P. Hohenstein, J. Am. Chem. Soc., 68 (1946) 2464.
162 M. F. Margaritova and G. D. Berezhnov, Tr. Mosk. Inst. Tonkoi Khim. Tekhnol., 4 (1953) 46.
163 C. S. Marvel and J. F. Porter, J. Org. Chem., 24 (1959) 137.
164 C. S. Marvel, S. L. Jacobs, W. K. Taft and B. G. Labbe, J. Polym. Sci., 19 (1956) 59.
165 E. Grzywa and Z. Jedlinski, Vysokomol. Soedin., 8 (1966) 1653; Polymer Sci., U.S.S.R., 8 (1966) 1824.
166 R. J. Orr and H. L. Williams, J. Polym. Sci., 32 (1958) 89.
167 S. N. Ushakov and S. S. Ivanov, Izv. Akad. Nauk S.S.S.R., Otd. Khim. Nauk, (1957) 1465.
168 V. Arendt and S. Kaizerman, J. Polym. Sci., Part A-1, 7 (1969) 2741.

169 H. Kamogawa, R. Murase and T. Sekiya, Kogyo Kagaku Zasshi, 62 (1959) 1749.
170 W. Cooper and E. Catterall, Can. J. Chem., 34 (1956) 387.
171 B. L. Funt and E. A. Ogryzlo, J. Polym. Sci., 25 (1957) 279.
172 M. F. Sarokin and M. M. Babkina, Vysokomol. Soedin., 7 (1965) 737; Polym. Sci., U.S.S.R., 7 (1965) 816.
173 S. S. Spasskii, A. V. Tokarev, S. A. Mikhailova, T. V. Molchanova and M. E. Mat'kova, Zh. Obshch. Khim., 30 (1960) 250, Engl. translation, p. 268.
174 A. L. Klebanskii and O. A. Timofeev, Zh. Prikl. Khim., 32 (1959) 2294; J. Polym. Sci., 52 (1961) 23.
175 N. V. Rakityanskii and R. L. Rabinovich, Report of the All Union Scientific Institute of Synthetic Rubber, 1951.
176 G. E. Krybekyan, E. G. Sinanyan and A. N. Akopyan, Izv. Akad. Nauk Arm. S.S.R., Khim. Nauki, 15 (1962) 527.
177 S. N. Ushakov and L. B. Trukhmanova, Izv. Akad. Nauk S.S.S.R. Otd. Khim. Nauk, (1957) 980.
178 S. P. Mitsengendler, V. N. Krasulina and L. B. Trukhmanova, Izv. Akad. Nauk S.S.S.R. Otd. Khim. Nauk, (1956) 1120.
179 S. Iwatsuki, M. Shin and Y. Yamashita, Makromol. Chem., 102 (1967) 232.
180 K. Tsuda, S. Kobayashi and T. Otsu, J. Polym. Sci., Part A-1, 6 (1968) 41.
181 J. Pellon, R. L. Kugel, R. Marcus and R. Rabinowitz, J. Polym. Sci., Part A, 2 (1964) 4105.
182 M. Imoto, M. Kinoshita and M. Nishigaki, Makromol. Chem., 94 (1966) 238.
183 Chas. Pfizer and Co., Product News, quoted in ref. 1.
184 C. Walling, E. R. Briggs, K. B. Wolfstirn and F. R. Mayo, J. Am. Chem. Soc., 70 (1948) 1537.
185 M. Ragazzini, D. Carcano, M. Modena and G. Serboli, Eur. Polym. J., 6 (1970) 763.
186 J. F. Bork and L. E. Coleman, J. Polym. Sci., 43 (1960) 413.
187 T. Alfrey, L. Arond and C. G. Overberger, quoted in ref. 189, p. 35.
188 T. Alfrey and B. Magel, quoted in ref. 189.
189 T. Alfrey, Jr., J. Bohrer and H. Mark, Copolymerization, High Polymer Series, Vol. VIII, Interscience, New York, 1952, p. 40.
190 T. Alfrey, C. G. Overberger and S. H. Pinner, J. Am. Chem. Soc., 75 (1953) 4221.
191 E. I. Egorova and A. F. Dokukina, Vysokomol. Soedin. Karbotsepnye Vysokomol. Soedin., Sb. Statei (1963) 40.
192 G. B. Butler, G. Van Laaren and M. F. Ramadier, J. Polym. Sci., Part A-1, 5 (1967) 1265.
193 K. I. Beynon, J. Polym. Sci., Part A, 1 (1963) 3343.
194 K. I. Beynon, J. Polym. Sci., Part A, 1 (1963) 3357.
195 K. I. Beynon and E. J. Hayward, J. Polym. Sci., Part A, 3 (1965) 1793.
196 C. C. Unruh and T. M. Laakso, J. Polym. Sci., 33 (1958) 87; Ind. Eng. Chem., 50 (1958) 1124.
197 H. Ito and S. Suzuki, Kogyo Kagaku Zasshi, 60 (1957) 341.
198 D. A. Kangas and R. R. Pelletier, J. Polym. Sci., Part A-1, 8 (1970) 3543.
199 H. Hopff and H. Gutenberg, Makromol. Chem., 60 (1963) 129.
200 L. Bogetich, G. A. Mortimer and G. W. Daues, J. Polym. Sci., 61 (1962) 3.
201 F. H. C. Edgecombe, Nature (London), 198 (1963) 1085.
202 G. A. Mortimer, J. Polym. Sci., Part B, 3 (1963) 343.
203 N. Ariya, M. Uchiyama, T. Kurosaki and I. Iwakura, Makromol. Chem., 99 (1966) 126.
204 G. A. Mortimer, J. Polym. Sci., Part B, 8 (1970) 323.
205 P. Colombo, L. E. Kakacka, J. Fontana, R. N. Chapman and M. Steinberg, Polym. Prepr. Am. Chem. Soc., Div. Polym. Chem., 5 (1964) 914.
206 P. Colombo, M. Steinberg and J. Fontana, J. Polym. Sci., Part B, 1 (1963) 447.

207 E. E. Brando and A. I. Dintses, Neftekhimiya, 4 (1964) 68.
208 D. D. Coffman, P. S. Kinkney, F. T. Wall, W. H. Wood and H. S. Young, J. Am. Chem. Soc., 75 (1952) 3391.
209 M. M. Brubaker, D. D. Coffman and H. H. Hoehn, J. Am. Chem. Soc., 74 (1951) 1509.
210 M. Ragazzini, C. Garbuglio, D. Carcano, B. Minasso and Gb. Cevidalli, Eur. Polym. J., 3 (1967) 129.
211 R. D. Burkhart and N. L. Zutty, J. Polym. Sci., Part A, 1 (1963) 1137.
212 J. T. Khamis, Polymer, 6 (1965) 99.
213 Y. Tabata, K. Ishigure, K. Oshima and H. Sobue, J. Polym. Sci., Part A, 2 (1964) 2445.
214 R. J. Orr and H. L. Williams, Can. J. Chem., 30 (1952) 108.
215 W. H. Daley, C.-D. S. Lee and C. G. Overberger, J. Polym. Sci., Part A-1, 9 (1971) 1723.
216 G. Akazome, S. Sakai and K. Murai, Kobunshi, 17 (1960) 449.
217 M. C. de Wilde and G. Smets, J. Polym. Sci., 5 (1950) 253.
218 F. M. Lewis and F. R. Mayo, J. Am. Chem. Soc., 70 (1948) 1533.
219 B. Yamada, H. Yoneno and T. Otsu, J. Polym. Sci., Part A-1, 8 (1970) 2021.
220 M. Kato and Y. Nakano, J. Polym. Sci., Part B, 10 (1972) 157.
221 V. A. Orlov and O. G. Tarakanov, Plast. Massy, (1964) 6.
222 K. Crauwels and G. Smets, Bull. Soc. Chim. Belg, 59 (1950) 182.
223 S. H. Pinner, J. Polym. Sci., 10 (1953) 379.
224 G. G. Cameron, D. H. Grant, N. Grassie, J. E. Lamb and I. C. McNeill, J. Polym. Sci., 36 (1959) 173.
225 N. Grassie and I. C. McNeill, J. Polym. Sci., 27 (1958) 207.
226 G. S. Kolesnikov, A. S. Trevlina and A. B. Alovitdinov, Vysokomol. Soedin., 7 (1965) 1913; Polym. Sci., U.S.S.R., 7 (1965) 2097.
227 A. Ulinska and Z. Mankowski, Stud. Soc. Sci. Torun., Sect. B, 2 (1960) 16.
228 A. V. Ryabov, Yu. D. Semchikov and N. N. Slavnitskaya, Tr. Khim. Khim. Tekhnol. (1963) 334.
229 G. Smets, N. Deval and P. Hous, J. Polym. Sci., Part A, 2 (1964) 4835.
230 H. Lüssi, Makromol. Chem., 110 (1967) 100.
231 H. K. Johnston and A. Rudin, Macromolecules, 4 (1971) 661.
232 Y. Inaki, S. Hirose, S. Nozakura and A. Murahashi, Polym. J., 3 (1972) 159.
233 J. Roovers and G. Smets, Makromol. Chem., 60 (1963) 89.
234 R. M. Joshi and S. L. Kapur, J. Sci. Ind. Res. (India), Sect. B, 16 (1957) 379.
235 M. K. Saha, P. Ghosh and S. R. Palit, J. Polym. Sci., Part A, 2 (1964) 1365.
236 E. de Witte and E. J. Goethals, Makromol. Chem., 115 (1968) 234.
237 C. G. Overberger, D. E. Baldwin and H. P. Gregor, J. Am. Chem. Soc., 72 (1950) 4864.
238 M. Shima and A. Kotera, J. Polym. Sci., Part A, 1 (1963) 1115.
239 F. Leonard, W. P. Hohenstein and E. Merz., J. Am. Chem. Soc., 70 (1948) 1283.
240 S. L. Aggarwal and F. A. Long, J. Polym. Sci., 11 (1953) 127.
241 T. Takahashi, J. Polym. Sci., Part A-1, 8 (1970) 739.
242 C. G. Overberger and E. Sarlo, J. Polym. Sci., Part A, 2 (1964) 1017.
243 T. Tamikado, J. Polym. Sci., 43 (1960) 489.
244 J. C. Lai, T. D. Rounsefell and C. U. Pittman, Macromolecules, 4 (1971) 155.
245 K. Crauwels and G. Smets, Bull. Soc. Chim. Belg., 59 (1950) 443.
246 F. C. Baines and J. C. Bevington, Polymer, 11 (1970) 647.
247 C. C. Price and J. Zomlefer, J. Am. Chem. Soc., 72 (1950) 14.
248 E. D. Holly, J. Polym. Sci., 36 (1969) 329.
249 B. A. Zaitsev, A. G. Zak, R. G. Luchko and G. A. Shtraikhman, Eur. Polym. J., 8 (1972) 1121.
250 C. C. Price and H. Morita, J. Am. Chem. Soc., 75 (1950) 4747.
251 R. Hart, Makromol. Chem., 49 (1961) 33.

252 A. N. Akopyan, G. E. Krbekyan and E. G. Sinanyan, Vysokomol. Soedin., 5 (1963) 681; Polym. Sci., U.S.S.R., 4 (1963) 1383.
253 I. Simek and L. Komora, Chem. Zvesti, 17 (1963) 757.
254 N. G. Kulkarni, N. Krishnamurti, P. C. Chatterjee and M. A. Sirasamban, Makromol. Chem., 139 (1970) 165.
255 R. Hart, Makromol. Chem., 47 (1961) 143.
256 G. V. Tkachenko, L. V. Stupen, L. P. Kofman and L. Z. Frolova, Zh. Fiz. Khim., 31 (1957) 2676.
257 G. V. Tkachenko, L. V. Stupen, L. P. Kofman and L. A. Karacheva, Zh. Fiz. Khim., 32 (1958) 2492.
258 H. Kamogawa and H. G. Cassidy, J. Polym. Sci., Part A, 1 (1963) 1971.
259 R. J. Kern, J. Polym. Sci., 43 (1960) 549.
260 C. C. Price, B. D. Halpern and S. T. Voong, J. Polym. Sci., 11 (1953) 575.
261 W. Hahn and A. Fischer, Makromol. Chem., 21 (1956) 77.
262 M. Jacob, G. Smets and F. de Schryver, J. Polym. Sci., Part B, 10 (1972) 669.
263 G. Saini, A. Leono and S. Franco, Makromol. Chem., 144 (1971) 235.
264 R. L. Whistler and J. L. Goatley, J. Polym. Sci., 50 (1961) 127.
265 R. Kerber and H. Glaman, Makromol. Chem., 144 (1971) 1.
266 G. Smets and R. Van Gorp, Eur. Polym. J., 5 (1969) 15.
267 A. Guyot and J. Guillot, J. Macromol. Sci., Chem., 2 (1968) 889.
268 N. L. Zutty and F. J. Welch, J. Polym. Sci., 43 (1960) 447.
269 U. Johnsen and K. Kolbe, Makromol. Chem., 116 (1968) 173.
270 F. M. Lewis, F. R. Mayo and W. F. Hulse, J. Am. Chem. Soc., 67 (1945) 1701.
271 R. D. Burkhart and N. L. Zutty, J. Polym. Sci., 57 (1962) 793.
272 N. L. Zutty and R. D. Burkhart, Advances in Chemistry Series, Vol. 34 Polymerization and Condensation Processes, American Chemical Society, Washington, D.C., 1962 p. 52.
273 G. Smets, A. Poot and G. L. Duncan, J. Polym. Sci., 54 (1961) 65.
274 E. J. Gothals and E. de Witte, J. Macromol. Sci., Chem., 4 (1970) 63.
275 M. F. Sorokin, V. K. Latov, Zh. T. Korkishko and Z. A. Kochnova, Plast. Massy, (1963) 11.
276 B. N. Rutovskii and A. M. Shur, Zh. Prikl. Khim., 24 (1951) 1173.
277 M. F. Shostakovskii, G. G. Skvortsova and M. Ya. Samoilova, Vysokomol. Soedin., 5 (1963) 966.
278 G. A. Petrova, G. A. Shtraikman and A. A. Vansheidt, Zh. Fiz. Khim., 33 (1959) 1240.
279 L. E. Coleman and W. S. Durrell, J. Org. Chem., 23 (1958) 1211.
280 H. Yamaguchi and Y. Minoura, J. Polym. Sci., Part A-1, 8 (1970) 1467.
281 M. F. Margaritova and V. A. Raiskaya, Tr., Mosk. Inst. Tonkoi Khim. Tekhnol., 4 (1953) 37.
282 N. Grassie, B. J. D. Torrance, J. D. Fortune and J. D. Genell, Polymer, 6 (1965) 653.
283 J. C. Bevington and D. O. Harris, J. Polym. Sci., Part B, 5 (1967) 799.
284 L. E. Coleman and J. E. Conrady, J. Polym. Sci., 38 (1959) 241.
285 A. M. Rozhkov, Isv. Sib. Otd. Akad. Nauk S.S.S.R., Ser. Khim. Nauk, (1963) 103.
286 S. Nagai, T. Uno and K. Yoshida, Kobunshi, 15 (1958) 550.
287 C. Walling, E. R. Briggs and K. B. Wolfstirn, J. Am. Chem. Soc., 55 (1948) 1543.
288 W. M. Thomas and M. T, O'Shaughnessy, J. Polym. Sci., 11 (1953) 455.
289 T. Otsu and B. Yamada, Makromol. Chem., 111 (1967) 297.
290 R. W. Roth and R. F. Church, J. Polym. Sci., 55 (1961) 41.
291 S. Hashimoto and I. Furukawa, Kobunshi, 22 (1965) 231.
292 S. Hashimoto and I. Furukawa, Kobunshi, 21 (1964) 647.
293 J. F. Bork, D. P. Wyman and L. E. Coleman, J. Appl. Polym. Sci., 7 (1963) 451.
294 P. Agron, T. Alfrey, J. Bohrer, H. Haas and H. Wechsler, J. Polym. Sci., 3 (1948) 157.

295 W. T. Bengough, D. Goldrich and R. A. Young, Eur. Polym. J., 3 (1965) 117.
296 G. M. Burnett, J. M. Pearson and J. D. B. Smith, J. Polym. Sci., Part A, 4 (1966) 2024.
297 A. Kh. Usbekova and V. V. Razumovskii, Vysokomol. Soedin., Ser. A, 14 (1972) 856.
298 C. L. Sandberg and F. A. Bovey, J. Polym. Sci., 15 (1955) 553.
299 F. W. Knobloch, J. Polym. Sci., 25 (1957) 453.
300 R. Rabinowitz, R. Marcus and J. Pellon, J. Polym. Sci., Part A, 2 (1964) 1233.
301 W. A. P. Black, J. A. Colquhoun and E. T. Dewar, Makromol. Chem., 102 (1967) 266.
302 H. Kamogawa and H. G. Cassidy, J. Polym. Sci., Part A, 2 (1964) 2409.
303 C. M. Burns and W. Y. Lau, J. Appl. Polym. Sci., 16 (1972) 1309.
304 A. M. North and A. M. Scallan, Polymer, 5 (1964) 447.
305 G. Saini, A. Leoni and S. Franco, Makromol. Chem., 146 (1971) 165.
306 B. M. Culbertson, E. A. Sedor and R. C. Slagel, Macromolecules, 1 (1968) 254.
307 L. E. Coleman, J. F. Bork and P. D. Wyman, Am. Chem. Soc. Polym. Prepr., 5 (1964) 250.
308 A. T. Coscia, R. L. Kugel and J. Pellon, J. Polym. Sci., 55 (1961) 303.
309 R. Rabinowitz, R. Marcus and J. Pellon, J. Polym. Sci., Part A, 2 (1964) 1241.
310 C. E. Scott and C. C. Price, J. Am. Chem. Soc., 81 (1959) 2672.
311 M. F. Shostakovskii, A. M. Khomutov and A. P. Alimov, Izv. Akad. Nauk S.S.S.R., Otd. Khim. Nauk, (1961) 707.
312 T. Tamikado, T. Sakai and K. Sagisaka, Makromol. Chem., 50 (1961) 244.
313 M. F. Shostakovskii, E. N. Prilezhaeva and J. M. Karavaeva, Vysokomol. Soedin., 1 (1959) 781; J. Polym. Sci., 40 (1959) 598.
314 M. Matsuda, K. Oshima and N. Tokura, J. Polym. Sci., Part A, 2 (1964) 4271.
315 C. C. Price and R. D. Gilbert, J. Polym. Sci., 8 (1952) 580.
316 M. S. Gluckman, M. J. Kampf, J. L. O'Brien, T. G. Fox and R. K. Graham, J. Polym. Sci., 37 (1959) 411.
317 Y. Iwakura, T. Kurosaki, N. Ariya and T. Ito, Makromol. Chem., 97 (1966) 128.
318 M. F. Sorokin, K. A. Lyalyushko, R. A. Dudakova, V. S. Vasil'ev and A. N. Shuvalova, Plast. Massy, (1963) 3.
319 Y. Iwakura and N. Nakabayashi, Makromol. Chem., 66 (1963) 142.
320 M. Kato and H. Kamogawa, J. Polym. Sci., Part A, 4 (1966) 2771.
321 R. Hart and G. Smets, J. Polym. Sci., 5 (1950) 55.
322 Y. Iwakura, M. Sato, T. Tamikado and S. Mimashi, Kobunshi Kagaku, 13 (1956) 125.
323 T. Kagiya, T. Matsuda and Y. Zushi, J. Macromol. Sci., Chem., 5 (1971) 1349.
324 R. H. Wiley and J.-I. Jin, J. Macromol. Sci., Chem., 3 (1969) 835.
325 J. Exner and M. Bohdanecky, Chem. Listy, 48 (1954) 483.
326 T. Higuchi and H. Imoto, Kogyo Kagaku Zasshi, 61 (1958) 1053.
327 A. M. North and D. Postlethwaite, Polymer, 5 (1964) 237.
328 D. C. Blackley and H. W. Melville, Makromol. Chem., 18 (1956) 16.
329 G. van Paesschen and D. Timmerman, Makromol. Chem., 78 (1964) 112.
330 G. Saini, A. Leoni and S. Franco, Makromol. Chem., 147 (1971) 213.
331 E. Müller, K. Dinges and W. Graulich, Makromol. Chem., 57 (1962) 27.
332 A. V. Ryabov, Yu. D. Semchikov and N. N. Slavnitskaya, Dokl. Akad. Nauk S.S.S.R., 145 (1962) 822.
333 E. P. Bonsall, L. Valentine and H. W. Melville, Trans. Faraday Soc., 48 (1952) 763.
334 J. C. Bevington, D. O. Harris and M. Johnson, Eur. Polym. J., 1 (1965) 235.
335 N. Grassie, J. Thomson and A. Walton, Eur. Polym. J., 4 (1968) 139.
336 J. C. Bevington and B. W. Malpass, Trans. Faraday Soc., 60 (1964) 1268.
337 M. Izu and K. F. O'Driscoll, J. Polym. Sci., Part A-1, 8 (1970) 1687.
338 S. Murahashi, S. Nozakura, A. Umehara and K. Obata, Kobunshi, 21 (1964) 625.

339 G. Smets and A. Reckers, Rec. Trav. Chim. Pays-Bas, 68 (1949) 983.
340 T. Alfrey and W. H. Ebelke, J. Am. Chem. Soc., 71 (1949) 3235.
341 A. Kh. Usbekova and V. V. Razumovskii, quoted in ref. 249.
342 F. Ida, K. Uemura and S. Abe, Kagaku To Kogyo (Osaka), 38 (1964) 215.
343 J. K. Stille and L. D. Gother, Macromolecules, 2 (1969) 468.
344 M. Gordon, B. M. Grieveson and I. D. McMillan, J. Polym. Sci., 18 (1955) 497.
345 C. U. Pittman, P. J. Grube, O. E. Ayers, S. P. McManus, M. D. Rausch and G. A. Moser, J. Polym. Sci., Part A-1, 10 (1972) 379.
346 Y. Minoura, Y. Suzuki, Y. Sakanaka and H. Doi, J. Polym. Sci., Part A, 4 (1966) 2757.
347 Y. Minoura and Y. Sakanaka, J. Polym. Sci., Part A-1, 7 (1969) 3287.
348 S. Murahashi, S. Nozakura, A. Umehara and F. Nagoshi, Kobunshi, 22 (1965) 451.
349 J. N. Atherton and A. M. North, Trans. Faraday Soc., 58 (1962) 2049.
350 (Szu-Kwei) Min and Chen Ho Chu, Hua Hsueh Hsueh Pao, 23 (1957) 262.
351 G. G. Cameron, G. P. Karr and D. A. Russell, Eur. Polym. J., 7 (1971) 1029.
352 J. C. Bevington and M. Johnson, Eur. Polym. J., 4 (1968) 668.
353 G. Blauer and L. Goldstein, J. Polym. Sci., 25 (1957) 19.
354 T. Alfrey and S. L. Kapur, J. Polym. Sci., 4 (1949) 215.
355 M. F. Shostakovskii, E. N. Prilezhaeva and J. M. Karavaeva, J. Polym. Sci., 40 (1959) 598.
356 R. Hart and D. Timmerman, Bull. Soc. Chim. Belg., 67 (1958) 123.
357 K. Hayashi and G. Smets, J. Polym. Sci., 27 (1958) 275.
358 V. L. Folt, Canadian Patent 509,259 (1955).
359 E. Husemann, quoted in ref. 2. See also G. Greber and E. Reese, Makromol. Chem., 55 (1962) 96.
360 R. Hart and D. Timmerman, Makromol. Chem., 31 (1959) 223.
361 S. Fujii and K. Maeda, J. Appl. Polym. Sci., 16 (1972) 1331.
362 W. J. Burlant and D. H. Green, J. Polym. Sci., 31 (1958) 227.
363 V. D. Bezuglyi, T. A. Alekseyeva, L. I. Dmitriyevskaya and V. I. Grigor'eva, Vysokomol. Soedin., 9 (1967) 889; Polym. Sci., U.S.S.R., 9 (1967) 995.
364 R. G. Fordyce and G. E. Ham, J. Am. Chem. Soc., 73 (1951) 1186.
365 C. C. Price and J. G. Walsh, J. Polym. Sci., 6 (1951) 239.
366 M. Tanaka, S. Asai and S. Takeya, Kogyo Kagaku Zasshi, 62 (1959) 1786.
367 C. G. Overberger and F. W. Michelotti, J. Am. Chem. Soc., 80 (1958) 988.
368 A. B. Davankov, L. B. Zubakova and A. A. Gurov, Vysokomol. Soedin., 6 (1964) 237; Polym. Sci., U.S.S.R., 6 (1964) 274.
369 M. E. Mat'kova and S. S. Spasskii, Vysokomol. Soedin., 3 (1961) 93.
370 M. E. Mat'kova and S. S. Spasskii, Vysokomol. Soedin., 2 (1960) 879.
371 K. Ueberreiter and W. Krull, Z. Phys. Chem. (Frankfurt am Main), 12 (1957) 303.
372 A. Leoni and S. Franco, Macromolecules, 4 (1971) 355.
373 R. Kerber, Makromol. Chem., 96 (1966) 30.
374 M. Goodstein, quoted in ref. 189, p. 35.
375 A. Chapiro and A.-M. Jendrychowska-Bonamour, J. Polym. Sci., Part C, 1 (1963) 1211.
376 A. Guyot and J. Guillot, J. Macromol. Sci., Chem., 1 (1967) 793.
377 T. Alfrey and J. G. Harrison, J. Am. Chem. Soc., 68 (1946) 299.
378 T. Otsu, T. Ito and M. Imoto, J. Polym. Sci., Part B, 3 (1965) 113.
379 M. Kreisel, U. Garbatski and D. H. Kohn, J. Polym. Sci., Part A, 2 (1964) 105.
380 S. H. Ranel and D. H. Kohn, J. Polym. Sci., Part A-1, 7 (1969) 2209.
381 H. Ringsdorf and G. Greber, Makromol. Chem., 31 (1959) 27.
382 M. Akiyama, Y. Yanagisawa and M. Okawara, J. Polym. Sci., Part A-1, 6 (1969) 1905.
383 I. Iwakura, F. Toda and H. Suzuki, J. Polym. Sci., Part A-1, 5 (1967) 1599.

384 L. F. Vanderburgh and C. E. Brockway, J. Polym. Sci., Part A, 3 (1965) 575.
385 R. H. Wiley and G. L. Mayberry, J. Polym. Sci., Part A, 1 (1963) 217.
386 M. Imoto, T. Otsu, K. Tsuda and T. Ito, J. Polym. Sci., Part A, 2 (1964) 1407.
387 D. Braun and H.-G. Keppler, Makromol. Chem., 78 (1964) 100.
388 W. Kern and D. Braun, Makromol. Chem., 27 (1958) 23.
389 Polymer Corporation Ltd., Sarnia, Canada, quoted in ref. 2.
390 E. J. Arlman and H. W. Melville, Proc. R. Soc. London, Ser. A, 203 (1950) 301.
391 J. H. Bradbury and H. W. Melville, Proc. R. Soc. London, Ser. A, 222 (1954) 456.
392 G. M. Burnett, P. Evans and H. W. Melville, Trans. Faraday Soc., 49 (1953) 1096.
393 J. W. Vanderhoff, in E. H. Riddle (Ed.), Monomeric Acrylic Esters, Reinhold, New York, 1954, p. 94, ref. 154.
394 J. P. Fischer and W. Lüders, Makromol. Chem., 155 (1972) 239.
395 H.-G. Elias and W. Lengweiller, Makromol. Chem., 113 (1968) 155.
396 B. Yamada and T. Otsu, J. Macromol. Sci., Chem., 3 (1969) 1551.
397 A. R. Gantmakher, Yu. L. Spirin and S. S. Medvedev, Vysokomol. Soedin., 1 (1959) 1526.
398 G. Greber and J. Tölle, Makromol. Chem., 67 (1963) 98.
399 F. M. Lewis, C. Walling, W. Cummings, E. R. Briggs and F. R. Mayo, J. Am. Chem. Soc., 70 (1948) 1519.
400 C. A. Barson, J. Polym. Sci., 62 (1962) S128.
401 E. T. Borrows, R. N. Haward, J. Porges and J. Street, J. Appl. Chem., 5 (1955) 379.
402 C. S. Marvel, V. Sziraky and J. P. Economy, J. Org. Chem., 21 (1956) 1314.
403 D. G. L. James and T. Ogawa, J. Polym. Sci., Part B, 2 (1964) 991.
404 Y. Minoura, T. Tadokoro and Y. Suzuki, J. Polym. Sci., Part A-1, 5 (1967) 2641.
405 A. I. Volkova, P. A. Yel'tsova, M. M. Koton and A. V. Mikhailova, Vysokomol. Soedin., 8 (1966) 1997; Polym. Sci., U.S.S.R., 8 (1966) 2206.
406 A. Winston and F. L. Hamb, J. Polym. Sci., Part A, 3 (1965) 583.
407 D. Braun and T.-O. Ahn, Kolloid Z., 188 (1963) 1.
408 F. Sakurai and C. Huang, Kogyo Kagaku Zasshi, 61 (1958) 1629.
409 K. Uno, M. Ohara and H. G. Cassidy, J. Polym. Sci., Part A-1, 6 (1968) 2729.
410 R. C. P. Cubbon and J. D. B. Smith, Polymer, 10 (1969) 489.
411 G. S. Kolesnikov, E. F. Rodionova and I. G. Safaralieva, Izv. Akad. Nauk S.S.S.R., Ser. Khim., (1963) 2028.
412 J. W. Breitenbach, A. Schindler and C. Pflug, Monatsh. Chem., 81 (1950) 21.
413 T. Alfrey and S. Greenburg, J. Polym. Sci., 3 (1948) 297.
414 Y. Iwakura, F. Toda, M. Kosugi and Y. Torii, J. Polym. Sci., Part A-1, 9 (1971) 2423.
415 T. Alfrey, E. Merz and H. Mark, J. Polym. Sci., 1 (1946) 37.
416 R. Hess, quoted in ref. 189 pp. 36, 37.
417 R. H. Wiley and B. Davis, J. Polym. Sci., 62 (1962) S140.
418 J. C. Bevington, M. Johnson and J. P. Sheen, Eur. Polym. J., 7 (1971) 1147.
419 H. Ringsdorf, N. Weinshenker and C. G. Overberger, Makromol. Chem., 64 (1963) 126.
420 T. Tsuda and Y. Yamashita, Kogyo Kagaku Zasshi, 65 (1962) 811.
421 C. L. Arcus and R. J. S. Matthews, J. Chem. Soc., (1956) 4607.
422 R. H. Wiley, L. K. Heidemann and B. Davis, J. Polym. Sci., Part B, 1 (1963) 521.
423 J. P. Fischer, Makromol. Chem., 155 (1972) 227.
424 R. H. Wiley and E. E. Sale, J. Polym. Sci., 42 (1960) 491.
425 R. H. Wiley, S. P. Rao, J.-I. Jin and K. S. Kim, J. Macromol. Sci., Chem., 4 (1970) 1453.
426 R. H. Wiley, W. K. Matthews and K. F. O'Driscoll, J. Macromol. Sci., Chem., 1 (1967) 503.
427 R. H. Wiley and B. Davis, J. Polym. Sci., Part B, 1 (1963) 463.

428 R. G. Fordyce and G. E. Ham, J. Am. Chem. Soc., 69 (1947) 695.
429 G. Akazome, S. Sakai and K. Murai, Kobunshi Kagaku, 17 (1960) 482.
430 N. N. Rozovskaya and A. D. Abkin, Lakokras. Mater. Ikh Primen., (1961) 9.
431 T. Kimura and K. Yoshida, Kagaku To Kogyo (Osaka), 29 (1955) 43, 288.
432 V. D. Bezuglyi, T. A. Alekseeva, L. I. Dmitrievskaya, A. V. Chernobai and L. P. Kruglyak, Vysokomol. Soedin., 6 (1964) 125.
433 D. T. Mowry, U.S. Patent 2,417,607 (1947), values calculated by F. R. Mayo and C. Walling, ref. 21.
434 L. Rodriguez, Makromol. Chem., 12 (1954) 110.
435 M. F. Sorokin and I. M. Kochnov, Plast. Massy, (1963) 7.
436 M. F. Sorokin and I. M. Kochnov, Vysokomol. Soedin., 6 (1964) 798.
437 J. A. Simms, J. Appl. Polm. Sci., 5 (1961) 58.
438 M. F. Sorokin and I. M. Kochnov, Polym. Sci., U.S.S.R., 6 (1964) 877.
439 Y. Matsubara, M. Yoshihara and T. Maeshima, Kogyo Kagaku Zasshi, 74 (1971) 2163.
440 D. Braun, T. Ahn and W. Kern, Makromol. Chem., 53 (1962) 154.
441 M. P. Zverev and M. F. Margaritova, Ukr. Khim. Zh., 24 (1958) 626.
442 R. D. Gilbert and H. L. Williams, J. Am. Chem. Soc., 74 (1952) 4114.
443 R. H. Wiley and B. Davis, J. Polym. Sci., Part A, 1 (1963) 2819.
444 Y. Iwakura, F. Toda, N. Kusakawa and H. Suzuki, J. Polym. Sci., Part B, 6 (1968) 5.
445 Y. Iwakura, F. Toda, Y. Torii and R. Sekii, J. Polym. Sci., Part A-1, 6 (1968) 2681.
446 R. Hart and A. E. van Dormael, Bull. Soc. Chim. Belg. 65 (1956) 571.
447 J. Drougas and R. L. Guile, J. Polym. Sci., 55 (1961) 297.
448 V. V. Shneiderova, Lakrokras. Mater. Ikh Primen., (1962) 33.
449 G. Enomoto, Y. Ogo and T. Imoto, Makromol. Chem., 138 (1970) 19.
450 T. L. Ang and H. J. Harwood, Polym. Prepr. Am. Chem. Soc., Div. Polym. Chem., 5 (1964) 306.
451 K. Noma, M. Niwa and K. Iwasaki, Kobunshi, 20 (1963) 646.
452 M. B. Huglin, Polymer, 3 (1962) 335.
453 C. H. Bamford and W. G. Barb, Discuss. Faraday Soc., 14 (1953) 208.
454 T. Alfrey and E. Lavin, J. Am. Chem. Soc., 67 (1945) 2044.
455 A. Rudin and R. G. Yule, J. Polym. Sci., Part A-1, 9 (1971) 3009.
456 J. E. Mulvaney, J. G. Dillon and J. L. Laverty, J. Polym. Sci., Part A-1, 6 (1968) 1841.
457 O. Wichterle and V. Gregor, J. Polym. Sci., 34 (1959) 309.
458 T. M. Frunze, V. V. Korshak, Ye. L. Baranov and B. V. Lokshin, Vysokomol. Soedin., 8 (1966) 455; Polym. Sci., U.S.S.R., 8 (1966) 498.
459 G. S. Kolensnikov, S. L. Davydova and N. V. Klimentova, Vysokomol. Soedin., 4 (1962) 1098.
460 K. W. Doak and C. Walling, quoted in ref. 461.
461 F. R. Mayo, F. M. Lewis and C. Walling, J. Am. Chem. Soc., 70 (1948) 1529.
462 H. W. Melville and L. Valentine, Trans. Faraday Soc., 51 (1955) 1474.
463 K. Ito and Y. Yamashita, J. Polym. Sci., Part B, 3 (1965), 637.
464 K. Chikanishi and T. Tsuruta, Makromol. Chem., 81 (1965) 211.
465 A. D. Jenkins and M. G. Rayner, Eur. Polym. J., 8 (1972) 221.
466 E. J. Arlman, H. W. Melville and L. Valentine, Rec. Trav. Chim. Pays-Bas, 68 (1949) 945.
467 F. T. Wall, R. E. Florin and C. J. Delbecq, J. Am. Chem. Soc., 72 (1950) 4769.
468 A. F. Nikolayev and V. M. Gal'perin, Vysokomol. Soedin., 9 (1967) 2467; Polymer Sci., U.S.S.R., 9 (1967) 2793.
469 H. Asai and T. Imoto, J. Polym. Sci., Part B, 2 (1964) 553.

470 T. Ito and T. Otsu, J. Macromol. Sci., Chem., 3 (1969) 197.
471 V. E. Meyer, J. Polym. Sci., Part A, 4 (1966) 2819.
472 K. Ito and Y. Yamashita, J. Polym. Sci., Part B, 3 (1965) 625.
473 R. H. Wiley and E. E. Sale, J. Polym. Sci., 42 (1960) 479.
474 R. H. Wiley and B. Davis, J. Polym. Sci., 62 (1962) S132.
475 F. R. Mayo and F. M. Lewis, J. Am. Chem. Soc., 66 (1944) 1594.
476 V. E. Meyer, J. Polym. Sci., Part A-1, 5 (1967) 1289.
477 R. H. Wiley and B. Davis, J. Polym. Sci., 46 (1960) 423.
478 A. V. Golubeva, N. F. Usmanova and A. A. Vansheidt, J. Polym. Sci., 52 (1961) 63.
479 E. C. Chapin, P. C. Hamm and R. G. Fordyce quoted in ref. 2.
480 J. P. Fischer, Makromol. Chem., 155 (1972) 211.
481 I. Sakurada, Chem. High Polym. (Tokyo), 18 (1961) 496.
482 V. G. Ostroverkhov, I. S. Vakarchuk and V. G. Sinyavskii, Vysokomol. Soedin., 3 (1961) 1197.
483 C. C. Price and R. D. Gilbert, J. Am. Chem. Soc., 74 (1952) 2073.
484 M. Kreisel, U. Garbatski and D. H. Kohn, J. Polym. Sci., Part B, 2 (1964) 81.
485 N. N. Rozovskaya and A. D. Abkin, Lakokras. Mater. Ikh Primen., 1 (1961) 9.
486 K. Murata and A. Terada, J. Polym. Sci., Part A-1, 6 (1968) 2945.
487 A. V. Tokarev and S. S. Spasskii, Zh. Fiz. Khim., 33 (1959) 554.
488 M. A. Bulatov and S. S. Spasskii, Vysokomol. Soedin., 2 (1960) 658.
489 G. G. Cameron and N. Grassie, Makromol. Chem., 51 (1962) 130.
490 R. H. Wiley and W. A. Trinler, J. Polym. Sci., 28 (1958) 163.
491 K. W. Doak, J. Am. Chem. Soc., 70 (1948) 1525.
492 H. S. Kolesnikov, A. P. Suprun, T. A. Soboleva, V. A. Ershova and V. B. Bondareva, Vysokomol. Soedin., 4 (1962) 743; Polym. Sci., U.S.S.R., 3 (1962) 977.
493 C. W. Lewis and D. W. Lewis, J. Polym. Sci., 36 (1959) 325.
494 S. R. Sandler, J. Dannin and K. C. Tsau, J. Polym. Sci., Part A, 3 (1965) 3199.
495 R. H. Wiley and T.-O. Ahn, J. Polym. Sci., Part A-1, 6 (1968) 1293.
496 S. Yui-u, S. S. Skorokhodov and A. A. Vansheidt, Vysokomol. Soedin., 6 (1964) 1291; Polym. Sci., U.S.S.R., 6 (1964) 1426.
497 S. Murahashi, S. Nozakura, A. Umehara and F. Nagoshi, Kobunshi, 22 (1965) 451.
498 D. Katz, J. Polym. Sci., Part A, 1 (1963) 1635.
499 S. Banerjee and M. S. Muthana, J. Polym. Sci., 35 (1959) 292.
500 D. Braun, W. Neumann and J. Faust, Makromol. Chem., 85 (1965) 143.
501 M. Imoto, T. Otsu and W. Fukuda, J. Polym. Sci., Part B, 1 (1963) 225.
502 G. V. Tkachenko, V. S. Etlis, L. V. Stupen and L. P. Kofman, Zh. Fiz. Khim., 33 (1959) 25.
503 M. F. Shostakovskii, A. M. Khomutov and A. P. Alimov, Izv. Akad. Nauk S.S.S.R., Ser. Khim. (1963) 1839.
504 B. M. Culbertson, S. Dietz and R. D. Wilson, J. Polym. Sci., Part A-1, 9 (1971) 2727.
505 H. S. Kolesnikov, A. S. Tevlina, I. I. Grandberg, S. Ye. Vasyukov and G. I. Sharova, Vysokomol. Soedin., 9 (1967) 2492; Polym. Sci., U.S.S.R., 9 (1967) 2820.
506 V. D. Bezuglyi, T. A. Alekseyeva, L. I. Dmitriyevskaya, A. V. Chernobai and L. P. Kruglyak, Vysokomol. Soedin., 6 (1964) 125; Polym. Sci., U.S.S.R., 6 (1964) 148.
507 D. Braun, K. H. Tio and W. Neumann, Makromol. Chem., 123 (1969) 29.
508 F. Miller, Canadian Patent, 516,532 (1955).
509 G. N. Freidlin and M. N. Adamova, Vysokomol. Soedin., 9 (1967) 1531; Polym. Sci., U.S.S.R., 9 (1967) 1718.

510 S. Loshaek and E. Broderick, J. Polym. Sci., 39 (1959) 241.
511 D. Braun, H. Daimon and G. Becker, Makromol. Chem., 62 (1963) 183.
512 G. Greber and E. Reese, Makromol. Chem., 55 (1962) 96.
513 A. F. Nikolaev and M. N. Tereschenko, Vysokomol. Soedin., 6 (1964) 379; Polym. Sci., U.S.S.R., 6 (1964) 419.
514 A. F. Nikolaev and M. N. Tereschenko, Vysokomol. Soedin., 6 (1964) 379.
515 A. V. Ryabov, Yu. D. Semchikov and V. N. Vakhrusheva, Tr. Khim. Khim. Tekhnol., (1963) 188.
516 R. M. Fuoss and G. I. Cathers, J. Polym. Sci., 4 (1949) 97.
517 C. L. Schilling and J. E. Mulvaney, Macromolecules, 1 (1968) 445.
518 D. W. Brown and L. A. Wall, J. Polym. Sci., Part A-1, 6 (1968) 1367.
519 M. Yamada and I. Takase, Kobunshi, 22 (1965) 626.
520 I. Sakurada and G. Takahashi, Kobunshi, 11 (1954) 286.
521 M. Sadamichi and K. Noro, J. Macromol. Sci., Chem., 3 (1969) 845.
522 I. Sakurada, G. Takahashi and H. Mata, Kobunshi, 12 (1955) 362.
523 M. Kinoshita, T. Irie and M. Imoto, Makromol. Chem., 110 (1967) 47.
524 A. M. Khomutov, Vysokomol. Soedin., 5 (1963) 1121.
525 A. Mitsutani and M. Yano, Kogyo Kagaku Zasshi, 67 (1964) 935.
526 G. E. Ham. J. Polym. Sci., 24 (1954) 87.
527 S. N. Ushakov and E. M. Lavrent'eva, Zh. Prikl. Khim., 31 (1958) 1686.
528 S. N. Ushakov and L. B. Trukhmanova, Vysokomol. Soedin., 1 (1959) 1754.
529 S. D. Ross, M. Markarian, H. H. Young and M. Nazzewski, J. Am. Chem. Soc., 72 (1950) 1133.
530 G. Takahashi, Kobunshi, 14 (1957) 151.
531 A. A. Durgaryan, A. S. Grigoryan and O. A. Chaltykyan, Izv. Akad. Nauk Arm. S.S.R., Khim. Nauki, 15 (1962) 455.
532. T. L. Dawson, R. D. Ludenberg and F. J. Welch, J. Polym. Sci., Part A-1, 7 (1969) 173.
533 M. Yamada and I. Takase, Chem. High Polym. (Tokyo), 18 (1961) 85.
534 S. G. Matsoyan, M. G. Voskanyan and A. A. Cholakyan, Vysokomol. Soedin., 5 (1963) 1035.
535 G. Akazome, S. Sakai and K. Murai, Kobunshi, 17 (1960) 449.
536 B. Erussalinsky, N. Tumarkin, F. Duntoff, S. Lyubetzky and A. Goldenberg, Makromol. Chem., 104 (1967) 288.
537 M. Takebayashi, and Y. Ito, Bull. Chem. Soc. Jpn., 29 (1956) 287.
538 E. Imoto and H. Horiuchi, Kobunshi, 8 (1951) 463.
539 E. W. Moffett and R. E. Smith, U.S. Patent 2,356,871 (1944), values calculated by F. R. Mayo and C. Walling, ref. 21.
540 C. Pichot, J. Guillot and A. Guyot, J. Macromol. Sci., Chem., 5 (1971) 753.
541 J. W. L. Fordham, G. H. McCain and L. E. Alexander, J. Polym. Sci., 39 (1959) 335.
542 H. L. Marder and C. Schuerch, J. Polym. Sci., 44 (1960) 129.
543 H. C. Haas and N. W. Schuler, J. Polym. Sci., 31 (1958) 237.
544 C. J. Kurian and M. S. Muthana, Makromol. Chem., 29 (1959) 26.
545 A. Adicoff and A. Buselli, J. Polym. Sci., 21 (1956) 340.
546 W. S. Port, E. F. Jordan, J. E. Hansen and D. Swern, J. Polym. Sci., 9 (1952) 493.
547 A. F. Nikolaev, S. N. Ushakov, L. P. Vishnevetskaya, N. A. Voronova and E. I. Rodina, Vysokomol. Soedin., 4 (1962) 1053.
548 S. Imoto, J. Ukida and T. Kominami, Kobunshi, 14 (1957) 101.
549 S. Okamura and K. Uno, Kobunshi, 8 (1951) 467.
550 J. W. Breitenbach and H. Edelhauser, Ric. Sci., 25 (1955) 242.
551 D. J. Kahn and H. Horowitz, J. Polym. Sci., 54 (1961) 363.
552 A. F. Nikolayev, S. N. Ushakov and L. S. Mishkileyeva, Vysokomol. Soedin., 6 (1964) 287; Polym. Sci., U.S.S.R., 6 (1964) 331.

553 A. F. Nikolaev, S. N. Ushakov and L. S. Mishkileyeva, Vysokomol. Soedin., 6 (1964) 287.
554 H. C. Haas, E. S. Emerson and N. W. Schuler, J. Polym. Sci., 22 (1956) 291.
555 L. Ghosez and G. Smets, J. Polym. Sci., 35 (1959) 215.
556 F. P. Sidel'kovskaya, M. F. Shostakovskii, F. Ibragimov and M. A. Askarov, Vysokomol. Soedin., 6 (1964) 1585; Polym. Sci., U.S.S.R., 6 (1964) 1757.
557 M. Imoto and S. Shimizu, Kobunshi, 18 (1961) 747.
558 Z. Janovic, C. S. Marvel, M. J. Diamond, D. J. Kertesy and G. Fuller, J. Polym. Sci., Part A-1, 9 (1971) 2639.
559 A. G. Kronman, I. V. Pasmanik, B. I. Fedoseyev and V. A. Kargin, Vysokomol. Soedin., 9 (1967) 2503; Polym. Sci., U.S.S.R., 9 (1967) 2834.
560 R. C. L. Chow and C. S. Marvel, J. Polym. Sci., Part A-1, 6 (1968) 1515.
561 G. S. Park, D. G. Smith and A. A. Rich, Eur. Polym. J., 7 (1971) 335.
562 G. V. Tkachenko, P. H. Khomikovskii, A. D. Abkin and S. S. Medvedev, Zh. Fiz. Khim., 31 (1957) 242.
563 N. Kliman and M. Lazar, Chem. Prum., 9 (1959) 668.
564 T. Kimura and K. Yoshida, Kagaku To Kogyo (Osaka), 28 (1954) 158.
565 S. Nagai and K. Yoshida, Kobunshi, 17 (1960) 77.
566 G. Akazome, S. Sakai, Y. Choshi and K. Murai, Kobunshi, 17 (1960) 478.
567 B. Erussalimsky, F. Duntoff and N. Tumarkin, Makromol, Chem., 66 (1963) 205.
568 F. I. Duntov, A. L. Gol'denberg, M. A. Litvonova and B. L. Yerusalimskii, Vysokomol. Soedin., 9 (1967) 1920; Polym. Sci., U.S.S.R., 9 (1967) 2166.
569 A. P. Sheinker, M. K. Yakovleva, B. V. Kristal'nyi and A. D. Abkin, Dokl. Akad. Nauk S.S.S.R., 124 (1959) 632; Resins, Rubbers, Plastics, 13 (1959) 659.
570 S.-P. Chang, T. K. Miwa and W. H. Tallent, J. Polym. Sci., Part A-1, 7 (1969) 471.
571 R. C. Reinhardt, Ind. Eng. Chem., 35 (1943) 422, calculated by F. R. Mayo and C. Walling ref. 21.
572 W. I. Bengough and R. G. W. Norrish, Proc. R. Soc. London, Ser. A, 218 (1953) 155.
573 S. Enomoto, J. Polym. Sci., 55 (1961) 95.
574 V. L. Folt, Canadian Patent 510,354 (1955).
575 R. Büring and H.-J. Kotzsch, Angew. Makromol. Chem., 13 (1970) 89.
576 C. G. Overberger, H. Biletch and R. G. Nickerson, J. Polym. Sci., 27 (1958) 381.
577 Z. Machacek, Chem. Listy, 48 (1954) 477.
578 G. Dolgin and P. Gordon, quoted in ref. 189.
579 O. Wichterle and J. Zelinka, Chem. Listy, 51 (1957) 2146.
580 E. C. Britton, C. W. Davis and F. L. Taylor, U.S. Patent 2,160,940 (1939), calculated by F. R. Mayo and C. Walling quoted in ref. 21.
581 C. S. Marvel and E. B. Mano, J. Polym. Sci., 31 (1958) 165.
582 S. Kamenar, I. Simek and E. Regensbogenova, Chem., Zvesti, 14 (1960) 581.
583 A. Ardis, Canadian Patent 516,315 (1955).
584 C. C. Price and C. E. Greene, J. Polym. Sci., 6 (1951) 111.
585 F. P. Sidel'kovskaya, M. A. Askarov and F. Ibragimov, Vysokomol. Soedin., 6 (1964) 1810; Polym. Sci., U.S.S.R., 6 (1964) 2005.
586 C. U. Pittman and P. L. Grube, J. Polym. Sci., Part A-1, 9 (1971) 3175.
587 R. H. Wiley and J.-I. Jin, J. Macromol. Sci., Chem., 2 (1968) 1097.
588 G, E. Ham, J. Polym. Sci., 24 (1957) 349.
589 G. E. Ham. J. Polym. Sci., 45 (1960) 183.
590 W. G. Barb, J. Polym. Sci., 11 (1953) 117.
591 G. E. Ham, J. Polym. Sci., 45 (1960) 177.
592 G. E. Ham, J. Polym. Sci., Part B, 3 (1965) 187.
593 R. Simha and L. A. Wall, J. Res. Nat. Bur. Stand., 41 (1948) 521.
594 H. Hopff and P. Schlumbom, Makromol. Chem., 43 (1961) 173.
595 S. N. Ushakov, S. P. Mitsengendler and B. M. Polyatskina, Zh. Prikl. Khim., 24 (1951) 289.

596 S. N. Ushakov, S. P. Mitsengendler and B. M. Polyatskina, Zh. Prikl. Khim, 23 (1950) 512.
597 S. N. Ushakov, S. P. Mitsengendler and B. M. Polyatskina, Khim. Fiz. Khim. Vysokomol. Soedin. Dokl. 7-ya Konf. Vysokomol. Soedin., (1952) 59.
598 M. Vrancken and G. Smets, Makromol. Chem., 30 (1959) 197.
599 A. M. Khomutov, Vysokomol. Soedin., 5 (1963) 1121; Polym. Sci., U.S.S.R., 5 (1964) 181.
600 G. Hardy, J. Varga, K. Nytrai, I. Tsajlik and L. Zubonyai, Vysokomol. Soedin., 6 (1964) 758; Polym. Sci., U.S.S.R., 6 (1964) 832.
601 S. N. Ushakov and A. F. Nikolaev, Izv. Akad. Nauk S.S.S.R., Otd. Khim. Nauk, (1956) 83.
602 S. N. Ushakov and A. F. Nikolaev, Bull. Acad. Sci. U.S.S.R., Div. Chem. Sci., 1 (1956) 79.
603 C. S. Marvel, D. G. Jones, T. W. Mastin and G. L. Schertz, J. Am. Chem. Soc., 64 (1942) 2356.
604 N. Grassie, I. C. McNeill and I. F. McLaren, J. Polym. Sci., Part B, 3 (1965) 897.
605 W. Ring, Angew. Makromol. Chem., 21 (1972) 149.

Chapter 5

The Reactivity of Monomers and Radicals

N. C. BILLINGHAM and A. D. JENKINS

1. Introduction

1.1 BASIC POLYMERIZATION

The free-radical addition polymerization of unsaturated molecules by the action of a radical-generating initiator is a very powerful method for the production of a wide variety of polymers, especially those derived from vinyl, vinylidene and diene monomers; it is the most widely used industrial method for the manufacture of addition polymers. Radical polymerizations are typical chain processes, and indeed, our understanding of radical chain reactions is probably greatest in the field of polymerization, since polymer-forming reactions normally occur at rates which are readily measured experimentally; the study of reaction rates, molecular weight and copolymer composition usually involves rather simple techniques. The kinetic scheme usually applicable to the radical propagated polymerization of a single monomer may be formulated as follows

(i) *Initiation*

$$\left.\begin{array}{lcl} \text{Initiator} & \xrightarrow{k_D} & 2R\cdot \\ R + M & \xrightarrow{k_i} & X_1\cdot \end{array}\right\} \text{Overall rate} = I$$

In the initiation step, *primary radicals* are generated by some means, usually involving the thermal or photochemical cleavage of an added initiator, such as benzoyl peroxide or azo-bis-isobutyronitrile; radicals may also be generated by direct thermal, photochemical or radiation induced homolysis of the monomer, although such reactions are much less common. Where radicals are generated from an added initiator, the rate coefficient for initiator decomposition is denoted by k_D. Following their generation, some fraction of the primary radicals will react with the monomer to produce a new radical X_1, containing one monomer unit; the rate coefficient for this process is denoted by k_i and the overall rate of generation of X_1 radicals is given the symbol I. The fraction, f, of primary radicals which react with the monomer is termed the *initiator efficiency*.

(ii) *Propagation*

$$X_n\cdot + M \xrightarrow{k_p} X_{n+1}^{\cdot}$$

Chain propagation proceeds by the rapid addition of monomer molecules to the active radicals generated in the initiation step. Each propagation step increases the polymer chain length by one unit without destroying the radical activity. The propagation rate coefficient is denoted by k_p; the nature of this coefficient is discussed below.

(iii) *Termination*

$$X_n\cdot + X_m\cdot \xrightarrow{k_t} P_{n+m}$$

$$\text{or} \xrightarrow{k_t'} P_n + P_m$$

Termination of growing chains involves a bimolecular reaction of two polymeric radicals. Two termination processes may occur, *combination* (rate coefficient k_t) in which two radicals couple to form a single polymer chain, and *disproportionation* (rate coefficient k_t') in which one radical transfers a hydrogen atom to a second, producing two polymer molecules.

From the foregoing exposition it can readily be seen that both the rate of polymerization and the molecular weight of the polymer are governed by the relative rates of the elementary steps which constitute the overall reaction; the rates of the elementary reactions are, in their turn, determined by the reactivity of the radicals involved. Thus, the efficiency of initiation is determined by the reactivity of the primary radical towards monomer, relative to its reactivity towards any other species (including other primary radicals) present; efficient initiation occurs only when the primary radical is most reactive towards the monomer. Furthermore, both rate and molecular weight are governed by the competition between propagation and termination rates and hence by the relative reactivity of the $X_n\cdot$ radical towards monomer and towards other radicals.

In order to place the kinetics of polymerization on a quantitative basis it is necessary to make certain simplifying assumptions, of which the most important is the assumption that the value, k_p, for the rate coefficient of the propagation reaction is independent of the chain length n. Implicit in this statement is one of the fundamental assumptions of almost all treatments of polymerization behaviour, i.e. that all radicals have the same reactivity irrespective of chain length; the justification for this assumption rests mainly on the fact that kinetic equations derived by its use are normally satisfactorily applicable to experimental results. On this basis, separate discussion of the reactivity of polymeric radicals in isolation from their non-polymeric counterparts arises purely as a matter of convenience; size is of no special importance and the segregation of large and small radicals occurs largely because they have been studied in different ways and in reaction with different substrates.

Application of the stationary-state assumption to the simple reaction scheme [1] allows the following expressions for the rate of polymerization R_p and the degree of polymerization, $\bar{P}$, to be derived.

$$R_p = k_p[M]\left[\frac{I}{(k_t + k_t')}\right]^{1/2} \tag{1}$$

$$\bar{P} = k_p[M]\,[I(k_t' + k_t/2)]^{-1/2} \tag{2}$$

From these equations it follows that, for a given monomer concentration, [M], and initiation rate, the rate of polymerization and the molecular weight of the product are dependent upon the value of $k_p/(k_t + k_t')^{1/2}$, which quantity may be regarded as a measure of the relative reactivity of the chain carrying radical towards propagation and termination. Although $k_p/(k_t + k_t')^{1/2}$ is a most useful quantity for practical evaluation of polymerizability, too many factors are involved in determining the individual rate coefficients for the overall quantity to be simply related to radical structure in any meaningful way. Methods are available for the determination of absolute velocity coefficients for polymerization processes [2] but these are usually difficult to apply and of somewhat limited accuracy; relatively few such determinations have been made. Fortunately a great deal of useful information is available in the form of ratios of rate coefficients from studies of two additional reactions which may occur in a polymerization process and which will now be considered in some detail.

1.2 CHAIN TRANSFER REACTIONS

The general chain transfer process may be formulated as

$$X_n\cdot + T \xrightarrow{k_{tr}} P_n + T\cdot$$

In such a process the growth of a polymer chain is terminated by abstraction of a hydrogen or other atom from the monomer, polymer, initiator or any other species present in the system; the reaction terminates the growing polymer chain whilst simultaneously generating a new radical capable of starting a new growing chain.

Qualitatively it can readily be seen that transfer reactions may have two effects: the degree of polymerization is reduced by an amount dependent on the relative reactivity of the monomer and the transfer agent T towards $X_n\cdot$, whilst the rate of polymerization is altered by an amount which depends on the ability of T· radicals to reinitiate polymerization, and hence on the relative reactivity of $X_n\cdot$ and T· radicals towards the monomer. Quantitatively the effect of transfer reactions on the number-average degree of polymerization may be expressed by the equation [3]

$$\frac{1}{\bar{P}} = \sum \frac{k_{tr}}{k_p}\frac{[T]}{[M]} + \frac{1}{\bar{P}_0} \tag{3}$$

where $\bar{P}_0$ is the degree of polymerization expected in the absence of transfer and is given by eqn. (2). The summation term is included in this equation to take account of the fact that transfer may occur to several species simultaneously. The ratio k_{tr}/k_p which will be denoted as C_T is termed the *chain transfer constant*; it is normally evaluated by measurement of the variation of number-average molecular weight with the concentration of the transfer agent T, although care may be needed in the application of eqn. (3) due to the interdependence of the concentrations of the species present [4]. Clearly C_T is a most useful parameter since it is a measure of the reactivity of a propagating radical towards a substrate T, relative to its reactivity towards monomer; in cases where absolute propagation rate coefficients are known C_T provides a basis for direct comparison of reactivities.

If the radical T·, produced by transfer, is less reactive towards monomer than the propagating radical X_n· then transfer reactions will result in a decrease in the rate of polymerization and *retardation* will be observed. The limiting case of this type occurs when every kinetic chain is interrupted by transfer and the transfer radicals are incapable of initiation of a new polymer chain. Under these conditions the polymerization will be unobservably slow and *inhibition* will occur. Suitable kinetic equations are available for both retarded and inhibited polymerizations and their application leads to the evaluation of C_T from rate measurements.

Some typical examples of atom transfer reactions include

Hydrocarbons	$X_n\cdot + RH$	$\longrightarrow$	$X_n{-}H + R\cdot$
Halides	$X_n\cdot + CCl_4$	$\longrightarrow$	$X_n{-}Cl + CCl_3\cdot$
Mercaptans	$X_n\cdot + RSH$	$\longrightarrow$	$X_n{-}H + RS\cdot$
Amines	$X_n\cdot + (C_2H_5)_3N$	$\longrightarrow$	$X_n{-}H + (C_2H_5)_2N\dot{C}HCH_3$

Polymeric radicals can also enter into electron transfer reactions with transition metal ions, for example, ferric and cupric ions [5—8],

$$X_n\cdot + FeCl_3 \longrightarrow X_n{-}Cl + FeCl_2$$

This reaction is actually a termination process since no possible chain carrier is produced; its occurrence to any significant extent is invariably accompanied by either inhibition or retardation.

1.3 COPOLYMERIZATION

When a mixture of two polymerizable monomers is treated with a free radical initiator the product polymer will be expected to contain repeat units derived from both monomers. Suppose we consider a system containing two monomers A and B and we assume that the reactivity of a chain radical depends only upon the nature of the terminal monomer unit, i.e. radicals ending in —A—A· and —B—A· are assumed to be equally reactive towards the monomer. We may then consider that the system contains only two types of chain radical, $X_a\cdot$ radicals having end-groups derived from A and $X_b\cdot$ radicals whose end groups are derived from B. It follows that we may distinguish four propagation steps

$$X_a\cdot + A \longrightarrow X_a\cdot \quad k_p^{aa}$$

$$X_a\cdot + B \longrightarrow X_b\cdot \quad k_p^{ab}$$

$$X_b\cdot + A \longrightarrow X_a\cdot \quad k_p^{ba}$$

$$X_b\cdot + B \longrightarrow X_b\cdot \quad k_p^{bb}$$

For a low conversion, where the concentrations of all reactants may be assumed constant it follows that the ratio n of the concentrations of A and B units in the polymer is given by

$$n = (-\mathrm{d}[A]/\mathrm{d}t)/(-\mathrm{d}[B]/\mathrm{d}t) \tag{4}$$

Application of the stationary state approach to this equation allows the relation between the copolymer composition and the composition of the starting monomer mixture to be expressed in a form known as the *copolymer composition equation* [9–11]

$$n = \frac{[A]}{[B]}\left\{\frac{r_a[A] + [B]}{r_b[B] + [A]}\right\} \tag{5}$$

where

$$r_a = \frac{k_p^{aa}}{k_p^{ab}} \quad \text{and} \quad r_b = \frac{k_p^{bb}}{k_p^{ba}}$$

r_a and r_b are termed the *monomer reactivity ratios* and they may, in principle, be determined by analysis of two copolymers prepared with different molar feed ratios of A to B. In practice their determination is usually more complex [12], but even so a large number of values have been measured and extensive tabulations are available [13–15].

Monomer reactivity ratios from copolymerization experiments are extremely important parameters in any discussion of radical reactivity. The value of r_a, for example, is a direct measure of the relative rates at which the radical $A\cdot$ reacts with monomers A and B, that is to say it is an inverse measure of the reactivity of the $A\cdot$ radical towards the second monomer B, relative to its reactivity towards its own monomer A. Reactivity ratios are thus comparable to transfer constants, the former measuring reactivity towards a polymerizable monomer and the latter reactivity towards a non-polymerizable substrate.

It is evident that, if the relative reactivity of any radical towards the two monomers A and B were independent of the radical structure, we would always find that $r_a = r_b^{-1}$ or $r_a r_b = 1$. The condition defines an *ideal copolymerization* [10] and Fig. 1 shows the dependence of copolymer composition on monomer feed composition for a number of systems of

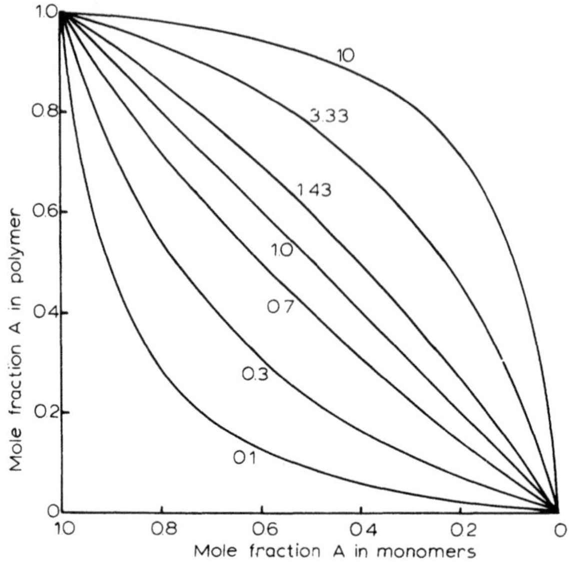

Fig. 1. Relationship between polymer and monomer compositions for ideal copolymerization, $r_a r_b = 1$. Values of r_a indicated.

this kind. In practice ideal copolymerization is a rare phenomenon since the condition that $r_a r_b = 1$ is not often satisfied. Generally two types of behaviour are found in real systems.

(i) r_a is greater than unity and r_b is small. This condition implies that both A· and B· radicals react preferentially with monomer A, causing the

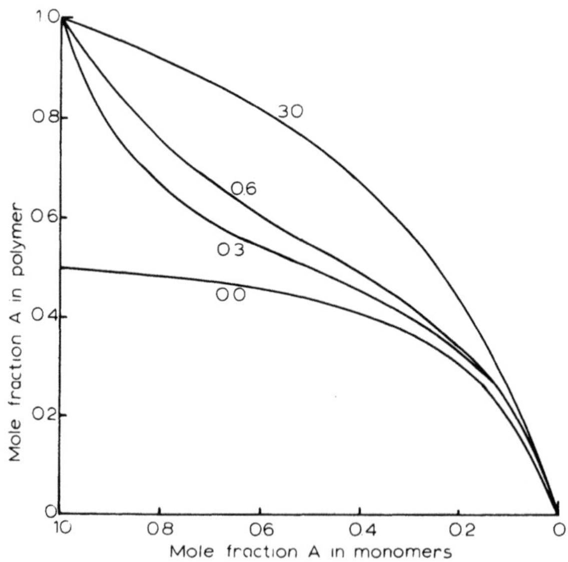

Fig. 2. Relationship between polymer and monomer composition for non-ideal copolymerization; $r_b = 0.30$, r_a indicated.

copolymer to be richer in repeat units derived from A than would be predicted from ideal behaviour.

(ii) r_a and r_b are both less than unity. This implies that each radical reacts preferentially with the other monomer, giving rise to a situation in which the two monomers tend to enter the growing chain alternately; for this reason this condition is sometimes referred to as the *alternating tendency*. The limit of this situation occurs when $r_a = r_b = 0$ so that each radical can only react with the other monomer and the polymer has a regular alternating structure.

A third condition, in which r_a and r_b are both greater than unity, so that the monomers tend to simultaneously homopolymerize, is unknown and most monomer pairs copolymerize with reactivity ratios given by $0 < r_a r_b < 1$. Some examples of non-ideal copolymerization are shown in Fig. 2.

From the foregoing discussion it should be clear that from relatively simple measurements of polymerization kinetics, molecular weights and copolymer composition, it is possible to extract rate coefficients for the reactions of polymeric radicals with themselves (in termination), with non-polymerizing substrates (transfer agents) and with other monomers (in copolymerization), the rate coefficients in each case being measured relative to the homopropagation rate coefficient. In the subsequent sections we shall attempt to illustrate the way in which meaningful correlations of radical and substrate reactivity with structure can be achieved.

2. Radical—radical reactions: the termination process

2.1 HOMOPOLYMERIZATION

The combination reactions of small radicals in the gas phase are usually processes of high kinetic efficiency, i.e. they have very low activation energies and steric factors which are close to unity. In the closely packed structure of a liquid, radical mobilities are very much lower than in the gas phase so that the time taken for two radicals to diffuse together, and the length of the period for which they remain in contact are both increased. It is thus expected that mutual combination of two radicals in a liquid medium will be a process of very high efficiency whose rate is controlled by the rate of diffusion together of the radicals. Effects of this kind are particularly probable with polymeric radicals which, by virtue of their large size, have low mobility and low diffusion coefficients in the liquid phase. North and Reed [16] have shown that the termination reaction in methyl methacrylate polymerization is diffusion controlled under all normal conditions, whilst Allen and Patrick [17], comparing known termination rate coefficients with measured diffusion coefficients for polymers in solution, concluded that all termination processes are most probably controlled by diffusion.

The control of termination by the rate of diffusion of macro-radicals has important consequences for the polymerization behaviour, particularly in heterogeneous or highly viscous systems; these consequences are discussed more fully in the next chapter. In view of the fact that termination is diffusion controlled and that the diffusion coefficient will be determined by the size of the polymer radical, rather than the structure of its terminal unit, it is not surprising to find that termination rate coefficients usually depend on the viscosity of the reaction medium and often vary very significantly with conversion even in the early stages of the polymerization. Equally, it is not surprising to find that there is no meaningful correlation between the termination rate coefficient and the structure of the polymeric radical.

TABLE 1
Termination mechanism for polymer radicals

Monomer	Temperature (°C)	Mechanism	Ref.
Styrene	60—100	Combination	19, 20
Acrylonitrile	60	Combination	21—23
Methyl methacrylate	0	Mainly combination	19, 24
	60 and above	Mainly disproportionation	19, 24
Methyl acrylate	90	Mainly disproportionation	20
Vinyl acetate	90	Mainly disproportionation	20

When two polymer radicals terminate by mutual reaction they may do so either by combination or by disproportionation and it is not immediately obvious which mode of reaction will be preferred in any given system. Experimental determination of termination mechanism is often difficult; Table 1 shows the results in some systems where reasonable evidence for the termination mechanism has been established. Bevington [18] has applied a thermodynamic treatment based on estimates of the enthalpy and entropy changes in the two modes of reaction. Although such studies cannot be made with great accuracy it is broadly concluded that the enthalpy term favours combination and the entropy term disproportionation; it is thus expected that there will be a "threshold" temperature for each radical, below which termination is likely to be by combination and above which disproportionation is the favoured mode. The only useful correlation between radical structure and termination mechanism is that an α-methyl substituent will offer more sites for double bond formation and hence will favour disproportionation.

2.2 COPOLYMERIZATION

Using the simple picture of a copolymerization process given in Sect. 1.3, we can readily see that three termination reactions are possible, viz.

$$A\cdot + A\cdot \longrightarrow \text{Polymer} \quad k_{ta}$$

$$A\cdot + B\cdot \longrightarrow \text{Polymer} \quad k_{tx}$$

$$B\cdot + B\cdot \longrightarrow \text{Polymer} \quad k_{tb}$$

These termination steps do not make any significant contribution to the composition of the copolymer, but their relative rates sometimes have an influence on the kinetics of the reaction. If it is assumed that each termination rate coefficient is composed of two terms, each characteristic of the reactivity of one of the participating radicals, then we may write

$$k_{ta} = k_a^2 \quad (6)$$

$$k_{tx} = k_a \cdot k_b \quad (7)$$

$$k_{tb} = k_b^2 \quad (8)$$

and hence

$$k_{tx} = (k_{ta} \cdot k_{tb})^{1/2} \quad (9)$$

This is known as the *geometric mean assumption* and there is ample evidence that it is valid for the rate coefficients for mixed radical reactions in the gas phase where all three reactions are chemically controlled [25]. In copolymerization of two monomers, which are chemically very similar, such as styrene and a nuclear substituted styrene [26] the experimental data is in good agreement with the geometric mean assumption, but in other cases [27, 28] there are divergences from this assumption, the observed cross-termination rate coefficient being higher than the predicted value, so that $\phi > 1$ in the expression

$$k_{tx} = \phi(k_{ta} \cdot k_{tb})^{1/2} \quad (10)$$

This condition implies that each growing radical preferentially terminates by reaction with the opposite radical and is thus analogous to the alternating tendency in propagation reactions. However, unlike reactivity ratios, the ϕ-factor is not a characteristic of any given monomer pair but varies with the ratio of the monomers present [29]. This effect has been taken by some workers to imply that the reactivity of a radical in termination reactions is influenced not only by the terminal unit but also by the repeat units preceding the terminal position, the nature of these units being determined by the monomer composition. Ham [30—32] has generalized this *penultimate unit effect* to include contributions to termination reactivity from units other than the terminal unit; this procedure leads to better curves to fit the relationship between monomer and polymer composition, if only because additional variable parameters are introduced.

Whilst the penultimate unit effect provides a satisfactory explanation for the deviation of measured and calculated cross-termination rates it is very difficult to see valid reasons for expecting the reactivity of a polymeric radical to be dependent upon the nature of units other than the terminal unit. The geometric-mean postulate implicitly contains the assumption that the three termination reactions are chemically controlled, k_{ta}, k_{tx} and k_{tb} being true rate coefficients for chemical reactions. However, we have already seen in the previous section that current thinking regards radical termination reactions as being diffusion controlled processes; in this case k_{ta}, k_{tb} and k_{tx} are regarded as being essentially diffusion coefficients. If this is true then the geometric-mean assumption of eqn. (9) can no longer be regarded as valid; it would for instance imply that a radical trapped in a solid matrix, where its diffusion coefficient is zero, could not be terminated by a mobile radical no matter how high its diffusion coefficient. It is qualitatively obvious, and assumed in all quantitative treatments of diffusion control, that the rate of diffusion together of two different species is proportional to the sum of their diffusion rates through the medium. Under these conditions k_{tx} is given by an arithmetic mean expression

$$k_{tx} = (k_{ta} + k_{tb})/2 \tag{11}$$

On this basis the effect of monomer composition on termination rate is attributed to the different viscosity (and hence diffusion) characteristics of polymers containing different proportions of the two monomer units so that the effective rate coefficients for all termination processes should vary with polymer (and hence monomer) composition. Whilst this supposition would seem to be more acceptable than the penultimate unit effect and more consistent with an overall view of termination as a diffusion controlled process, it remains to be established that copolymers with varying composition do indeed have viscosity parameters which vary in a manner consistent with observed cross-termination rate coefficients.

3. Radical—substrate reactions: copolymerization and chain transfer

3.1 FACTORS INFLUENCING RADICAL REACTIVITY

3.1.1 Inherent reactivity of radicals and substrates

In Sect. 1 it was shown that studies of copolymerization and transfer reactions yield useful basic information on reactivity of radicals and substrates. In attempting to correlate such reactivity data with structure it is initially tempting to look for some simple order of reactivity of radicals towards substrates which is independent of the nature of the substrate, and vice-versa. The concept of an *inherent reactivity* of radicals and substrates has been given considerable importance and is worthy of detailed consideration. Let us first examine the information available from reactivity ratios in copolymerization. Consider the copolymerization reactions of a monomer A with a series of other monomers. In this case the reciprocal of the reactivity ratio r_a is a direct measure of the reactivity of the series of monomers towards the radical A· since

$$\frac{1}{r_a} = \frac{k_p^{ab}}{k_p^{aa}} = \frac{\text{Specific rate of } (\text{A}\cdot + \text{B})}{\text{Specific rate of } (\text{A}\cdot + \text{A})}$$

and the rate coefficients for all reactions of radical A· are measured with reference to a common standard, the homopropagation rate coefficient for monomer A. Using $1/r_a$ values as a measure of reactivity, Walling and Mayo [33, 34] tabulated the reactivity of a number of monomers with respect to a series of reference radicals. Table 2 shows the resulting data for six common monomers which, for reasons which will be outlined later, we shall use throughout as a basis for the discussion of general theories of radical reactivity. The six monomers are arranged in the columns of this Table in what Walling and Mayo consider to be the decreasing order of their inherent reactivity. Clearly there is some evidence to support this order of reactivity although it will be noted that the order is much better defined in the right-hand columns of the Table. However, there are some marked inconsistencies, for example the relative reactivity of styrene and methacrylonitrile is 0.3 with respect to the styryl radical but 10.6 with respect to the acrylonitrile radical, leaving a factor of 35 to be explained.

An alternative presentation of the data of Table 2 is given in Table 3, in which all the values in each column have been normalized to 1.00 for reaction with the styryl radical. From the lowest row of this Table it is seen that the reactivity of vinyl acetate monomer relative to styrene monomer is virtually independent of the radical concerned whereas it is clear that, on the same standard, the behaviour of the remaining monomers is very dependent upon the structure of the radical with which they are reacting. Thus although the order of monomer reactivity is qualitatively valid, the concept of inherent reactivity of monomers does not hold quantitatively.

If we wish to extend this approach to examine the relative reactivity of a series of radicals towards the same monomer then the situation becomes more complex. Thus if we measure the values of r_a for the copolymerization of a series of monomers A with a reference monomer B we have again

$$\frac{1}{r_a} = \frac{k_p^{ab}}{k_p^{aa}} = \frac{\text{Specific rate of } (\text{A}\cdot + \text{B})}{\text{Specific rate of } (\text{A}\cdot + \text{A})}$$

TABLE 2
Reactivities of radical—monomer combinations relative to homopolymerization

Monomer	Radical					
	Styrene	Vinyl acetate	Methyl methacrylate	Methyl acrylate	Methacrylonitrile	Acrylonitrile
Styrene	1.0	100	2.17	5.56	6.25	33.3
Methyl methacrylate	2.0	66.7	1.0	2.38	1.43	6.67
Methacrylonitrile	3.33	100	1.43	3.33	1.0	3.13
Acrylonitrile	2.22	16.7	0.83	1.05	3.73	1.0
Methyl acrylate	1.33	10	1.09	1.0	0.50	0.71
Vinyl acetate	0.02	1.0	0.05	0.11	0.08	0.20

TABLE 3
Reactivities of radical—monomer combinations relative to reactions of styrene monomer

Monomer	Radical					
	Styrene	Vinyl acetate	Methyl methacrylate	Methyl acrylate	Methacrylonitrile	Acrylonitrile
Styrene	1.00	1.00	1.00	1.00	1.00	1.00
Methyl methacrylate	2.00	0.67	0.46	0.43	0.23	0.20
Methacrylonitrile	3.33	1.00	0.66	0.60	0.16	0.09
Acrylonitrile	2.22	0.17	0.38	0.19	0.60	0.03
Methyl acrylate	1.33	0.10	0.50	0.18	0.08	0.02
Vinyl acetate	0.02	0.01	0.02	0.02	0.01	0.01

TABLE 4
Homopropagation rate coefficients at 60 °C

Monomer	k_p ($l\ mole^{-1}\ sec^{-1}$)
Styrene	176
Methyl methacrylate	734
Methacrylonitrile	201
Acrylonitrile	2458
Methyl acrylate	2090
Vinyl acetate	3700

so that the value of $1/r_a$ measures the reactivity of the radical A· with respect to its own homopropagation coefficient. In order to make realistic comparisons it is necessary to know the value of the homopropagation coefficient for each of the monomers A. It is for this reason that we shall confine our attention to the six monomers of Tables 2 and 3 since the propagation rate coefficients for these monomers have been established with a reasonable degree of accuracy. Table 4 shows these values; using this information, together with r_a values we are able to construct Table 5, which shows the reactivity of the radicals derived from our six monomers, relative to each monomer, the figures in each row being the rate coefficients relative to the appropriate homopropagation rate coefficient. Again we can see that qualitiatively there is a definite order of reactivity, the vinyl acetate radical being most reactive and styryl radical the least reactive with the complete list being VA > AN > MA > MM > MAN > S. However, there are again considerable discrepancies in the order and large variations in the actual ratios of reactivities.

It will be observed that the order of inherent reactivity of radicals is approximately the inverse of the order for monomers, the most reactive monomers yielding the least reactive radicals and vice versa. This is very easily understood since a substituent which decreases the reactivity of a radical by some stabilizing influence would be expected to increase the reactivity of the double bond precursor because these terms refer to two conjugate species

$$X_n\cdot + CH_2{=}CHY \longrightarrow X{-}CH_2{-}\dot{C}HY$$

Walling [34] has shown that a given substituent will stabilize a radical to a much greater extent that its conjugate monomer, an effect which is readily confirmed by examination of Tables 2 and 5. Thus the vinyl acetate radical is about 1000 times more reactive than the styryl radical towards any monomer, although styrene monomer is only about 50 times more reactive than vinyl acetate towards any given radical.

TABLE 5
Reactivities of radical—monomer combinations relative to the homopropagation of the radical

Radical	Monomer					
	Styrene	Methacrylonitrile	Methyl methacrylate	Methyl acrylate	Acrylonitrile	Vinyl acetate
Vinyl acetate	2100	1840	336	17.8	24.8	1.0
Acrylonitrile	484	38.3	23.3	0.88	1.0	0.14
Methyl acrylate	65.9	34.6	6.8	1.0	0.9	0.06
Methyl methacrylate	9.07	5.23	1.0	0.4	0.25	0.07
Methacrylonitrile	7.1	1.0	0.4	0.05	0.03	0.005
Styrene	1.0	2.92	0.50	0.11	0.16	0.0009

TABLE 6
Absolute rate coefficients for transfer reactions (l $mole^{-1}$ sec^{-1})

Radical	Transfer agent			
	Chloroform	*n*-Butyl mercaptan	Carbon tetrabromide	Ferric chloride
Styrene	8.80×10^{-3}	3870	387	9.41×10^{4}
Methacrylonitrile	—	—	14.1	619
Methyl methacrylate	3.30×10^{-2}	492	198	5080
Methyl acrylate	0.26	3530	857	6770
Acrylonitrile	0.75	1006	211	7020
Vinyl acetate	46.4	1.78×10^{5}	$>1.44 \times 10^{5}$	2.35×10^{6}

TABLE 7
Relative rate coefficients for propagation and transfer[a]

Radical	Substrate				
	Ferric chloride	Acrylonitrile	Vinyl chloride	Styrene	Triethylamine
Vinyl acetate	100	500	1000	5×10^3	25×10^3
Styrene	1	1	1	1	1
Acrylonitrile	10^{-2}	0.5	20	500	5×10^3
Ratio of reactivity VA·/AN·	10^4	10^3	50	10	5
Ratio of reactivity S·/AN·	100	2	5×10^{-2}	2×10^{-3}	2×10^{-4}

[a] In each column the figures have been "normalized" relative to that for attack by the styryl radical. The figures in different columns are therefore not comparable.

If the general conclusion that there is a qualitative order of inherent reactivity is valid then a similar order of reactivity of radicals should be observed in their chain transfer reactions with substrates other than monomers. For the six radicals so far considered absolute chain transfer rate coefficients are listed in Table 6 for a variety of substrates, the radicals being tabulated in the order of increasing inherent reactivity derived from Table 5. It is immediately clear that the proposed inherent order of reactivity is not maintained in transfer processes since there are glaring inconsistencies in the data of Table 6. To illustrate this fact rather better, consider the data of Table 7 in which both chain transfer and propagation rate coefficients are given. From this Table it is immediately obvious that an order of inherent reactivity does not apply. Thus the ratio of reactivities for the radicals derived from vinyl acetate and acrylonitrile changes from 10^4 to 5 between ferric chloride and triethylamine, whilst the ratio for styrene and acrylonitrile radicals inverts across the Table, changing from 10^2 with ferric chloride to 2×10^{-4} with triethylamine. Thus it is seen that, although copolymerization data might suggest a definite order of reactivity radicals, this order is not supported when transfer data are considered; in this case the order of reactivity is not unique but depends upon the particular substrate. The more extreme type of behaviour can be seen to occur with highly polar substrates such as $FeCl_3$ and NEt_3 which might suggest that polar factors have an influence upon radical reactivity.

From the foregoing discussion we can see that, although one might initially be tempted to conclude that radicals and substrates have a unique inherent reactivity, this view is not supported by a closer examination of the data, especially in the case of highly polar substrates. The basic problem is thus to derive a theory of reactivity which will take account of the fact that rate coefficients for radical—substrate reactions include contributions which depend on the precise structure of both radical and substrate. Before attempting to discuss the development of such theories it is pertinent to examine some of the factors which may influence radical—substrate reactions and to show how some of these factors dominate particular systems.

3.1.2 Polar effects

It was pointed out in Sect. 1.3 that, in a copolymerization reaction for which r_a and r_b are both less than unity each propagating radical reacts preferentially with the other monomer, leading to an alternating tendency; the product $r_a r_b$ is frequently used as a measure of the inverse of the alternating tendency. Walling and Mayo [33] have shown that it is possible to tabulate monomers, in the form of Table 8, so that the product $r_a r_b$ decreases with increasing separation of the monomers in the Table. Examination of this Table suggests that it represents a Table of

TABLE 8

Alternating tendencies in copolymerization reactions

	α-Methyl styrene	Butadiene	Styrene	Isopropenyl acetate	Vinyl acetate	Vinyl chloride	2-Chlorobutadiene	2,5-Dichlorostyrene	Methyl methacrylate	Vinylidene chloride	Methyl acrylate	Methyl vinyl ketone	β-Chloroethyl acrylate	Methacrylonitrile	Acrylonitrile	Diethyl fumarate
α-Methyl styrene																
Butadiene																
Styrene																
Isopropenyl acetate																
Vinyl acetate				1.0												
Vinyl chloride			0.34	0.55	0.63											
2-Chlorobutadiene			0.2	0.4												
2,5-Dichlorostyrene			0.2	0.16												
Methyl methacrylate	0.07	0.19	0.24	0.5			0.7	1.0								
Vinylidene chloride		0.1	0.16		0.12				0.61							
Methyl acrylate		0.04	0.14			0.75	0.9	0.5		0.8						
Methyl vinyl ketone			0.10													
β-Chloroethyl acrylate			0.06								0.9					
Methacrylonitrile	0.006	0.01	0.06		0.24				0.43							
Acrylonitrile	0.006	0.02	0.02		0.25	0.13	0.24	0.015	0.24	0.34		1.1				
Diethyl fumarate			0.02		0.004	0.06	0.17			0.56						
Maleic anhydride				0.00006	0.00002	0.002			0.13		0.6					

electrical polarities, species differing widely in polarity having a greater tendency to react with each other. The monomers are found to follow the same order of polar influence as that which is found in organic reactions generally and which is attributed to the polar influence of substituent groups, the electronegative influence increasing through the series hydrocarbons < chlorinated hydrocarbons < esters < nitriles. A direct demonstration of the polar influence on radical reactivity has been given by Walling et al. [35], who showed that there is a good correlation between polymerization reactivity and the Hammett σ function for the substituent, in a series of nuclear substituted styrenes.

In the light of the obvious influence of polar factors on the alternating tendency in copolymerization and the fact, illustrated in Sect. 3.1, that the proposed order of inherent reactivity of radicals is not maintained in reactions with highly polar substrates it seems not unreasonable to suggest that the reactivity of a given radical—substrate combination is made up of two components, the inherent reactivity of radical and substrate and the modification of this inherent reactivity by polar factors specific to each radical substrate pair; this concept will be examined more fully in Sect. 4.

There has been some discussion as to whether the alternating tendency in copolymerization of two monomers of different polarity results from the interaction of permanent charges or from the temporary charges arising only during the formation of the transition complex. Walling et al. [33, 35, 36] consider that polar effects arise from temporary charges, since the electrostatic forces involved in the interaction of permanent charges would be dependent upon the dielectric constant of the reaction medium and no such effect is observed. However, this argument would apply equally to any situation in which the polarity of the transition state differs from that of the initial state, even if the latter is non-polar. It seems more probable that the insensitivity of reactivity ratios to solvent dielectric constant is due to a fortuitous mutual compensation of solvent effects on energy and entropy changes.

3.1.3 Resonance stabilization effects

If we are to adopt the view that the reactivity of monomers and radicals depends upon the perturbation of their inherent reactivity by polar influences then we should expect to find that resonance stabilization of monomers and radicals is an important factor in determining the inherent or non-polar component of reactivity. It has already been pointed out in Sect. 3.1.1 that such is indeed qualitatively the case; the general order of monomer and radical reactivity corresponds roughly to the degree of resonance stabilization of monomers and radicals. Thus, for example, the low reactivity of the styrene radical and the high reactivity of styrene monomer are attributable to the resonance delocalization of the unpaired electron of the radical into the orbitals of the benezene ring.

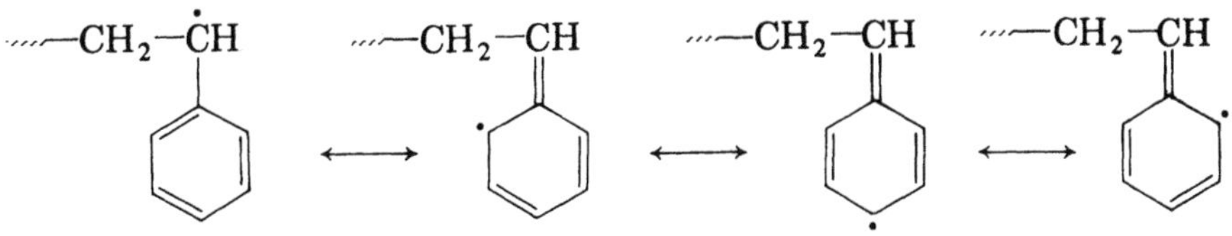

The importance of resonance contributions to radical reactivity is greatest where polar contributions are at a minimum, i.e. for reactions with weakly polar substrates. It will be shown in Sect. 4.4.1 that transfer constants for reactions with hydrocarbons can be used as a measure of resonance contributions to radical reactivity.

3.1.4 Steric effects

Although resonance and polar contributions are important in determining the reactivity of radical—monomer pairs there are many cases in which steric hindrance of the propagation reaction is an important factor in determining the rate of reaction. Thus, for example, 1,2 disubstituted monomers homopolymerize only with difficulty although they often copolymerize with other monomers. This generalization remains valid even when the equivalent 1,1-disubstituted olefin is readily polymerized. For example, although maleic anhydride copolymerizes fairly readily it can only homopolymerize very slowly [37, 38] due to steric hindrance to the transition state of the propagating radical.

Some typical values of reactivity ratios for copolymerization of maleic anhydride are given in Table 9; in view of the low reactivity of this monomer in homopolymerization, its reactivity in copolymerization is surprisingly high. Walling et al. [36] consider that this is due to an extreme example of polar interaction between a propagating radical and maleic anhydride monomer. However, Bartlett and Nozaki [40] have proposed that the enhanced reactivity is due to a donor—acceptor interaction between the two monomers; in a sense the copolymerization

TABLE 9
Reactivity ratios for copolymerization of maleic anhydride [13, 39]

Monomer A	Monomer B	r_a	r_b
Maleic anhydride	Methyl acrylate	0.02	2.8
Maleic anhydride	Methyl methacrylate	0.02	6.7
Maleic anhydride	Styrene	0.02	0.04
Maleic anhydride	Vinyl acetate	0.003	0.06
Maleic anhydride	Vinyl chloride	0.008	0.30

TABLE 10[a]
Effect of chlorine substitution on reactivity

Monomer	Relative rate of addition of	
	Styrene radical	Vinylidene cyanide radical
$CH_2{=}CHCl$	1.0	1.0
$CH_2{=}CCl_2$	9.7	12.9
trans-$CHCl{=}CHCl$	0.56	0.02
cis-$CHCl{=}CHCl$	0.21	0.02

[a] Based on average data from ref. 14.

may be regarded as a homopolymerization of a complex of the two monomers. This concept of monomer—monomer interactions in reactivity is further discussed in Section 3.1.6.

The steric effect of 1,2 disubstitution may also be illustrated by reference to Table 10, which shows the effect of substitution of a second chlorine atom into vinyl chloride on the reactivity of the resulting monomers towards the radicals derived from styrene and vinylidene chloride. Although the effect of a second chlorine atom in the 1-position is to increase the reactivity of the monomer by stabilization of the active centre, substitution in the 2-position decreases the reactivity by a factor which depends on whether the two chlorine atoms are *cis* or *trans*; the *trans* isomer is usually the more reactive. Steric effects of this type are by no means restricted to 1,2-disubstituted monomers in which both substituents are identical. Table 11 shows reactivity ratios for copolymerization of methacrylic and crotonic acids with vinyl acetate and vinylidene chloride [14]; this data illustrates clearly that both vinylidene chloride and vinyl acetate radicals react much more slowly with 1,2 substituted compounds than with their 1,1 substituted counterparts. In this context it is interesting to examine the following data [14]

Monomer A	Monomer B	r_a
$CH_2{=}CH_2$	$CF_2{=}CF_2$	0.15
$CH_2{=}CHCl$	$CF_2{=}CFCl$	2.53

It can be seen that, although the ethylene radical adds more rapidly to tetrafluoroethylene than to ethylene, the vinyl chloride radical is less reactive towards it perfluorinated analogue than towards vinyl chloride. Since, in the latter case, there may be more important polar effects arising from the presence of the chlorine atoms it seems not unreasonable to suggest that the presence of two fluorine atoms does not appreciably hinder the addition of radicals; this view is in accord with the fact that the

TABLE 11
Reactivity ratios for copolymerization of crotonic and methacrylic acids

Monomer A	Monomer B	
	Crotonic acid	Methacrylic acid
Vinyl acetate	$r_a = 0.33$ $r_b = 0.01$	$r_a = 0.01$ $r_b = 20$
Vinylidene chloride	$r_a = 35$ $r_b = 0.065$	$r_a = 0.15$ $r_b = 3$

fluorine atom is much smaller than the chlorine atom or the methyl group and would be expected to exert much less steric hindrance.

The effect of steric factors may be more complex in situations where radical stabilization results from electron delocalization in the transition state, steric effects may then inhibit resonance. For example, addition of a radical R· to a monomer of the maleic or fumaric ester series presumably produces a radical which may be formally represented as

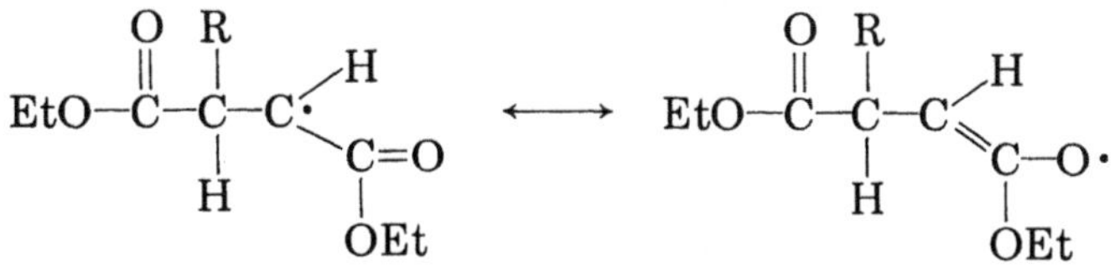

in which delocalization of the unpaired electron can only be effective if all of the atoms attached to the doubly-bonded carbon in the right hand canonical form are coplanar. This condition cannot be met in the maleic esters, where it is found that there is very considerable steric hindrance to the required coplanarity; the fumarate esters are not subject to this restriction so that stabilization of the transition state is effective and reduces the activation energy for propagation relative to that for the maleates.

3.1.5 Radical depropagation

Under the conventional conditions of constant temperature and pressure any polymerization process can only proceed spontaneously if the free energy change, ΔG, is negative. The free energy change is defined by

$$\Delta G = \Delta H - T\Delta S$$

and its magnitude depends upon the relative importance of the ΔH and ΔS terms. Since polymerization is an association process, its entropy change must be negative, because the reaction involves combination of independent monomer molecules into the more ordered polymer chain. Thus it follows that, for polymerization to proceed, the enthalpy change ΔH must be negative, and sufficiently so to overcome the effect of the

$T\Delta S$ term; obviously as the temperature rises a point must be reached where ΔG becomes positive so that depropagation is thermodynamically favoured. The critical temperature T_c ($=\Delta H/\Delta S$) at which $\Delta G = 0$ is termed the *ceiling-temperature* for the polymerization. Clearly quite misleading results will be obtained if polymerization rates or copolymerization reactivity ratios are measured at temperatures close to T_c.

In practice the ceiling-temperatures for most common monomers are considerably higher than normal working temperatures although there are several systems in which the ceiling-temperatures are quite low. For example, the ceiling-temperature for polymerization of α-methyl styrene is 61 °C so that this monomer is not susceptible to radical polymerization. Ceiling-temperature effects are also well known in the copolymerization of olefins with sulphur dioxide. Dainton and Ivin [41] and Sawada [42] have reviewed the thermodynamics of addition polymerization, whilst O'Driscoll [43] has considered the effect of depropagation on copolymerization reactions.

3.1.6 Complexing of radicals and monomers

The free-radical initiated polymerization of polar monomers containing pendant carbonyl or nitrile groups (such as methyl methacrylate or acrylonitrile) is characterized by an increased rate of polymerization in the presence of metal halides [44—46]. Bamford et al. [44] have attributed this effect to complex formation between the propagating radical and the metal halide, whilst Imoto et al. [45] consider that the enhanced reactivity is due to complex formation between the halide and monomer. This salt effect extends to copolymerization with non-complexed monomers such as vinylidene chloride, typically with increased r_a values for the complexed monomer. Monomers of the α-olefin type, which are not readily polymerized by radical means are more readily copolymerized with complexed polar monomers than with the uncomplexed monomers [47, 48]. A particularly interesting and potentially valuable aspect of the copolymerization of complexed polar monomers with uncomplexed non-polar monomers is the fact that such reactions frequently yield highly regular alternating copolymers, often irrespective of the mole ratio of the two monomers [46, 48].

Although the polymerizations of metal salt-complexed monomers are undoubtedly complex reactions, there is good evidence to suggest that their copolymerization behaviour is frequently dominated by monomer—monomer interactions. This is in complete contrast with the behaviour of uncomplexed monomers where the important factor is the interaction between a propagating radical and the incoming monomer molecule. It appears that the effect of the complexing of a polar monomer with Lewis acids is to increase the electron-accepting ability of the monomer, relative to its uncomplexed state; the complexed monomer may then participate in a one-electron transfer process with uncomplexed monomer or another electron donor monomer to form a charge-transfer complex. Copolymerization under such conditions may be regarded as the radical-induced homopolymerization of a Lewis acid activated charge-transfer complex of the two monomers, thus producing a regular alternating copolymer. It is found that if a sufficiently strong Lewis acid, such as diethyl aluminium chloride, is employed the resulting charge-transfer complex may induce polymerization even in the absence of radical initiators [49, 50]. Gaylord and Takahashi [46] have extensively reviewed the polymerization of complexed monomers. In the present context we shall employ the concept of radical—monomer interaction as the dominant influence upon radical reactivity but it must be borne in mind that this approach is probably inadequate to describe reactivity in very highly polar systems.

3.1.7 Penultimate and more remote units

In some cases [29, 51] analysis of copolymerization reactivity data shows that the monomer reactivity ratios are dependent upon monomer composition. This effect has been attributed to a breakdown of the usual assumption that radical reactivity is dependent only upon the terminal unit of the radical; if radicals of the form ~~AA· and ~~BA· are not identical in reactivity then r_a will be dependent upon the monomer ratio since the relative concentrations of the two radicals will vary with [A]/[B]. Table 12 [52] shows some examples of this behaviour. It will be noted that in all of these examples there is a large polarity difference between the monomers and, in general, penultimate unit effects are only significant when this is so. A generalized treatment of the effect of penultimate and more remote units has been given by Ham [30—32], who

TABLE 12
Effects of penultimate unit on radical reactivity

Monomer A	Monomer B	Reactivity ratio for radical with terminal structure	
		AA·	BA·
Styrene	Maleic anhydride	0.017	0.063
Styrene	Citraconic anhydride	0.07	0.25
Styrene	Fumaronitrile	0.07	1.0
Styrene	Acrylonitrile	0.30	0.45
α-Methyl styrene	Acrylonitrile	0.055	0.093
Chloroprene	Vinylidene cyanide	0	0.057
Methyl methacrylate	Vinylidene cyanide	0.04	0.08
Styrene	Oxygen	4.4×10^{-8}	$0.33—1.8 \times 10^{-6}$

has considered the effect of monomer units as far back as four units from the radical end. Although this approach is satisfactory (possibly due to the large number of variable parameters available) it is difficult to see how remote unit effects can operate; it seems possible, in view of the large polarity difference between monomers which display this effect, that it may arise at least in part from monomer—monomer interactions of the type described in Sect. 3.1.6.

4. Interpretation of radical reactivity data

4.1 POTENTIAL ENERGY SURFACE APPROACH

Evans et al. [53] have attempted to examine the interaction of radicals with monomers in terms of the potential energy surfaces for reaction. Typical examples of such curves are shown in Fig. 3. The curves represent, respectively, the repulsion energy between the radical and the atoms of the monomer which will form the new bond, and the potential energy of the new bond when formed (since the double bond has a much higher force constant than the final single bond it is a reasonable simplification to regard the latter bond length as the only parameter). The point of intersection of the two curves, where the energies of the bonded and unbonded states are equal, corresponds to the transition state of the addition reaction and the energy difference between this point and the

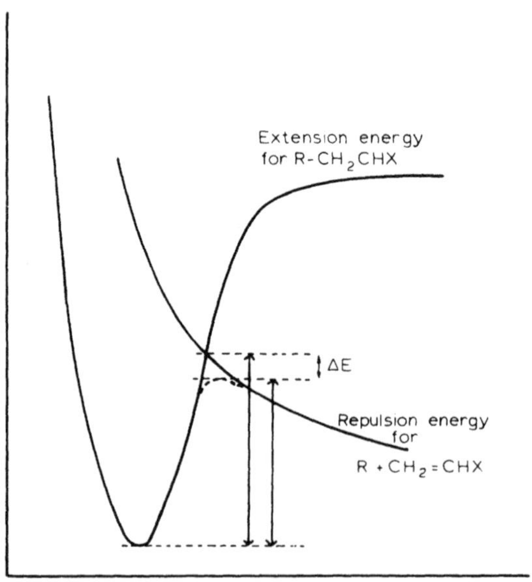

Fig. 3. Potential energy diagram for addition of a radical to a monomer.

potential energy minimum of the bonded state is the activation energy for propagation. However, this simplified view takes no account of the fact that the transition state may be stabilized by both resonance and polar effects which will lead to a lowering of the activation energy; such effects will be equivalent to a rounding off of the intersection and the omission of any estimate of the extent of rounding off is equivalent to ignoring the effect of stabilizing influences. The lack of any basis for estimating the effect of transition-state stabilization precludes the general application of the energy surface approach.

4.2 FRONTIER ELECTRON DENSITY APPROACH

Yonezawa et al. [54, 55] have shown that a theoretical molecular orbital approach can yield useful explanations of radical reactivity, provided that certain assumptions are made about the nature of the transition state for the addition reaction. The procedure adopted is essentially an LCAO—MO calculation of the activation energy for addition, on the assumption that π conjugation between the approaching radical and the monomer is the most important stabilizing factor determining the energy of the transition state. It is assumed that conjugation between the rth atomic π orbital in the monomer and the sth atomic π orbital in the radical constitutes a perturbation of the system; the resulting stabilization, which is computed by the LCAO—MO approach, is equated to the second-order perturbation energy $(\Delta E)_{r,s}$ which is assumed to contribute to the overall activation energy in the form

$$E = C - (\Delta E)_{r,s}$$

where C is a constant related to the σ contribution to the activation energy. On this basis the larger $(\Delta E)_{r,s}$ the more easily reaction will occur at the r and s positions of monomer and radical, respectively.

It is found, from a study of the complete LCAO equation for $(\Delta E)_{r,s}$, that this quantity is determined primarily by the electron densities in the frontier orbitals, which are the highest bonding molecular orbitals for electrophilic attack and the lowest bonding molecular orbitals for nucleophilic attack. Although there is no rigorous justification for this approach it gives results which are closely analogous to those obtained in calculations of delocalization energy. It will be noted that this treatment takes into account the influence of substituents without any need to divide their effects into polar and resonance contributions; in this sense the frontier orbital approach contrasts with all of the semi-empirical techniques to be discussed in later sections. Clearly, the prediction of reactivity in radical addition reactions involves the calculation of frontier electron densities at the various possible reacting sites in the monomer and radical. Some electron densities, calculated for two monomer—radical pairs are shown below

	Monomer	Radical
Styrene	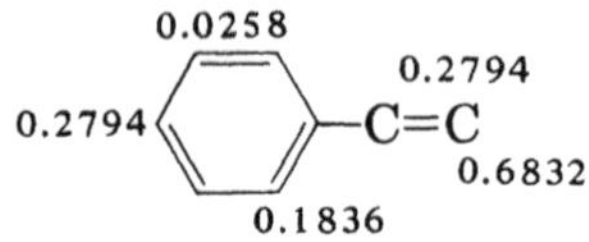	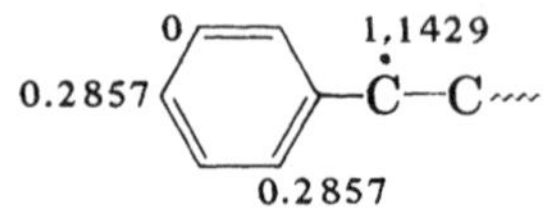
Acrylonitrile	N≡C—C=C (0.3333, 0.7644 above; 0.5679, 0.3852 below)	N≡C—Ċ—C〰 (0.2152 above; 0.6986, 1.0862 below)

These results are in agreement with the experimental observation that radical—monomer reactions are normally in the configuration represented by

$$\sim\!CH_2\dot{C}H(X) + CH_2{=}CH(Y) \longrightarrow \sim\!CH_2{-}CH(X)CH_2\dot{C}H(Y)$$

In principle in a homopolymerization reaction there are four possible modes of addition of the propagating radical to the monomer, viz.

$$\sim\!CH_2\dot{C}HX + CH_2{=}CHX \longrightarrow CH_2CHXCH_2\dot{C}HX \qquad (a)$$

$$\sim\!CH_2\dot{C}HX + CHX{=}CH_2 \longrightarrow CH_2CHXCHX\dot{C}H_2 \qquad (b)$$

$$\sim\!CHX\dot{C}H_2 + CH_2{=}CHX \longrightarrow CHXCH_2CH_2\dot{C}HX \qquad (c)$$

$$\sim\!CHX\dot{C}H_2 + CHX{=}CH_2 \longrightarrow CHXCH_2CHX\dot{C}H_2 \qquad (d)$$

For all monomers to which they have been applied, frontier electron density calculations indicate that the reactivity in propagation should follow the sequence (c) > (a) ≈ (d) > (b). Thus if a radical of the type $\sim\!CH_2\dot{C}HX$ is generated it will react by mode (a) rather than (b), tending to replicate its structure, whereas a radical of the type $\sim\!CHX\dot{C}H_2$ will react preferentially by mode (c) producing the isomeric $\sim\!CH_2{-}\dot{C}HX$ radical. The result is that, whichever type of radical is generated, it will very quickly form the $\sim\!CH_2\dot{C}HX$ radical which will thereafter perpetuate its own kind; this conclusion implies that head-to-tail addition will be predominant, a conclusion which is amply supported by experimental results.

It is not claimed that frontier electron density calculations can give accurate values for reactivity ratios but there is found to be reasonable correlation between reactivity ratios and the $(\Delta E)_{rs}$ values for competitive reactions. If a radical A can react with its own monomer A or a second monomer B, the reactivity ratio r_a is equal to k_p^{aa}/k_p^{ab} ·; on an activation energy basis r_a should be proportional to log Δ where

$$\Delta = (\Delta E)_{ab} - (\Delta E)_{aa}$$

Table 13 [55] and Fig. 4 illustrate that the predicted correlation is observed with only one serious exception. However, this approach is rather limited since relatively few $(\Delta E)_{rs}$ values have been calculated.

An unusual result of frontier electron density calculations arises in homopolymerization of some 1,2 disubstituted olefins where it appears that there is a critical value of $(\Delta E)_{rs}$ below which addition will not take place; this effect may provide an alternative to steric hindrance in explaining the low reactivity of 1,2-disubstituted monomers, although it is not possible to separate the two factors and, indeed, both may be operative.

TABLE 13
Comparison of frontier electron density calculations with experimental reactivity ratios

Radical A	Monomer B	2 + log Δ	r_a
Acrylonitrile	Butadiene	1.881	0
Acrylonitrile	Styrene	1.836	0.04
Butadiene	Acrylonitrile	1.560	0.35
Styrene	Acrylonitrile	1.395	0.41
Vinyl chloride	Acrylonitrile	1.375	0.02
Butadiene	Styrene	1.133	1.39
Styrene	*o*-Chlorostyrene	0.679	0.56
Styrene	Butadiene	0.394	0.78
Styrene	*m*-Chlorostyrene	0.344	0.64

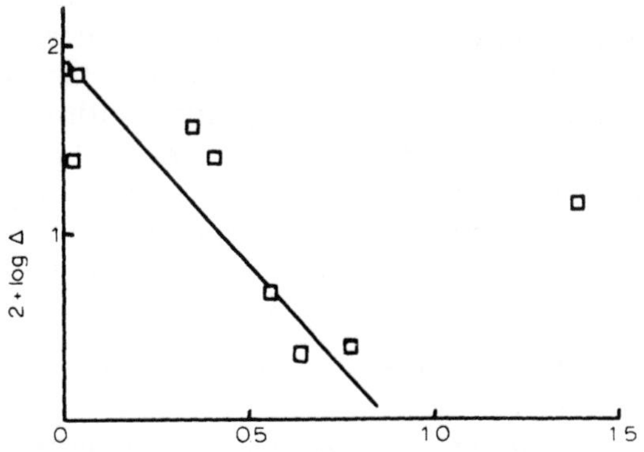

Fig. 4. Correlation of frontier electron energy calculations and measured reactivity ratios.

An alternative, but similar, approach to the direct computation of radical reactivities has been proposed by Kawabata et al. [56, 57] based on the calculation of conjugation stabilization energies between monomer and radical; the results obtained are comparable with the frontier orbital approach. Although theoretical treatments of this kind, based upon the use of calculable properties of monomers and radicals and their interaction, are attractive, Dewar and Thompson [58] consider that there are serious criticisms of many such methods and that their success is due solely to fortuitous correspondencies between the quantities used (such as frontier electron density, charge density and delocalization energy) and the energy differences which in fact determine the reaction rate. These authors consider that the only valid theoretical approach is via attempted calculation of the energy differences between the reactants and the transition state, and they, and other authors [58—60], have attempted to apply the SCF—MO approach to such calculations. Whilst the results obtained are extremely successful in some systems, such as estimation of partial rate factors in hydrocarbon substitution, relative rates of base deprotonation of arylmethanes and relative rates of solvolysis of arylmethyl chlorides, it is unfortunate that the method has not yet been applied to the rather more difficult problem of computation of radical reactivities in polymerization processes. It therefore seems that at the present time theoretical predictions of radical reactivity should be regarded with extreme caution.

4.3 THE *Q—e* SCHEME

In discussing the factors which may influence reactivity of radicals and monomers, we have seen that, in the absence of specific effects such as steric hindrance or monomer—monomer interaction, two main factors appear to be important in determining reactivity, the inherent reactivity of monomers and radicals and the alternating tendency. The latter effect becomes more apparent with monomers bearing polar groups, a fact which led Price [61] to suggest that the alternating tendency arises entirely from electrostatic interactions of permanent charges due to the mutual polarization of radicals and monomers. This concept was extended by Alfrey and Price [62] in an attempt to provide a quantitative interpretation of radical reactivity data by means of a method now known as the *Q—e* scheme.

The *Q—e* approach makes three basic assumptions

(i) The inherent reactivities of monomers and radicals can be described by parameters Q and P, respectively.

(ii) The polar properties of monomers and radicals can be denoted by a parameter e, the e values for a given monomer and its own radical being identical.

(iii) The rate coefficient for reaction of a radical A with a monomer B is given by

$$k_{ab} = P_a Q_b \exp[-e_a e_b] \tag{12}$$

If these assumptions are accepted it follows that the monomer reactivity ratios are given by the expressions

$$r_a = (Q_a/Q_b) \exp[-e_a(e_a - e_b)] \tag{13}$$

$$r_b = (Q_b/Q_a) \exp[-e_b(e_b - e_a)] \tag{14}$$

$$r_a r_b = \exp[-(e_a - e_b)^2] \tag{15}$$

Thus if the Q and e values are known for a series of monomers, the reactivity ratios of copolymerization of any pair of monomers can be predicted. In practice, the assignment of Q and e values is quite empirical, the values being derived from copolymerization data in such a way as to lead to a self-consistent set of figures. It is clearly necessary to adopt arbitrary Q and e values for one monomer which is to act as a reference standard; once this has been done values can be assigned to other monomers to give correct reactivity ratios.

It is possible to criticize the Q—e approach on a number of grounds. In particular there is no justification for the assumption of equal effective charges on radical and monomer and, indeed, such permanent charges which may exist may be much less important in determining reactivity than polarization effects arising only in the transition state of the radical addition reaction. Furthermore the Q—e scheme assumes that e values are dependent on the dielectric constant of the medium in the form

$$e_a = C_a/(rDkT)^{1/2} \tag{16}$$

where k is Boltzmann's constant, r the separation between charges, T the absolute temperature, D the dielectric constant and C_a the actual charge on monomer A; reactivity ratios are experimentally found to be independent of D. However, although there is no doubt that these criticisms are valid it is equally difficult to deny that the Q—e scheme has proved to be one of the most generally useful theories of radical reactivity as judged by its ability to predict reactivity ratios for copolymerization of a very wide range of monomer pairs. Provided that it is accepted that the approach is empirical and that the Q and e parameters do not have a sound theoretical basis, the predictive utility of the Q—e scheme is remarkably good and in good agreement with experimental observations.

Clearly a prerequisite of any tabulation of Q and e values is the assignment of arbitrary values to a reference monomer; the usual reference is styrene, for which a Q value of 1.0 has been universally adopted. Assignment of an e value is rather more difficult since e values may be both positive and negative and a change in the reference e may change

TABLE 14
Q—*e* Parameters

Monomer	*e*	*Q*
p-Methoxystyrene	−1.11	1.36
Butadiene	−1.05	2.39
Styrene	(−0.80)[a]	(1.00)[a]
2-Vinylpyridine	−0.50	1.30
Vinyl acetate	−0.22	0.026
Vinyl chloride	0.20	0.044
Methyl methacrylate	0.40	0.74
Methyl acrylate	0.60	0.42
Methacrylic acid	0.65	2.34
Acrylic acid	0.77	1.15
Methacrylonitrile	0.81	1.12
Vinyl trifluoroacetate	1.06	0.50
Acrylonitrile	1.20	0.60
Vinyl fluoride	1.28	0.012
Maleic anhydride	2.25	0.23
Vinylidene cyanide	2.58	20.13

[a] These figures are assumed and all others are derived from them.

both the magnitude and the sign of the values for other monomers. The originators of the scheme used the value $e = -0.8$ for styrene and it is this value which is used in Table 14, which lists *Q*—*e* parameters for a number of monomers.

Although the *Q*—*e* scheme is a satisfactory empirical approach to the prediction of reactivity, attempts have been made to correlate its parameters with more fundamental factors, having some real significance in terms of structure; in this context the choice of reference point for assignment of *Q* and *e* values becomes important and there have been many discussions concerned with establishing the best values for *Q* and *e* for styrene [52, 56, 57, 63—66]. For example, Kawabata et al. [56] presented a revised list of *Q* and *e* parameters based on the values $Q = 1.0$ $e = 0$ for styrene; the values thus obtained were considered to have some relevance in terms of structure. Thus the *Q* values for many monomers correlate in a linear fashion with methyl affinities and the *P* values for radicals are found to be constant in copolymerization of a given radical with a wide range of monomers. Unfortunately these factors are partially offset by the fact that the proposed *e* values behave in a much less regular fashion and the authors propose that it is the product $e_r e_m$ (termed e_{rm}) which has physical significance.

Although the empirical value of the Q—e approach cannot be disputed it is clear that the difficulty of assigning absolute values of Q and e is a limiting factor in attempting to find a satisfactory basis for interpretation of the parameters in terms of structure. Comparison with theoretical work, in so far as the results of such work are regarded as valid, suggests that for styrene $Q = 1.0$, $e = 0$ are the preferred values and we shall show later that there are other good reasons for this choice. However, the majority of the available Q—e values are computed on the basis of $Q = 1.0$, $e = -0.8$ for styrene and are not readily amenable to interpretation in terms of structure. It is therefore desirable to consider alternative approaches.

4.4 "PATTERNS OF REACTIVITY" APPROACH

The methods of making structure-reactivity correlations which have been discussed so far fall into two types. The first group are those based upon attempts to correlate reactivity with theoretical parameters, such as charge density or delocalization energy, which are calculable on the basis of valence theory, whilst the second approach, typified by the Q—e scheme, is essentially an empirical method based on the self-consistent use of arbitrary parameters with little theoretical justification. Although both of these methods and their comparison, lead to interesting and useful results, their use can be criticized in many respects and it is of interest to consider alternative approaches to the problem of structure—reactivity correlation. One such approach is the semi-empirical scheme described by its authors Bamford, Jenkins and Johnston [63, 67, 68] under the title "Patterns of Free Radical Reactivity"; for convenience in discussion we shall refer to it simply as "Patterns". This scheme is very successful in interpreting radical reactivity in a number of monomers and is worthy of more detailed examination.

4.4.1 The basis of the scheme

The Patterns approach to reactivity adopts essentially similar assumptions to the Q—e scheme in that it is considered that reactivity may be ascribed, in the absence of specific steric effects, to a combination of the effects of polar and non-polar contributions. However, the two methods diverge considerably in that the Q—e scheme assigns arbitrary parameters as a measure of polar and non-polar influences, whereas the Patterns scheme assesses their relative importance on the basis of experimentally determined parameters. Furthermore, the Patterns approach does not distinguish addition and transfer processes; both atom and electron transfer reactions are considered and may be represented by the equations

$$X_n\cdot + YQ \longrightarrow X_n{-}Y + Q\cdot \qquad \text{Atom transfer} \qquad (a)$$

$$\left.\begin{array}{l} X_n\cdot + M^{Z+} \longrightarrow X_n^{+} + M^{(Z-1)+} \\ X_n\cdot + M^{Z-} \longrightarrow X_n^{-} + M^{(Z-1)-} \end{array}\right. \qquad \text{Electron transfer} \qquad (b)$$

in which $X_n\cdot$ is a propagating polymer radical, Y is an atom, usually hydrogen or a halogen, which is transferred from the molecule YQ and M is a metal or other ion. The propagation reaction in homo- or copolymerization can formally be represented as

$$X_n\cdot + \rangle C{=}C\langle \longrightarrow X_n{-}\overset{|}{\underset{|}{C}}{-}\overset{|}{\underset{|}{C}}\cdot$$

In the Patterns scheme it is postulated that the rates of addition and transfer processes are determined by a combination of resonance and polar effects. Thus in atom transfer reactions the relative dissociation energies of the $X_n{-}Y$ and $Q{-}Y$ bonds are important in determining reactivity; these energies are in their turn determined not only by the nature of the atoms involved, but also by the resonance energies of the radicals $X_n\cdot$ and $Q\cdot$. Similarly the rate of propagation will be partially determined by the difference in resonance energy between $X_n\cdot$ and $X_n{-}\overset{|}{\underset{|}{C}}{-}\overset{|}{\underset{|}{C}}\cdot$. It is clear from the foregoing discussion that polar contributions to transition state energy will also be important factors but, if their effect is to be quantitatively assessed, it is necessary to have some means of estimating the effect of non-polar contributions, which will be formally analogous to the *P* and *Q* parameters of the *Q*—*e* scheme whilst being more soundly based. We thus require a quantitative measure of the inherent reactivities of radicals in reactions in which there are no polar contributions to the transition state; the Patterns method uses as such a parameter the rate coefficient for the atom transfer reaction with toluene, a reaction in which it is reasonable to assume that polar effects will be at a minimum. Table 15 lists the appropriate rate coefficients in order of increasing reactivity for the radicals of the six monomers considered earlier. In order to retain consistency with the original literature we adopt the symbol k_3 for the specific rate coefficient of a general atom transfer reaction and the symbol $k_{3,\mathrm{T}}$ for the reaction with toluene. Since any other influences are to be regarded as modifying the transition-state energy, with a resulting change in the logarithm of the rate coefficient, double logarithmic plots may be used to compare the reactivity of radicals in Table 15 with respect to toluene and any other substrate. Figure 5

TABLE 15
Transfer constants and Hammett σ functions for radicals

Radical	$k_{3,T}$ (l mole^{-1} sec^{-1})	4 + log $k_{3,T}$	σ[a]
Styrene	2.11×10^{-3}	1.324	−0.01
Methacrylonitrile	2.0×10^{-2}	2.301	0.49
Methyl methacrylate	1.25×10^{-2}	2.097	0.28
Methyl acrylate	0.56	3.748	0.45
Acrylonitrile	0.785	3.895	0.66
Vinyl acetate	7.73	4.888	0.31

[a] σ-Values are taken from ref. 70.

shows such plots for a series of hydrocarbons; in each case it can be seen that, within experimental error, there is a linear relation between log $k_{3,T}$ and log $k_{3,hc}$ (where hc denotes an hydrocarbon).

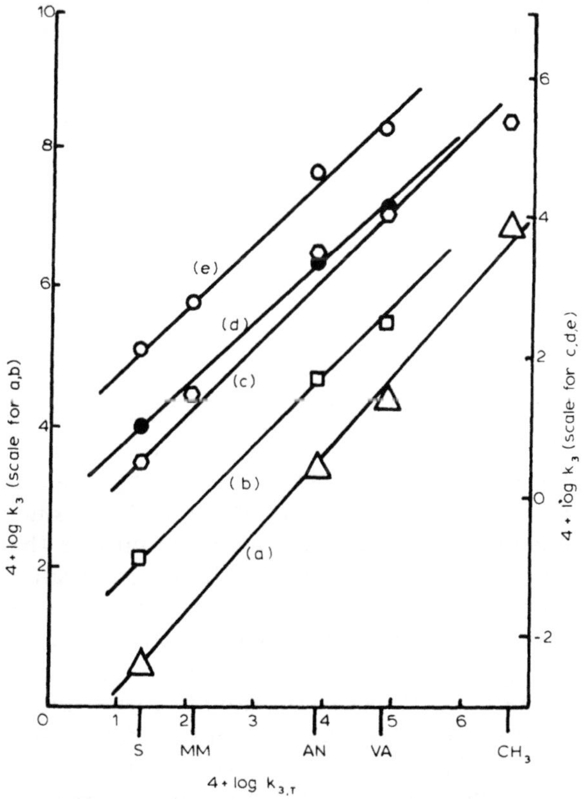

Fig. 5. Double logarithmic plot of the rate coefficients for chain transfer with various hydrocarbons as functions of the rate coefficients for chain transfer with toluene. All reactions at 60 °C. (a) Cyclohexane; (b) isopropylbenzene; (c) benzene; (d) *t*-butylbenzene; (e) ethylbenzene. Abbreviations for radicals as in Fig. 6.

TABLE 16

γ-Values for transfer reactions of hydrocarbons

Hydrocarbon	γ
Benzene	0.15
Toluene	1.0
Ethyl benzene	4.2
Isopropyl benzene	5.6
t-Butyl benzene	0.24
Cyclohexane	0.42

Furthermore, since the slopes of these lines are all unity it follows that there is a direct proportionality

$$k_{3,hc} = \gamma k_{3,T} \tag{17}$$

This result implies that the dominant influences in transfer reactions with hydrocarbons are independent of the hydrocarbon, a result which is particularly striking, since reference to Fig. 5 shows that the range of $k_{3,T}$ values is five orders of magnitude and the range of $k_{3,hc}$ values is eight orders of magnitude. Thus the choice of toluene as a reference standard for the measure of non-polar reactivity may be justified on the grounds that any other hydrocarbon would be equally valid. The main advantages of toluene lie in the fact that $k_{3,T}$ values are usually high enough to be measured with reasonable accuracy and more values have been determined than for any other hydrocarbon. Some values of γ for typical substrates are shown in Table 16.

4.4.2 *Extension to polar substrates*

It will be noted that the order of reactivity of the six standard radicals in their transfer reactions with toluene given in Table 15 is identical to the proposed order of inherent reactivity of these radicals in copolymerization (see Sect. 3.1.1 and Table 5). The linear correlations of $k_{3,T}$ values with rate coefficients for transfer with other hydrocarbons implies that this order of reactivity is maintained in all reactions with non-polar substrates. We have seen that the concept of inherent reactivity of radicals breaks down when applied to their reactions with polar substrates and it is of interest to examine the way in which this behaviour is reflected in the attempted correlation of rate coefficients by double logarithmic plots. Figures 6—14 show the results of such correlations for a series of substrates of increasing polarity. For vinyl acetate and styrene, both substrates of low polarity, straight lines of unit slope are observed (Fig. 6) indicating that these monomers are similar to hydrocarbons and that their reactivity is governed by the same factors. However, with increasing polarity through the series methyl methacrylate, methacrylonitrile,

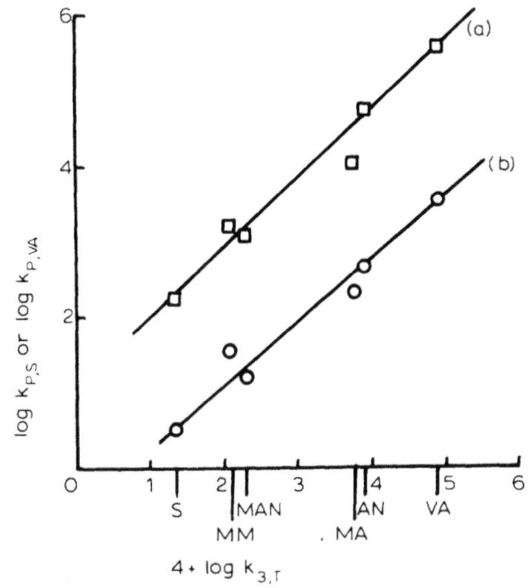

Fig. 6. Double logarithmic plots of the rate coefficients for addition of transfer reactions with the substrate indicated as functions of the rate coefficients for chain transfer with toluene. All reactions at 60 °C. (a) Styrene; (b) vinyl acetate. Abbreviations for radicals: S, styrene; MM, methyl methacrylate; MAN, methacrylonitrile; MA, methyl acrylate; AN, acrylonitrile; VA, vinyl acetate. These abbreviations are also used in Figs. 7—14.

methyl acrylate and acrylonitrile (Figs. 7—10) there is increasing deviation of the results from a straight line although the points form a pattern which, with minor modifications, is the same in each case. Furthermore, as the points become more widely scattered the dotted line joining the points for styrene and vinyl acetate radicals becomes progressively lower in slope. (This line is not intended to have any fundamental significance

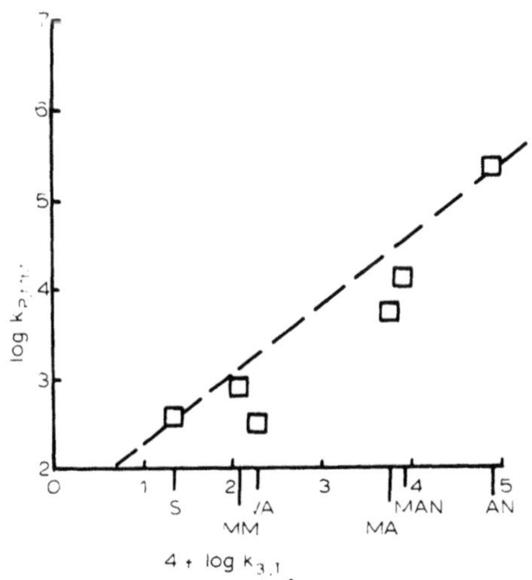

Fig. 7. Methyl methacrylate.

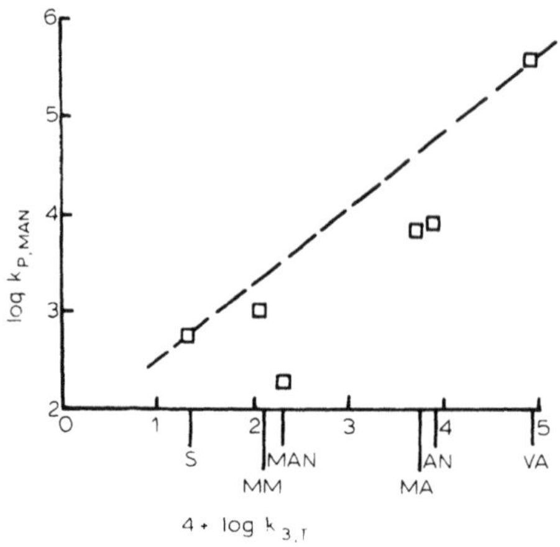

Fig. 8. Methacrylonitrile.

but is included in order to facilitate comparison of the plots.) With transfer reactions to polar substrates, carbon tetrabromide, *n*-butyl mercaptan and ferric chloride (Figs. 11—13) the same pattern is observed but it becomes progressively more exaggerated. In some exceptional cases such as triethylamine (Fig. 14) a different type of pattern is observed in which the points lie above the datum line between styrene and vinyl acetate radicals.

Walling and Mayo and their collaborators [35, 36, 39] suggested that the transition states for radical propagation and transfer reactions may contain contributions from polar structures such as $X^+Y \cdot Q^-$ and $X^+[>C{-}C]^{\dot{-}}$, and this supposition has been confirmed by further work on atom transfer processes. If such polar contributions are important then

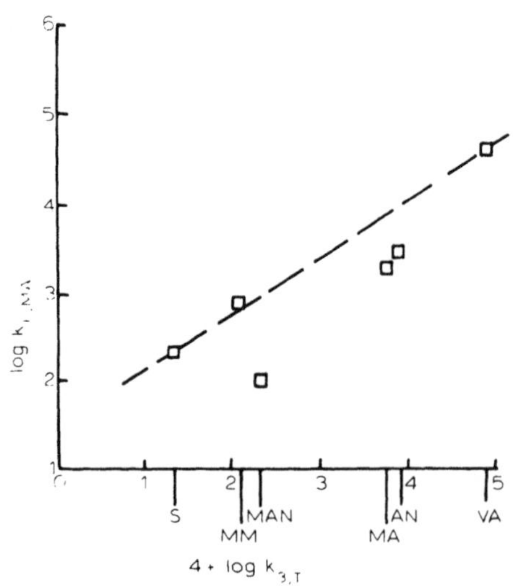

Fig. 9. Methyl acrylate.

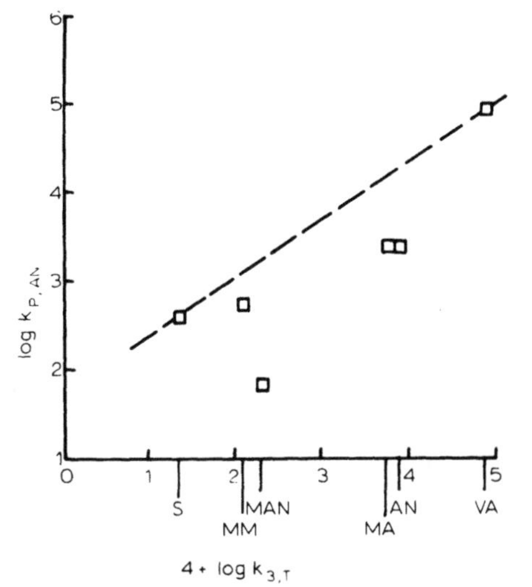

Fig. 10. Acrylonitrile.

the electron donating or accepting capacities of the reactants will have a considerable influence on the reaction rate, an influence which will be expected to reach a maximum in reactions with highly polar substrates such as metal ions. Examination of the data in Figs. 6—14 will show that, qualitatively, the deviations from the straight line due to inherent reactivity increase with increasing substrate polarity. Clearly if these deviations can be placed upon a quantitative footing they may be used as

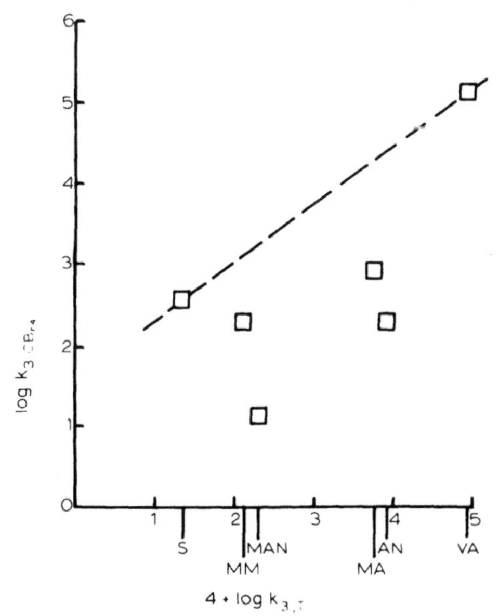

Fig. 11. Carbon tetrabromide.

References pp. 471—473

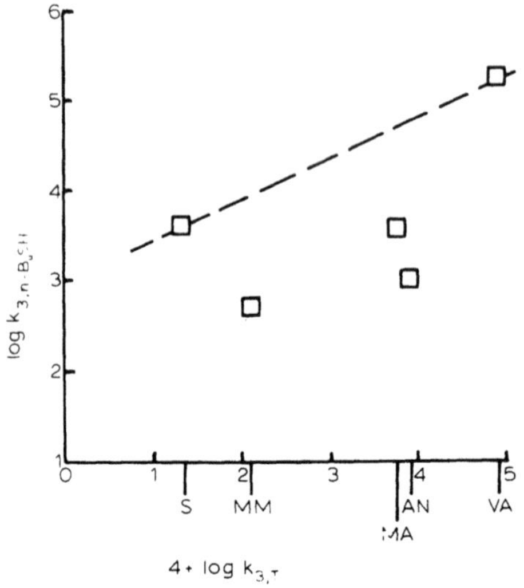

Fig. 12. *n*-Butyl mercaptan.

a basis for assessment of the influence of radical and substrate polarity on reactivity.

From a kinetic viewpoint the effect of polar contributions to radical reactivity may be interpreted in terms of the perturbation of the

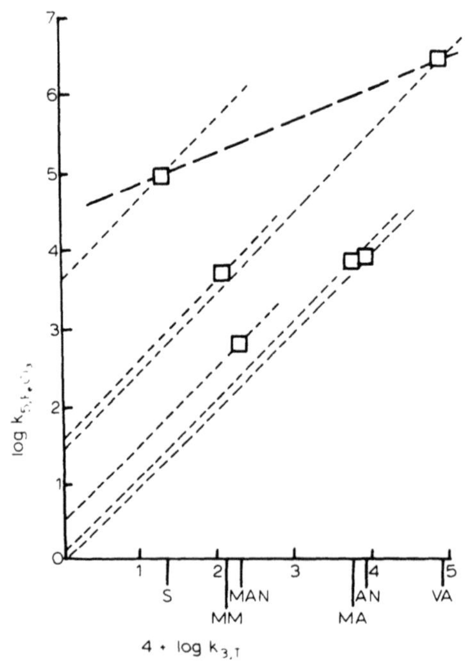

Fig. 13. Ferric chloride.

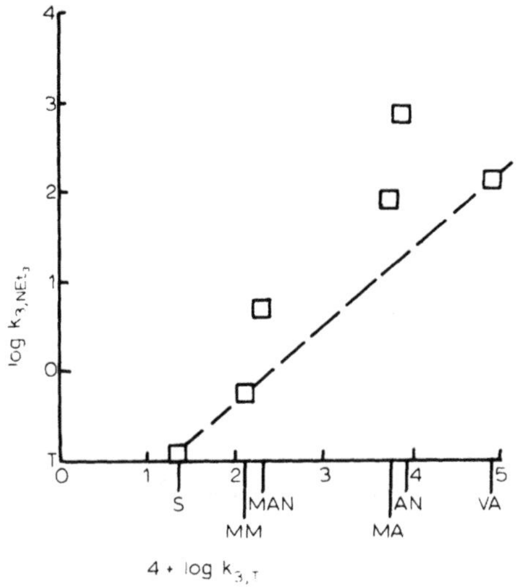

Fig. 14. Triethylamine.

transition state for reaction, with a consequent variation in the activation energy. Let us consider the reaction of a radical X· with toluene (T) and any other substrate (S) and let the changes in activation energy resulting from polar contributions to these reactions be ΔE_T and ΔE_s respectively. The additional change in activation energy for reaction with S, relative to reaction with T, is then

$$\Delta E = \Delta E_s - \Delta E_T \quad (18)$$

and we may consider that $P = \Delta E/RT$ is a measure of the polar contribution to the reaction X· + S relative to that for X· + T. Extending this argument to the reactions of two different radicals with the same substrate S we may write

$$k_s = A_s \exp[-(E_s - \Delta E_s)/RT] = k_s^* \exp \Delta E_s/RT \quad (19)$$

$$k_s' = A_s' \exp[-(E_s' - \Delta E_s')/RT] = k_s'^* \exp \Delta E_s'/RT \quad (20)$$

where E_s, E_s' are the hypothetical activation energies for reactions of X· and X'· with S, in the absence of any polar contribution, k_s^* and $k_s'^*$ are the rate coefficients for these two hypothetical reactions and k_s, k_s', A_s and A_s' are the observed rate coefficients and pre-exponential factors for the "real" reaction. Using the same notation we may consider the reactions of X· and X'· with toluene and write

$$k_{3,T} = A_T \exp[-(E_T - \Delta E_T)/RT] = k_{3,T}^* \exp \Delta E_T/RT \quad (21)$$

$$k_{3,T}' = A_T' \exp[-(E_T' - \Delta E_T')/RT] = k_{3,T}^* \exp \Delta E_T'/RT \quad (22)$$

From these four equations we can derive the relationship

$$\log k'_s - \log k_s = \log k'_{3,T} - \log k_{3,T} + \frac{0.434}{RT}(\Delta E'_s - \Delta E_s - \Delta E'_T + \Delta E_T)$$

$$+ (\log k'^*_s - \log k^*_s - \log k'^*_{3,T} + \log k^*_{3,T}) \quad (23)$$

For hydrocarbon and other non-polar substrates the results given in Figs. 5 and 6 imply a relationship of the form

$$\log k'_s - \log k_s = \log k'_{3,T} - \log k_{3,T} \quad (24)$$

so that for such substrates

$$\frac{0.434}{RT}(\Delta E'_s - \Delta E_s - \Delta E'_T + \Delta E_T) = -(\log k'^*_s - \log k^*_s - \log k^*_{3,T}$$

$$+ \log k^*_{3,T}) \quad (25)$$

Now, in reactions with non-polar substrates the ΔE_s values will be expected to be small and the difference between them even smaller so that the left-hand term in eqn. (25) must be close to zero. Thus the right-hand term will also be zero for non-polar substrates; in the present treatment we shall assume that this term remains zero for polar substrates, since the k^* values refer to hypothetical reactions in which there is no polar contribution. Equation 23 then reduces to

$$\log k'_s - \log k_s = \log k'_{3,T} - \log k_{3,T} + P' - P \quad (26)$$

If $P' - P = 0$ the points for any two radicals will lie on a straight line of unit slope; in other cases it follows from eqn. (26) that $P' - P$ is given by the vertical separation of two lines of unit slope passing through the points for radicals X′ and X. Since we are only concerned with differences in polarity we can separate eqn. (26) into two components and write the general equation

$$\log k_s = \log k_{3,T} + P \quad (27)$$

For any radical, P can thus be estimated from the double logarithmic plot of log k_s against log $k_{3,T}$ as the intercept, on the ordinate, of a straight line of unit slope through the point corresponding to the particular radical. For convenience we use decadic logarithms and the abscissa ($4 + \log k_{3,T}$), which gives P values close to unity but does not alter the differences in P values for different species.

Using this approach we may estimate P values for every point in Figs. 6—14 by the simple procedure of drawing a line of unit slope through each point in turn and determining its intercept. The spread of such lines must be a measure of polar contributions to the reaction; where the lines

TABLE 17
Polarity parameters for radical—substrate reactions

Radical	Substrate												
	$FeCl_3$	*n*-BuSH	CBr_4	AN	MA	MAN	$CHCl_3$	MM	VC	VA	S	C_6H_6	NEt_3
Styrene	3.64	2.26	1.26	1.30	1.04	1.44	0.61	1.20	−0.32	−0.82	0.92	−4.82	−2.23
Methyl methacrylate	1.61	0.59	0.20	0.64	0.80	0.94	0.42	0.77	−0.35	−0.53	1.10	−4.63	−2.31
Vinyl acetate	1.48	0.36	0.27	0.08	−0.32	0.68	0.78	0.50	−0.83	−1.32	0.68	−4.85	−2.75
Methacrylonitrile	0.49	—	−1.15	−0.43	−0.30	0.0	—	0.19	—	−1.08	0.90	—	−1.57
Methyl acrylate	0.08	−0.20	−0.82	−0.35	−0.43	0.09	−0.33	−0.05	−1.38	−1.38	0.31	—	−1.83
Acrylonitrile	−0.06	−0.89	−1.57	−0.50	−0.42	−0.01	−0.01	0.25	−1.02	−1.20	0.90	−4.39	−1.02
Intrinsic polarity	1.25	1.06	0.96	0.66	0.62	0.55	0.41	0.41	0.40	0.30	0.24	0.17	−0.56

coincide the polar contribution is taken as zero and in strongly polar cases the separation is large. It should be noted at this stage that the values of *P* assigned in this way may be regarded as a quantitative measure of polar contributions to a given radical—substrate reaction; *P* values will contain contributions from both radical and substrate reactivity and we shall show in Sect. 4.4.3 how these contributions may be separated.

Table 17 shows the values of *P* for the reactions of our six standard radicals with a variety of substrates in both copolymerization and chain transfer. The radicals are listed in decreasing order of the values of *P* for reactions with ferric chloride, the compound showing the greatest spread of values. It will be seen that essentially the same order is maintained with all substrates, although as might be expected any real semblance of order disappears in reactions with comparatively non-polar substrates such as styrene and vinyl acetate.

It will be recalled that the absolute values of rate coefficients for both copolymerization and transfer reactions are determined from a knowledge of the propagation rate coefficient for homopolymerization of the monomer concerned, so that any error in measurement of this latter quantity will reflect in the accuracy with which the rate coefficients for radical substrate reactions are known; this error will apply to all values including $k_{3,\mathrm{T}}$. In the present context the affect of such an error will be to displace the points concerned along a line of unit slope since the error will reflect equally in the values of log k_s and log $k_{3,\mathrm{T}}$; for this reason the derived values of the polarity parameter *P* will be independent of any errors in the propagation rate coefficient.

4.4.3 Separation of radical and substrate reactivities

In the foregoing discussion we have shown how the rate coefficient for any radical—substrate reaction can be split into two components, a non-polar contribution measured by the value of $k_{3,\mathrm{T}}$ and a polar contribution measured by *P*. Furthermore, the order of polarity parameters for a given series of radicals is constant and independent of the substrate although the spread of the *P* values about the mean varies with the substrate concerned, the spread being largest for highly polar substrates such as $FeCl_3$ and smallest for non-polar substrates such as styrene. It has been proposed that this degree of spread of the *P* values can be used to assign to each substrate a figure which represents its ability to involve polar transition states in its reactions; this so-called *intrinsic polarity* (IP) is defined as the root mean square deviation of the *P* values from the mean, for a series of reactions of a given substrate. For this value to be of significance it must clearly be deduced for each substrate from its reactions with the same series of radicals and only when data for all of the radicals is available. Values of IP are given in Table 17; it is notable that for triethylamine the order of *P* values is the reverse of that for ferric chloride and the IP for the former compound correspondingly has a negative value.

Although intrinsic polarity is a useful concept its derivation and the limitations on its value are in distinct contrast to the declared aim of the Patterns approach, which is the assessment of radical and substrate reactivities in terms of experimentally measured parameters. Accordingly it is desirable to seek an interpretation of the values of *P* in terms of some such quantity. The most successful quantity is found to be the Hammett σ parameter for the substituent(s) on the α-carbon atom of the radical; the values of this function are accepted as being a measure of the extent to which substituents affect the electron density at the reaction site and they have been widely used with great success for the correlation of reactivity data in physical-organic chemistry.

Increasing σ values correspond to an increasing electron-withdrawing capacity on the part of the substituent; with the exception of styrene (for which σ is virtually zero) all of our six radicals contain substituents which are electron-withdrawing so that the importance of polar transition states in which the radical bears a positive charge will increase as σ increases.

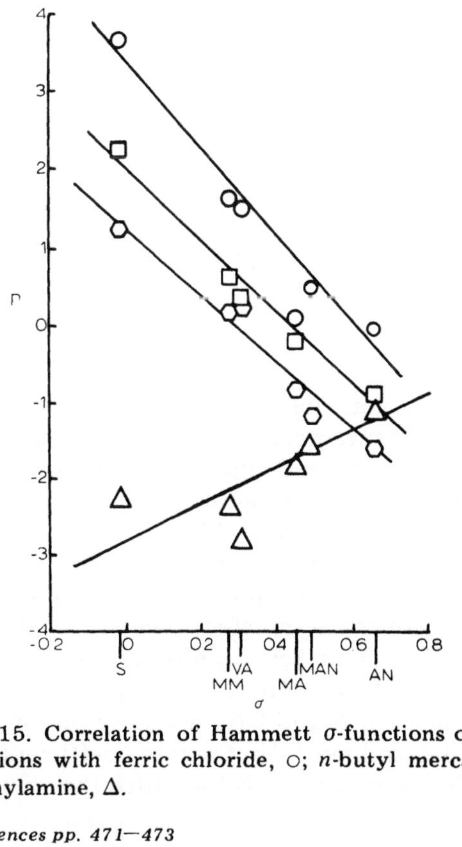

Fig. 15. Correlation of Hammett σ-functions of radicals and polarity parameters for reactions with ferric chloride, ○; *n*-butyl mercaptan, □; carbon tetrabromide, ⬡; and triethylamine, Δ.

Figure 15 shows the results of plotting P against the *para*-σ-values for the α substituents in the six radicals, for reactions with carbon tetrabromide, ferric chloride and triethylamine. Where the radical carries two α substituents the algebraic sum of the σ values has been used. The linear correlation exhibited in this Fig. is found to hold not only for these three cases but for virtually every reaction for which data is available; for triethylamine the correlation with *para*-σ-values is good (with the exception of the anomalous point for styrene) and indeed it is, perhaps surprisingly, considerably better than the correlation with σ^+ values which are especially applicable to electrophilic reactions.

Although all plots of P against σ are straight lines, these lines do not coincide. It therefore follows that the values of P can be expressed by an equation of the form

$$P = \alpha\sigma + \beta \tag{28}$$

where α and β are constants for a given substrate and represent the slope and intercept of the P—σ plot, respectively. It thus follows that eqn. (27) can be expressed in the form

$$\log k_s = \log k_{3,T} + \alpha\sigma + \beta \tag{29}$$

in which $k_{3,T}$ and σ are parameters characteristic of the radical and α and β are parameters characteristic of the substrate. This equation is the final form of the Patterns expression for the rate coefficient of a radical—substrate reaction and it fulfils the objective of separating reactivity into contributions which may be independently assigned to the radical and the substrate. Clearly the term β is a measure of the contribution of the substrate towards reactivity in reactions with a non-polar radical for which $\sigma = 0$; it follows by comparison with eqn. (17) that, under these conditions

$$\beta = \log \gamma \tag{30}$$

The term $\alpha\sigma$ represents the additional polar contribution to the rate coefficient k_s which occurs when both radical and substrate exert polar influences; the magnitude of this term depends upon both the extent of the contribution of polar structures and the direction of charge transfer between radical and substrate. Reference to Fig. 15 will show that for triethylamine the value of α is positive whilst for carbon tetrabromide and ferric chloride it is negative. These values are qualitatively in agreement with the proposed transition states for these transfer processes in which it is believed that carbon tetrabromide and ferric chloride show charge transfer in the opposite sense to triethylamine [5—7, 71, 72].

Table 18 shows a list of the values of α and β for a variety of common substrates. These values may be determined from the rate coefficients for the reaction of the substrate with any two radicals for which the values of

TABLE 18
α and β values for substrates

Substrate	α	β
Triethylamine	+2.4	1.8
N,N-Dimethylformamide	+1.0	−0.2
p-Methoxystyrene	+1.0	4.86
Butadiene	0	5.03
Styrene	0	4.85
Vinyl acetate	0	3.00
2-Vinylpyridine	−1.0	5.18
Chloroform	−1.4	0.90
Vinyl chloride	−1.5	3.65
Methyl methacrylate	−1.5	4.90
Vinyl trifluoroacetate	−2.0	3.22
Methacrylonitrile	−2.5	5.40
Methyl acrylate	−3.0	5.20
Acrylic acid	−3.0	5.52
Acrylonitrile	−3.0	5.30
Vinyl fluoride	−3.5	3.77
Carbon tetrabromide	−4.3	5.25
Maleic anhydride	−4.5	5.75
n-Butyl mercaptan	−4.8	6.05
Vinylidene cyanide	−5.0	7.22
Ferric chloride	−5.65	7.40

$k_{3,\mathrm{T}}$ and σ are established, although in practice it is usual to derive the values from data relating to more than two radicals wherever possible.

In the reactions of radicals with polymerizable monomers there is a close relationship between the structure of a radical $\sim\!CH_2-\dot{C}HY$ and the structure of its parent monomer $CH_2=CHY$. In the Patterns treatment the polar properties of a monomer and its derived radical would be expressed by the parameters α and σ, respectively. Since these polar contributions arise from the same source, the modification of electron density by the substituent Y, in each case, it is reasonable to enquire whether there is any relationship between α and σ for monomers and their derived radicals. Such a relationship is indeed found as shown in Fig. 16, in which the straight line corresponds to the relation

$$\alpha = -5.3\sigma \qquad (31)$$

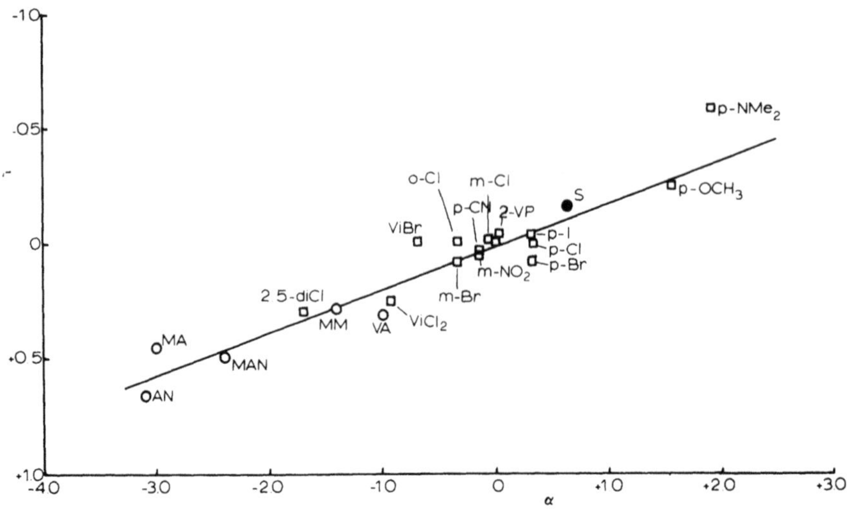

Fig. 16. Correlation of α-values for monomers and Hammett σ-values for conjugate radicals. The monomers include substituted styrenes and vinylbromide (ViBr), 2-vinylpyridine (2-VP) and vinylidene chloride ($ViCl_2$).

Although it is tempting to suggest that this is a rigorous relationship such is not the case since the degree of scatter on this plot is far greater than the range of experimental error; the relationship should only be used to obtain α values in cases where the experimental value is not accessible and in such cases it should be used with great care.

It was stated earlier that where a radical bears two substituent groups the σ value was taken as the algebraic sum of the values for the two groups. Clearly if α and β and $k_{3,T}$ are known, eqn. (29) can be used to assess the value of σ. In the case of disubstituted monomers the results are instructive. Where the σ values are of opposite sign there is good agreement between measured and experimental values (methyl methacrylate, calculated = experimental = 0.28; methacrylonitrile, calculated = 0.45, experimental = 0.47). However, where the σ values are the same sign it is found that the effective σ is always less than the algebraic sum (vinylidene chloride, experimental = 0.38, 2 x vinyl chloride = 0.56; vinylidene cyanide, experimental = 0.94, 2 x acrylonitrile = 1.12). In the latter case the second substituent may be regarded as exerting its influence on a molecule which is already polarized in the same sense by the first substituent; the observed effect is thus not surprising but it does point to the need for caution in applying the Patterns approach to disubstituted monomers.

4.4.4 The graphical representation of radical reactivity

It has been pointed out several times during the course of this review that the concept of an inherent order of radical reactivity is at best a poor approximation since, experimentally, the order of radical reactivity is found to depend upon the substrate. This effect arises directly from the term $\alpha\sigma$ in eqn. (29) which represents the effect of polar interaction between a radical and substrate. Examination of eqn. (29) will show that the quantity $\log k_s - \beta$ is a linear function of α, with slope equal to σ and an intercept at $\alpha = 0$ equal to $\log k_{3,T}$. Thus this line can be constructed for any radical for which $k_{3,T}$ and σ are known; the effect of substrate polarity on the order of radical reactivity is readily demonstrated by consideration of a collection of such plots such as that shown in Fig. 17 for the six standard radicals. A vertical line drawn through the α value for any given substrate will cross the lines for the radicals in the order of their reactivity towards the substrate concerned. This conclusion is valid since for any given substrate the value of β is a constant which enters similarly into each ordinate value, hence affecting all k_s values equally. Clearly a single order of radical reactivity can only be observed for all substrates if the plots in Fig. 17 do not cross, a condition which is not fulfilled in practice. Whenever two radical lines cross there will be an inversion in the order of reactivity as α varies on either side of its value at the crossing point. At sufficiently positive and negative values of α there would be two orders of radical reactivity which would be the inverse of each other; so far the extremes of α encountered are +3 and −6 but even so the styrene radical changes its position from lowest to second highest as the α values changes within this range. On this basis it is clear that the Patterns approach provides an excellent quantitative interpretation of the way in which the order of radical reactivity is modified by substrate polarity.

4.4.5 The graphical representation of substrate reactivity

Figure 17 illustrates the way in which eqn. (29) may be plotted in order to display graphically the order of reactivity of a series of radicals towards

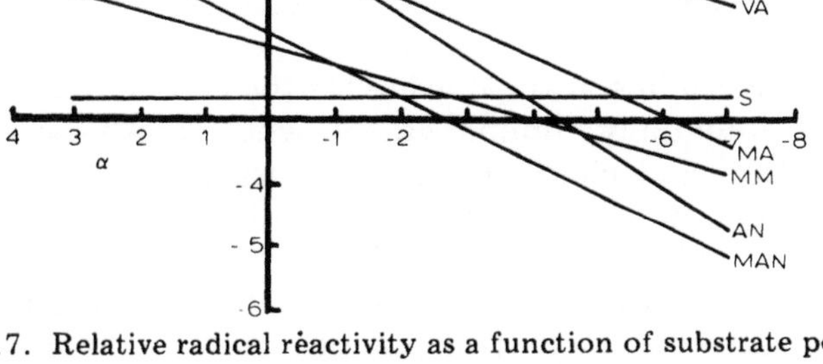

Fig. 17. Relative radical reactivity as a function of substrate polarity.

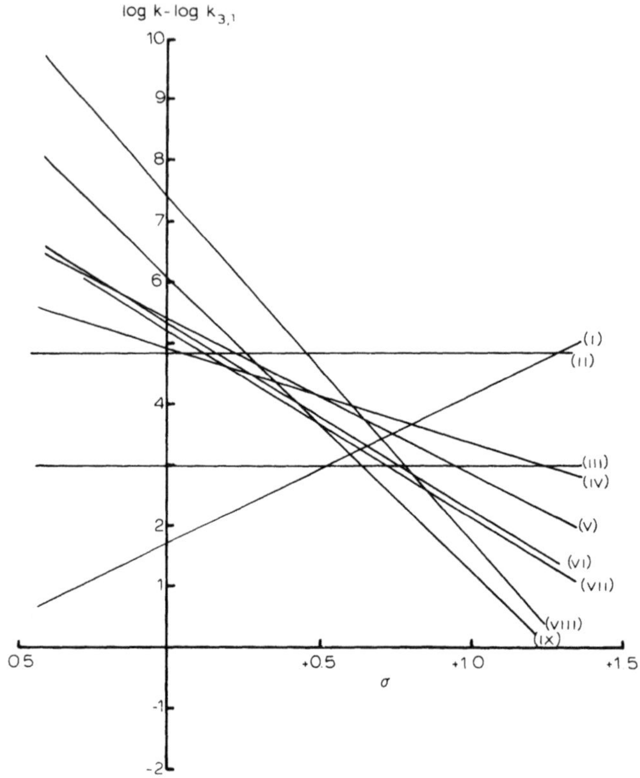

Fig. 18. Relative substrate reactivity as a function of radical polarity: (i) triethylamine; (ii) styrene; (iii) vinyl acetate; (iv) methyl methacrylate; (v) methacrylonitrile; (vi) acrylonitrile; (vii) methylacrylate; (viii) ferric chloride; (ix) *n*-butyl mercaptan.

any substrate whose α value is known; if the β value is also known the same plot allows values of k_s to be calculated. An alternative treatment of eqn. (29) allows us to display the way in which the order of reactivity of a series of substrates towards a given radical varies with the radical structure. Thus, examination of this equation will show that the function $\log k_s - \log k_{3,T}$ is linearly related to σ, with a slope of α and an ordinate intercept equal to β; this line can thus be constructed for any substrate for which α and β are known. Figure 18, which is analogous to Fig. 17, shows a series of such plots for each substrate for which α and β are known. The order of reactivity of any radical towards these substrates is readily assessed by drawing a vertical line at the σ value for the chosen radical.

Figures 17 and 18 potentially summarize the whole of the available data on copolymerization and chain transfer; by using the appropriate plot it is possible to assess the order of reactivity and the values of k_s for reaction of any substrate with a series of radicals or for reaction of any radical with a series of substrates. Strictly speaking we require a pair of plots for each temperature considered; in practice most of the available reactivity ratios and transfer constants have been measured over a narrow temperature range so that one pair of plots is sufficient for all normal requirements.

4.5 COPOLYMERIZATION REACTIONS AND THE PATTERNS SCHEME

4.5.1 Estimation of reactivity ratios

The Patterns scheme, unlike many others, is universal in its application, being equally valid for both polymerizable and non-polymerizable substrates. However, at this point it is instructive to consider its application in the field of copolymerization behaviour. Since Fig. 18 allows us to estimate the order of reactivity of any series of substrates towards a given radical, it is obviously possible to use this diagram as a means of determining copolymerization reactivity ratios. Thus if we have two monomers A and B with corresponding σ values σ_a and σ_b we may draw a vertical line through σ_a which will intersect the lines for monomers A and B at points with a vertical separation equal to $\log r_a$; similarly a vertical line through σ_b gives $\log r_b$. Note that it is necessary to take into account the order in which the vertical line intersects the plots for the two monomers in order to assign the correct sign to $\log r$. It will readily be seen that the same approach can be applied to the estimation of chain transfer constants since a vertical line at σ for the radical will intersect the lines for monomer and substrate with a separation equal to $\log C_s$.

The graphical procedure for evaluation of reactivity ratios may be expressed algebraically, the corresponding equations being

$$\log r_a = \sigma_a(\alpha_a - \alpha_b) + (\beta_a - \beta_b) \tag{32}$$

$$\log r_b = -\sigma_b(\alpha_a - \alpha_b) - (\beta_a - \beta_b) \tag{33}$$

If we recall the equivalent equation of the Q—e scheme, viz.

$$\ln r_a = -e_a(e_a - e_b) + \ln(Q_a/Q_b) \tag{34}$$

$$\ln r_b = e_b(e_a - e_b) - \ln(Q_a/Q_b) \tag{35}$$

then it is obvious that these two sets of equations are formally similar, a similarity which is emphasized if we rewrite eqns. (34) and (35), using the approximate correlation of α and σ expressed in eqn. (31) to give

$$\log r_a = -5.3\sigma_a(\sigma_a - \sigma_b) + (\beta_a - \beta_b) \tag{36}$$

$$\log r_b = 5.3\sigma_b(\sigma_a - \sigma_b) - (\beta_a - \beta_b) \tag{37}$$

Figure 19 illustrates the correlation between α and e values suggested by these equations; it can be seen that there is a qualitative correlation

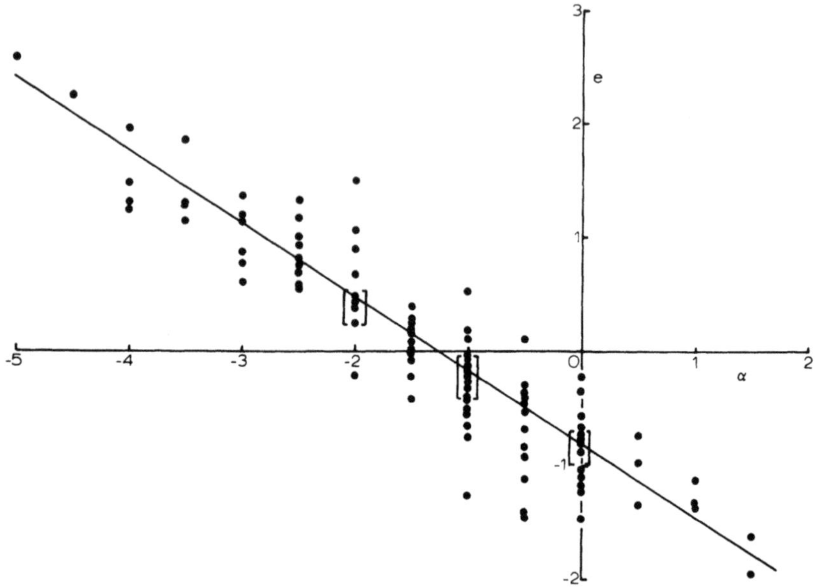

Fig. 19. Correlation between α- and e-values. In the regions in which a number of points are enclosed in brackets there are many more available experimental data than can be clearly plotted, hence these regions should be given extra weighting in any statistical analyses.

which may be expressed in the equations

$$\alpha = -1.55e - 1.25 \tag{38}$$

or

$$e + 0.8 = -0.645\alpha \tag{39}$$

The Patterns and Q—e schemes would thus formally correspond completely if the e value for styrene were 0 instead of —0.8, a conclusion which adds support to the proposal of this basic adjustment of e values by Kawabata et al. [56, 57].

4.5.2 Copolymerization of three monomers

Any three monomers, A, B and C can be used to produce three binary copolymerization mixtures which will be characterized by a total of six reactivity ratios, viz. r_{ab} and r_{ba} from A + B, r_{ac} and r_{ca} from A + C and r_{bc} and r_{bc} from B + C. Ham [73] has proposed that in all such cases the reactivity ratios should be related by

$$r_{ab} \cdot r_{bc} \cdot r_{ca} = r_{ac} \cdot r_{cb} \cdot r_{ba} \tag{40}$$

This relationship follows directly from the Q—e eqn. (13) and (14) [74] and thus, by definition, it is also directly derivable from the simplified

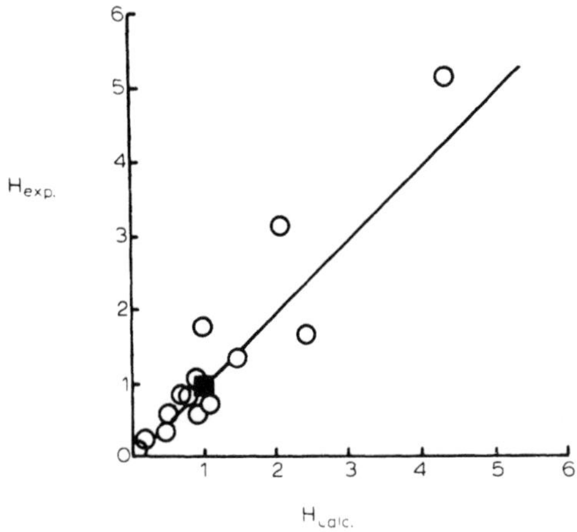

Fig. 20. Comparison of observed and calculated *H*-factors, ○, points corresponding to *H*-values calculated from the Patterns treatment: line of unit slope denotes complete agreement between experiment and calculations of this type. *Q—e* scheme predicts that all points coincide at *H* (experimental) = *H* (calcuiated) = 1, denoted by ■.

Patterns eqns. (36) and (37). However, Jenkins [75] has shown that application of the unsimplified Patterns equations leads to a different result.

Following Mayo [74] we may define a parameter *H*, such that

$$H = \frac{r_{ab} \cdot r_{bc} \cdot r_{ca}}{r_{ac} \cdot r_{cb} \cdot r_{ba}} \tag{41}$$

According to both the *Q—e* scheme and the simplified Patterns equations *H* is necessarily unity. However, application of the full Patterns equations gives

$$-\log H = \sigma_a(\alpha_b - \alpha_c) + \sigma_b(\alpha_c - \alpha_a) + \sigma_c(\alpha_a - \alpha_b) \tag{42}$$

so that the parameter *H* can be computed from a knowledge of the appropriate σ and α values. Figure 20 shows the excellent correlation between *H* values measured experimentally for fourteen systems of the three monomers and the values calculated from eqn. (42). This correlation is an excellent illustration of the power of the Patterns approach to account for radical and monomer reactivity in a situation where the *Q—e* scheme is totally inadequate.

4.6 COMPUTER-BASED ASSIGNMENT OF REACTIVITY PARAMETERS

Recently Hoyland [76—78] has made a new approach to the rationalization of radical reactivity by using computer methods to assign the best values of parameters which he regards as determining reactivity.

The treatment has been presented in two forms, but in both cases the division into polar and non-polar contributions is made.

In the first variant [76], the non-polar or inherent reactivity parameter is defined as the relative localization energy L of the monomers whilst the polar factor is taken as the Pauling electronegativity X, specified as the difference between the value appropriate to the radical (X_R) and that appropriate to the monomer (X_M). For a binary copolymerization, the reactivity ratio r_A is then thought to be represented quantitatively by the equation

$$\log r_a = [L(A) - L(B)] + [X_R(A) - X_M(A)] - [X_R(A) - X_M(B)] \quad (43)$$

This equation implies that, for any radical R and monomer M, the specific rate coefficient is given by

$$\log k_{RM} = [L_R - L_M] + [X_R - X_M] + \text{Constant} \quad (44)$$

As with the Q—e scheme, to clothe the skeleton with flesh we must first assign parameter values for one species, and Hoyland has chosen the values $L = 0$, $X_R = 0$, for styrene. On this basis the known reactivity data for five selected monomers — styrene, 2-vinyl pyridine, 4-vinyl pyridine, methyl acrylate and methyl methacrylate — were used to determine the best values for the compound parameters L, X_R and X_M by a computer-based least-squares numerical fit. Data for the five primary monomers being settled, a further twelve monomers were similarly appraised using copolymerization data relating to their reactions with the standard substances. The reactivity parameters derived in this way are said to give good agreement with the experimental data; that is to say that a set of parameters can be derived by the computer which, on resubstitution into eqn. (44), regenerate reactivity ratios close to those used in the analysis. An exception is recorded in the case of acrylonitrile, for which the recalculated reactivity data are not satisfactory.

Apart from the arbitrary assignment of values of parameters for one monomer, one might feel that it would have been better to include acrylonitrile among the five basic monomers; it is the most polar of the monomers considered and it would obviously avoid extrapolation if it were used at the primary stage of the treatment. Another point of criticism is that styrene and the two vinyl pyridines are so similar in character that they are not well suited to be used as three of the five primary monomers.

The alternative formulation adopted by Hoyland [77] identifies as the polar parameter in reactivity the difference in the charge transfer energy between the radical and each of the two monomers, whilst retaining the relative localization energy as the parameter characteristic of inherent reactivity. This approach leads again to a three-parameter equation from which a set of self-consistent reactivity parameters can be computed by the least-squares procedure; although the method is more complex than the electronegativity approach, the principle is similar and the results are again poor for the most polar monomer, acrylonitrile.

The use of computer methods is an interesting pointer towards ways of increasing the efficiency of data-handling in a field in which so much experimental work is now on record. However, it is doubtful whether any procedure which relies on the arbitrary assignment of basic parameters can be regarded as entirely satisfactory for fundamental purposes, particularly since the technique has not yet been made applicable to transfer processes.

The manuscript for this chapter was completed in 1972.

REFERENCES

1 C. H. Bamford, W. G. Barb, A. D. Jenkins and P. F. Onyon, The Kinetics of Vinyl Polymerization by Radical Mechanisms, Butterworths, London, 1958, Chap. 1.
2 C. H. Bamford, W. G. Barb, A. D. Jenkins and P. F. Onyon, ref. 1, Chap. 2.
3 C. H. Bamford, W. G. Barb, A. D. Jenkins and P. F. Onyon, ref. 1, p. 232.
4 C. H. Bamford, W. G. Barb, A. D. Jenkins and P. F. Onyon, ref. 1, pp. 235—238.
5 E. Collinson, F. S. Dainton and G. S. McNaughton, J. Chem. Phys., 52 (1955) 565.
6 C. H. Bamford, A. D. Jenkins and R. Johnston, Proc. R. Soc. London, Ser. A 239 (1957) 214.
7 E. Collinson, F. S. Dainton, D. P. Smith, G. J. Trudel and S. Tazuke, Discuss. Faraday Soc., 29 (1960) 188.
8 W. I. Bengough and W. H. Fairservice, Trans. Faraday Soc., 61 (1965) 1206.
9 R. Simha and H. Branson, J. Chem. Phys., 12 (1944) 253.
10 F. T. Wall, J. Am. Chem. Soc., 66 (1944) 2050.
11 F. R. Mayo and F. M. Lewis, J. Am. Chem. Soc., 66 (1944) 1594.
12 P. W. Tidwell and G. A. Mortimer, J. Macromol. Sci., Rev. Macromol. Chem., 4(2) (1970) 281.
13 L. J. Young, J. Polym. Sci., 54 (1961) 411.
14 H. Mark, B. Immergut, E. H. Immergut, L. J. Young and K. I. Beynon, in G. E. Ham (Ed.), Copolymerization, Interscience, New York, 1964.
15 H. Mark, B. Immergut, E. H. Immergut, L. J. Young and K. I. Beynon, in J. Brandrup and E. H. Immergut (Eds.), Polymer Handbook, Interscience, New York, 2nd edn., 1975.
16 A. M. North and G. A. Reed, Trans. Faraday Soc., 57 (1961) 859.
17 P. E. M. Allen and C. R. Patrick, Makromol. Chem., 57 (1961) 154.
18 J. C. Bevington, Trans. Faraday Soc., 48 (1952) 1045.
19 J. C. Bevington, H. W. Melville and R. P. Taylor, J. Polym. Sci., 12 (1954) 449; 14 (1954) 463.
20 C. H. Bamford and A. D. Jenkins, Nature (London), 176 (1955) 78; Proc. R. Soc. London, Ser. A 228 (1955) 220.
21 C. H. Bamford, A. D. Jenkins and R. Johnston, Trans. Faraday Soc., 55 (1959) 418.
22 B. E. Bailey and A. D. Jenkins, Trans. Faraday Soc., 56 (1960) 903.
23 J. C. Bevington and D. E. Eaves, Trans. Faraday Soc., 55 (1959) 1777.
24 G. A. Ayrey and C. G. Moore, J. Polym. Sci., 36 (1959) 41.
25 A. F. Trotman-Dickenson, Ann. Rep. Prog. Chem., 55 (1948) 40.
26 E. P. Bonsall, L. Valentine and H. W. Melville, J. Polym. Sci., 7 (1951) 39.
27 G. M. Burnett and H. R. Gersmann, J. Polym. Sci., 28 (1958) 655.

28 J. H. Bradbury and H. W. Melville, Proc. R. Soc. London, Ser. A, 222 (1954) 456.
29 W. G. Barb, J. Polym. Sci., 11 (1953) 117.
30 G. E. Ham, J. Polym. Sci., 45 (1960) 169, 177, 183.
31 G. E. Ham, J. Polym. Sci., 54 (1961) 1.
32 G. E. Ham, J. Polym. Sci., 61 (1962) 9.
33 C. Walling and F. R. Mayo, J. Polym. Sci., 3 (1948) 895.
34 C. Walling, Free Radicals in Solution, Wiley, New York, 1957, p. 123.
35 C. Walling, E. R. Briggs, K. B. Wolfstern and F. R. Mayo, J. Am. Chem. Soc., 70 (1948) 1537, 1544.
36 C. Walling, D. Seymour and K. B. Wolfstern, J. Am. Chem. Soc., 70 (1948) 2559.
37 J. L. Lang, W. A. Pavelich and A. D. Clarey, J. Polym. Sci., 55 (1961) 531.
38 R. M. Joshi, Makromol. Chem., 53 (1961) 33.
39 M. C. DeWilde and G. Smeto, J. Polym. Sci., 5 (1950) 253.
40 P. D. Bartlett and K. Nozaki, J. Am. Chem. Soc., 68 (1946) 1495.
41 F. S. Dainton and K. J. Ivin, Q. Rev., 12 (1958) 61.
42 H. Sawada, J. Macromol. Sci., Rev. Macromol. Chem., 3(2) (1969) 314.
43 K. F. O'Driscoll, Am. Chem. Soc. Adv. Chem. Ser., 91 (1969) 439.
44 C. H. Bamford, A. D. Jenkins and R. Johnston, Proc. R. Soc. London, Ser. A, 239 (1957) 214.
45 M. Imoto, T. Otsu and S. Shimizu, Makromol Chem., 65 (1963) 174, 180, 194.
46 N. G. Gaylord and A. Takahashi, Am. Chem. Soc. Adv. Chem. Ser., 91 (1969) 94.
47 G. E. Serniuk and R. M. Thomas, U.S. Pat. 3,183,217.
48 G. E. Serniuk and R. M. Thomas, U.S. Pat. 3,278,503.
49 M. Hirooka, Kobunshi, 16 (1967) 1172.
50 M. Hirooka, H. Yabuuchi, S. Morita, S. Kawasumi and K. Nakaguchi, J. Polym. Sci., Part B 5 (1967) 47.
51 E. Mertz, T. Alfrey and G. Goldfinger, J. Polym. Sci., 1 (1946) 75.
52 F. R. Mayo, Chem. Ber. 70 (1966) 233.
53 M. G. Evans, J. Gergely and E. S. Seamen, J. Polym. Sci., 3 (1948) 866.
54 T. Yonezawa, K. Hayashi, C. Nagata, S. Okamura and K. Fukui, J. Polym. Sci., 14 (1954) 312.
55 K. Hayashi, T. Yonezawa, C. Nagata, S. Okamura and K. Fukui, J. Polym. Sci., 20 (1956) 537.
56 T. Kawabata, T. Tsuruta and J. Furukawa, Makromol. Chem., 51 (1962) 70.
57 T. Kawabata, T. Tsuruta and J. Furukawa, Makromol. Chem., 51 (1962) 80.
58 M. J. S. Dewar and C. C. Thompson, J. Am. Chem. Soc., 87 (1965) 4414.
59 A. L. H. Chung and M. J. S. Dewar, J. Chem. Phys., 42 (1965) 756.
60 M. J. S. Dewar and G. J. Gleicher, J. Am. Chem. Soc., 87 (1965) 685, 692.
61 C. C. Price, J. Polym. Sci., 1 (1946) 83.
62 T. Alfrey and C. C. Price, J. Polym. Sci., 2 (1947) 101.
63 C. H. Bamford and A. D. Jenkins, Trans. Faraday Soc., 59 (1963) 530.
64 C. H. Bamford and A. D. Jenkins, J. Polym. Sci., Part B, 1 (1963) 609.
65 R. D. Burkhart and N. C. Zutty, J. Polym. Sci., Part A, 1 (1963) 1137.
66 C. C. Price, J. Polym. Sci., Part B, 1 (1963) 433.
67 C. H. Bamford, A. D. Jenkins and R. Johnston, Trans. Faraday Soc., 55 (1959) 418.
68 C. H. Bamford and A. D. Jenkins, J. Polym. Sci., 53 (1961) 149.
69 C. Walling, J. Am. Chem. Soc., 70 (1948) 2561.
70 C. D. Ritchie and W. F. Sager, Prog. Phys. Org. Chem., 2 (1964) 323.
71 C. H. Bamford and E. F. T. White, Trans. Faraday Soc., 52 (1956) 716.
72 N. Fuhrman and R. B. Mesrobian, J. Am. Chem. Soc., 76 (1954) 3281.
73 G. E. Ham, J. Polym. Sci., Part A, 2 (1964) 4169.
74 F. R. Mayo, J. Polym. Sci., Part A, 2 (1964) 4207.
75 A. D. Jenkins, Eur. Polym. J., 1 (1965) 177.

76 J. R. Hoyland, J. Polym. Sci., Part A-1, 8 (1970), 885.
77 J. R. Hoyland, J. Polym. Sci., Part A-1, 8 (1970), 901.
78 J. R. Hoyland, J. Polym. Sci., Part A-1, 8 (1970), 1863.

Chapter 6

Free-Radical Polymerization in Heterogeneous Systems

N. C. BILLINGHAM and A. D. JENKINS

1. Introduction

The polymerization of vinyl and related monomers by the action of radical-producing initiators is one of the best studied examples of a radical chain reaction, displaying the features of initiation, propagation, termination and chain transfer characteristic of such reactions. In homogeneous reaction systems at low conversions simple kinetic relationships are expected from a chain mechanism and are normally observed; first-order dependence of rate on monomer concentration and half-order dependence on initiator concentration are both predicted by a chain mechanism and are found in many cases. However, homogeneous polymerization reactions differ from their simple analogues in that the presence and production of high-molecular-weight polymer in the reaction system leads to a high and changing viscosity of the system; the mobility of long chain radicals is thus reduced so that the effects of diffusion control of the rates of elementary steps may become both significant and variable during the reaction. Thus, in discussing the reactivity of monomers and radicals in polymerization, it is important to recognize that both chemical and physical factors must be taken into account. Among the reactions which are dominated by physical factors are the so-called heterogeneous polymerizations, in which the propagating radicals pass from one phase to another, usually as a consequence of precipitation of solid polymer from a liquid- or vapour-phase reaction. The purpose of this chapter is to discuss the role of physical factors in the behaviour of heterogeneous polymerizations. However, it is pertinent to precede such discussion by a brief summary of the role of diffusion in the simpler homogeneous polymerization reactions.

2. Diffusion control in homogeneous polymerization

2.1 HOMOPOLYMERIZATION

The bimolecular interaction of two polymeric radicals is the main termination reaction in homogeneous polymerization; it leads to the disappearance of the radical activity and the formation of one or two molecules of "dead" polymer. In the gas phase, radical combination reactions are processes of high kinetic efficiency, i.e. they occur at rates close to the collision frequency. In the closely packed environment of a liquid, radical mobilities are very much less than they are in the gas phase so that the time taken for two radicals to diffuse together and the length of time for which they remain in contact are both increased. This latter effect, the "cage-effect", implies that efficient chemical reaction will occur during the relatively long contact time of two radicals, so that the rate of reaction is effectively determined by the rate at which the reacting radicals diffuse together. Rate coefficients of the order of magnitude predicted from diffusion coefficients of small molecules in liquids have been observed in many reactions between highly reactive species, supporting the view that diffusion control is dominant in such reactions [1, 2]. Although free-radical polymerizations frequently show simple kinetic behaviour at low conversions, the known behaviour of small species suggests that termination of polymeric radicals must be a diffusion controlled process under all conditions and indeed, the magnitude of the termination rate coefficient in solution polymerization [3] (around 10^8 l mole^{-1} sec^{-1}) is consistent with the known diffusivities of large molecules in solvents [2, 4].

Several studies [5, 6] and notably those of North and co-workers [1, 7—9], have shown that the termination rate coefficient in many polymerization reactions is strongly dependent on the viscosity of the solvent even at low conversions, where the concentration of polymer is effectively zero. It is thus established that the termination is diffusion controlled from the start of polymerization, even though the viscosity may be many times lower than that achieved at high conversions. Since termination is diffusion controlled under such conditions, it must remain so with increasing conversion. However, there are difficulties in interpreting kinetic results based on diffusion models under conditions where the viscosity of the polymerizing medium is increased by the effect of polymer molecules [1, 8]. Thus, except for very dilute solutions, the viscosity of a polymer solution contains contributions from chain entanglement so that the macroscopic viscosity may grossly overestimate the resistance to motion of the chain ends; the use of diffusion coefficients or viscosity data relating to translational motion of the centre of gravity of a polymer molecule will seriously underestimate the mobility of the chain-ends, especially at low separation. Several theoretical approaches have been used in an attempt to derive kinetic equations for the diffusion controlled reactions of macro-radicals, the most successful being one in which the overall reaction is visualized as involving three stages: translational diffusion of the centres of gravity of the reactants, segmental rearrangement, by diffusion, to bring the radical centres into proximity and, finally, chemical reaction. Of these steps, the rate of segmental rearrangement of the radical chain end in the coiled polymeric free radical appears to be rate controlling [7—9] and good correlations between termination rate coefficients and chain "stiffness" have been observed [10—12]. This model has been extended by Burkhart [13].

On the basis of a quantitative evaluation of the chain-end mobility in relation to the flexibility of the total chain [11, 12] it is suggested that the segment which rearranges during termination includes no more than three to five carbon—carbon bonds, which is in accord with the observation that in many cases the diffusion controlled termination rate coefficient at low conversions is inversely proportional to solvent viscosity but independent of the molecular weight of the radical.

Borgwardt et al. [14] have demonstrated the truth of Benson and North's theory [7] of chain-end mobility by the direct measurement of radical recombination rates, following the creation of radicals by pulse-radiolysis.

The simple kinetic behaviour shown by radical polymerization at low conversions implies that, although termination is controlled by diffusion, the termination rate coefficient remains essentially constant. However, with increasing conversions, particularly in bulk polymerization reactions, the increase in viscosity caused by the production of high-molecular-weight polymer affects the rate of termination, causing it to decrease. The consequence of a decrease in termination rate is autoacceleration of the polymerization reaction which may lead in extreme cases to an explosively fast reaction. This effect, known as the "gel-effect", has been extensively studied [8, 15—17] and is undoubtedly due to the decrease in termination rate or to complete suppression of termination in the high viscosity of a bulk polymerization. Figure 1 illustrates the gel-effect in a typical case. It is interesting to note that in certain polymerizations carried to high conversion, the polymerization stops well short of

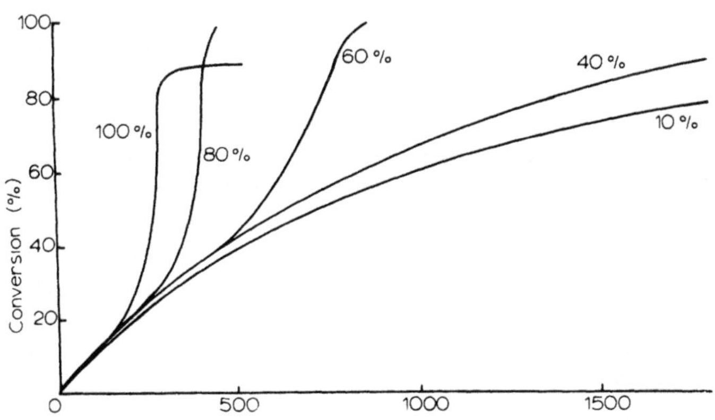

Fig. 1. Autoacceleration in the polymerization of methyl methacrylate by benzoyl peroxide in benzene solution at 50 °C. Concentrations of monomer as indicated. (After Schulz and Harborth [15].)

complete reaction; the limiting conversion is temperature dependent and complete conversion is achieved above the glass transition temperature of the polymer [18, 19]. These results indicate that monomer molecules may become completely immobilized in the glassy polymer, suggesting that diffusion control of chain propagation is an important factor in high conversion systems. This suggestion is confirmed by measurements of the propagation rate coefficient at high conversions; this decreases as complete conversion is approached [17, 20].

Accurate kinetic measurements on bulk polymerization reactions are complicated by the high viscosity and the difficulty of maintaining isothermal conditions in an exothermic autoaccelerating reaction. Horie et al. [21] have studied the bulk polymerization of methyl methacrylate under isothermal conditions by using differential scanning calorimetry. The results indicate that diffusion control of propagation becomes significant at around 40 % conversion. Comparison of the limiting conversions obtained at several temperatures with the glass transition temperatures calculated for monomer—polymer mixtures shows that propagation ceases at the conversion at which the polymerization temperature is equal to the glass transition temperature of the mixture; polymerizations performed at temperatures above the glass transition of the pure polymer proceed to complete conversion. Changes in the rate of polymerization over the complete conversion range were interpreted in terms of the effects of diffusion control of propagation and termination on both the propagation rate coefficient and the concentration of free radicals in the system, the latter being increased by a factor of 10^3—10^4 at high conversions, in accord with ESR measurements [22]. The implications of the results obtained in this study and that of Schulz [18] is that, at temperatures below the glass transition of a polymer containing small amounts of monomer, not only are the polymer radicals immobilized in the glass but the monomer molecules are also unable to diffuse to the radical centres. Chapiro [23] has questioned this view and produced evidence that the radiation induced bulk polymerization of methyl methacrylate can go to completion at temperatures 40 °C below the glass transition. However, the same author [24] has indicated that, with other systems, polymerization in vitreous media at low temperatures does not occur. In addition to the propagation and bimolecular termination reactions it is important to recognize that, as the interaction of two polymeric radicals becomes less favourable with increasing viscosity, termination by the much more mobile primary radicals generated by initiator decomposition may become important. This factor has been included by Ito [25, 26] in a treatment of diffusion controlled termination rates at both high and low conversions, although the treatment does not allow complete analysis of the gel-effect since it does not include contributions from diffusion control of propagation.

In summary, it is well established that bimolecular reaction of polymeric radicals is a diffusion controlled process under all conditions, the rate controlling process being the mobility of the chain-end rather than the translational mobility of the whole polymer molecule. With increasing medium viscosity termination is retarded, and in the vitreous state below the glass transition temperature radical—radical reactions are impossible. The extent to which chain propagation can proceed in polymeric glasses remains open to further investigation although on balance the present information suggests that diffusion control of chain propagation is an important factor at high conversions. These conclusions are of importance for the discussion of heterogeneous polymerization as will be shown in subsequent discussions.

2.2 COPOLYMERIZATION

In a binary copolymerization reaction the termination process will be diffusion controlled in a similar manner to that of a simple homopolymerization. However, the existence of different types of propagating radical introduces added complications which are worth consideration. Since the recombination reactions of small radicals are essentially collision controlled it is not surprising to find that there is no specific effect or preferred reaction between unlike radicals [27]. Since the activation energies are low the rate coefficient for cross-combination of unlike radicals is related to the coefficients for combination of like radicals by the geometric mean assumption, such that

$$k_{12} = (k_{11} \cdot k_{22})^{0.5}$$

For polymeric free radicals highly specific cross-termination reactions have been proposed to account for anomalies in copolymerization rates and copolymer compositions. The so called ϕ factor defined as

$$\phi = k_t^{12}/(k_t^{11} \cdot k_t^{22})^{0.5}$$

has been widely used in discussion of copolymerization behaviour despite the fact that it is highly composition dependent for many systems and not very soundly based in theory [28, 29]. (In this context the termination rate coefficient is defined by $-\mathrm{d}[\mathrm{X}\cdot]/\mathrm{d}t = k_t[\mathrm{X}\cdot]^2$, which is the definition commonly employed in discussing free-radical polymerization; the definition widely used in other branches of free-radical chemistry is $-\mathrm{d}[\mathrm{X}\cdot]/\mathrm{d}t = 2k_t[\mathrm{X}\cdot]^2$ leading to the formulation of ϕ as $\phi = k_t^{12}/2(k_t^{11} \cdot k_t^{22})^{0.5}$.) The geometric mean assumption implies that the three termination rate coefficients are those for true chemical reactions, whereas it has already been shown that radical termination reactions are diffusion controlled.

The geometric mean assumption cannot be applied to diffusion controlled processes; for example, it would imply that radicals trapped in a solid matrix (where $k_t^{11} = 0$) could never be terminated by mobile radicals no matter how high their diffusion coefficients. Since the rate of diffusion together of two species in a solution is proportional to the sum of the diffusion rates it is more reasonable to write the cross-termination rate coefficient as an arithmetic mean of the rate coefficients for the mutual termination radicals so that $k_t^{12} = (k_t^{11} + k_t^{22})/2$. However, it has been shown in the previous section that the rate determining step in the diffusion controlled termination of two similar polymer radicals is the rate of rearrangement of a segment of the chain-end encompassing several carbon atoms. Since the composition of the chain-end in a copolymer radical is a function of the initial monomer feed composition, the reactivity ratios and the fractional conversion, it is expected that the three diffusion controlled termination rate coefficients will also be dependent upon these variables. North et al. [8, 30, 31] suggested that an "ideal" diffusion controlled termination rate coefficient might be defined in relation to the homopolymerization coefficients by

$$k_t^{12} = k_t^{11} x_1 + k_t^{22} x_2$$

where x_1 and x_2 represent the mole fractions of monomer 1 and monomer 2 units in the polymer, respectively. Expressions for the rate of copolymerization derived on this basis agree well with experiment for the copolymerization of vinyl acetate and methyl methacrylate [30]. Russo and Munari [11, 12] have used a model for chain termination which includes contributions from both chemical and chain flexibility factors; this scheme, which takes into account the dependence of termination rate coefficient on the composition of the chain-end, gives good agreement with experiment for the copolymerization of styrene with methyl methacrylate [32].

3. Heterogeneous polymerization

A heterogeneous polymerization may be defined as a polymerization reaction in which the propagating free radical passes from one phase to another during its growth. Phase separation may occur as a result of insolubility of the polymer in the reaction medium either naturally or induced by the presence of a precipitant added to the system, such reactions being termed precipitation polymerizations. Phase separation also occurs in the polymerization of monomers in the vapour phase, polymeric radicals and molecules separating as a mist which deposits as solid polymer on the surface of the reaction vessel. Emulsion polymerization, in which free radicals, generated in an aqueous phase, induce polymerization of monomer contained in surfactant micelles, also falls into this category of heterogeneous polymerization; reactions of this category show special features which are outside the scope of the present chapter but which have been extensively reviewed elsewhere [33—35]. In contrast, suspension polymerization is best regarded as the subdivision of a bulk homogeneous polymerization into a series of small volume elements each behaving as an independent homogeneous reaction; as such these reactions are not properly heterogeneous and will not be considered further. It is reasonable to include as heterogeneous polymerizations the "popcorn" polymerizations, which consist of the separation from the liquid or vapour phase of small nodules of polymer which proliferate rapidly producing a highly cross-linked material. Breitenbach [36] has given an excellent review of these reactions which will not be considered further. Although heterogeneous polymerizations occur over a wide range of reaction conditions these reactions have many features in common. Among precipitation polymerizations the polymerization of acrylonitrile has received most attention and, accordingly, this system will be discussed first, followed by an examination of other reactions.

3.1 PRECIPITATION POLYMERIZATION OF ACRYLONITRILE

3.1.1 Overall features of the polymerization reaction

Polyacrylonitrile is totally insoluble in its own monomer, even at high temperatures; only a few simple solvents will dissolve polyacrylonitrile, the most common being *N,N*-dimethylformamide (DMF) and dimethyl sulphoxide (DMSO). For this reason precipitation of polymer is a common feature of acrylonitrile polymerization. In the presence of sufficient DMF to maintain a homogeneous solution, the free-radical polymerization exhibits normal homogeneous reaction kinetics; in particular the initiator exponent is 0.5 and the after-effect in photo-initiated polymerization is dependent upon the intensity but not the duration of irradiation, provided that the latter is not too short or too long. Acrylonitrile does not polymerize thermally, evén in the total absence of oxygen, which is a powerful inhibitor for the polymerization; direct photoinitiation is possible but inefficient.

Figure 2 illustrates the conversion—time behaviour observed by Bamford and Jenkins [37] in a bulk polymerization of acrylonitrile initiated by dibenzoyl peroxide. The reaction is clearly autoaccelerative and this behaviour is unaffected by the use of residual monomer from a reaction which has been allowed to proceed to 60 % conversion; autoacceleration is therefore not due to gradual consumption of a retarding impurity. As soon as polymerization begins in the bulk monomer, the polymer precipitates as a white granular solid without any increase in the viscosity of the liquid monomer, and the precipitate normally remains dispersed causing the reaction mixture to become opalescent. Although the autocatalytic behaviour of the polymerization makes quantitative kinetic

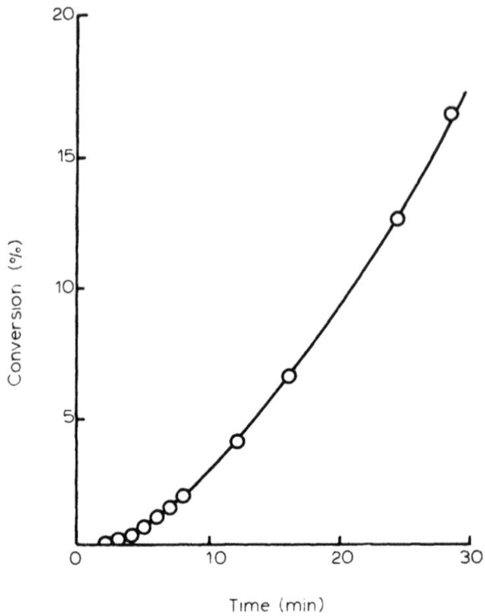

Fig. 2. Conversion—time curve for bulk polymerization of acrylonitrile at 60 °C. Initiator is benzoyl peroxide at 2.7 × 10^{-2} mole %. (After Bamford and Jenkins [37].)

statements difficult, the initiator exponent appears to be around 0.9. The most striking anomalies appear in studies of the photochemical after-effect. In contrast to the homogeneous system the duration of the after-effect increases progressively as the time of irradiation increases [24]. Further, successive irradiations of a given reaction give progressively larger after-effects for a given period of irradiation, unless the system is allowed to stand in darkness for long periods between irradiations. In this latter case the after-effect is found to be the same for each irradiation. The only conclusion which can be drawn from these observations is that the precipitated polymer retains its propagating activity for periods of several hours at room temperature [37].

When polyacrylonitrile, produced from a bulk polymerization, is added to the reaction system it does not display catalytic effects. However, in view of the strong inhibiting action of oxygen it is probable that any catalytic activity possessed by the polymer is destroyed by exposure to air. In an attempt to examine the apparent initiating activity of polyacrylonitrile, Bamford and Jenkins prepared polymers in bulk monomer by photopolymerization at 25 °C with benzoyl peroxide as sensitizer. After a photochemical polymerization to about 20 % conversion the mixture was allowed to stand in the dark for a while, then transferred to a thermostat bath at 60 °C. The results are illustrated in Fig. 3, which shows that, as the reaction temperature approached 60 °C, an extremely fast

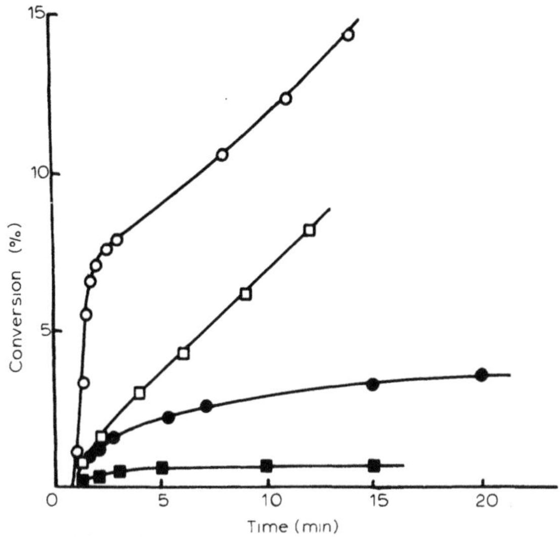

Fig. 3. The "fast-reaction" in the polymerization of acrylonitrile at 60 °C, with 2.7 × 10^{-2} mole % benzoyl peroxide as initiator. ○, After 45 min photopolymerization to 18 % conversion at 25 °C; ●, after photopolymerization to 18 % conversion without initiator; □, after photopolymerization to 18 % conversion with initiator present but initiation rate adjusted to be identical to ○; ■, as in ○, but monomer replaced with ethanol after photopolymerization. (After Bamford and Jenkins [37].)

reaction occurred in which approximately 10 % of the available monomer was consumed in about two minutes. After this initial "fast reaction" the reaction rate fell and normal behaviour with some autoacceleration was resumed. No significant reaction was observed in an experiment in which the residual monomer was replaced by ethanol before heating the photopolymer, thus demonstrating that the fast reaction is not simply due to a contraction of photopolymer in the dilatometer on heating.

The results shown in Fig. 3 also indicate that the fast reaction is observed if the photopolymer is prepared by direct irradiation of the monomer with no added sensitizer; in this case the reaction rate falls to near zero after the fast reaction period. Similar results were obtained when the fast reaction was isolated by using a volatile initiator such as di-*tert*-butyl peroxide [38]. This initiator was used to sensitize the photopolymerization, then the residual monomer and initiator were removed by vacuum distillation and replaced with fresh monomer. Under these conditions any reaction observed at 60 °C must arise from the influence of the photopolymer since the pure monomer does not polymerize thermally. Figure 4 shows the results of such an experiment, demonstrating that the fast reaction still takes place but that its rate rapidly declines to a very low value. Extensive studies of the fast reaction [38] have established that it is unaffected by allowing the photopolymer to remain in contact with monomer for long periods before heating;

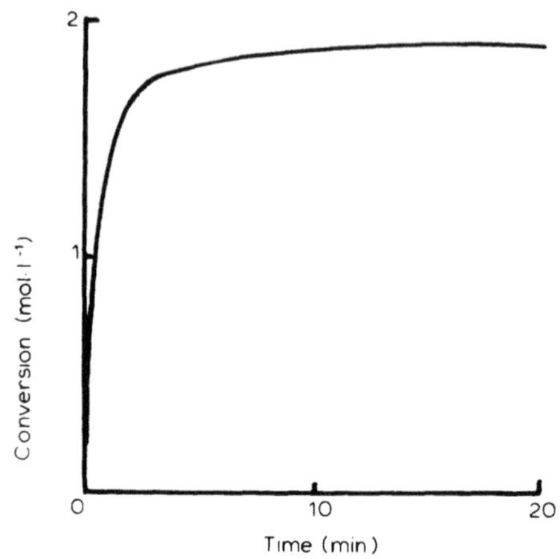

Fig. 4. Fast reaction in acrylonitrile polymerization at 60 °C in the absence of initiator. Di-*t*-butyl peroxide, used as a sensitizer for photopolymerization, was removed by distillation before heating to 60 °C. (After Bamford and Jenkins [38].)

replacement of the monomer after photoreaction and heating the dry photopolymer to 60 °C have no effect on the fast reaction although it is completely inhibited by contact of the photopolymer with oxygen. All of these results lead to the conclusion that the fast reaction is induced by the presence of active free radicals, trapped in the polymer and having extremely long lifetimes at temperatures below 60 °C. The presence of trapped radicals in the polyacrylonitrile photopolymer has been established beyond doubt by both chemical means and by the observance of an ESR signal from the photopolymer [39]. Thus, α,α-diphenyl picryl hydrazyl (DPPH) undergoes reaction with the photopolymer when heated with it under the same conditions as those used for the fast reaction. Although the chemistry and stoichiometry of the reaction of DPPH with polymeric radicals is not always simple [40, 41], radical counting experiments using this inhibitor indicate that trapped radicals may be present at concentrations up to 10^{-4} mole l^{-1}, as a result which is in good agreement with the concentration of spins measured by ESR [42]. From the concentration of trapped radicals measured in this way the mean value of the termination rate coefficient in the fast reaction at 60 °C can be estimated as around 200 l $mole^{-1}$ sec^{-1}. This value is about seven orders of magnitude lower than the value for the homogeneous reaction in DMF solution, showing that motion of the trapped radicals is still very restricted even at 60 °C. Comparison of the trapped radical concentration with the known rate of radical generation shows that between 1 and 10 % of the total number of radicals produced become trapped in the polymer.

Since the radicals trapped in the photopolymer at 25 °C are able to terminate when the temperature is raised to 60 °C it is expected that photopolymerization at 60 °C would lead to negligible radical trapping.

TABLE 1
Concentration of radicals trapped in polyacrylonitrile as a function of photopolymerization temperature

Temp. (°C)	No. of radicals
20	3.7×10^{16}
25	2.3×10^{16}
40	0.8×10^{16}
60	$<0.1 \times 10^{16}$

Table 1 shows the variations of trapped radical concentration with the temperature of photopolymerization, demonstrating that the concentration declines with increasing temperature and that a high concentration can only be achieved at temperatures below 60 °C [43].

The fast reaction discussed above occurs when photopolymer produced at 25 °C is heated to 60 °C. Similar reactions are observed when the same photopolymer is heated to temperatures below 60 °C; the half-life and extent of the process at intermediate temperatures are summarized in Table 2 [39]. However, at temperatures below 60 °C virtually no decrease in radical concentration accompanies the fast reaction. For example, at 40 °C the rate of the fast reaction declines to near zero with a half-life of 40 min, but a further fast reaction occurs if the temperature is then raised to 60 °C; following this latter reaction, further increases in temperature have no effect. Although these observations of the behaviour of the fast reaction so far presented allow a reasonable interpretation in terms of a model for the heterogeneous polymerization, it is useful to consider some other relevant studies before discussing the detailed interpretation.

In a homogeneous polymerization there is an intimate connection between the rate of polymerization and the degree of polymerization of the product polymer; for constant initiator concentration the rate of initiation rises with increasing temperature and, since the activation energy for initiation is much greater than that for either propagation or termination, the degree of polymerization is expected to decrease with

TABLE 2
Rate and extent of fast reaction of acrylonitrile as a function of temperature

Temp. (°C)	Half-life of fast reaction (min)	Monomer consumption by fast reaction (mole l^{-1})
40	40	0.38
50	12	0.95
55	4.5	1.01
60	2.3	1.46
70	1.5	1.91

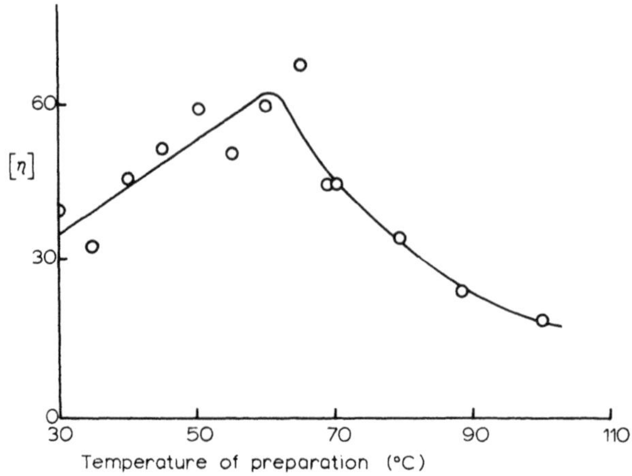

Fig. 5. Intrinsic viscosities of polyacrylonitrile samples as a function of temperature of preparation. Bulk monomer initiated by 2.7 x 10^{-2} mole % benzoyl peroxide. (After Bamford and Jenkins [37].)

increasing polymerization temperature. Figure 5 shows that this expectation is not found for the heterogeneous polymerization of acrylonitrile, which displays a pronounced maximum in the degree of polymerization at about 60 °C.

Figure 6 illustrates the results obtained when the photopolymerization at 20 °C is conducted in the presence of varying amounts of DMF which acts as a swelling agent when present in low concentrations [43]. The concentration of trapped radicals, as measured by ESR, increases with increasing DMF concentrations up to 8 mole %; above this concentration of solvent the trapped radical yield decreases, becoming zero at about 50%, when the polymer formed is a very swollen opaque gel. Similar

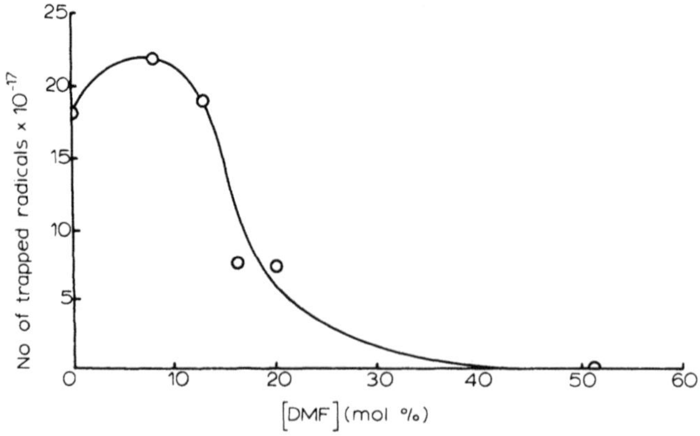

Fig. 6. Variation of trapped radical concentration with concentration of added DMF after photopolymerization to 20 % conversion at 20 °C. (After Bamford et al. [42].)

results are obtained when methyl acrylate is copolymerized with acrylonitrile; the concentration of trapped radicals apparently increases with up to about 2 % methyl acrylate and thereafter it decreases.

3.1.2 Occlusion theory of acrylonitrile polymerization

The observations of the behaviour of heterogeneous polymerization of acrylonitrile summarized in the previous section have been used by Bamford and Jenkins to develop a qualitative theory of the underlying processes based on the concept of an occluded radical. From the kinetic observations and radical trapping studies it is clear that many of the free radicals, produced from the liquid phase initiator breakdown, precipitate from the liquid phase before termination. Since the liquid phase is a poor solvent the precipitated polymer will be coiled to at least its unexpanded random coil dimensions [44] and the dimensions of the coiled molecule will be further reduced by intramolecular hydrogen bonding and by crystallization. Such coiling will reduce the reactivity of the radicals, since on a statistical basis the active chain-end will be shielded by the remainder of the molecule so that the reaction will only occur if the radical diffuses to a point near the surface of the polymer molecule or if a potential reactant can penetrate the diffusion barrier imposed by the polymer chain. In this sense the diffusion barriers to reaction of the radical are rather similar to those encountered in homogeneous reactions (see Sect. 2.1), except that the radical will be more tightly coiled in the precipitate phase than in a homogeneous phase. However, in heterogeneous polymerization the reactivity of the precipitated radical will be further reduced by coalescence of the radical with dead polymer particles to produce the observed precipitate particles which have dimensions considerably larger than those associated with a single molecule (electron microscopy of polyacrylonitrile precipitates produced in heterogeneous polymerization [45] suggests that approximately 50 polymer molecules are present in each particle, although this number will vary with the polymerization conditions). The combination of these two effects, whose relative importance may vary with reaction conditions, is termed occlusion. The importance of the morphology of the polymer particles in controlling the extent of occlusion has been stressed by Lewis and King [46].

Since radical occlusion imposes diffusion barriers to both the motion of the radical chain-end and the diffusion of small species in the polymer it is expected that all of the rate coefficients for the reactions of an occluded radical will be lower than the equivalent coefficients for reaction in a homogeneous phase. It is immediately obvious that bimolecular termination will be most seriously affected since this reaction depends upon two radicals diffusing into sufficiently close proximity for reaction to occur. Propagation and transfer, however, are impeded only to the extent that monomer and transfer agent must diffuse into the precipitate particle. Since the rate of polymerization depends on the relative rates of propagation and termination, the overall effect of occlusion is expected to lead to an increase in polymerization rate and an increase in the contribution of chain transfer relative to bimolecular termination.

Although bimolecular termination is expected to be drastically impeded by radical occlusion, other termination reactions may be involved. Thus two primary radicals or propagating radicals which have not become sufficiently long to precipitate, may react in the liquid phase. Such small radicals may also diffuse into the precipitate particles; although the diffusion barrier to entry of a radical may be high, the probability of termination within the particle is high since there will be a large barrier to diffusion out of the particle. Chain transfer within the polymer particle may produce a small radical which is able to diffuse out, although the diffusion barrier will be high and will be increased by any propagation of the radical within the particle. Finally a polymeric radical end may diffuse near to the surface of the particle where it may react with radicals in the solution; this latter process will be favoured by high temperatures where thermal agitation may tend to disperse the coalesced particles. However, the prolonged activity of acrylonitrile photopolymer at 25 °C shows that high degrees of occlusion must arise, preventing subsequent reactions even after long periods in contact with monomer. The alternative termination mechanisms must therefore be negligible under these conditions [37]. Since the polymerization behaviour is very dependent upon physical factors which control the degree of occlusion it is not surprising that quantitative kinetic analysis is difficult. However, the observed initiator exponent (ca. 0.9) is consistent with a first-order termination mechanism such as is radical occlusion. Thomas and Pellon [47] have attempted to analyse the kinetics of the heterogeneous polymerization at 60 °C on the basis of a reaction in which termination includes both bimolecular interaction of radicals and a trapping of radicals, the latter being a kinetically first-order process; initiator exponents in the range 0.5—1.0 were predicted. However, it has been shown [43] that the rate of termination by radical trapping at 60 °C is too small to be kinetically significant. Lewis and King [46] obtained initiator exponents of around 0.9 at low initiator concentrations but found that high initiator concentrations gave a value closer to 0.3; this value was attributed to a significant contribution from primary-radical termination at high initiator concentrations.

It follows that the behaviour of the heterogeneous polymerization is determined at least in part by both the number of particles present and their size, the former determining the probability of a radical becoming occluded by coalescence and the latter determining the diffusion barrier within the particle, i.e. the degree of occlusion. Since both the number and size of the particles will increase during the reaction the effective termination rate coefficient will fall progressively with conversion; the autoacceleration and the photochemical after-effect are both consequences of this. The dependence of molecular weight on polymerization temperature is also qualitatively explicable in terms of an occlusion model. It has been shown that, in studies of the reaction initiated by low-temperature photopolymer, a series of fast reactions can be observed as the temperature is raised. Up to a maximum of 60 °C the fast reaction occurs without loss of radical activity. Thus it appears that, in the simple heterogeneous polymerization, an increase in temperature from 25 °C causes swelling or agitation of the aggregates, which allows propagation to occur more readily without affecting termination, so that the molecular weight of the polymer increases with temperature. However, as the temperature approaches 60 °C occlusion is so diminished that the rate of bimolecular termination increases, accompanied by a reduction in molecular weight; the combination of these opposing influences causes the observed maximum in molecular weight near 40 °C.

The effect of increased temperature below 40 °C in allowing propagation to occur without termination is effectively duplicated by swelling agents. The slight but definite increase in trapped radical concentration at low concentrations of swelling agent indicates that some additional polymerization can occur under these conditions, without significant termination, thus leading to an even higher concentration of occluded radicals. This view is confirmed by observations of the rate of photopolymerization at 25 °C, the rate passing through a maximum at around 8 mole % DMF. Greater concentrations of swelling agents reduce both coiling and coalescence, increasing the radical mobility and causing a reduction in the heterogeneous characteristics of the polymerization.

In summary, therefore, it appears that the result of moderate degrees of radical occlusion is a decrease in the rate of termination with respect to the rate of propagation, leading to an increase in rate of polymerization and polymer molecular weight. Extreme degrees of occlusion cause complete trapping of radicals to the point where chain propagation is inhibited; in the chemical sense the intrinsic reactivity of the trapped radicals is unaltered but the rate coefficients for their reactions are totally controlled by physical constraints.

In the context of the occlusion model it is interesting to examine the effect of transfer agents and inhibitors on the heterogeneous polymerization. Bamford and White [48] studied the effect of *n*-butyl mercaptan on the bulk polymerization of acrylonitrile at 60 °C, with results summarized in Fig. 7. The rate of polymerization decreases markedly as the concentration of mercaptan is increased and, at the same time, the autoacceleration becomes less marked until the conversion—time plot becomes linear at mercaptan concentrations greater than 0.03 mole l^{-1}; above this concentration the plot remains linear and the rate is indepen-

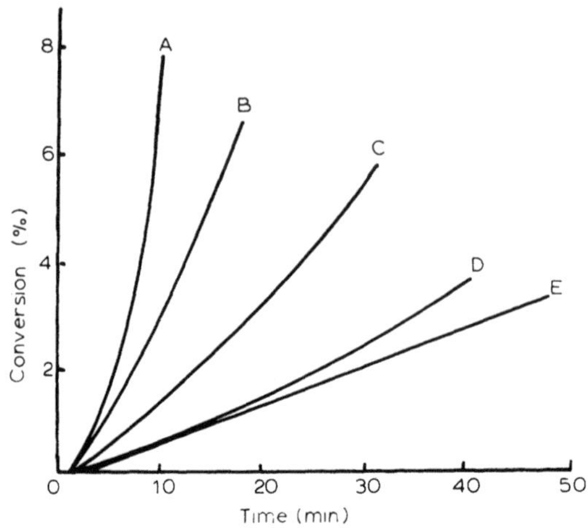

Fig. 7. Effect of *n*-butyl mercaptan on the bulk polymerization of acrylonitrile at 60 °C with 1.6 x 10^{-2} mole % azo-isobutyronitrile as initiator. Concentration of transfer agent (mole l^{-1}) A, 0; B, 3.74 x 10^{-3}; C, 2.53 x 10^{-2}; D, 2.95 x 10^{-2}; E, 3.14 x 10^{-1}. (After Bamford and White [48].)

dent of mercaptan concentration. This effect was explained in terms of the escape from polymer particles of the small mobile radicals produced by transfer, together with the reduction in molecular weight arising from transfer; this latter effect will reduce the degree of occlusion of the radicals and facilitate their termination. When transfer is abundant the overall effect is to eliminate the heterogeneous character of the reaction as shown in Fig. 7. These arguments are supported by observations of the effect of ferric chloride, which acts as a radical trap [49]. Under appropriate conditions the first-order (in radical concentration) termination process

$$R\cdot + Fe^{3+} \longrightarrow Fe^{2+} + R^{+}$$

completely dominates the bimolecular reaction [50]. Figure 8 shows the effect of ferric chloride on the polymerization at 60 °C in DMF solution. At concentrations of monomer above 6 mole l^{-1} the polymerization becomes heterogeneous and the balance between autoacceleration and retardation is determined by the relative rates of diffusion of monomer and inhibitor into the polymer particles. As Fig. 8 shows, the rate falls slightly, indicating that propagation is rather more adversely affected than is termination. In contrast, when propagation and bimolecular termination are in competition the onset of heterogeneous conditions is accompanied by marked acceleration.

Amdur [51] has studied the effect of ferric chloride on the bulk polymerization of acrylonitrile in the absence of solvent. With high concentrations of inhibitor the termination reaction is dominated by the

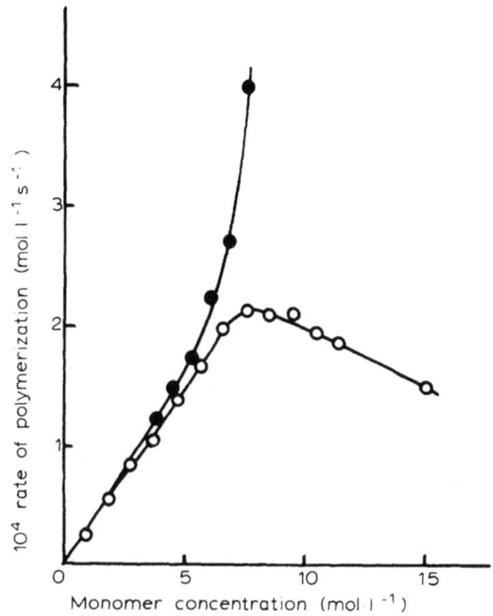

Fig. 8. Effect of ferric chloride on polymerization of acrylonitrile by azo-isobutyronitrile in DMF at 60 °C. ●, [initiator] = 7.90 x 10^{-3} mole l^{-1}, [$FeCl_3$] = 0; ○, [initiator] = 6.20 x 10^{-3} mole l^{-1}; [$FeCl_3$] = 6.8 x 10^{-3} mole l^{-1}. (After Bamford et al. [43].)

added inhibitor and steady-state kinetics were observed at low (<3 %) conversion. Assuming that stationary-state kinetic equations can be applied and that the ratio (k_p/k_t) of the propagation rate coefficient to the termination rate coefficient with the metal ion is the same as that measured with DMF solution [50], the effective monomer concentration at the propagating centres was calculated. The value obtained was approximately 3 mole l^{-1} as compared with the bulk concentration of 14.4 mole l^{-1} of pure acrylonitrile at 60 °C. On this basis, Amdur suggests that propagation occurs from a radical located in the surface of the polymer particles where the monomer concentration is significantly lower than that of the bulk monomer but much higher than within the polymer particles. This conclusion, which is in agreement with that of Lewis and King [46], depends upon the assumption that the ratio k_p/k_t is the same in both the heterogeneous bulk polymerization and the homogeneous polymerization in DMF. This conclusion may be criticized on two grounds. (a) There is no a priori reason to assume that k_p/k_t is constant since in the homogeneous reaction it is determined by chemical influences whilst in the heterogeneous reaction it is presumably determined by the relative diffusion coefficients of monomer and ferric chloride; indeed the results of Bamford et al. shown in Fig. 8 and discussed above suggest that the ratio is not constant. (b) It is well established [52, 53] that the rate coefficient, k_t, for reaction of ferric chloride with polymeric free radicals is not a simple constant but depends strongly on the nature of the metal complexes present in the solution: these complexes will undoubtedly differ very considerably in pure acrylonitrile and in acrylonitrile/DMF mixtures and a dependence of k_t on the monomer concentration has been demonstrated for polymerization of methacrylonitrile in DMF solution [52]. A good deal of further work is required before the exact locus of propagation in the heterogeneous polymerization of acrylonitrile can be properly defined.

3.1.3 Kinetics of the fast reaction

When acrylonitrile is photopolymerized at 25 °C, a proportion of the radicals which are generated become occluded during propagation, to such an extent that they are totally unable to undergo bimolecular termination. It is these radicals which are responsible for the fast reaction, observed when the photopolymer is heated above 25 °C in contact with monomer. When no other initiating species is present the fast reaction is thus initiated at a number of preferred active centres and it proceeds either until further occlusion causes propagation to become negligible (at temperatures below 60 °C) or until bimolecular termination destroys the radicals (reactions at temperatures of 60 °C or above). In this first case it is qualitatively obvious that a moderate increase in temperature to around 40 °C allows propagation to occur within the aggregates until a diffusion barrier is set up which prevents further polymerization. However, when the fast reaction is performed at 60 °C, at which temperature all radicals are destroyed quite rapidly, the locus of polymerization is not so well-defined, since it is not immediately obvious whether the polymer particles disintegrate instantaneously or whether reaction occurs within swollen aggregates. If the particles disintegrate instantaneously the propagation and termination rate coefficients k_p and k_t would be expected to assume the values found for homogeneous solution (k_p = 2458 and k_t = 9.8×10^8 l mole^{-1} sec^{-1}) [54]. This situation would then be equivalent to an ordinary after-effect which can be treated according to the equation [3]

$$\Delta M = \frac{k_p[M]}{k_t} \log\left(\frac{\theta + 1}{2}\right) \tag{1}$$

in which ΔM is the depletion of monomer concentration and θ is the ratio of the initial and final rates of the fast reaction. Application of this equation to the fast reaction at 60 °C leads to a calculated value of ΔM of 6.0×10^{-5} mole l^{-1} which is in marked contrast to the observed value 1.4 mole l^{-1}. The only reasonable explanation for this discrepancy is that k_t is very low and this is confirmed by direct measurements of radical concentration [42]; polymerization must therefore be regarded as taking place within the polymer particles.

With some reservations it is possible to treat the kinetics of the fast reaction by a semi-quantitative approach which can also be extended to cases in which the trapped radicals are used to initiate polymerization of a second monomer. Thus if the residual monomer is removed after photoreaction and replaced with another, polymerization at 60 °C will yield a block copolymer, provided that rapid transfer does not occur [55]. If the concentration of trapped radicals in the system is [X] and the mean effective monomer concentration is $[\overline{\mathrm{M}}]$ at the site of reaction, then the rate of polymerization is given by

$$r = k_p[\mathrm{X}][\overline{\mathrm{M}}] \tag{2}$$

and if we ignore any variation in k_p we can write

$$\frac{\mathrm{d}r}{\mathrm{d}t} = k_p[\overline{\mathrm{M}}]\frac{\mathrm{d}[\mathrm{X}]}{\mathrm{d}t} + k_p[\mathrm{X}]\frac{\mathrm{d}[\overline{\mathrm{M}}]}{\mathrm{d}t} \tag{3}$$

Although the assumption of constant k_p is questionable and a detailed analysis of the kinetics on the basis of this equation is impossible, a number of limiting cases can be described and are depicted in Fig. 9.

(i) If [X] and [M] are both constant so that $\mathrm{d}[\mathrm{X}]/\mathrm{d}t$ and $\mathrm{d}[\overline{\mathrm{M}}]/\mathrm{d}t$ are both zero the rate will be constant (Fig. 9(a)).

(ii) If the diffusion barrier to propagation increases with polymerization auto-occlusion will develop, $\mathrm{d}[\overline{\mathrm{M}}]/\mathrm{d}t$ will decrease as $[\overline{\mathrm{M}}]$ tends to zero and the fast reaction will follow a curve of type (b). If the new polymer is able to enhance swelling of the polymer particle the opposite effect may be observed giving rise to an autoacceleration as shown in curve (c), as swelling leads to increasing $[\overline{\mathrm{M}}]$.

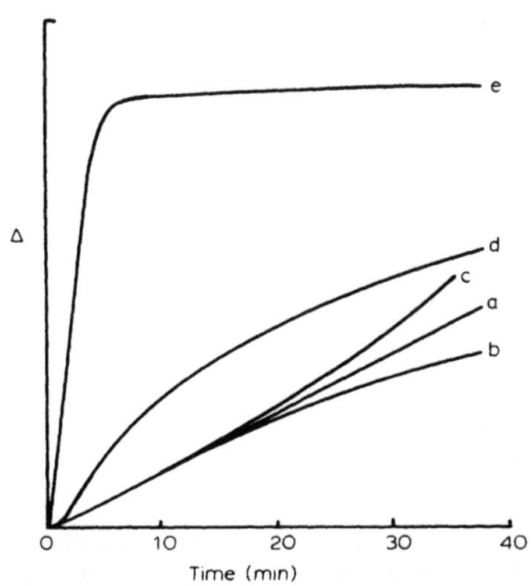

Fig. 9. Predicted form of kinetic behaviour for polymerizations initiated by trapped radicals. For identification of curves, see text. (After Bamford and Jenkins [55].)

(iii) If [$\overline{M}$] remains constant but bimolecular termination reduces [X] there will be no simple kinetic form unless it is possible to make some assumption about the rate of radical termination. If termination occurs by occlusion rather than bimolecular reaction then termination might reasonably be expected to be first-order, leading to a relation similar to curve (b).

(iv) If bimolecular termination occurs with constant [$\overline{M}$] the simplest situation corresponds to a normal after-effect with a constant rate coefficient for the termination process having a much lower value than it does in solution. This is equivalent to curve (d) of Fig. 9.

(v) In the more general case where the period required to produce swelling is comparable to the length of the experiment, [X] will decrease slowly and [$\overline{M}$] may decrease due to auto-occlusion or increase due to swelling. The kinetic behaviour cannot be easily be predicted, although extreme examples would give curves (e) and (d).

Figure 10 shows the results obtained in the polymerization of a series of monomers by polyacrylonitrile photopolymer. The concentrations of the trapped radicals which have survived at the time indicated are shown as a fraction of the initial concentration of polyacrylonitrile radicals. Behaviour typical of class (i) reactions is exhibited by *n*-propyl acrylate

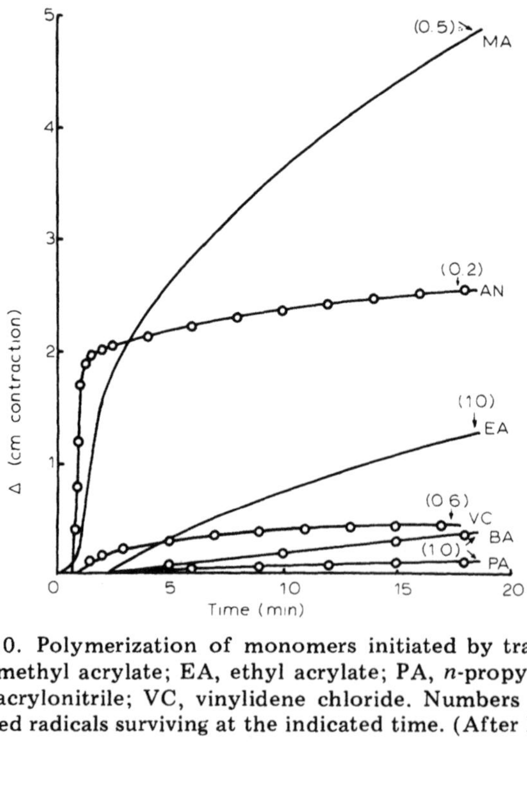

Fig. 10. Polymerization of monomers initiated by trapped polyacrylonitrile radicals. MA, methyl acrylate; EA, ethyl acrylate; PA, *n*-propyl acrylate; BA, *n*-butyl acrylate; AN, acrylonitrile; VC, vinylidene chloride. Numbers in brackets give the fraction of trapped radicals surviving at the indicated time. (After Bamford and Jenkins [55].)

and n-butyl acrylate both of which show a slow linear reaction with no decrease in radical concentration. Case (ii) is obeyed by ethyl acrylate, the rate decreasing with time without any consumption of radicals; this case is also typical of the fast reaction of acrylonitrile at 40 °C. With vinylidene chloride the situation is more complex, the rate decreasing according to a second-order rate law, with very little decline in radical concentration; this reaction probably belongs to case (v), the second-order rate law presumably arising fortuitously as a combination of effects. The reaction of acrylonitrile itself resembles curve (e) in Fig. 9; the reaction is accompanied by destruction of radical activity and is assigned to case (v). Reaction of polyacrylonitrile radicals with methyl acrylate is exceptional in that it can reach almost complete conversion with very slow radical decay. The final polymer is insoluble in DMF which is a good solvent for both polyacrylonitrile and poly(methyl acrylate) suggesting that cross-linking occurs. This reaction has some of the characteristics of the "popcorn" polymerization [36].

From the foregoing discussion it can be seen that the concept of radical occlusion is able, at least qualitatively, to explain all of the characteristic features of the heterogeneous polymerization of acrylonitrile. In view of the fact that there is as yet no complete quantitative interpretation of the behaviour of high conversion homogeneous polymerization it is not surprising that the much more complex heterogeneous systems, even one as widely studied as is acrylonitrile, have not yet succumbed to quantitative description. However, in the light of the discussion so far presented it is relevant to examine how far the qualitative occlusion model can be extended to the description of heterogeneous polymerization of other monomers.

3.2 HETEROGENEOUS POLYMERIZATION OF VINYL CHLORIDE

3.2.1 General features

The heterogeneous polymerization of vinyl chloride was amongst the earliest heterogeneous reactions to be studied. The autocatalytic nature of the reaction was observed by Prat [56] and subsequently confirmed by several workers [57—60]. Several studies have established that the initiator exponent is close to 0.5 in contrast to the value close to 0.9 found for acrylonitrile polymerization [57, 58, 60]. Bengough and Norrish [57] showed that the precipitated polymer acts as a cocatalyst for the radical polymerization, producing a rate increase proportional to the $\frac{2}{3}$ power of the weight of the polymer; a similar effect was observed when pre-formed polymer was added to the reaction mixture. These authors suggested that chain transfer reactions with the polymer surface produced radicals immobilized in the surface and thus less likely to become involved in bimolecular termination; on this basis the cocatalytic effect of the polymer would be expected to be a function of the surface area of the polymer particles, thus accounting for the $\frac{2}{3}$ power dependence on the weight of polymer. Bengough and Norrish observed that the conversion—time curves for polymerization in the presence of added preformed polymer display an inflexion point at conversions in the range 2—10 %; above this inflexion the rate of polymerization and the autoacceleration were both reduced. This critical point was attributed to the conversion of the dispersed polymer precipitate into a glass state with a concomitant decrease in surface area. Similar results were obtained by Talamini and Vidotto [61] and attributed to coagulation of the polymer particles.

Since the mechanism proposed by Bengough and Norrish requires a chain transfer reaction to the polymer it is a consequence of this mechanism that long branches should be formed on the polymer. Cotman [62] treated poly(vinyl chloride) with lithium aluminium hydride to replace chlorine atoms by hydrogen. Infrared determination of the concentration of CH_3— end-groups in the resulting polyethylene indicated that commercial poly(vinyl chloride) can contain up to one branch per fifty monomer units. Batzer and Nisch [63] established the presence of tertiary chlorine atoms in commercial polymer samples by observing a reduction in molecular weight on treatment of the polymer with $LiAlH_4$ in the presence of oxygen. George et al. [64] used the method of Cotman and showed the presence of branching in polymer prepared at 45 °C, but were unable to detect branching in polymer prepared at —40 °C. However, Bovey and Tiers [65] suggest that the branching in poly(vinyl chloride) arises mainly from a back-biting chain transfer process, producing short-chain branches; this type of branching is also consistent with the results of a molecular-weight distribution analysis performed by Talamini and Vidotto [66]. Since back-biting chain transfer might be expected to be independent of conversion it is expected that such reactions would produce an extent of branching which does not depend on conversion. Krozer and Czlonkowska [67] suggest that the extent of branching, in fact, decreases with increasing conversion and conclude that branching occurs by chain transfer to polymer, producing long-chain branches. In contrast, Boccato et al. [68] studied branching in poly(vinyl chloride) prepared under both heterogeneous and homogeneous conditions and concluded that the amount of branching is independent of conversion in both cases although it decreases as the temperature of the polymerization is lowered. De Vries et al. [69] have made an extensive study of the dilute solution properties of poly(vinyl chloride) and of the solution and melt properties of polyethylenes prepared by $LiAlH_4$ reduction of poly(vinyl chloride); these authors conclude that commercial polymer samples contain no more than one long branch per polymer molecule, in agreement with the findings of Hahn and Müller [70]. A similar conclusion was reached by Lyngaae-Jorgensen, [71, 72], who has published an extensive bibliography on chain branching mechanisms in poly(vinyl chloride). Thus, although the presence of short chain branches in poly(vinyl chloride) is well established there is little evidence to support the view that extensive long-chain branching occurs as would be required by the mechanism of Bengough and Norrish.

In the bulk polymerization of vinyl chloride, the molecular weight of the polymer is found to be independent of both initiator concentration and conversion over a wide range of both parameters [56—60, 73]. Most authors have attributed this effect to the occurrence of chain transfer to the monomer although Talamini and Vidotto [66] have questioned this assumption on the basis of an analysis of molecular-weight distribution and the fact that similar kinetic behaviour has been found for acrylonitrile, for which transfer to monomer is absent [45]. Danusso and Sianesi [73] obtained a value of 1.48×10^{-3} l mole^{-1} sec^{-1} for the chain transfer coefficient with monomer at 60 °C; this result, which corresponds to one transfer step in every 680 propagation steps, was obtained by application of the Mayo equation and later confirmed by the use of a labelled initiator [74]. The existence of a facile transfer reaction with the monomer implies that polymeric radicals produced on the surface of a polymer particle are unlikely to have the very long lifetimes characteristic of the occluded radicals in polyacrylonitrile since transfer reactions of surface radicals with monomer would destroy the immobilized radical.

The bulk polymerization of vinyl chloride shows a number of features similar to those found in the bulk polymerization of acrylonitrile. Thus Talamini and Vidotto [61] showed that poly(vinyl chloride) prepared by bulk polymerization has a molecular weight which increases with temperature in the range —78 to —30 °C, the molecular weight passing through a maximum at —30 °C and then decreasing with increasing temperature; these results are qualitatively similar to those obtained with acrylonitrile (Fig. 5) although in the latter case the molecular weight maximum occurs some 70 °C higher, at 40 °C. The origin of this maximum is discussed further in Sect. 3.3.

Bawn et al. [75] studied the post-polymerization initiated by γ-radiation. Acrylonitrile gave a "fast reaction" of the type observed by Bamford and Jenkins and described in Sect. 3.1.1. However, only a very slight and irreproducible fast reaction was observed when poly(vinyl chloride) pre-polymer was heated to 56 °C although a more reproducible fast reaction was found when polymer produced at —23 °C was heated to —19 °C. The absence of a significant fast reaction at high temperature together with the low temperature maximum in the molecular weight/temperature relationship suggests that complete radical occlusion, such as is observed in polyacrylonitrile, can only occur in poly(vinyl chloride) at much lower temperatures.

Breitenbach and Schindler [58] showed that the addition of transfer agents markedly reduces the autoacceleration in the bulk polymerization reaction (Fig. 11) although the presence of low concentrations of transfer

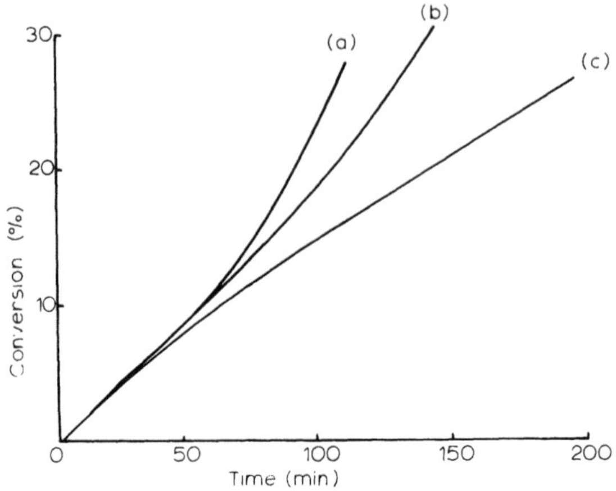

Fig. 11. Effect of carbon tetrabromide on kinetics of polymerization of vinyl chloride in bulk at 50 °C, initiated by lauryl peroxide. (a) [CBr_4] = 5 x 10^{-4} mole l^{-1}; (b) [CBr_4] = 5 x 10^{-1} mole l^{-1}; (c) [CBr_4] = 5 x 10^{-1} mole l^{-1}. (After Vidotto et al. [77].)

agent causes an increase in polymerization rate (Fig. 12) in contrast with the reduction observed for acrylonitrile polymerization. Ryska et al. [76] and Vidotto et al. [77] have obtained similar results.

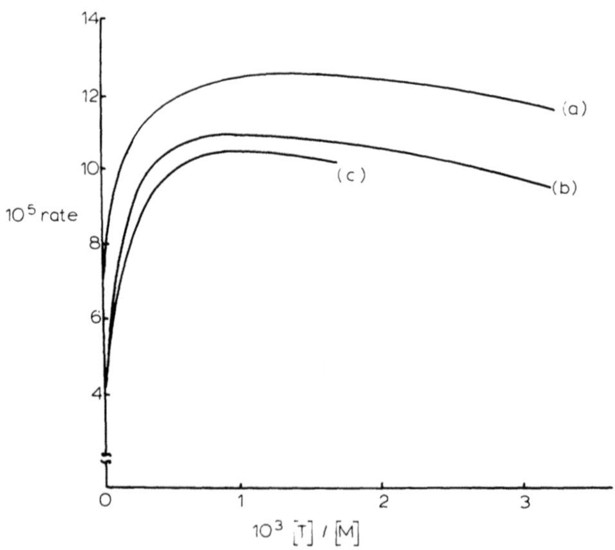

Fig. 12. Effect of chain transfer agents on rate of bulk polymerization of vinyl chloride at 50 °C. (a) Initiator azo-isobutyronitrile, transfer agent CBr_4 (6.9 x 10^{-3} mole l^{-1}); (b) initiator benzoyl peroxide, transfer agent CBr_4 (13.6 x 10^{-3} mole l^{-1}); (c) initiator benzoyl peroxide, transfer agent dodecyl mercaptan (13.6 x 10^{-3} mole l^{-1}). (After Breitenbach and Schindler. [58].)

TABLE 3
Particle size and number in bulk polymerization of vinyl chloride at 50 °C [78]

Fractional conversion	Mean particle diameter x 10^5 (cm)	$N \times 10^{-11}$ [a]
0.0017	<1.0	>23.2
0.009	2.90	5.10
0.045	5.10	4.65
0.096	5.90	6.41
0.148	7.20	5.40
0.300	9.60	4.70

[a] N = number of particles present per gm of vinyl chloride initially present.

Cotman et al. [78] studied the concentration of particles present in bulk polymerization of vinyl chloride. These authors showed that the number of particles present remained constant between 1 and 30 % conversion and was independent of initiator concentration, although the mean particle diameter increased with conversion (Table 3). Between 0.2 and 1 % conversion there was a sharp decrease in the number of particles; this was associated with the sharp decrease in rate at low conversions described previously. These authors conclude that the absence of secondary nucleation implies that radicals produced in the solution are predominantly absorbed into existing particles. Bort et al. [79, 80] studied the concentration and morphology of the particles produced under varying polymerization conditions. At relatively high rates of radical production, single particles with a relatively smooth surface were formed in about the same concentration as that found by Cotman (approximately 4.5×10^{14} particles per litre of monomer initially present). At low rates of radical production aggregates were formed consisting of a number of globular particles. However, irrespective of the particle morphology, both the number and morphology of the particles present at high conversions were found to be determined only by the conditions of the reaction at less than 1 % conversion; Bort, in agreement with Cotman concludes that the polymerization mainly takes place in the particles produced at low conversion and that secondary particle production is negligible. These authors also investigated the effect of particle size on polymerization rate [81] and showed that the polymerization rate is a function of the surface area of the particles, although this function we found to be linear only at high conversions; the initial stages in the formation of particles have been studied by electron microscopy [82] and the importance of the initial rate of polymerization and the rate of nucleation of the primary particles in determining the overall rate of polymerization has been discussed [83].

From the foregoing discussion it is clear that the bulk polymerization of vinyl chloride has several features in common with that of acrylonitrile, notably the autoaccelerating nature of the reaction, the observed maximum in the molecular weight/temperature relationship and the effect of transfer agents in reducing the tendency towards autoacceleration. However, there are several very significant differences between the two systems. In particular pre-formed poly(vinyl chloride) is a catalyst for vinyl chloride polymerization, whereas the same is only true of acrylonitrile in the very special circumstances when trapped radicals are present in the polymer. Although the effect of chain transfer agents in reducing autoacceleration is common to both vinyl chloride and acrylonitrile polymerization, low concentrations of transfer agent increase the polymerization rate of vinyl chloride but reduce the rate of polymerization of acrylonitrile. Finally the initiator exponent for acrylonitrile is around 0.9 whilst the value for vinyl chloride is 0.5. The origin of the differences between these two polymerization systems can be attributed partly to chemical and partly to physical factors. Thus in the polymerization of vinyl chloride, chain transfer to monomer is apparently relatively facile so that a radical can only become occluded completely if both monomer and other radicals are excluded from its vicinity. Furthermore, the free energy associated with the interaction of vinyl chloride monomer with the polymer is much higher than is the corresponding value for acrylonitrile; poly(vinyl chloride) is relatively readily swollen by its own monomer, which is a good plasticizer for the polymer [84], whilst polymer has finite solubility in the monomer. According to Gerrens et al. [85] the Flory—Huggins interaction constant χ_1 for the vinyl chloride—poly(vinyl chloride) system is 0.88 at 50 °C, which corresponds to a concentration of monomer in the polymer particles of 6 mole l^{-1}; once the particle density is established in the early stages of the reaction the polymerization system thus consists of two different phases each having a constant composition, at least up to around 70 % conversion. The bulk polymerization of vinyl chloride must therefore be regarded as a two-phase reaction, propagation proceeding in both the high monomer concentration phase and the high polymer concentration phase. The two phases may exchange radicals by precipitation of radicals into the polymer particles and by the escape from the particles of mobile radicals, produced by chain transfer with monomer. It is important to recognize that most published studies of this polymerization have been performed using reaction temperatures well above room temperature; since the molecular weight/temperature maximum is at —30 °C and no significant fast reaction is observed at high temperatures it is reasonable to suggest that the characteristic features of occluded polymerization observed with acrylonitrile will only be observed for vinyl chloride polymerization if the latter is performed at temperatures below —30 °C. In contrast, polyacrylonitrile is totally insoluble in its own monomer so that propagating radicals can undergo virtually no reaction before precipitating and the monomer concentration within the polymer particles is low; polymerization external to the particles can thus be essentially neglected.

3.2.2 Kinetic analysis of the bulk polymerization of vinyl chloride

The bulk polymerization of vinyl chloride has been the subject of many attempts to derive kinetic expressions which are capable of rationalizing the overall behaviour of the reaction as described in the previous section. Talamini and Peggion [86] have reviewed the literature on this aspect of vinyl chloride polymerization up to 1965. Following these authors we may summarize the reactions which have been proposed by various workers as follows.

Reaction in the liquid phase

$I \longrightarrow R\cdot$	Initiation	$R_i = k_d[I]$	(i)
$R\cdot + M \longrightarrow R\cdot$	Propagation	$R_p = k_p[R\cdot][M]$	(ii)
$R\cdot + M \longrightarrow P + M\cdot$	Transfer to monomer	$R_{tr} = k_{tr}[R\cdot][M]$	(iii)
$R\cdot + R\cdot \longrightarrow P$	Termination	$R_t = k_t[R\cdot]^2$	(iv)

Reactions within the polymer particles

$R\cdot + P \longrightarrow T\cdot + P$	Radical trapping	$R_c = k_c[R\cdot]F([P])$	(v)
$T\cdot + M \longrightarrow T\cdot$	Propagation	$R'_p = k'_p[T\cdot][M']$	(vi)
$T\cdot + M \longrightarrow P + M\cdot$	Transfer to monomer	$R'_{tr} = k'_{tr}[T\cdot][M']$	(vii)
$T\cdot + T\cdot \longrightarrow P$	Termination	$R'_t = k'_t[T\cdot]^2$	(viii)
$T\cdot + M\cdot \longrightarrow P$	Termination	$R''_t = k''_t[T\cdot][M'\cdot]$	(ix)

Reactions of M· radicals within the particles

$M\cdot + M \longrightarrow T\cdot$	Propagation	$R''_p = k''_p[M'\cdot][M']$	(x)
$M\cdot + P \longrightarrow T\cdot$	Transfer to polymer	$R^p_{tr} = k^p_{tr}[M'\cdot][P]$	(xi)
$M\cdot \longrightarrow R\cdot$	Escape to liquid phase	$R_e = k_e[M'\cdot]$	(xii)

Reactions (i)—(iv) are characteristic of the homogeneous free-radical polymerization, whilst reaction (v) indicates the formation of trapped radicals in the polymer with a rate which is a function of polymer concentration. Reactions (vi)—(ix) indicate the reactions, analogous to reactions (ii)—(iv), which may be undergone by trapped radicals, with rate coefficients which may be different from those of the homogeneous reactions. Reactions (x), (xi) and (xii) represent the reactions undergone by monomeric radicals M· within the polymer particles. The simultaneous possibility of reactions (v) and (xii) allows interchange of radicals between the two phases of the system.

Bengough and Norrish [57] assumed a steady state concentration for all radicals present in the system, equivalent in the present nomenclature to the expression

$$d([R\cdot] + [T\cdot] + [M\cdot])/dt = 0 \tag{4}$$

As described previously these authors assume that the radical trapping process represented by reaction (v) corresponds to a chain transfer reaction with the polymer surface so that the rate of this reaction becomes

$$R_c = k_c[R\cdot]F([P]) = Kk_c[P]^{2/3} = k_cA \tag{5}$$

where k_c is the rate coefficient for transfer to polymer; A is the surface area of the polymer particles and $A = K[P]^{2/3}$. It is further assumed (unjustifiably) that the concentration of monomer is the same in both phases and that reactions (viii) and (ix) do not occur (i.e. $k_t' = k_t'' = 0$). On this basis it is possible to derive an equation for the rate of polymerization having the form

$$-d[M]/dt = \alpha[M][I]^{1/2} + \beta[P]^{2/3}[I]^{1/2} \tag{6}$$

where α and β are constants. This equation contains two terms, the first of which is typical of the homogeneous reaction whilst the second describes the autoacceleration resulting from the presence of polymer. However, it has already been shown that the absence of long-chain branching in poly(vinyl chloride) implies that this mechanism is invalid and, furthermore, the $\frac{2}{3}$ power dependence of rate on particle area over a wide conversion range is doubtful [81].

The steady-state approximation is also assumed by Breitenbach and Schindler [58]. However, these authors assume that polymer radicals are present in the particles due both to precipitation and to the presence of initiator in the monomer swollen particles; it is further assumed that all propagation steps proceed equally efficiently in both phases so that $k_p = k_p'$ and $[M] = [M']$. In contrast to the assumption of Bengough and Norrish that bimolecular termination of polymeric radicals is absent in the concentrated phase, Breitenbach and Schindler adopt a partial occlusion model in which it is assumed that $k_t' < k_t$ and that the overall termination rate k_t^0 decreases with conversion according to the equation

$$k_t^0 = k_t/(1 + ac) \tag{7}$$

where c is the fractional conversion and a is a constant. These assumptions allow the conversion—time relation for an initial monomer concentration $[M]_0$ to be expressed as

$$c = \alpha[I]^{1/2}\cdot t\cdot[M]/[M]_0 + a\cdot\alpha^2[I][M]^2t^2/4[M]_0 \tag{8}$$

where α is a constant. This expression again contains two terms, of which the first is characteristic of the homogeneous phase and the second represents the modifying effect of the heterogeneous phase. According to this equation plots of c/t against t should be linear with slope and intercept both depending on the initiator concentration and such was found to be the case at least at low conversions. The predicted conversion—time curves gave good agreement with experiment at least up to $c = 0.2$; in particular the first-order termination process resulting from radical trapping contributes to the deviation of the initiator exponent from 0.5 and the authors showed that measurements of the initiator exponent at zero time gave a value of exactly 0.5. According to Breitenbach and Schindler the effect of transfer agents in increasing the polymerization rate may be ascribed to the conversion of relatively unreactive polymer radicals into active radicals by the chain transfer reaction.

Although the steady-state assumption gives reasonable agreement between theory and experiment for the results of Breitenbach and Schindler, Magat [87] has offered a quantitative treatment of the reaction by regarding autoacceleration as an extended pre-effect due to an abnormally low termination rate coefficient. This treatment essentially ignores both chain transfer to monomer and the two-phase nature of the reaction system which is equivalent, in the present reaction scheme, to the assumptions that $k_t = k_t' = k_t''$, $k_{tr} = k_{tr}'$ and $k_p = k_p'$; the conversion—time curve is then of the form expected for a homogeneous reaction with a long pre-effect and can be expressed as

$$c = \frac{k_p}{k_t} \ln \cosh \{(k_t k_d [\mathrm{I}])^{\frac{1}{2}} t\} \tag{9}$$

which, for low values of t, can be transformed by series expansion of the logarithm to give

$$c = k_d k_p [\mathrm{I}] t^2/2 \tag{10}$$

This treatment regards the individual specific rate coefficients as constant, unlike the treatment of Breitenbach and Schindler; although a reasonable fit to experimental data was obtained at low conversions an extension to the polymerization of vinylidene chloride was much less successful. Furthermore, this equation predicts that the plot of c/t against t is linear and passes through the origin; in practice as shown by Breitenbach and Schindler, such plots have finite intercepts.

Of the kinetic treatments so far described, that due to Breitenbach and Schindler is the most successful although it fails at conversions greater than 20 % and requires a purely empirical relationship between the overall termination rate coefficient and conversion. Mickley et al. [88] have attempted to obtain kinetic expressions usable over a wide conversion range by taking into account the particulate nature of the reaction and attempting to define the function F([P]) for reaction (v) in a precise manner. These authors assume that bimolecular termination reactions are absent in the polymer phase so that $k_t = k_t' = 0$, and that monomer radicals produced by chain transfer in the polymer particles, reaction (vii) may either propagate, reaction (x) undergo chain transfer, reaction (xi) or escape into the liquid phase, reaction (xii). The parameter k_cF(P) characterizing reaction (v) was set equal to $4\pi ND$, where N is the number of particles and D the diffusion coefficient of polymer radicals in the liquid phase. k_e, the rate coefficient for escape of M radicals was set equal to $4\pi rD$, where r is the average particle radius and D is the diffusion coefficient of monomer radicals in the liquid phase. From a kinetic analysis based on these assumptions Mickley et al. suggested that the rate of polymerization can be described by an equation of the form

$$-\mathrm{d[M]}/\mathrm{d}t = \alpha[\mathrm{M}][\mathrm{I}]^{1/2} + \alpha[\mathrm{I}]^{1/2}\mathrm{F}([\mathrm{P}]) \qquad (11)$$

When [P] is low (<30 g l^{-1}) the particles are small so that radical activity involves the whole of the polymer particles and F([P]) can be written in the form

$$([\mathrm{P}]) = Dk_{\mathrm{tr}}^{\mathrm{p}}[\mathrm{T}\cdot]\,V/D'\gamma k_{\mathrm{tr}} \qquad (12)$$

where V is the total volume of the particles in unit volume of reaction mixture and γ is the ratio of the monomeric radical concentration at the particle surface to the average concentration of radicals in the active volume. However, when [P] is large (>60 g l^{-1}) the particles are large and it is considered that radical activity is confined to a thin surface layer of thickness δ on each particle; F([P]) is then written in the form

$$\mathrm{F}([\mathrm{P}]) = Dk_{\mathrm{tr}}^{\mathrm{p}}[\mathrm{T}\cdot]\,\delta\cdot S/D'\delta k_{\mathrm{tr}} \qquad (13)$$

where S is the total surface area of the particles. It is interesting to note that this model predicts that the second term in eqn. (11) is proportional to the particle volume when the particles are small, but becomes proportional to surface area (in agreement with Bengough and Norrish) when the particles are large. The homogeneous component of eqn. (11) arises from the propagation reactions of polymer radicals in the homogeneous phase before they reach a critical size for coiling into primary polymer particles. By application of Smoluchowski's treatment of particle flocculation it was shown that virtually every polymer particle is incorporated into an aggregate as soon as it is formed thus maintaining the total number of particles constant but causing them to increase in size. Whilst the polymer particles remain small enough for radical activity to be uniform a dependence of the heterogeneous component of the rate on volume is observed which gradually becomes a surface area dependence as the particles are enlarged and their inner regions become inaccessible to the liquid phase. Since it is assumed that termination within the particles is absent, this model is essentially a modified occlusion model and requires the chain transfer to monomer as a means of maintaining the assumed steady-state radical concentration.

Danusso and Sianesi [73] observed that the conversion—time curves obtained over the full conversion range with different concentrations of initiator but at the same temperature are accurately superimposable if the fractional conversion is plotted against $[I]^{1/2}t$ instead of t. This observation leads to an empirical equation for the conversion—time behaviour of the form

$$c = F([I]^{1/2}t) \tag{14}$$

According to Talamini and Peggion [86] this equation may be developed in the form

$$c = \sum_{i=1}^{n} \alpha_i [I]^{i/2} t \tag{15}$$

where α_i is a constant for low values of c. In the particular case where $n = 2$ the summation may be expanded to yield

$$c = \alpha_1 [I]^{1/2} t + \alpha_2 [I] t^2 \tag{16}$$

This equation has the exact form of that due to Breitenbach and Schindler, eqn. (8), but does not agree with the expressions proposed by any of the other workers.

Although eqn. (8) is consistent with low conversion experimental data and with the expansion of Danusso's empirical equation (16), Talamini et al. [89—91] have criticized the assumption, implicit in the derivation of eqn. (2), that the overall termination rate coefficient is a function of conversion. These authors emphasize that the bulk polymerization takes place in two phases from low conversion onwards; the two phases have a constant composition in the whole range of their coexistence, so that although the values of the termination rate coefficients may be different in the two phases they will be expected to be independent of conversion up to high values. Furthermore, Talamini and Vidotto [66] consider that chain transfer to monomer is relatively unimportant and that bimolecular radical termination is able to occur in both the dilute and concentrated phases. According to Talamini, if it is assumed that k_p has the same value for both phases, the conversion—time behaviour can be written in the form

$$\frac{dc}{dt} = k_p [R\cdot] \frac{[M]}{[M]_0} + k_p [T\cdot] \frac{[M']}{[M]_0} \tag{17}$$

where $[M]_0$ is the initial monomer concentration, $[M]$ is the monomer concentration in the liquid phase and $[M']$ is the monomer concentration in the precipitate phase. If the specific rates of reaction in the liquid and precipitate phases (expressed as conversion per unit time) are defined as r and r', respectively, we may write

$$\frac{dc}{dt} = r\frac{[M]}{[M]_0} + r'\frac{[M']}{[M]_0} \tag{18}$$

Talamini assumes that, since the two phases have constant composition, the ratio of the specific rates r' and r will remain constant so that

$$r' = Qr \tag{19}$$

which is equivalent to the assumption that the ratio of radical concentrations in the two phases remains constant at all conversions. It is further assumed that r can be expressed in the form

$$r = k[I]^{1/2} = k_p\left(\frac{k_d}{k_t}\right)^{1/2}[I]^{1/2} \tag{20}$$

Analysis on this basis leads to the result that the conversion

$$c = [\exp(qrt) - 1]/q \tag{21}$$

where q is a constant dependent only on Q and the composition of the particles. Expansion of the exponential term as a series and substitution of r from eqn. (20) yields

$$c = \frac{1}{q}\sum_{n=1}^{\infty}\frac{(qk)^n}{n!}[I]^{n/2}t^n \tag{22}$$

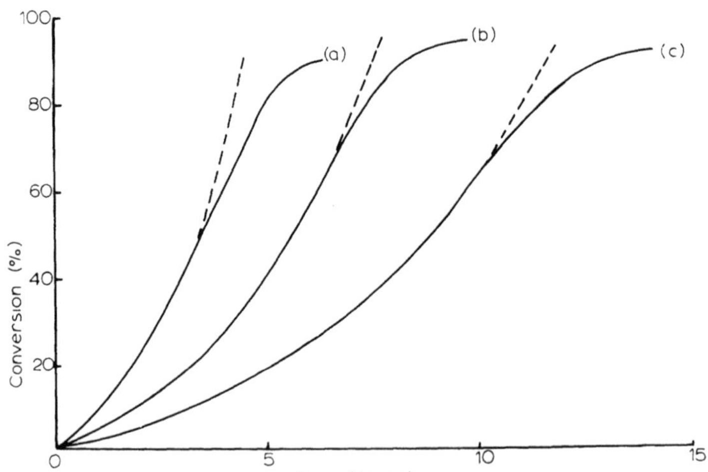

Fig. 13. Bulk polymerization of vinyl chloride at 50 °C, initiated by benzoyl peroxide. Dotted lines are curves derived from eqn. (22). (a) $[I]_0 = 10.3 \times 10^{-3}$ mole l^{-1}; (b) $[I]_0 = 5.16 \times 10^{-3}$ mole l^{-1}. (c) $[I]_0 = 2.1 \times 10^{-3}$ mole l^{-1}. (After Talamini et al. [91].)

This equation has the precise form of the empirical equation derived from the observations of Danusso and Sianesi (eqn. (15)), and agrees well with experiment up to high conversion as illustrated in Fig. 13. By limiting eqn. (22) to two terms it may be reduced to the form

$$c = \alpha_1 [\mathrm{I}]^{1/2} t + \alpha_2 [\mathrm{I}] t^2 \tag{23}$$

which is identical to the equation of Breitenbach and Schindler (eqn. (8)). This equation correctly predicts that the initiator exponent should be dependent on conversion, being 0.5 only at zero time; however, it predicts a constant value of 0.5 for the exponent determined at constant conversion and such is indeed found to be the case.

According to this theory the effect of chain transfer agents in increasing the rate arises from the fact that a reduction in polymer molecular weight will effect the equilibrium distribution of radicals between the two phases to the extent that lower-molecular-weight species will have a higher preference for the liquid phase where the reaction rate is high.

The model used by Talamini gives good agreement with experiment at low conversion but only poor agreement at high conversion. Abdul-Alim and Hamielec [92] applied a model based on that of Talamini to kinetic studies of the polymerization at 30, 50 and 70 °C and extended their model to include analysis of the average molecular weights and molecular-weight distributions of the polymers produced. Following Talamini, these authors essentially assume an equilibrium distribution of radicals between the liquid and precipitate phases so that the specific polymerization rates in both phases are related by eqn. (19). By making allowance for the changes in volume of the two phases and the initiator consumption they derived the equation

$$\frac{dc}{dt} = \frac{1 + Px}{\sqrt{1 - Bx}} k_{\mathrm{L}} [\mathrm{I}]_0^{1/2} \exp(-k_{\mathrm{d}} t/2) \tag{24}$$

where B is a constant determined by the relative densities of polymer and monomer, P is a constant determined by the parameter Q of eqn. (19) and by the composition of the precipitate phase, $[\mathrm{I}]_0$ is the initial concentration of initiator, which decomposes with a rate coefficient k_{d}, and k_{L} $(=k_{\mathrm{p}}(k_{\mathrm{d}}/k_{\mathrm{t}})^{1/2})$ is the constant characterizing the simple homogeneous polymerization rate. Equation (24) describes the polymerization up to the point where the conversion is such that the overall composition of the mixture is identical to that of the precipitate phase. Beyond this point the authors assume that both k_{p} and k_{t} will fall with increasing conversion and they further assume that this fall can be empirically expressed by the relation $k_{\alpha}(1 - c)$. With this assumption a rate equation is derived which gives an excellent fit to both the conversion—time curves (see Fig. 14) and the molecular-weight distribution curves.

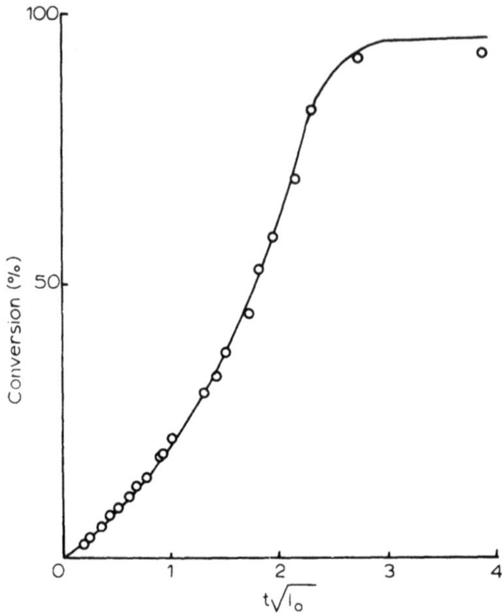

Fig. 14. Bulk polymerization of vinyl chloride at 30 °C initiated by azo-isobutyronitrile. Solid line is generated from eqn. (24), ○ = experimental points. (After Abdul-Alim and Hamielec [92].)

The main assumption made both by Talamini and by Abdul-Alim and Hamielec is the constancy of the factor Q in eqn. (19), which is equivalent to the assumption of a thermodynamic equilibrium distribution of radicals between the two phases. Furthermore it will readily be seen from the combination of eqns. (19) and (20) that it is implicitly assumed that the concentration of radicals in the precipitate phase is determined by the rate of termination in the liquid phase; this is equivalent to the assumption that bimolecular radical termination is absent from the precipitate phase. In an extensive discussion, Ugelstad et al. [84] have criticized these assumptions on several grounds. Without reference to the detailed chemistry of the reaction these authors consider that it is reasonable to assume a quasi-stationary state in radical concentration in both phases, so that, with an appropriate change in notation to emphasize the generality of the proposed equations, we may write stationary-state equations for the numbers of moles of radicals in both phases as follows

$$\frac{d[R\cdot]_L}{dt} = k_i[I]_L V_L - k_a[R\cdot]_L = k_{de}[R\cdot]_P - k_{t_L}[R\cdot]_L^2 V_L = 0 \qquad (25)$$

$$\frac{d[R\cdot]_P}{dt} = k_i[I]_P V_P + k_a[R\cdot]_L - k_{de}[R\cdot]_P - k_{tP}[R\cdot]_P^2 V_P = 0 \qquad (26)$$

where the subscripts L and P denote the liquid and precipitate phases, k_a is the rate coefficient for entry of radicals into the precipitate phase, k_{de} is the rate coefficient for desorption of radicals from the precipitate phase and the formation of particles is accompanied by a significant decrease in rate.

Ugelstad et al. have used eqns. (25) and (26) to calculate the form of the conversion—time curve from first principles. Known values of k_i and k_t were used and it was assumed, following Mickley et al. [88], that the rate coefficients for radical adsorption and desorption are determined by diffusion rates and may be expressed in the form

$$k_a = 4\pi D_L N r \tag{27}$$

$$k_{de} = 4\pi D_P N r \tag{28}$$

where N is the number of particles, r is the average particle radius and D_L and D_P are the radical diffusion coefficients in the liquid and precipitate phases, respectively. Values taken for k_{t_P} (3.6×10^8 l mole^{-1} sec^{-1}) and D_P (2×10^{-11} dm^3 sec^{-1}) were experimentally derived from emulsion polymerization data and it was found that the best fit of the calculated curve to the experimental data was obtained with a value of D_L equal to 10^{-7} dm^3 sec^{-1} and the ratio k_{t_L}/k_{t_P} set equal to 20. Although this approach gives a good fit with experiment for conversions above 30 %, it consistently predicts polymerization rates at low conversions which are significantly lower than those observed; this discrepancy is particularly marked at conversions below 10 % and is in contrast with the approach used by Talamini, which gives curves deviating from experiment at high conversions.

Ugelstad considers that the deviation from experimental data of the conversion curves predicted from eqns. (25) and (26) arises directly from the particulate nature of the reaction, and is due to the restriction of the number of radicals which are able to coexist in a polymer particle when the particle is small. This situation is very familiar in kinetic treatments of emulsion polymerization, the classic Smith—Ewart theory being based on the view that, if the polymer particles are sufficiently small, only one radical may exist in a particle, the entry of a second radical resulting in immediate termination [33, 34]. The effect of subdivision of a polymerization reaction into discrete particles may be treated by the methods developed by Stockmayer [93] and O'Toole [94]. Using a modification of this treatment, Ugelstad et al. [84, 85] have calculated the effect of subdivision on the rate of vinyl chloride polymerization as a function of conversion. When both subdivision, and the possibility of both desorption and adsorption of radicals, were taken into account it was found that the effect is to decrease the chance of termination in the particles at low conversion leading to a marked increase in the number of moles of radicals within the particles and a consequent increase in the rate of polymerization at low conversions, bringing the predicted behaviour into agreement with experiment. According to these authors the rate of the termination reaction within the particles is the dominant factor in determining the radical concentrations in both the liquid and precipitate phases which is exactly opposite to the conclusion of Talamini who considers that the radical concentration in the dilute phase is determined by the termination rate in that phase. The effect of transfer agents is considered by Ugelstad to arise from an increase in the value of D_P caused by lowering of the molecular weight of polymer radicals. The treatment developed by Ugelstad and co-workers gives the best fit of all available theories to experimental data over the full conversion range although it suffers from the problem that several of its parameters are not directly accessible but must be varied to obtain the best fit.

In summary, it is reasonable to conclude that the polymerization of vinyl chloride in bulk at elevated temperatures (up to 60 °C) is a two-phase reaction proceeding both in the virtually pure liquid monomer and in the monomer-swollen polymer particles. Within the particles the effective viscosity is high so that the rate of bimolecular termination is lower than that in the liquid phase. However, because of the relatively high solubility of monomer in the polymer particles the mobility of polymer radicals remains high and complete radical occlusion, such as is observed with acrylonitrile, can only be expected at low temperatures.

3.3 THE HETEROGENEOUS POLYMERIZATION OF VINYLIDENE CHLORIDE

Polymerization of both acrylonitrile and vinyl chloride leads to a polymer in which each monomer unit contains a pseudo-asymmetric carbon atom so that both polymers are capable of exhibiting stereoisomerism. Poly(vinyl chloride) prepared at temperatures above 25 °C contains short-chain branches produced by intramolecular chain transfer reactions. Both polymers precipitate from their parent monomers as globular particles of low crystallinity. In contrast, vinylidene chloride is a symmetrical monomer giving rise to a polymer chain which cannot exhibit stereoisomerism. Poly(vinylidene chloride) is able to crystallize to a very high extent and precipitates from the bulk monomer in the form of reasonably well defined lamellar crystals. The glass transition temperature of the polymer (—10 °C) is below the temperatures employed in most polymerization studies [96, 97] and solubility studies suggest that precipitation is entirely due to crystallization since the monomer is predicted to be a good solvent for the amorphous polymer [98].

Burnett and Melville [99] studied the photoinitiated polymerization of vinylidene chloride in the temperature range 13—35 °C. Although the reaction is heterogeneous many of their observations were typical of homogeneous kinetics, there being no autoacceleration and no unusual relationship between the rate of reaction and the incident light intensity.

However, measurement of the absolute values of the rate coefficients for propagation and termination as a function of temperature gave rather surprising values of the activation parameters. It was found that the two rate coefficients were given by

$$k_p = 10^{16} \exp[-25{,}000/RT] \text{ l mole}^{-1} \text{ sec}^{-1} \quad (29)$$

$$k_t = 10^{30} \exp[-40{,}000/RT] \text{ l mole}^{-1} \text{ sec}^{-1} \quad (30)$$

the subscripts p and t indicating propagation and termination, respectively. Although the values of k_p and k_t are normal in the narrow temperature range studied, it is apparent from the high activation energies that a relatively small increase in the polymerization temperature would cause the rate coefficients to assume impossibly high values; furthermore the very high pre-exponential factors are not interpretable by any currently accepted theory of absolute reaction rates. Burnett and Melville suggested that the high activation energies for both propagation and termination might result from incorporation of the propagating radicals into the crystalline polymer precipitate.

In a later study Bengough and Norrish [100] examined the kinetics of the reaction initiated by benzoyl peroxide in the temperature range 47—75 °C, and found that the reaction displays autoaccelerative behaviour similar to that of acrylonitrile and vinyl chloride. This apparent discrepancy is in agreement with the observation by Bamford and Jenkins [37] that the photoinitiated polymerization of acrylonitrile shows a square root dependence on intensity of irradiation and is not autoaccelerative; these latter authors suggest that the normal kinetic behaviour in photoinitiated polymerization is due to the fact that scattering of light by the precipitating polymer progressively reduces the rate of initiation in opposition to the autoaccelerating tendency. Bengough and Norrish suggested that the autoacceleration arises from the immobilization of polymer radicals on the polymer particle surfaces in a manner similar to that described by the same authors for vinyl chloride polymerization [57]. Magat [87] attempted to apply the extended pre-effect treatment, described for vinyl chloride, to the data of Bengough and Norrish but found a poorer correlation for vinylidene chloride than for vinyl chloride.

Bamford et al. [43] were unable to detect trapped radicals by ESR in poly(vinylidene chloride) prepared by photoreaction at 20 °C, and Bawn et al. [75] were unable to detect any significant fast reaction. Talamini and Vidotto [66] studied the variation of molecular weight and dissolution time with polymerization temperature in the bulk heterogeneous polymerization of acrylonitrile, vinyl chloride and vinylidene chloride. In each of the first two cases a pronounced maximum was observed in the molecular weight/temperature curve and was accompanied by a sharp increase in dissolution time at the temperature of the molecular weight maximum. In contrast the molecular weight of poly(vinylidene chloride) decreased continuously with increasing temperature in the range −50 to +15 °C. Talamini and Vidotto suggest that the maximum observed with vinyl chloride and acrylonitrile can be ascribed to a structural change in the polymers as the preparation temperature is lowered; these authors suggest that it is the stereoregularity (and therefore crystallinity) of the polymers which changes with temperature leading to greater crystallinity and a higher probability of radical trapping within the polymer particles. Although superficially attractive, this theory is unsupported by experimental evidence. Bovey [101, 102] has reviewed the extensive NMR studies of stereoisomerism in poly(vinyl chloride). The 220-MHz spectrum of the polymer, produced by free-radical polymerization, is interpretable in terms of pentad resonances for the α-protons and tetrad resonances for the β-protons. The tetrad and pentad relative intensities are in good agreement with Bernouillian chain propagation statistics and the value of P_m (the probability of isotactic placement) is around 0.4 (a value of 0.5 corresponds to production of completely atactic polymer). The value of P_m is remarkably insensitive to polymerization temperature, P_m values of 0.46 at 100 °C and 0.37 at −78 °C being reported [103]. Even if it is assumed that syndiotactic stereoblocks of only four or five monomer units are able to crystallize, with a P_m value of 0.4 such blocks would constitute, at most, less than 10 % of the total monomer units and, furthermore this fraction would be insensitive to polymerization temperature. Thus the changes in the properties of poly(vinyl chloride) which are associated with lower polymerization temperatures cannot be attributed to the small increase in syndiotactic content of the polymer. Rather, it is probable that the reduction in branching caused by lower polymerization temperatures [65, 66, 104] is responsible for the enhancement of crystallinity. Rather similar conclusions apply to polyacrylonitrile [101] although there are considerable difficulties in interpretation of the ^1H NMR spectrum [105]. Svegliado et al. [106] showed that the 100-MHz β-methylene spectrum of polyacrylonitrile is unaffected over a polymerization temperature range from −78 to +150 °C, implying that stereoregularity is essentially independent of temperature, in agreement with potential energy calculations of Krigbaum and Tokita [107]. Although differences in crystallinity with polymerization temperature have been observed [66, 108] these cannot be attributed to variations in stereoregularity; in the case of polyacrylonitrile the effect is also not readily attributed to branching, since Ham [109] has shown that the chain transfer coefficients for models of the polyacrylonitrile chain are so low that significant chain branching is unlikely. It seems most probable that the increase in crystallinity with reduction in polymerization temperature is a consequence rather than a cause of molecular weight changes. Alternatively, it is possible that the maximum in molecular weight is associated with the glass transition temperature of the polymer—monomer mixture. The glass transition temperature of polyacrylonitrile is in the range

80—100 °C depending on the method of preparation and the care taken to avoid degradation of the polymer during, or prior to, measurement [110]. For poly(vinyl chloride) the glass transition temperature is 80 °C [111]. The interaction of monomer with polymer would be expected to lower the glass transition temperature, this effect being much greater with poly(vinyl chloride) than with polyacrylonitrile, owing to the much higher equilibrium monomer concentration in the polymer particles in the former case. The absence of a molecular weight maximum for poly-(vinylidene chloride) at temperatures above —50 °C would then imply either that crystallization of the polymer is near complete or that the solubility of monomer in the amorphous regions of the polymer is sufficient to depress the glass transition temperature below —50 °C.

Bort et al. [83, 112, 113] studied the morphology of the particles produced by free-radical bulk polymerization of vinylidene chloride and found that, even at conversions as low as 0.1 %, two types of particle morphology were present, globular particles and crystalline platelets. Similar results were obtained by Wessling and Harrison [114] who found that a variety of particle morphologies could be obtained by polymerization in poor solvents. However, the globule plus platelet morphology of the bulk reaction could not be duplicated by solution crystallization of the polymer. Wessling and Harrison suggest that polymer precipitation is due to crystallization and that, because of this fact, radical reactions within the polymer particles are unlikely. This conclusion is partially based on the view that vinylidene chloride is a solvent for the amorphous polymer [97]. If this is true then it is difficult to account for the presence of the, apparently amorphous, globules particularly since they seem to increase in size with increasing conversion [113]. Wessling and Harrison treated the polymerization kinetics by assuming a model in which the normal steps of the homogeneous polymerization are accompanied by precipitation of polymeric radicals followed by propagation and bimolecular termination of the radicals on the surface of the polymer particles. With this model the form of the kinetic equation is dependent upon the shape of the polymer particles and these authors derived a kinetic equation of the form

$$-\mathrm{d}[\mathrm{M}]/\mathrm{d}t = k[\mathrm{I}]^{1/2}([\mathrm{M}]/[\mathrm{M}]_0)^{1/4} \tag{31}$$

where k is a constant determined partly by the rate coefficients of the elementary steps and partly by the morphology of the particles. The best fit with experimental data was obtained when it was assumed that all propagating radicals precipitate, so that liquid phase termination is negligible, and that growth occurs on the edges of lamellar crystals. It is possible to envisage two models of this type; in the first it is assumed that a polymeric radical, whose chain is involved in a folded chain lamellar crystal, propagates in the liquid phase until the length of the chain in the liquid corresponds approximately to a single fold length of the crystal when the chain is deposited on the crystal face and the growth process is repeated. Alternatively, the process may involve radicals which are completely crystallized so that the chain-end is on the crystal surface; propagation is then envisaged as involving adsorption of monomer onto the crystal surface followed by coupling with the radical. Although present evidence is insufficient to distinguish these two mechanisms, Wessling and Harrison consider that the former is more probable since their kinetic scheme requires bimolecular radical termination which is unlikely to occur if the propagating radicals are immobilized on a crystal lattice surface.

3.4 OTHER VINYL HALIDES AND RELATED MONOMERS

Blauer et al. [115] examined the polymerization of vinyl bromide in 1,2-dibromethane solution and found a first-order dependence of rate on initiator concentration and a second-order dependence on monomer concentration. Although the reaction system is not obviously heterogeneous, light scattering studies indicated that polymer association occurs in the solvent medium used and there was evidence that the system is close to precipitation. The authors suggest that some radical occlusion is occurring since it is only on this basis that a satisfactory explanation of the kinetic results can be derived. Minsker et al. [116] studied the polymerization of vinyl bromide in bulk and in benzaldehyde solution and found that the polymer produced in the latter case is significantly more crystalline. Thomas and O'Shaughnessy [117] showed that the polymerization of chlorotrifluoroethylene is zero-order in monomer up to 80% conversion with an initiator exponent of between 0.7 and 0.8. They concluded that this reaction is heterogeneous, the autoaccelerating tendency being compensated by depletion of monomer. Qualitatively similar results were obtained using diluents which are non-solvents for the polymer. Volkova et al. [118] have reported that the rates of the γ-initiated polymerizations of $CF_2{:}CFCl$, $CF_2{:}CH_2$ and $CF_2{:}CF_2$ pass through maxima with polymerization temperature, the position of the maximum being a function of dose rate and monomer structure. Lazar [119] found that the kinetics of polymerization of chlorotrifluoroethylene are sensitive to the nature of the diluent; with chlorinated hydrocarbons a maximum was observed in the polymerization rate at a critical ratio of diluent and monomer concentration. Although very little detailed kinetic work has been published it seems very probable that the heterogeneous polymerizations of vinyl bromide and chlorotrifluoroethylene involve at least partial radical occlusion. Wall [120] has edited a comprehensive review of the synthesis, polymerization and polymer properties of fluorinated monomers.

3.5 THE COPOLYMERIZATION OF STYRENE WITH MALEIC ANHYDRIDE

In the context of heterogeneous polymerization, the free radical induced copolymerization of styrene (S) with maleic anhydride (MA) represents an interesting case, since the reaction may be homogeneous or heterogeneous depending only on the ratio of the reactant concentrations. Although the copolymerization in homogeneous solution has been extensively studied [121—123] because of its ability to yield alternating copolymers via the complexed monomers mechanism [124], relatively few publications have dealt specifically with the homogeneous—heterogeneous transition. However, this aspect of the reaction has been studied in some detail by Bamford et al. [125, 126], who found that the critical ratio [MA]/[S] at which the homogeneous—heterogeneous transition occurred was approximately 10^{-2}, both for bulk polymerization and for polymerization in the presence of a diluent such as *o*-dichlorobenzene. For [MA]/[S] $< 10^{-2}$, the polymer remains in solution, but above this value it precipitates in a swollen form. Figure 15 shows how the rate of reaction, as measured by the rate of consumption of the anhydride, varies with the anhydride concentration for the reaction in *o*-dichlorobenzene at 60 °C. Assuming normal copolymerization behaviour and neglecting the homopropagation of maleic anhydride, which is extremely slow, it is easily shown that the order of reaction with respect to MA should change from zero to unity as [MA] increases. The early part of the observed curve follows this prediction, until the critical concentration of anhydride is reached, when polymer precipitation occurs; at this point there is a

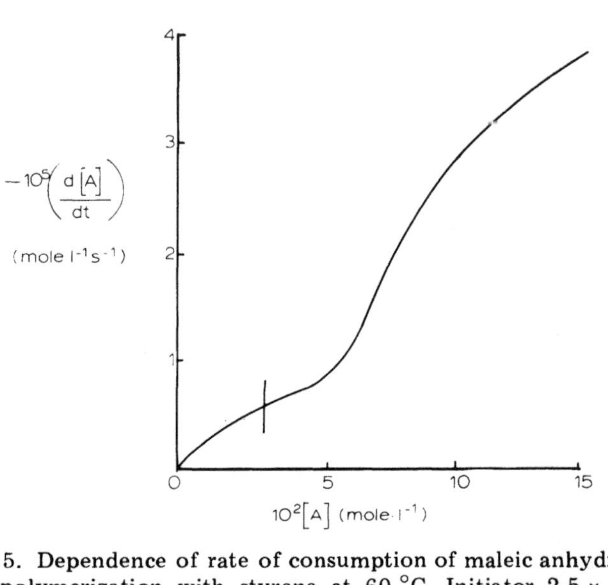

Fig. 15. Dependence of rate of consumption of maleic anhydride on its concentration in copolymerization with styrene at 60 °C. Initiator 2.5×10^{-2} w/v azo-isobutyronitrile. Homogeneous—heterogeneous transition is indicated by a vertical bar.

TABLE 4
Conversion dependence of the rate of copolymerization of maleic anhydride with styrene in the absence of solvents

Reaction time (min)	$[MA]_0 \times 10^2$ (mole l^{-1})	Mean $[MA] \times 10^2$ (mole l^{-1})	Conversion (%)	[MA] in polymer (%)	10^5 Mean rate ($= -10^5\, d[MA]/dt$) (mole l^{-1} sec^{-1})
Initiator, 0.014 % azo-isobutyronitrile; polymer precipitates					
7	29.1	28.3	0.4	36.5	3.62
10	29.8	28.4	0.7	37.1	4.37
15	30.9	28.5	1.2	37.0	5.15
20	32.0	28.2	2.0	36.8	6.40
Initiator, 0.025 % azo-isobutyronitrile; polymer remains in solution					
8	9.30	8.92	0.35	22.1	1.53
15	9.65	8.95	0.60	22.9	1.55
30	10.15	8.88	1.10	22.7	1.44

sharp increase in rate, following which the order of reaction with respect to MA continues to diminish. This increase in rate is attributed to an increase in the termination rate coefficient associated with precipitation of the polymer. Table 4 shows the results obtained for a series of polymerizations, both homogeneous and heterogeneous, in which the initial concentration of maleic anhydride was varied to maintain a constant mean concentration over varying reaction periods. The results show an essentially constant mean rate under homogeneous conditions, but clearly indicate that autoacceleration occurs under heterogeneous conditions. In the homogeneous case the initiator exponent is 0.5 rising to 0.7 in the heterogeneous case. The behaviour under heterogeneous conditions is thus very similar to that of acrylonitrile. The change from homogeneous to heterogeneous conditions has surprisingly little effect on the composition of the copolymer produced in this reaction. If it is assumed that the reactivity ratio r_1 for maleic anhydride radicals is zero, which is true to a good approximation, then standard copolymerization theory predicts that the mole ratio, n, of S to MA in the polymer, varies with the mole ratio in the feed according to the relationship

$$n - 1 = r_2[\mathrm{S}]/[\mathrm{MA}] \qquad (32)$$

where r_2 is the reactivity ratio for styryl radicals. Figure 16 illustrates the fact that the copolymerization obeys eqn. (32) closely in several reaction systems with no detectable change associated with the homogeneous–heterogeneous transition. The implication of these results is that the ratio of the monomer concentrations within the precipitate phase is the same as that in the bulk phase, implying that there is no difference in the availability of the two monomers within the precipitate phase.

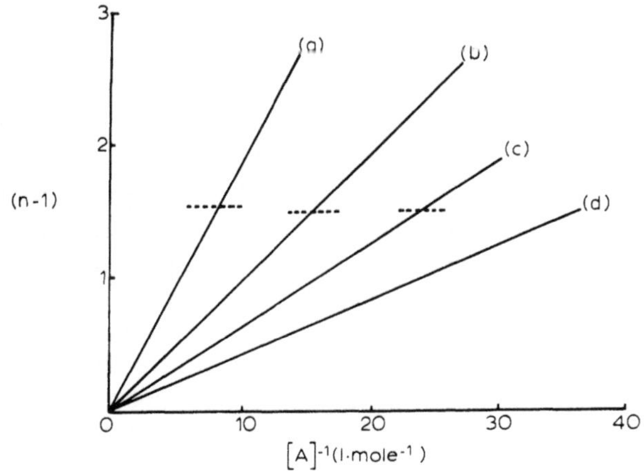

Fig. 16. Test of eqn. (32) for copolymerization of maleic anhydride and styrene at 60 °C. Solvent (a), bulk styrene; (b) 1:1 styrene: *o*-dichloro benzene; (c) 1:3 styrene; *o*-dichlorobenzene; (d) 1:1 styrene; decane. The horizontal bars indicate the homogeneous-heterogeneous transition. (After Bamford et al. [125].)

3.6 POLYMERIZATION IN THE PRESENCE OF PRECIPITANTS

In all of the heterogeneous polymerization reactions so far considered, the precipitate phase is formed because the polymer (or copolymer) is insoluble in the parent monomer. However, many normally homogeneous reactions in which the polymer remains soluble throughout the reaction, can be rendered heterogeneous by the addition of a diluent which acts as a non-solvent for the polymer and induces precipitation. The addition of a precipitant frequently results in increases in both rate and degree of polymerization [127—129] and, although most studies of reactions of this type are semi-quantitative, it is reasonable to conclude that diffusional limitations of the termination rate are responsible for most of the observed effects.

Abere et al. [130] studied the polymerization of styrene induced by benzoyl peroxide at 60 °C in the presence of methanol. Initial rates were measured as a function of monomer concentration and it was found that reduction in the monomer concentration in the range 100—20 % monomer caused the expected decrease in the initial rate of polymerization. However, when the monomer concentration is further decreased below 20 % the initial rate of polymerization increases and there is a concomitant increase in the degree of polymerization. This increase in rate is associated with the precipitation of polymer and it is reasonable to postulate that diffusion limitation of chain termination is responsible. Similar results were obtained by Chapiro [131] for the polymerization initiated by γ-radiation at 15 °C, although the results are complicated by the fact that radical production by γ-irradiation is more facile for methanol than for styrene so that the rate of initiation is a function of the ratio of monomer and methanol concentrations. Provided due allowance is made for this the results are similar to those of Abere; in particular, a rapid increase in both rate and degree of polymerization is observed at low monomer concentrations where polymer precipitation occurs. Individual kinetic runs in this region of monomer concentration show autoacceleration and the initiator exponent is greater than 0.5.

It has already been shown in Sect. 2.1, that the bulk polymerization of methyl methacrylate at high conversions displays autoacceleration associated with diffusion control of termination, even though precipitation does not occur. In the presence of sufficient cyclohexane, precipitation of the polymer occurs leading to autoacceleration at low conversion and an initiator exponent greater than 0.5. Under these conditions the calculated rate coefficients for both propagation and termination decrease, and the behaviour of the reaction under these conditions is entirely consistent with a radical occlusion mechanism [17].

A particularly important industrial use of heterogeneous polymerization occurs in those cases where a water soluble monomer is polymerized in aqueous solution from which the polymer precipitates. Dainton et al.

[132—135] have published an extensive study of the polymerization of acrylonitrile in aqueous solutions. Initiation was achieved either by the use of Fenton's reaction at 25 °C

$$Fe^{2+} + H_2O_2 \longrightarrow Fe^{3+} + OH^- + \cdot OH$$

or photochemically with Fe^{2+}, Fe^{3+} or I^- at temperatures in the range 15—50 °C. In studies using Fenton's reaction it was found that a brief period of acceleration was followed by a prolonged period of constant rate extending to greater than 5 % conversion, provided care was taken to remove ferric ions produced in the initiation process; removal of ferric ions was effected by complexing with fluoride ions and was necessary to eliminate a termination reaction between the propagating radicals and ferric ions, similar to that observed in dimethyl formamide solutions [50]. The linear initial polymerization rate was followed by a gradual deceleration which was always slower than would have been predicted from the known decrease in initiator and monomer concentrations. Kinetic studies of varying concentrations of monomer showed that the linear rate is proportional to $[M]^2$ at low monomer concentrations and to [M] at high monomer concentrations, the transition from second- to first-order behaviour occurring over a relatively narrow range of monomer concentration. The mean monomer concentration $[M]_c$ at the transition point is strongly dependent on the initiator concentration. In the range of first-order dependence on monomer the polymerization rate was proportional to $([Fe^{2+}][H_2O_2])^{0.5}$ whilst in the second-order range the exponent varied with initiation rate, from 0.9 at low initiation rate to 0.2 at high initiation rate. When the polymerization was initiated photochemically it was found that the reaction exhibits a long photochemical after-effect equivalent to the bimolecular decay of chain centres with a mean lifetime of 150—800 sec, depending upon temperature. This lifetime is independent of the intensity and duration of the illumination and rotating sector measurements confirmed the presence of radicals with long lifetimes. Thomas et al. [136] observed that the aqueous polymerization of acrylonitrile produces a constant number of particles in the early stages, the particles being uniform in size and about 50 A in diameter — coagulation of the particles occurs only when their volumes have increased by 10^4 or 10^5 times.

Dainton et al. have attempted to interpret all of these observations on the basis of a model of the polymerization process involving reaction in both solution and precipitate phases. These authors consider that, in the early stages of the reaction, precipitating polymer molecules and polymer radicals will coalesce to produce particles which will increase in size until they are stabilized by surface charge at a critical size corresponding to about 50 A diameter; particles which are at or above this critical size are regarded as being sufficiently stabilized by their surface charge that they are unable to coalesce. If radicals contained within sub-critical and post-critical particles are denoted by S· and P·, respectively, then it is expected that the concentration of S· will increase rapidly in the early stages, leading to a stationary value. The sub-critical particles formed at this stage will coalesce to form large polymer particles (denoted by P) at a rate which is assumed to increase with the degree of polymerization of the S· radicals prior to their precipitation. Since the degree of polymerization is proportional to the monomer concentration, it is qualitatively expected that the initial rate of formation of P particles will be proportional to monomer concentrations at low conversions. As the polymerization proceeds and the number and size of the P particles increases, the flocculation arguments, employed earlier in the discussion of vinyl chloride polymerization, imply that sub-critical particles will increasingly coagulate with P particles rather than with each other; in the limiting case all sub-critical particles are captured by P particles leading to a constant particle number for the P particles. The coalescence of a sub-critical and a P type particle may then lead to two reactions

$$\mathrm{S\cdot + P \longrightarrow P\cdot}$$

$$\mathrm{S\cdot + P\cdot \longrightarrow P}$$

depending on whether or not the S· radical terminates on an existing P· radical on its entry into the P particles. The reaction scheme adopted for this stage of the reaction is formally similar to that describing emulsion polymerization, except that in the latter case propagation of the S· radical prior to its entry into a polymer particle is normally insignificant in terms of its contribution to the polymerization rate. In order to derive kinetic expressions it is necessary to assume a reaction of equilibrium adsorption of monomer onto the polymer particles. If it is assumed that the overall polymerization rate is the sum of two terms corresponding to the consumption of monomer by S· and P· radicals then we may write

$$R_\mathrm{p} = R_\mathrm{p}^\mathrm{s} + R_\mathrm{p}^\mathrm{p} = K_\mathrm{p}[\mathrm{M}]\,[\mathrm{S\cdot}] + k'_\mathrm{p}[\mathrm{M}]_\mathrm{ads}[\mathrm{P\cdot}] \qquad (33)$$

and a kinetic scheme based on this equation was shown to give reasonable agreement with experimental data, although the form of the adsorption isotherm chosen to describe $[\mathrm{M}]_\mathrm{ads}$ is very critical and the scheme is much less successful in predicting the variation of $[\mathrm{M}]_\mathrm{c}$ with the rate.

As has already been pointed out, polyacrylonitrile is completely insoluble in its own monomer and, conversely, the monomer concentration within a precipitate particle must be very low. Napper and Parts [137] have examined the polymerization of vinyl acetate in aqueous solution, a system in which the polymer is insoluble in water but soluble in the monomer. In this system polymerization proceeds in monomer swollen polymer particles and the reaction kinetics were found to resemble those for emulsion polymerization. Comparison of rate coefficients for the propagation and termination reactions in the precipitating system and the homogeneous solution reaction showed that the propagation coefficient is unaffected by precipitation but that the termination coefficient is reduced by a factor of 25 in comparison with the homogeneous reaction.

An interesting example of a homogeneous polymerization having kinetics very similar to heterogeneous systems is the polymerization of acrylamide in aqueous solutions. Although both monomer and polymer are water soluble and the reaction displays essentially homogeneous kinetics, rotating-sector measurements indicate very long radical lifetimes and there is a strong photochemical after-effect corresponding to radical lifetimes of several hours. Dainton and Tordoff [138] consider that simple radical occlusion cannot be responsible for lifetimes of this magnitude in an essentially homogeneous system. They suggest that the long lifetimes are due to slow imidization of the terminal radical, especially at low pH values, so that for long periods the radical has the glutarimide structure (II) rather than the acetamide structure (I)

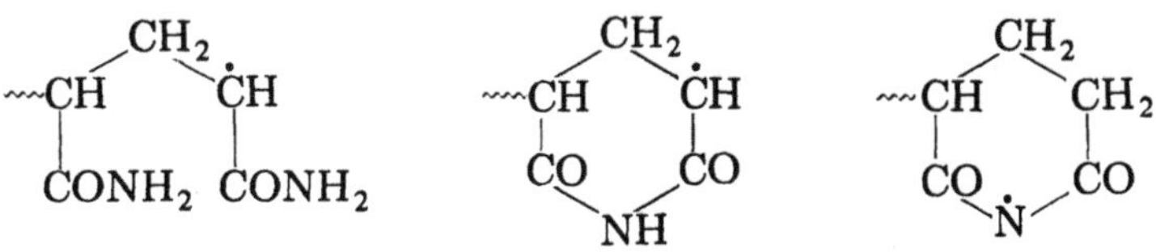

(I) (II) (III)

Radicals (I) and (II) would not differ greatly in reactivity, but (II) would be in equilibrium with its much more stable tautomer (III). On this basis the velocity coefficient for radical "burial" would be that for the tautomerization (II) $\rightleftharpoons$ (III); the detailed kinetic observations were shown to be consistent with this postulate and supporting evidence was derived from the fact that the after-effect is not observed at high pH, where tautomerism would be restricted.

The manuscript for this chapter was completed in 1972.

REFERENCES

1 A. M. North, Q. Rev., 20 (1966) 421.
2 C. R. Patrick and P. E. M. Allen, Makromol. Chem., 47 (1961) 154.
3 C. H. Bamford, W. G. Barb, A. D. Jenkins and P. F. Onyon, Kinetics of Vinyl Polymerization by Free Radical Mechanisms, Butterworths, London, 1958.
4 H. Kobayashi, J. Polym. Sci., 17 (1955) 81.
5 C. H. Bamford and S. Brumby, Makromol. Chem., 105 (1967) 122.
6 Y. Yokota and M. Itoh, J. Polym. Sci., Part B, 6 (1968) 825.
7 S. W. Benson and A. M. North, J. Am. Chem. Soc., 81 (1959) 1339.
8 A. M. North and D. Postlethwaite, in T. Tsuruta and K. F. O'Driscoll (Eds.), Structure and Mechanism in Vinyl Polymerization, Dekker, New York, 1969, Chap. 4.
9 J. Hughes and A. M. North, Trans. Faraday Soc., 60 (1964) 960; Proc. Chem. Soc., (1964) 404.

10 A. M. North and G. A. Reed, J. Polym. Sci., Part A, 1 (1963) 1311; Trans. Faraday Soc., 57 (1961) 859.
11 S. Russo and S. Munari, J. Makromol. Sci., Chem., 2 (1968) 1321.
12 S. Russo and S. Munari, Chim. Ind. (Milan), 51 (1969) 25.
13 R. D. Burkhart, J. Polym. Sci., Part A, 3 (1965) 883.
14 U. Borgwardt, W. Schnabel and A. Henglein, Makromol. Chem., 127 (1969) 176.
15 G. V. Schulz and G. Harborth, Makromol. Chem., 1 (1947) 106.
16 E. Trommsdorf, H. Köhle and P. Lagally, Makromol. Chem., 1 (1947) 169.
17 P. Hayden and H. Melville, J. Polym. Sci., 43 (1960) 201.
18 G. V. Schulz, Z. Phys. Chem., 8 (1956) 284, 290.
19 G. M. Burnett and G. L. Duncan, Makromol. Chem., 51 (1962) 154.
20 W. I. Bengough and H. W. Melville, Proc. R. Soc. London, Ser. A, 230 (1955) 429.
21 K. Horie, I. Mita and H. Kambe, J. Polym. Sci., Part A-1, 6 (1968) 2663.
22 N. M. Atherton, H. Melville and D. H. Whiffen, J. Polym. Sci., 34 (1959) 199.
23 A. Chapiro, Eur. Polym. J., Supplement (1969) 43.
24 A. Chapiro, Chimica, 21 (1967) 454.
25 K. Ito, J. Polym. Sci., Part A-1, 7 (1969) 827, 2247, 2707, 2995, 3387.
26 K. Ito, J. Polym. Sci., Part A-1, 10 (1972) 3159.
27 J. A. Kerr and A. F. Trotman-Dickenson, in G. Porter (Ed.), Progress in Reaction Kinetics, Pergamon, London, 1961, Chap. 4.
28 Kh. S. Bagdasar'yan, Theory of Free Radical Polymerization, Israel Programme for Scientific Translations, Jerusalem, 1968, p. 146.
29 W. B. Barb, J. Polym. Sci., 11 (1953) 117.
30 J. N. Atherton and A. M. North, Trans. Faraday Soc., 58 (1962) 2049.
31 A. M. North and D. Postlethwaite, Polymer, 5 (1964) 237.
32 G. Benta, B. M. Gallo and S. Russo, J. Chem. Soc., Faraday Trans., 2 (1973) 328.
33 J. W. Vanderhoff, in G. E. Ham (Ed.), Vinyl Polymerization, Vol. 1, Part II, Dekker, New York, 1969, pp. 1—138.
34 B. M. E. van der Hoff, Adv. Chem. Ser., 34 (1962) 6.
35 J. L. Gardon, J. Polym. Sci., Part A-1, 6 (1968), 623, 643, 665, 687, 2853, 2859.
36 J. W. Breitenbach, Adv. Macromol. Chem., 1 (1968) 139.
37 C. H. Bamford and A. D. Jenkins, Proc. R. Soc. London, Ser. A, 216 (1953) 515.
38 C. H. Bamford and A. D. Jenkins, J. Polym. Sci., 20 (1956) 405.
39 C. H. Bamford and A. D. Jenkins, Proc. R. Soc. London, Ser. A, 228 (1955) 220.
40 G. S. Hammond, J. N. Sen and C. F. Boozer, J. Am. Chem. Soc., 77 (1955) 3244.
41 J. C. Bevington, J. Chem. Soc. (1956) 1127.
42 C. H. Bamford, D. J. E. Ingram, A. D. Jenkins and M. C. R. Symons, Nature (London), 175 (1955) 894.
43 C. H. Bamford, A. D. Jenkins, M. C. R. Symons and M. G. Townsend, J. Polym. Sci., 34 (1959) 181.
44 P. J. Flory, Principles of Polymer Chemistry, Cornell, New York, 1953, p. 495.
45 W. M. Thomas, Adv. Polym. Sci., 2 (1961) 401.
46 O. G. Lewis and R. M. King, Adv. Chem. Ser., 91 (1969) 25.
47 W. M. Thomas and J. Pellon, J. Polym. Sci., 13 (1954) 329.
48 C. H. Bamford and E. F. T. White, unpublished results, see A. D. Jenkins, in G. E. Ham (Ed.), Vinyl Polymerization Vol. 1, Part I, Dekker, London, 1964, p. 369.
49 C. H. Bamford, A. D. Jenkins and R. Johnston, Nature (London), 177 (1956) 992.
50 C. H. Bamford, A. D. Jenkins and R. Johnston, Proc. R. Soc. London, Ser. A, 239 (1957) 214.
51 S. Amdur, J. Polym. Sci., Part A-1, 9 (1971) 175.
52 A. Chapman, D. Phil. Thesis, University of Sussex, 1971.

53 N. N. Dass and M. H. George, J. Polym. Sci., Part A-1, 7 (1969) 269.
54 C. H. Bamford, A. D. Jenkins and R. Johnston, Trans. Faraday Soc., 55 (1959) 418.
55 C. H. Bamford and A. D. Jenkins, J. Chim. Phys., 56 (1959) 798.
56 J. Prat, C. Rend. J. Inst. Plast., (1949) 58.
57 W. I. Bengough and R. G. W. Norrish, Proc. R. Soc. London, Ser. A, 220 (1950) 301.
58 J. W. Breitenbach and A. Schindler, Monatsh. Chem., 80 (1949) 429; Ric. Sci., Suppl., 25 (1955) 34.
59 E. J. Arlman and W. M. Wagner, J. Polym. Sci., 9 (1951) 581.
60 F. Danusso and G. Perugini, Chim. Ind. (Milan), 35 (1953) 881.
61 G. Talamini and G. Vidotto, Makromol. Chem., 50 (1961) 129.
62 J. D. Cotman, J. Am. Chem. Soc., 77 (1955) 2750; Ann. N.Y. Acad. Sci., 57 (1953) 417.
63 H. Batzer and A. Nisch, Makromol. Chem., 22 (1957) 131.
64 M. H. George, R. J. Grisenthwaite and R. F. Hunter, Chem. Ind. (London) (1958) 1114.
65 F. H. Bovey and G. D. V. Tiers, Chem. Ind. (London), 42 (1962) 1826.
66 G. Talamini and G. Vidotto, Chem. Ind. (Milan), 46 (1964) 16.
67 Sz. Krozer and Z. Czlonkowska, J. Appl. Polym. Sci., 8 (1964) 1275.
68 G. Boccato, A. Rigo, G. Talamini and F. Zilic-Grandi, Corsi. Semin. Chim., 8 (1968) 126.
69 A. J. de Vries, C. Bonnebat and M. Carrega, Pure Appl. Chem., 26 (1971) 209.
70 W. Hahn and O. Müller, Makromol. Chem., 16 (1955) 71.
71 J. Lyngaae-Jorgenson, Seventh Int. Symp. Gel. Perm. Chromatogr., Monaco, 1969.
72 J. Lyngaae-Jorgenson, J. Chromatogr. Sci., 9 (1971) 331.
73 F. Danusso and D. Sianesi, Chim. Ind. (Milan), 37 (1955) 695.
74 F. Danusso, G. Pajaro and D. Sianesi, Chem. Ind. (Milan), 41 (1959) 1170,
75 C. E. Bawn, T. P. Hobin and W. J. McGarry, J. Chim. Phys., 56 (1959) 791.
76 M. Ryska, M. Kolinsky and D. Lim, J. Polym. Sci., Part C, 16 (1967) 621.
77 G. Vidotto, S. Brugnaro and G. Talamini, Makromol. Chem., 140 (1970) 249.
78 J. D. Cotman, M. F. Ganzalez and G. C. Claver, J. Polym. Sci., Part A-1, 5 (1967) 1137.
79 D. N. Bort, V. G. Marinin, A. J. Kalinin and V. A. Kargin, Vysokomol. Soedin., Ser. A, 10 (1968) 2574.
80 D. N. Bort, V. G. Marinin, A. J. Kalinin and V. A. Kargin, Dokl. Akad. Nauk S.S.S.R., 183 (1968) 1080.
81 D. N. Bort, V. G. Marinin, A. J. Kalinin and V. A. Kargin, Vysokomol. Soedin., Ser. B, 12 (1970) 391.
82 D. N. Bort, V. G. Marinin, A. J. Kalinin and V. A. Kargin, Kinet. Mech. Polyreact. Int. Symp. Macromol. Chem. Prepr., 3 (1969) 403.
83 D. N. Bort and A. S. Arzhakov, Vestn. Akad. Nauk S.S.S.R., 40 (1970) 79.
84 J. Ugelstad, H. Lervik, B. Gardinovacki and E. Sund, Pure Appl. Chem., 26 (1971) 121.
85 H. Gerrens, W. Fink and E. Köhnlein, J. Polym. Sci., Part C, 16 (1967) 2781.
86 G. Talamini and E. Peggion, in G. E. Ham (Ed.), Kinetics and Mechanism of Vinyl Polymerization, Vol. 1, Part I, Dekker, New York, 1967, p. 331.
87 M. Magat, J. Polym. Sci., 16 (1955) 491.
88 H. S. Mickley, A. S. Michaels and A. L. Moore, J. Polym. Sci., 60 (1962) 121.
89 G. Talamini, J. Polym. Sci., Part A-2, 4 (1966) 535.
90 G. Talamini and G. Vidotto, Chim. Macromol., (1968) 118.
91 A. Crosato-Arnaldi, P. Gasparini and G. Talamini, Makromol. Chem., 117 (1968) 140.

92 A. H. Abdul-Alim and A. E. Hamielec, J. Appl. Polym. Sci., 16 (1972) 783.
93 W. H. Stockmayer, J. Polym. Sci., 24 (1957) 314.
94 J. T. O'Toole, J. Polym. Sci., 9 (1965) 1291.
95 J. Ugelstad, P. C. Mark and J. Aasen, J. Appl. Polym. Sci., Part A-1, 5 (1967) 2281.
96 R. F. Boyer and R. G. Spencer, J. Appl. Phys., 15 (1944) 398.
97 R. A. Wessling, Am. Chem. Soc. Polym. Prepr., 14 (1973) 283.
98 R. A. Wessling, J. Appl. Polym. Sci., 14 (1970) 1531.
99 J. W. Burnett and H. W. Melville, Trans. Faraday Soc., 46 (1950) 976.
100 W. I. Bengough and R. G. W. Norrish, Proc. R. Soc. London, Ser. A, 218 (1953) 149.
101 F. A. Bovey, Prog. Polym. Sci., 3 (1971) 1.
102 F. A. Bovey, Pure Appl. Chem., 26 (1971) 193.
103 F. A. Bovey, F. P. Hood, E. W. Anderson and R. L. Kornegay, J. Phys. Chem., 71 (1967) 312.
104 L. Cavalli, G. C. Borsini, G. Carraro and G. Confalonieri, J. Polym. Sci., Part A-1, 8 (1970) 801.
105 T. Yoshino, H. Kenjo and K. Kuno, J. Polym. Sci., Part B, 5 (1967) 703.
106 G. Svegliado, G. Talamini and G. Vidotto, J. Polym. Sci., Part A-1, 5 (1967) 2875.
107 W. R. Krigbaum and S. Tokita, J. Polym. Sci., 43 (1960) 467.
108 R. Chiang, J. Polym. Sci., Part A, 3 (1965) 2019.
109 G. E. Ham, J. Polym. Sci., 21 (1956) 337.
110 R. B. Beevers, Macromol. Rev., 3 (1968) 113.
111 J. Brandrup and E. H. Immergut (Eds.), Polymer Handbook, Interscience, New York, 1967.
112 D. N. Bort, I. N. Vishnevskaya and V. A. Kargin, Vysokomol. Soedin., Ser. B, 12 (1970) 176.
113 D. N. Bort and I. N. Vishnevskaya, Vysokomol. Soedin., Ser. A, 13 (1971) 1950.
114 R. A. Wessling and I. R. Harrison, J. Polym. Sci., Part A-1, 9 (1971) 3471.
115 G. Blauer, M. Shenblat and A. Katchalsky, J. Polym. Sci., 38 (1959) 189.
116 K. S. Minsker, A. G. Kronman, Yu A. Sangalev, D. N. Bort and G. A. Razuvaev, Vysokomol. Soedin. Sb. Statei, (1963) 45.
117 W. M. Thomas and M. T. O'Shaughnessy, J. Polym. Sci., 11 (1953) 455.
118 E. V. Volkova, P. V. Zimakov, A. V. Fokin, A. D. Sorokin, V. M. Belikov, L. A. Bulygina, A. I. Skobina and L. A. Krasnousov, Radiats. Khim. Polim. Mater. Simp. Moscow (1964) 109; Chem. Abs., 67 (1967) 22304n.
119 M. Lazar, J. Polym. Sci., 29 (1958) 753.
120 L. A. Wall (Ed.), Fluoropolymers, Wiley, New York, 1972.
121 M. Hirooka, Kobunshi, 16 (1967) 1172.
122 E. Ehrenfeld and W. Sporysz, Chemik, 24 (1971) 49.
123 P. Shantorovitch and L. N. Sosnovskaya, Dokl. Akad. Nauk S.S.S.R., 191 (1970) 100; Izv. Akad. Nauk S.S.R., 2 (1970) 358.
124 N. G. Gaylord and A. Takahashi, Adv. Chem. Ser., 91 (1969) 94.
125 C. H. Bamford, W. G. Barb and A. D. Jenkins, Nature (London), 169 (1952) 1044.
126 C. H. Bamford and W. G. Barb, Discuss. Faraday Soc., 14 (1953) 208.
127 R. G. W. Norrish and R. R. Smith, Nature (London), 150 (1942) 336.
128 T. White and R. N. Haward, J. Chem. Soc., (1953) 25.
129 R. N. Haward, J. Polym. Sci., 3 (1948) 10.
130 J. Abere, G. Goldfinger, H. Naidus and H. F. Mark, J. Phys. Chem., 49 (1945) 211.
131 A. Chapiro, J. Chim. Phys., 47 (1950) 747,764.

132 F. S. Dainton, P. H. Seaman, D. G. L. James and R. S. Eaton, J. Polym. Sci., 34 (1959) 209.
133 F. S. Dainton and P. H. Seaman, J. Polym. Sci., 39 (1959) 279.
134 F. S. Dainton and D. G. L. James, J. Polym. Sci., 39 (1959) 299.
135 F. S. Dainton and R. S. Eaton, J. Polym. Sci., 39 (1959) 313.
136 W. M. Thomas, E. H. Gleason and G. Mino, J. Polym. Sci., 24 (1957) 43.
137 D. H. Napper and A. G. Parts, J. Polym. Sci., 61 (1962) 113.
138 F. S. Dainton and M. Tordoff, Trans. Faraday Soc., 53 (1957) 499, 666.

Chapter 7

The Calculation of Mole-Weight Distributions from Kinetic Schemes

H. TOMPA

1. Introduction

Only in a few very simple cases can the mole-weight (or chain-length) distribution of a polymer resulting from a polymerization reaction be obtained by intuitive reasoning; in all other cases it must be obtained by solving the set of simultaneous differential equations by which the rates of the different reactions are described. These differential equations can be solved as such in some cases to give analytical results; in some cases the solutions can be obtained by the appropriate classical methods, in others they can be solved successively, for increasing values of the chain length, and occasionally the solution can be guessed and then verified. These methods have been described for a number of kinetic schemes in several text books [1—3] and we shall use them only by way of illustration; it is the purpose of this chapter to describe a general method, employing moments of distributions and generating functions or equivalent concepts, and to show how to use it; the fundamentals of the method, with some applications, have been given previously [4—8]. A large number of applications have been described in a recent review article [9] and further applications have also been reported [10—14].

Another method, consisting in transforming the kinetic equations into a partial differential equation for the polymer concentration as a function of chain length and time, or chain length and conversion, has been proposed by Bamford and Tompa [5] and used with great success by Saito et al. [14], but in view of the mathematical difficulties likely to be encountered precisely in those cases which are not amenable to treatment by the methods described in this chapter, we shall not discuss the method in detail and refer the interested reader to the two papers quoted. We feel all the more justified in this attitude as the mathematical problems to be solved will vary from case to case since they depend on the type of differential equations obtained and this again depends on the kinetic scheme. No general treatment is therefore possible.

We shall not discuss here the question of the validity of the assumptions underlying the kinetic equations, e.g. the independence of rate coefficients of chain length and viscosity; this has been done elsewhere in this work.

2. Notation and definitions

2.1 THE MOMENTS OF THE DISTRIBUTION

We shall denote a polymer of chain length r by P_r and shall, for the sake of simplicity, always use the same symbol to denote the concentration of the polymer species or other substance in question. We shall usually use the chain length, i.e. the number of monomer units, to express the size of the polymer molecule rather than the mole weight; this cannot lead to any difficulty since one is a multiple of the other with the mole-weight of the repeating unit as a factor of proportionality. We shall denote quantities referring to time $t = 0$ by a superscript ^{0}.

The moment of the distribution of order n, or the n-th moment, is defined by

$$Q_n = \sum_{r=1}^{\infty} r^n P_r \tag{1}$$

It is sometimes convenient to regard P_r as a continuous function of the chain length; we shall define the independent variable ρ as an appropriate fraction of r

$$\rho = r/\bar{r} \tag{2}$$

where $\bar{r}$ is arbitrary and is to be chosen conveniently. The discrete set of numbers P_r is then replaced by a continuous function $P(\rho)$ or simply P. In terms of $P(\rho)$ the moments are

$$Q_n = \bar{r}^{n+1} \int_0^{\infty} \rho^n P(\rho)\, \mathrm{d}\rho \tag{3}$$

since not only must r be expressed in terms of ρ from eqn. (2) but on passing from summation to integration the implicitly understood Δr (which is of course equal to unity) must be transformed into $\bar{r}\,\mathrm{d}\rho$.

Clearly the moment of order zero, Q_0, equals the total number of polymer molecules and the first moment Q_1 the total weight of polymer present, relative to the weight of the monomer unit.

It follows from the well-known definitions of the number average and the weight average chain lengths r_n and r_w that

$$r_\mathrm{n} = Q_1/Q_0; \quad r_\mathrm{w} = Q_2/Q_1 \tag{4}$$

It is often convenient to take $\bar{r} = r_\mathrm{n}$ though this is by no means a necessity.

2.2 THE GENERATING FUNCTION

The generating function $R(z)$ is defined by

$$R(z) = \sum_{r=1}^{\infty} z^r P_r \tag{5}$$

and is a function of the auxiliary variable z; P_r is therefore the coefficient of z^r in a power-series expansion of R and it is preferable to regard z as being less than unity to avoid all problems of convergence.

To familiarize the reader with the notion of the generating function, we give the following example: since

$$\exp(az) - 1 = \sum_{r=1}^{\infty} \frac{(az)^r}{r!} \tag{6}$$

it is clear that $R(z) = \exp(az) - 1$ is the generating function of the distribution

$$P_r = \frac{a^r}{r!} \tag{7}$$

which is known as the Poisson distribution.

A mole-weight distribution of polymers which is fairly common is the so-called exponential distribution, which we write in the form

$$P_r = Aa^r = A \exp(r \ln a) \tag{8}$$

where the value of a has to be below unity if the series of P_r is to converge and just below unity if the values of P_r are not to decrease to extremely small values after the first few terms; the values of P_r form a geometric series. The first two moments are

$$Q_0 = \frac{Aa}{1-a}; \quad Q_1 = \frac{Aa}{(1-a)^2} \tag{9}$$

so that it appears to be more reasonable to write

$$P_r = Q_0(1-a)a^{r-1} = Q_1(1-a)^2 a^{r-1} \tag{10}$$

The generating function R is then

$$R(z) = Q_0 \frac{(1-a)z}{1-az} = Q_1 \frac{(1-a)^2 z}{1-az} \tag{11}$$

The moments Q_n can be obtained from the generating function directly, without first calculating P_r; clearly $Q_0 = R(1)$ and it is easy to show that in general

$$Q_n = \left(z \frac{\mathrm{d}}{\mathrm{d}z} \right)^n R(z)\ (z = 1) \tag{12}$$

where z is to be given the value unity after n differentiations and multiplications by z.

2.3 THE LAPLACE TRANSFORM

If the polymer distribution is regarded as a continuous function of the chain length, the equivalent of the generating function is the Laplace transform

$$\mathscr{L}(P) = L(p) = \int_0^\infty \exp(-p\rho)P(\rho)\,\mathrm{d}\rho \tag{13}$$

The generating function goes over into the Laplace transform, except for a constant factor, if z is replaced by $\exp(-p/\bar{r})$ and summation by integration. Even though there is correspondence between the two, it is preferable to distinguish them by different notations.

To obtain P when L has been obtained one must look up Tables of Laplace transforms or inverse Laplace transforms; short Tables are given in the Handbook of Chemistry and Physics [15], more complete ones in two other publications [16, 17]. The following simple rules of Laplace transforms help considerably in manipulating them.

If $L(p)$ is the transform of $P(\rho)$, $L(p + a)$ is the transform of $\exp(-a\rho)$. $P(\rho)$ and $pL(p) - P(0)$ that of $\mathrm{d}P/\mathrm{d}\rho$. Further, the Laplace transform of rP_r is $-\bar{r}\ \partial L/\partial p$.

P_{r-1} can be expressed in terms of P_r as

$$P_{r-1} = P_r - \frac{1}{\bar{r}} \frac{\partial P_r}{\partial \rho} \tag{14}$$

so that on forming the Laplace transform P_{r-1} goes over into $L + [P(0) - pL]/\bar{r}$. The Laplace transform of the convolution of two functions is the product of their transforms.

The moments of the distribution can be obtained from the Laplace transform by

$$Q_n = (-1)^n \bar{r}^{n+1} \frac{\mathrm{d}^n L}{\mathrm{d}p^n}\ (p = 0) \tag{15}$$

2.4 RADICALS

In radical polymerizations we have not only polymer molecules but also radicals present; we denote a radical of chain length r and its concentration by X_r, the n-th moment by Y_n, the generating function by $Z(z)$ and the Laplace transform by $S(p)$. Diradicals will be denoted by D_r and the corresponding functions by E_n, $F(z)$ and $T(p)$, respectively.

2.5 DISCRETE VERSUS CONTINUOUS DISTRIBUTIONS

It is impossible to say a priori whether in a given case it is preferable to regard the chain length as a discrete or a continuous variable. Very often one way will lead to a solution of the problem whereas the other yields equations which have no closed mathematical solutions. When the problem can be solved by either way, the two solutions obtained may differ by quantities of the order of $(\bar{r})^{-1}$ and this is due to the difference between a discrete and a continuous variable; if $\bar{r}$ is taken to be r_n, the number average chain length, with values of several hundreds or even several thousands, the difference is physically insignificant. However, these small differences can explain the fact that sometimes one way leads to a closed solution while the other does not.

2.6 THE USE OF MOMENTS TO OBTAIN THE DISTRIBUTION

There are a number of kinetic schemes for which it is not possible to solve the differential equations for R or for L analytically, though of course it is always possible to obtain solutions by numerical methods, but there does not seem to be any practicable way to obtain P from L or R in such cases. However, it is possible to obtain the distribution from the moments by a method which has been described for the case of an independent variable going from $-\infty$ to $+\infty$ by Margenau and Murphy [18] and for the case of a variable going from 0 to ∞ by Bamford and Tompa [4, 5].

We write the polymer distribution as an expansion into Laguerre polynomials, which are the orthogonal polynomials appropriate to the interval 0 to ∞, in the form

$$P(\rho) = \exp(-\rho) \sum_{j=0}^{\infty} c_j L_j(\rho) \tag{16}$$

where the Laguerre polynomial $L_j(\rho)$ is defined from

$$\frac{L_j(\rho)}{j!} = \sum_{i=0}^{j} \binom{j}{i} \frac{(-\rho)^i}{i!} \tag{17}$$

It is shown in text-books of analysis that due to the orthogonality of the Laguerre polynomials the coefficients c_j are given by

$$(j!)^2 c_j = \int_0^{\infty} P(\rho) L_j(\rho)\, \mathrm{d}\rho \tag{18}$$

If one substitutes for L_j in this equation its value from eqn. (17) and reverses the order of integration and summation one obtains

$$j!\, c_j = \sum_{i=0}^{j} \binom{j}{i} (-1)^i \frac{Q_i}{i!\bar{r}^{i+1}} \tag{19}$$

so that c_j can be calculated from the values of the moments of orders 0 to j. If the expansion (16) converges fairly rapidly a few values of Q_j will be sufficient to give P with an acceptable accuracy; this will be the more so the nearer P is to an exponential distribution and the better one chooses $\bar{r}$.

2.7 OTHER MATHEMATICAL TECHNIQUES

In statistics and probability theory (see e.g. Aitken [19], p. 21 ff.) the moment generating function G is used extensively. G is related to Q_n in the same way as R is to P_r, i.e. by

$$G(t) = \sum_{n=0}^{\infty} t^n Q_n = \sum_{r=1}^{\infty} \frac{P_r}{1 - tr} \tag{20}$$

where the latter equation follows from the definition of Q_n; as a matter of fact, the definition used in statistics contains a factor $n!$ but this does not affect the argument. No case is known to the author in which the use of the moment generating function was more advantageous than one of the techniques described previously, but it is conceivable that such a case will arise one day. It might also be advisable to try for a solution via the so-called cumulant generating function [19] when other techniques fail.

3. The use of the generating function

3.1 POLYADDITION

3.1.1 Monomer concentration constant

If a polymer molecule grows simply by addition of monomer, which we call M,

$$P_r + M \rightarrow P_{r+1}\,;\ k \tag{21}$$

where k is the rate coefficient, and the kinetic equation is

$$\frac{\mathrm{d}P_r}{\mathrm{d}t} = kMP_{r-1} - kMP_r \tag{22}$$

This equation can be taken to be formally valid for $r = 1$ as well if we put $P_0 = 0$. Assuming first that the monomer concentration can be regarded as constant, we can write

$$\frac{dP_r}{d\tau} = P_{r-1} - P_r \tag{23}$$

with $\tau = kMt$; for $r = 1$ we have

$$\frac{dP_1}{d\tau} = -P_1 \tag{24}$$

the solution of which is

$$P_1 = P_1^0 \, e^{-\tau} \tag{25}$$

where we recall that a superscript ^{0} denotes the value of a quantity at time $t = 0$, in other words its initial value. Substituting into eqn. (23) for $r = 2$ and solving for P_2 we obtain

$$P_2 = (P_1^0 \tau + P_2^0) \, e^{-\tau} \tag{26}$$

Proceeding we obtain

$$P_3 = \left(P_1^0 \frac{\tau^2}{2} + P_2^0 \tau + P_3^0 \right) e^{-\tau} \tag{27}$$

and "one can see" what the general solution will look like. If all $P_r^0 = 0$ except P_1^0, the general solution is

$$P_r = P_1^0 \frac{\tau^{r-1}}{(r-1)!} \, e^{-\tau} \tag{28}$$

which is easily confirmed by substitution into eqn. (23).

The general solution is easily obtained by the use of the generating function: multiplying eqn. (23) by z^r and summing over r we have

$$\frac{\partial R}{\partial \tau} = zR - R \tag{29}$$

(we write a partial differential coefficient since R is a function of z and of τ) since

$$\sum_{r=2}^{\infty} z^r P_{r-1} = z \sum_{r=2}^{\infty} z^{r-1} P_{r-1} = z \sum_{r=1}^{\infty} z^r P_r = zR \tag{30}$$

so that

$$R = R^0 \, e^{(z-1)\tau} = R^0 \, e^{-\tau} \, e^{z\tau} \tag{31}$$

References p. 576

If all $P_r^0 = 0$ except P_1^0, $R^0 = zP_1^0$ and

$$R = P_1^0 \, \mathrm{e}^{-\tau} \sum_{r=1}^{\infty} z \frac{z^{r-1} \tau^{r-1}}{(r-1)!} \tag{32}$$

which gives for P_r eqn. (28). The distribution of chain lengths is thus a Poisson distribution.

In the general case the distribution has to be obtained from eqn. (31). If the analytical form of R^0 is known, P_r can in principle be obtained by development into a series of powers of z. If a direct development is not possible, an answer can sometimes be obtained by using Cauchy's theorem, according to which P_r, the coefficient of z^r, is obtained by integration around the origin in the complex plane, viz.

$$P_r = \frac{1}{2\pi i} \oint \frac{R(z)}{z^{r+1}} \, \mathrm{d}z \tag{33}$$

To calculate the number average and weight average chain lengths we put $z = 1$ into eqn. (31) and obtain

$$Q_0 = Q_0^0 \tag{34}$$

which is physically obvious and can be obtained for the special case by summing eqn. (28) over r.

Differentiating eqn. (31) with respect to z we have

$$\frac{\partial R}{\partial z} = \frac{\partial R^0}{\partial z} \, \mathrm{e}^{(z-1)\tau} + R^0 \, \mathrm{e}^{(z-1)\tau} \tau \tag{35}$$

and putting $z = 1$ now gives

$$Q_1 = Q_1^0 + Q_0^0 \tau \tag{36}$$

Since $r_\mathrm{n} = Q_1/Q_0$, and obviously $r_\mathrm{n}^0 = Q_1^0/Q_0^0$,

$$r_\mathrm{n} = r_\mathrm{n}^0 + \tau = r_\mathrm{n}^0 + kMt \tag{37}$$

Multiplying eqn. (35) by z, differentiating, and then putting $z = 1$

$$Q_2 = Q_2^0 + 2Q_1^0 \tau + Q_0^0 \tau(\tau + 1) \tag{38}$$

so that

$$\frac{Q_2}{Q_0} = r_\mathrm{n} r_\mathrm{w} = r_\mathrm{n}^0 r_\mathrm{w}^0 + 2\tau r_\mathrm{n}^0 + \tau(\tau + 1) \tag{39}$$

For the special case of all $P_r^0 = 0$ except P_1^0, where $Q_1^0 = Q_0^0 = P_1^0$, $r_\mathrm{n}^0 = r_\mathrm{w}^0 = 1$, and this gives

$$r_\mathrm{n} = \tau + 1; \quad r_\mathrm{w} = \frac{\tau^2 + 3\tau + 1}{\tau + 1} = r_\mathrm{n} + \frac{\tau}{\tau + 1} \tag{40}$$

which are easily obtained directly from eqn. (28). With increasing τ r_w approaches $r_n + 1$ as limit.

3.1.2 *Monomer concentration not constant*

If the monomer concentration can no longer be assumed constant, eqn. (22) is still valid but we must express M by solving the equation giving its rate of change with time. Clearly

$$\frac{dM}{dt} = -kM \sum_{r=1}^{\infty} P_r = -kMQ_0 \tag{41}$$

and since $Q_0 = Q_0^0$ still holds,

$$M = M^0 \exp(-kQ_0^0 t) \tag{42}$$

We introduce

$$d\tau = kM\, dt = kM^0 \exp(-kQ_0^0 t)\, dt \tag{43}$$

so that

$$\tau = \frac{M^0}{Q_0^0}[1 - \exp(-kQ_0^0 t)] \tag{44}$$

The amount of monomer converted in time t is

$$\Delta M = M^0 - M = Q_0^0 \tau \tag{45}$$

which gives a physically significant meaning to τ.

The constant of integration has been chosen so that $\tau = 0$ for $t = 0$; for small values of t this definition of τ agrees with that used in Sect. 3.1.1. Equation (22) can now be written again in the form of eqn. (23) so that all the results of the preceding section remain valid, with the new definition of τ.

After an infinite lapse of time M has decreased to zero and $\tau^\infty = M^0/Q_0^0$ so that

$$Q_1^\infty = Q_1^0 + M^0 \tag{46}$$

which is physically obvious. The final chain-length distribution and the average chain lengths can be obtained with this limiting value of τ.

3.2 POLYCONDENSATION

In polycondensation polymer molecules of the type A—B having two different reacting groups at the two ends react with similar molecules to form longer molecules of the same type. If the two reaction partners consist of r and s monomer units, respectively, the resulting molecule will have $r + s$ monomer units (we neglect the incidence of a possible cyclization reaction, in which the two ends of the same molecule react with each other), viz.

$$P_r + P_s \rightarrow P_{r+s};\ k \tag{47}$$

The kinetic equations are

$$\frac{dP_r}{dt} = \frac{k}{2}\sum_{s=1}^{r-1} P_s P_{r-s} - kP_r \sum_{s=1}^{\infty} P_s = \frac{k}{2}\sum_{s=1}^{r-1} P_s P_{r-s} - kQ_0 P_r \tag{48}$$

where the factor 1/2 in the first term on the right-hand side allows for the fact that every product $P_s P_{r-s}$ has been included twice. For the simplest case where the reaction mixture contained originally only monomer the solution can be obtained by intuitive reasoning using statistical arguments [20]: if p is the probability that a unit A or B has reacted in time t, the concentration of molecules of chain length r will be proportional to $(1-p)p^{r-1}$ since $(r-1)$ units must have reacted and one must not have reacted. P_r is therefore given by an expression of the type $A(1-p)p^{r-1}$; the total weight of polymer Q_1 is then found to be

$$Q_1 = \frac{A}{1-p} \tag{49}$$

and since this quantity is constant (independent of time) we have

$$A = Q_1^0(1-p)$$

and

$$P_r = Q_1^0(1-p)^2 p^{r-1} \tag{50}$$

The mole-weight distribution is therefore exponential.

To obtain the solution mathematically by a direct method we need first Q_0. We sum all equations (48) and note that

$$\sum_{r=1}^{\infty}\sum_{s=1}^{r-1} P_s P_{r-s} = \sum_{s=1}^{\infty} P_s \sum_{r=s+1}^{\infty} P_{r-s} = \sum_{s=1}^{\infty} P_s \sum_{q=1}^{\infty} P_q = Q_0^2 \tag{51}$$

Therefore

$$\frac{dQ_0}{dt} = \frac{k}{2} Q_0^2 - kQ_0^2 = -\frac{k}{2} Q_0^2 \tag{52}$$

the solution of which is

$$\frac{1}{Q_0} - \frac{1}{Q_0^0} = \frac{k}{2} t \tag{53}$$

or

$$Q_0 = \frac{Q_0^0}{(k/2)Q_0^0 t + 1} = \frac{Q_0^0}{\tau + 1} \tag{54}$$

with an obvious definition of τ. We can then write the differential equations in the form

$$\frac{dP_r}{d\tau} = \frac{1}{Q_0^0} \sum_{s=1}^{r-1} P_s P_{r-s} - \frac{2}{\tau+1} P_r \tag{55}$$

These equations have to be solved for successive values of r. For $r = 1$

$$\frac{dP_1}{d\tau} = -\frac{2}{\tau+1} P_1 \tag{56}$$

the solution of which is

$$P_1 = \frac{P_1^0}{(\tau+1)^2} \tag{57}$$

where P_1^0 is as before the value of P_1 for $t = 0$. For $r = 2$ we have now

$$\frac{dP_2}{dt} = \frac{1}{Q_0^0} P_1^2 - \frac{2}{\tau+1} P_2 = \frac{P_1^{0\,2}}{Q_0^0(\tau+1)^2} - \frac{2}{\tau+1} P_2 \tag{58}$$

the solution of which is

$$P_2 = \frac{P_1^{0\,2}}{Q_0^0} \frac{\tau}{(\tau+1)^3} + \frac{P_2^0}{(\tau+1)^2} \tag{59}$$

Similarly, the solution for $r = 3$ is

$$P_3 = \frac{P_1^{0\,3}}{Q_0^{0\,2}} \frac{\tau^2}{(\tau+1)^4} + \frac{2P_1^0 P_2^0}{Q_0^0} \frac{\tau}{(\tau+1)^3} + \frac{P_3^0}{(\tau+1)^2} \tag{60}$$

Clearly the solution becomes unmanageable except for the simple and physically most interesting case when all initial polymer is monomer, i.e. when $P_r^0 = 0$ for $r \neq 1$ and $P_1^0 = Q_0^0 = Q_1^0$; one guesses that the solution will be

$$P_r = \frac{Q_0^0}{(\tau+1)^2} \left(\frac{\tau}{\tau+1}\right)^{r-1} \tag{61}$$

and confirms that this is correct by substitution. Comparison with eqn. (50) shows also that p, the probability of a unit having reacted, is

$$p = \frac{\tau}{\tau+1} = \frac{kQ_0^0 t}{kQ_0^0 t + 2} \tag{61a}$$

The solution for the general case, where all $P_r^0 \neq 0$, can be effected by the use of the generating function. Q_0 is still given by eqn. (53) so that we have to solve the set of equations (48) or (55). We multiply each by z^r and sum over r; we note that, similarly to eqn. (51),

$$\sum_{r=1}^{\infty} z^r \sum_{s=1}^{r-1} P_s P_{r-s} = \sum_{s=1}^{\infty} z^s P_s \sum_{r=s+1}^{\infty} z^{r-s} P_{r-s}$$

$$= \sum_{s=1}^{\infty} z^s P_s \sum_{q=1}^{\infty} z^q P_q = R^2 \qquad (62)$$

so that we obtain

$$\frac{\partial R}{\partial t} = \frac{k}{2} R^2 - kQ_0 R \qquad (63)$$

or

$$\frac{\partial R}{\partial \tau} = \frac{1}{Q_0^0} R^2 - \frac{2}{\tau + 1} R \qquad (64)$$

Alternatively, if we did not want to use eqn. (53), we could put $z = 1$ in eqn. (63) and obtain again eqn. (54) for Q_0.

Equation (64) is of the type known as Riccati's equation; its solution is effected [21] by the substitution

$$R = -\frac{Q_0^0}{u} \frac{\partial u}{\partial \tau} \qquad (65)$$

which gives

$$\frac{\partial^2 u}{\partial \tau^2} + \frac{2}{\tau + 1} \frac{\partial u}{\partial \tau} = 0 \qquad (66)$$

We obtain successively

$$\frac{\partial u}{\partial \tau} = -\frac{c_1}{(\tau + 1)^2}; \quad u = \frac{c_1}{\tau + 1} + c_2; \quad R = \frac{Q_0^0}{(\tau + 1)[1 + c(\tau + 1)]} \qquad (67)$$

where c_1 and c_2 are constants of integration, of which only the ratio c appears in the final answer. If for $t = 0$, i.e. $\tau = 0$, only "polymer" of chain length unity is present, $Q_0^0 = P_1^0$ and $R^0 = zQ_0^0$ so that

$$R = \frac{Q_0^0 z}{(\tau + 1)(\tau + 1 - z\tau)} = \sum_{r=1}^{\infty} \frac{Q_0^0}{(\tau + 1)^2} \left(\frac{\tau}{\tau + 1}\right)^{r-1} z^r \qquad (68)$$

which gives

$$P_r = \frac{Q_0^0}{(\tau + 1)^2} \left(\frac{\tau}{\tau + 1}\right)^{r-1} \qquad (69)$$

This confirms eqn. (61) and identifies p as $\tau/(\tau + 1)$.

If the initial polymer has a general composition, $R^0 = \Sigma z^r P_r^0$ and

$$R = \frac{R^0}{(\tau + 1)[\tau + 1 - (R^0/Q_0^0)\tau]} \qquad (70)$$

To obtain the average chain lengths directly we differentiate eqn. (70) with respect to z

$$\frac{\partial R}{\partial z} = \frac{\partial R^0/\partial z}{(\tau+1)[\tau+1-(R^0/Q_0^0)\tau]} + \frac{R^0}{(\tau+1)[\tau+1-(R^0/Q_0^0)\tau]^2} \cdot \frac{\tau}{Q_0^0} \cdot \frac{\partial R^0}{\partial z} \tag{71}$$

and $z = 1$ gives

$$Q_1 = \frac{Q_1^0}{\tau+1} + \frac{Q_0^0}{\tau+1} \cdot \frac{\tau}{Q_0^0} \cdot Q_1^0 = Q_1^0 \tag{72}$$

Multiplication of eqn. (71) by z and renewed differentiation gives an expression which on putting $z = 1$ goes over into

$$Q_2 = Q_2^0 + 2\frac{Q_1^{0\,2}}{Q_0^0}\tau \tag{73}$$

so that

$$r_\mathrm{n} = r_\mathrm{n}^0(\tau+1); \quad r_\mathrm{w} = r_\mathrm{w}^0 + 2r_\mathrm{n}^0\tau \tag{74}$$

If all the initial polymer is monomer, $r_\mathrm{n}^0 = r_\mathrm{w}^0 = 1$, and

$$r_\mathrm{n} = \tau + 1; \quad r_\mathrm{w} = 2\tau + 1 \tag{75}$$

so that $r_\mathrm{w}/r_\mathrm{n}$ goes to two for large τ.

The moments of the distribution could be obtained from the differential equation for R without solving it explicitly. We write eqn. (63) in the form

$$\frac{\partial R}{\partial \tau} = \frac{1}{Q_0^0}R^2 - 2\frac{Q_0}{Q_0^0}R \tag{76}$$

and $z = 1$ gives a differential equation the solution of which is evidently eqn. (54). Differentiation of eqn. (76) with respect to z gives

$$\frac{\partial^2 R}{\partial \tau\, \partial z} = \frac{1}{Q_0^0}2R\frac{\partial R}{\partial z} - 2\frac{Q_0}{Q_0^0}\frac{\partial R}{\partial z} \tag{77}$$

so that with $z = 1$, $\partial Q_1/\partial \tau = 0$ and $Q_1 = Q_1^0$. Multiplication by z and differentiation gives an expression which on putting $z = 1$ reduces to

$$\frac{\partial Q_2}{\partial \tau} = 2\frac{Q_1^2}{Q_0^0} \tag{78}$$

the solution of which is clearly eqn. (73).

3.3 RADICAL POLYMERIZATION, MONOMER CONCENTRATION CONSTANT

3.3.1 *Stationary state*

a. Termination by disproportionation

Vinyl compounds usually polymerize by a radical mechanism; a simple scheme consisting of thermal initiation, propagation, chain transfer to monomer, and termination by disproportionation is

$$\begin{aligned} &2M \rightarrow 2X_1 ; k_1 \\ &X_r + M \rightarrow X_{r+1} \quad ; k_2 \\ &X_r + M \rightarrow P_r + X_1 ; k_3 \\ &X_r + X_s \rightarrow P_r + P_s ; k_4 \end{aligned} \tag{79}$$

The kinetic equations are

$$\frac{dX_1}{dt} = k_1 M^2 + k_3 M \sum_s X_s - k_2 M X_1 - k_3 M X_1 - k_4 \sum_s X_s \cdot X_1 \tag{80a}$$

$$\frac{dX_r}{dt} = k_2 M X_{r-1} - k_2 M X_r - k_3 M X_r - k_4 \sum_s X_s \cdot X_r \tag{80b}$$

$$\frac{dP_r}{dt} = k_3 M X_r + k_4 \sum_s X_s \cdot X_r \tag{80c}$$

If we add the first of these equations to the sum of the second over all r and introduce Y_0, the moment of order zero of the radical distribution, we obtain

$$\frac{dY_0}{dt} = k_1 M^2 - k_4 Y_0^2 \tag{81}$$

In the stationary-state treatment it is assumed that the rate of change of the radical concentrations can be neglected against the rates of formation and disappearance of the radicals; this is equivalent to putting all dX_r/dt equal to zero, and therefore of course also all dY_n/dt. This leads to

$$Y_0 = \left(\frac{k_1}{k_4}\right)^{1/2} M \tag{82}$$

We assume here at first that the monomer concentration is constant, e.g. because we are working at low conversions, so that Y_0 is also constant.

We substitute for Y_0 into eqn. (80b) and introduce the abbreviations w, γ and α by

$$w = (k_1 k_4)^{1/2}; \quad \gamma = \left(\frac{k_1}{k_4}\right)^{1/2}; \quad \alpha = \frac{k_2}{k_2 + k_3 + w} \tag{83}$$

If we put dX_1/dt equal to zero we have

$$X_1 = \gamma M(1-\alpha) \tag{84}$$

Putting dX_r/dt equal to zero gives similarly

$$X_r = \alpha X_{r-1} = \alpha^2 X_{r-2} = \cdots = \alpha^{r-1} X_1 = \gamma M(1-\alpha)\alpha^{r-1} \tag{85}$$

so that the radical distribution is exponential.

Substitution into eqn. (80c) gives

$$\frac{dP_r}{dt} = (k_3 + w)MX_r = \gamma(k_3 + w)M^2(1-\alpha)\alpha^{r-1} \tag{86}$$

so that

$$P_r = \gamma(k_3 + w)M^2(1-\alpha)\alpha^{r-1}t \tag{87}$$

and the polymer distribution is also exponential, in other words, the concentrations P_r form a geometric series. We have assumed that the initial polymer concentration is zero.

The first moments of the polymer distribution and the average chain lengths are easy to calculate. They are

$$Q_0 = \gamma(k_3 + w)M^2 t; \quad Q_1 = \gamma(k_3 + w)M^2 t\frac{1}{1-\alpha};$$

$$Q_2 = \gamma(k_3 + w)M^2 t\frac{1+\alpha}{(1-\alpha)^2} \tag{88}$$

and

$$r_n = \frac{1}{1-\alpha}; \quad r_w = \frac{1+\alpha}{1-\alpha}; \quad \frac{r_w}{r_n} = 1+\alpha \tag{89}$$

Unless propagation is much faster than termination by transfer or disproportionation one does not obtain a high polymer; in other words k_2 in such cases is much larger than $k_3 + w$, so that α is close to unity and the ratio r_w/r_n is nearly equal to 2.

It is usually convenient to write P_r in the form

$$P_r = Q_1(1-\alpha)^2\alpha^{r-1} \tag{90}$$

where Q_1 could also be replaced by ΔM, the monomer consumed.

It is fairly easy and instructive to solve this kinetic scheme by the use of generating functions. We multiply eqn. (80a) by z and (80b) by z^r and add the first to the sum over r of the second; this gives

$$\frac{\partial Z}{\partial t} = k_1 M^2 z + k_3 MzY_0 + k_2 MzZ - k_2 MZ - k_3 MZ - k_4 Y_0 Z \tag{91}$$

where the third term arises from

$$\sum_{r=2}^{\infty} z^r X_{r-1} = z \sum_{r=2}^{\infty} z^{r-1} X_{r-1} = zZ \tag{92}$$

Putting $z = 1$ transforms Z into Y_0 and the stationary-state assumption gives again eqn. (82). Substituting for Y_0 and using the stationary-state assumption gives

$$Z = \gamma M(1-\alpha)\frac{z}{1-\alpha z} \tag{93}$$

We multiply eqn. (80c) by z^r and sum over r and obtain

$$\frac{\partial R}{\partial t} = (k_3 + w)MZ \tag{94}$$

so that

$$R = \gamma M^2 (1-\alpha)\frac{z}{1-\alpha z}t \tag{95}$$

From this equation we easily obtain the polymer distribution and all related quantities and since

$$\frac{z}{1-\alpha z} = \sum_{r=1}^{\infty} \alpha^{r-1} z^r \tag{96}$$

the results agree with those given above.

b. Termination by combination

If in a scheme of radical polymerization termination is by combination but the other conditions are all as in the preceding section, the last of eqns. (79) has to be replaced by

$$X_r + X_s \rightarrow P_{r+s};\ k_4 \tag{97}$$

The kinetic equations for the radicals are the same as before; the last equation is

$$\frac{\mathrm{d}P_r}{\mathrm{d}t} = k_3 M X_r + \tfrac{1}{2}k_4 \sum_{s=1}^{r-1} X_s X_{r-s} \tag{98}$$

While the generating function Z of the radical distribution remains unchanged, eqn. (93), the equation for R becomes

$$\frac{\partial R}{\partial t} = k_3 MZ + \tfrac{1}{2}k_4 Z^2 \tag{99}$$

or

$$R = (k_3 MZ + \tfrac{1}{2}k_4 Z^2)t \tag{100}$$

To obtain P_r we need the development of Z^2 into powers of z: by the binomial theorem

$$(1-\alpha z)^{-2} = \sum_{r=0}^{\infty} (r+1)\alpha^r z^r \tag{101}$$

so that

$$P_r = \gamma[k_3 + \tfrac{1}{2}w(1-\alpha)(r-1)]M^2(1-\alpha)\alpha^{r-1}t \tag{102}$$

which is not a simple exponential distribution but the sum of an exponential distribution and a second distribution which has been called a generalized exponential distribution.

To obtain the number average chain length we put $z = 1$ into eqn. (100) giving

$$Q_0 = (k_3 MY_0 + \tfrac{1}{2}k_4 Y_0^2)t = \gamma(k_3 + \tfrac{1}{2}w)Mt \tag{103}$$

If we first differentiate R with respect to z and then put $z = 1$ we obtain

$$Q_1 = (k_3 MY_1 + k_4 Y_0 Y_1)t = \gamma(k_3 + w)MY_1 t \tag{104}$$

and we need Y_1, which we get by differentiating eqn. (93) with respect to z and then putting $z = 1$, viz.

$$Y_1 = \gamma M \frac{1}{1-\alpha} \tag{105}$$

so that

$$Q_1 = \gamma(k_3 + w)M^2 \frac{1}{1-\alpha} t \tag{106}$$

It follows that

$$r_n = \frac{k_3 + w}{k_3 + \tfrac{1}{2}w} \frac{1}{1-\alpha} = \frac{k_2}{k_3 + \tfrac{1}{2}w} \tag{107}$$

This is not the same expression as for termination by disproportionation; in fact, if the values of the rate coefficients are the same, the ratio of the number average chain lengths for the two cases is between one and two, depending on the relative values of the various constants.

3.3.2 *Non-stationary state*

It can be of interest on occasion to study the progress of a vinyl polymerization when the formation of radicals by an initiation reaction is increased or decreased suddenly, e.g. by changing abruptly the intensity of the light which initiates the reaction via a photochemical process. In fact additional information on the kinetic rate coefficients can be obtained by measuring the so-called pre-effect and after-effect; these are due to the fact that after any sudden change in conditions the stationary state is not established within an infinitely small time and the differences of the observed values against those calculated with the stationary-state assumption can be expressed in terms of the rate coefficients.

In a simple case of vinyl polymerization with termination by disproportionation eqns. (80) have to be evaluated without the assumption of the stationary state. The generating function Z has now to be obtained by solving the differential eqn. (91). Equation (81) for Y_0 is still valid; now it does not give, of course, (82), but has to be solved as a differential equation. By writing it in the form

$$\frac{\mathrm{d}Y_0}{k_1 M^2 - k_4 Y_0^2} = \mathrm{d}t \tag{108}$$

the solution is seen to be

$$Y_0 = \gamma M \,\mathrm{th}[wM(t+\tau)] = \gamma M \,\mathrm{cth}[wM(t+\tau')] \tag{109}$$

where the value of the integration constants τ and τ' are determined by the initial value of Y_0

$$Y_0^0 = \gamma M \,\mathrm{th}(wM\tau) = \gamma M \,\mathrm{cth}(wM\tau') \tag{110}$$

If $Y_0^0 = 0$, $\tau = 0$; in any case the value of Y_0 at time $t = \infty$ is that given by eqn. (82), since the hyperbolic tangent and cotangent both approach unity as the argument goes to infinity. The form with the hyperbolic tangent is the appropriate one to use if Y_0 is an increasing function of time, the hyperbolic cotangent if it is a decreasing one; both forms are formally correct in either case but of the two quantities τ and τ' only one is real.

Substituting the first form for Y_0 into eqn. (91), the differential equation can be solved, though the algebra is heavy, the solution being

$$Z = \mathrm{e}^{-qMt} \frac{\mathrm{ch}(wM\tau)}{\mathrm{ch}[wM(t+\tau)]} \times \left\{ Z^0 - z\gamma M \frac{k_2(1-z)w + (k_3 q - w^2)\,\mathrm{th}(wM\tau)}{q^2 - w^2} \right\} + 2\gamma M \frac{k_2(1-z)w + (k_3 q - w^2)\,\mathrm{th}[wM(t+\tau)]}{q^2 - w^2} \tag{111}$$

where we have used the abbreviation

$$q = k_2(1-z) + k_3 \tag{112}$$

This value of Z has to be substituted into eqn. (94) or (99), according to whether termination is by disproportionation or by combination, with the known value of Z^0, and the resulting equation has to be integrated with respect to t.

If the non-stationary state is due to a sudden change of the initiation rate coefficient from k_1^0 to k_1, Y_0^0 and Z^0 are given by the expression for the stationary state with k_1^0; the integration constants τ and τ' are related to k_1^0 by

$$k_1^0 = k_1 \ \mathrm{th}^2(wM\tau) = k_1 \ \mathrm{cth}^2(wM\tau') \tag{113}$$

The generating function Z becomes then, using the form with the hyperbolic tangent,

$$Z = z\gamma M \frac{1}{q^2 - w^2} \left\{ k_2(1-z)w + (k_3 q - w^2)\ \mathrm{th}[wM(t+\tau)] - \frac{\mathrm{e}^{-qMt}}{\mathrm{ch}[wM(t+\tau)]} \cdot \frac{wqk_2(1-z)}{q\ \mathrm{ch}(wM\tau) + w\ \mathrm{sh}(wM\tau)} \right\} \tag{114}$$

We shall not carry the calculation further since the rest is a question of algebra; this has been done by the author [7] for the pre-effect and the after-effect in terms of the intrinsic viscosity, and graphs are given for different relative values of the various rate coefficients.

4. The use of the Laplace transform

4.1 RADICAL POLYMERIZATION, INITIATION TO DIRADICALS

4.1.1 Termination by disproportionation

We shall illustrate the use of the Laplace transform to obtain the mole-weight distribution and the relevant quantities derived from it in the case when in a vinyl polymerization the primary initiation step produces diradicals; the cases treated in this section and in the next can be solved also by the use of generating functions.

The kinetic scheme we shall consider is

$$\begin{aligned}
&2M \to D_2;\ k_1\\
&D_r + M \to D_{r+1};\ 2k_2\\
&X_r + M \to X_{r+1};\ k_2\\
&D_r + M \to X_r + X_1;\ 2k_3\\
&X_r + M \to P_r + X_1;\ k_3\\
&D_r + D_s \to X_r + X_s;\ 4k_4\\
&D_r + X_s \to X_r + P_s;\ 2k_4\\
&X_r + X_s \to P_r + P_s;\ k_4
\end{aligned} \tag{115}$$

Writing the factors 2 and 4 in the rate coefficients is equivalent to assuming that the reactivity of a radical end is independent of the presence or absence of another radical end. The kinetic equations are

$$\frac{dD_2}{dt} = \tfrac{1}{2}k_1M^2 - 2k_2MD_2 - 2k_3MD_2 - 4k_4E_0D_2 - 2k_4Y_0D_2 \tag{116a}$$

$$\frac{dD_r}{dt} = 2k_2MD_{r-1} - 2k_2MD_r - 2k_3MD_r - 4k_4E_0D_r - 2k_4Y_0D_r \tag{116b}$$

$$\frac{dX_1}{dt} = 2k_3ME_0 + k_3MY_0 - k_2MX_1 - k_3MX_1 - 2k_4E_0X_1 - k_4Y_0X_1 \tag{116c}$$

$$\frac{dX_r}{dt} = k_2MX_{r-1} - k_2MX_r - k_3MX_r - 2k_4E_0X_r - k_4Y_0X_r + 2k_3MD_r + 4k_4E_0D_r + 2k_4Y_0D_r \tag{116d}$$

$$\frac{dP_r}{dt} = k_3MX_r + 2k_4E_0X_r + k_4Y_0X_r \tag{116e}$$

where we have already written Y_0 and E_0 for the moments of order zero of the single radicals X_r and the diradicals D_r, respectively.

We sum eqn. (116b) over all r and add eqn. (116a), we sum eqn. (116d) over r and add eqn. (116c) and introduce the assumption of the stationary state. This gives

$$\frac{dE_0}{dt} = 0 = \tfrac{1}{2}k_1M^2 - 2k_3ME_0 - 4k_4E_0^2 - 2k_4E_0Y_0 \tag{117a}$$

$$\frac{dY_0}{dt} = 0 = 4k_3ME_0 + 4k_4E_0^2 - k_4Y_0^2 \tag{117b}$$

If we add the second equation to twice the first we obtain

$$2E_0 + Y_0 = \gamma M \tag{118}$$

i.e. the same expression for the total radical end concentration as in the case of single radicals. We substitute this into the set of equations (116) and obtain in particular

$$\frac{dD_r}{dt} = 2k_2MD_{r-1} - 2(k_2 + k_3 + w)MD_r \tag{119a}$$

$$\frac{dX_r}{dt} = k_2MX_{r-1} - (k_2 + k_3 + w)MX_r + 2(k_3 + w)MD_r \tag{119b}$$

We now assume D_r and X_r to be continuous variables of $\rho = r/\bar{r}$, with $\bar{r}$ undefined at first, and denote them by $D(\rho)$ and $X(\rho)$, or simply D and X, respectively. We can assume formally that eqn. (119b) is valid down to $r = 0$, if we define X_0 in such a way that (116d) becomes identical with (116c) for $r = 1$, viz.

$$X_0 = \gamma \frac{k_3}{k_2} M \tag{120a}$$

Similarly (116b) remains formally correct to $r = 1$ if we put

$$D_1 = \frac{k_1}{4k_2} M \tag{120b}$$

We multiply eqn. (119) by $\exp(-p\rho)\,\mathrm{d}\rho$ and integrate over ρ, and reverse the order of integration over ρ and differentiation with respect to t; we write partial differential coefficients, since the Laplace transforms T and S are also functions of p, and obtain

$$\frac{\partial T}{\partial t} = 0 = 2k_2 M\left(T + \frac{D_0 - pT}{\bar{r}}\right) - 2(k_2 + k_3 + w)MT \tag{121a}$$

$$\frac{\partial S}{\partial t} = 0 = k_2 M\left(S + \frac{X_0 - pS}{\bar{r}}\right) - (k_2 + k_3 + w)MS + 2(k_3 + w)MT \tag{121b}$$

We identify D_0 with D_1, in fact we would have made an error only of the order of $(k_3 + w)/k_2$ if we had calculated D_2 and X_1 from eqn. (116a) and (116c) and identified them with D_0 and X_0, respectively. We obtain first for T and then for S

$$\left(k_3 + w + k_2 \frac{p}{\bar{r}}\right) T = \frac{k_1}{4\bar{r}} M \tag{122a}$$

$$\left(k_3 + w + k_2 \frac{p}{\bar{r}}\right) S = \frac{k_1}{2\bar{r}} \frac{k_3 + w}{k_3 + w + k_2 p/\bar{r}} M + \gamma \frac{k_3}{\bar{r}} M \tag{122b}$$

The Laplace transform of P is from (116e)

$$\frac{\partial L}{\partial t} = (k_3 + w)MS \tag{123}$$

and since S is independent of t,

$$L = \gamma \frac{k_3 + w}{k_3 + w + k_2 p/\bar{r}} \cdot \frac{M}{\bar{r}^2} \cdot \left[k_3 + \tfrac{1}{2} w \frac{k_3 + w}{k_3 + w + k_2 p/\bar{r}}\right] t \tag{124}$$

assuming that the initial polymer concentration is zero.

Before deriving the .polymer distribution we calculate the number average chain length of the polymer. From eqn. (15) we get

$$Q_0 = \gamma(k_3 + \tfrac{1}{2}w)M^2 t; \quad Q_1 = \gamma k_2 M^2 t \tag{125}$$

so that

$$r_n = \frac{k_2}{k_3 + \tfrac{1}{2}w} \tag{126}$$

Since L is the Laplace transform of P, $\mathscr{L}(P) = L$, we have, by a well-known theorem mentioned in Sect. 2.3,

$$\mathscr{L}\{P \exp[(k_3 + w)/k_2 \cdot \bar{r}\rho]\} = \gamma \frac{k_3 + w}{k_2 p} \left[k_3 + \tfrac{1}{2}w \frac{k_3 + w}{k_2 p} \bar{r}\right] M^2 t \tag{127}$$

Since the Laplace transform of ρ^n is $n!/p^{n+1}$, $1/p$ is the transform of unity and $1/p^2$ that of ρ, so that

$$P = \gamma \frac{k_3 + w}{k_2} \left[k_3 + \tfrac{1}{2}w \frac{k_3 + w}{k_2} r\right] M^2 \exp\left(-\frac{k_3 + w}{k_2} r\right) t \tag{128}$$

It is easy to show that this is practically the same distribution as that for initiation by single radicals and termination by combination, eqn. (102), since

$$\exp\left(\frac{k_3 + w}{k_2}\right) \doteq 1 + \frac{k_3 + w}{k_2} = 1 + \frac{1 - \alpha}{\alpha} = \frac{1}{\alpha} \tag{129}$$

so that the last term in P is α^r.

4.1.2 *Termination by combination*

If termination is by combination, other conditions being as in the preceding section, the last three equations of the set (115) have to be replaced by

$$\begin{aligned} &D_r + D_s \rightarrow D_{r+s};\ 4k_4 \\ &D_r + X_s \rightarrow X_{r+s};\ 2k_4 \\ &X_r + X_s \rightarrow P_{r+s};\ k_4 \end{aligned} \tag{130}$$

The kinetic equations are now

$$\frac{dD_2}{dt} = \tfrac{1}{2}k_1 M^2 - 2k_2 M D_2 - 2k_3 M D_2 - 4k_4 E_0 D_2 - 2k_4 Y_0 D_2 \tag{131a}$$

$$\frac{dD_r}{dt} = 2k_2 M D_{r-1} - 2k_2 M D_r - 2k_3 M D_r - 4k_4 E_0 D_r - 2k_4 Y_0 D_r + 2k_4 \sum_{s=2}^{r-2} D_s D_{r-s} \tag{131b}$$

$$\frac{\mathrm{d}X_1}{\mathrm{d}t} = 2k_3ME_0 + k_3MY_0 - k_2MX_1 - k_3MX_1 - 2k_4E_0X_1 - k_4Y_0X_1 \tag{131c}$$

$$\frac{\mathrm{d}X_r}{\mathrm{d}t} = k_2MX_{r-1} - k_2MX_r - k_3MX_r - 2k_4E_0X_r - k_4Y_0X_r + 2k_3MD_r + 2k_4\sum_{s=2}^{r-1} D_sX_{r-s} \tag{131d}$$

$$\frac{\mathrm{d}P_r}{\mathrm{d}t} = k_3MX_r + \tfrac{1}{2}k_4\sum_{s=1}^{r-1} X_sX_{r-s} \tag{131e}$$

We obtain again eqn. (118) and in substituting it into the second and fourth equations of (131) we have

$$\frac{\mathrm{d}D_r}{\mathrm{d}t} = 2k_2MD_{r-1} - 2(k_2 + k_3 + w)MD_r + 2k_4\sum_{s=2}^{r-2} D_sD_{r-s} \tag{132a}$$

$$\frac{\mathrm{d}X_r}{\mathrm{d}t} = k_2MX_{r-1} - (k_2 + k_3 + w)MX_r + 2k_3MD_r + 2k_4\sum_{s=2}^{r-1} D_sX_{r-s} \tag{132b}$$

We now repeat the steps of the preceding section and form the Laplace transforms. The sums which constitute the last terms of (132) change into convolutions as the discrete variable is changed into a continuous one, e.g.

$$\sum_{s=2}^{r-1} D_sX_{r-s} = \int_0^{\rho} D(\sigma)X(\rho-\sigma)\bar{r}\,\mathrm{d}\sigma \tag{133}$$

with $\sigma = s/\bar{r}$ (assuming the sum to go from $s = 0$ to $s = r$, which does not introduce a noticeable error); the transform of this is $\bar{r}TS$, as mentioned in Sect. 2.3. We have finally

$$\frac{\partial T}{\partial t} = 0 = 2k_2M\left(T + \frac{D_0 - pT}{\bar{r}}\right) - 2(k_2 + k_3' + w)MT + 2k_4\bar{r}T^2 \tag{134a}$$

$$\frac{\partial S}{\partial t} = 0 = k_2M\left(S + \frac{X_0 - pS}{\bar{r}}\right) - (k_2 + k_3 + w)MT + 2k_3MT + 2k_4\bar{r}ST \tag{134b}$$

These equations reduce to

$$\left(k_3 + w + k_2\frac{p}{\bar{r}}\right)MT = k_4\bar{r}T^2 + \frac{k_1}{4\bar{r}}M^2 \tag{135a}$$

$$\left(k_3 + w + k_2\frac{p}{\bar{r}}\right)MS = 2k_3MT + 2k_4\bar{r}ST + \gamma\frac{k_3}{\bar{r}}M^2 \tag{135b}$$

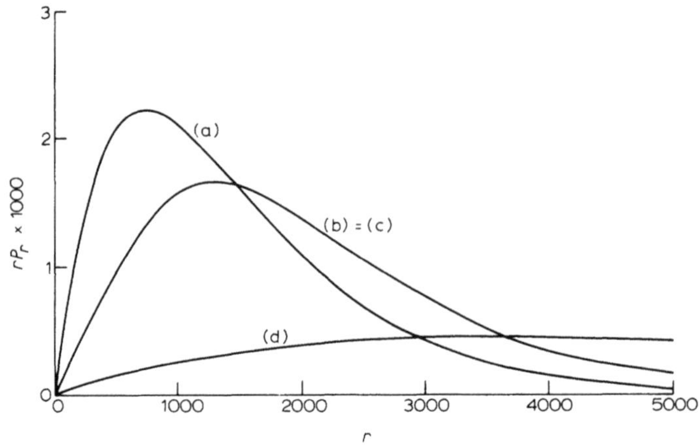

Fig. 1. Weight distributions of polymers formed under the following conditions: (a) single radicals, termination by disproportionation; (b) single radicals, termination by combination; (c) diradicals, termination by disproportionation; (d) diradicals, termination by combination; $k_1 = 2 \times 10^{-9}$, $k_2 = 10^2$, $k_3 = 2.7 \times 10^{-2}$, $k_4 = 6 \times 10^6$ l mole^{-1} sec^{-1}; $M = 8$ mole l^{-1} (constant), $\Delta M = 5\ M/9$ (from Bamford and Tompa [5]).

Equation (131e) transforms into

$$\frac{\partial L}{\partial t} = k_3 MS + \tfrac{1}{2} k_4 \bar{r} S^2 \tag{136}$$

If we eliminate T from eqns. (135) by substituting for T from the second equation into the first, we obtain a function of S which is exactly that on the right-hand side of eqn. (136) so that

$$\frac{\partial L}{\partial t} = \gamma k_3^2 \frac{M}{\bar{r}} \frac{1}{k_3 + k_2 p/\bar{r}} \tag{137}$$

If the initial polymer concentration is zero, Q_0, Q_1 and r_n are obtained easily from

$$Q_0 = \gamma k_3 Mt; \quad Q_1 = \gamma k_2 Mt; \quad r_n = \frac{k_2}{k_3} \tag{138}$$

and the polymer distribution is

$$P_r = \gamma \frac{k_3^2}{k_2} M \exp\left(-\frac{k_3}{k_2} r\right) t = \gamma \frac{k_3^2}{k_2} M \left(\frac{k_2}{k_2 + k_3}\right)^r t \tag{139}$$

This distribution is of the same functional form as regards chain length as that for initiation with single radicals and termination by disproportionation, but is broader when compared with smaller k_3 to w ratios. Figure 1, taken from Bamford and Tompa [5], shows the four distributions obtained for initiation by single radicals or diradicals and termination by disproportionation or combination, but otherwise under the same conditions and with the same values of the rate coefficients.

4.2 RADICAL POLYMERIZATION WITH CHAIN TRANSFER TO DEAD POLYMER, MONOMER CONCENTRATION CONSTANT

We consider a scheme of vinyl polymerization in which there is chain transfer to dead polymer producing branched molecules. The kinetic scheme, assuming termination to be by disproportionation, is

$$\begin{aligned} 2M &\rightarrow 2X_1\,;\, k_1 \\ X_r + M &\rightarrow X_{r+1}\,;\, k_2 \\ X_r + M &\rightarrow P_r + X_1\,;\, k_3 \\ X_r + X_s &\rightarrow P_r + P_s;\, k_4 \\ X_r + P_s &\rightarrow P_r + X_s\,;\, sk_5 \end{aligned} \tag{140}$$

where we have assumed that the rate coefficient for the reactivation of dead polymer is proportional to its chain length. We shall treat the case that the monomer concentration M is constant; further, in most applications to physical cases k_2 is again much larger than the other coefficients so that, for example, $k_3 + w$ can be neglected with regard to it.

The kinetic equations are

$$\frac{\mathrm{d}X_1}{\mathrm{d}t} = k_1M^2 + k_3MY_0 + k_5Y_0P_1 - k_2MX_1 - k_3MX_1 - k_4Y_0X_1 - k_5\Sigma\, sP_s \cdot X_1 \tag{141a}$$

$$\frac{\mathrm{d}X_r}{\mathrm{d}t} = k_2MX_{r-1} + k_5Y_0rP_r - k_2MX_r - k_3MX_r - k_4Y_0X_r - k_5\Sigma\, sP_s \cdot X_r \tag{141b}$$

$$\frac{\mathrm{d}P_r}{\mathrm{d}t} = k_3MX_r + k_4Y_0X_r + k_5\Sigma\, sP_s \cdot X_r - k_5Y_0rP_r \tag{141c}$$

Summation of the first and second equations again gives eqn. (82) for Y_0; we substitute this into eqn. (141b), multiply it by $\exp(-p\rho)\,\mathrm{d}\rho$, again leaving $\bar{r}$ undefined, and integrate to give

$$\frac{\partial S}{\partial t} = k_2M\left(S + \frac{X_0 - pS}{\bar{r}}\right) - \gamma k_5M\bar{r}\,\frac{\partial L}{\partial p} - \left(k_2 + k_3 + w + k_5\,\frac{Q_1}{M}\right)MS \tag{142}$$

We can assume (141b) to be valid for $r = 1$ if we put, formally,

$$X_0 = \gamma \frac{k_3 + w}{k_2} M \tag{143}$$

so that the integration can be extended down to $\rho = 0$. Equation (141c) gives

$$\frac{\partial L}{\partial t} = (k_3 + w)MS + k_5 Q_1 S + \gamma k_5 M \bar{r} \frac{\partial L}{\partial p} \tag{144}$$

If we multiply eqns. (141b) and (141c) by r and sum, adding (141a), we obtain, using the stationary-state assumption,

$$\frac{dQ_1}{dt} = \gamma(k_2 + k_3 + w)M^2 \tag{145}$$

so that

$$Q_1 = \gamma(k_2 + k_3 + w)M^2 t \tag{146}$$

Equation (146) could have been obtained at once from the physically obvious observation that the rate of increase of the total weight of polymer must equal the rate of decrease of monomer, but it is our aim to derive all expressions as far as possible mathematically and not by appealing to physical intuition.

We make use of the assumption of the stationary state by putting $\partial S/\partial t = 0$, compute S from eqn. (142) and substitute into eqn. (144); we also introduce a new time variable τ defined by

$$\tau = \gamma k_5 \frac{k_2 + k_3 + w}{k_3 + w} Mt \tag{147}$$

so that

$$k_5 \frac{Q_1}{M} = (k_3 + w)\tau \tag{148}$$

and obtain

$$\left[(k_3 + w)(\tau + 1) + k_2 \frac{p}{\bar{r}}\right] \frac{\partial L}{\partial \tau} = \frac{(k_3 + w)^3}{k_5(k_2 + k_3 + w)} (\tau + 1) \frac{M}{\bar{r}} + k_2 \frac{k_3 + w}{k_2 + k_3 + w} P \frac{\partial L}{\partial p} \tag{149}$$

It is convenient now to identify $\bar{r}$ with the value of r_n in the absence of chain transfer to dead polymer, i.e.

$$\bar{r} = \frac{k_2 + k_3 + w}{k_3 + w} \doteq \frac{k_2}{k_3 + w} \tag{150}$$

and eqn. (149) then simplifies to

$$(\tau + 1 + p)\frac{\partial L}{\partial \tau} - p\frac{\partial L}{\partial p} = \frac{k_2}{k_5}\frac{M}{\bar{r}^3}(\tau + 1) \tag{151}$$

This is a partial differential equation of the form of Lagrange's linear equation; if the initial polymer concentration is zero, i.e. $L^0 = 0$, its solution is

$$L = \frac{k_2}{k_5}\frac{M}{\bar{r}^3}\{p + \tau + 1 - [p^2 + 2p(\tau + 1) + 1]^{1/2}\} \tag{152}$$

This is the Laplace transform of P_r; it follows from a theorem mentioned in Sect. 2.3 that

$$\mathscr{L}\{P_r \exp[\rho(\tau + 1)]\} = \frac{k_2}{k_5}\frac{M}{\bar{r}^3}\{p - [p^2 - \tau(\tau + 2)]^{1/2}\} \tag{153}$$

We find from Tables of Laplace transforms (e.g. ref. 15, 43rd edn., p. 295, eqn. (58) with $a^2 = -\tau(\tau + 2)$; or ref. 17, p. 235, eqn. (28)) that

$$P_r = \frac{k_2}{k_5}\frac{M}{\bar{r}^3}\frac{1}{\rho}[\tau(\tau + 2)]^{1/2}\exp[-\rho(\tau + 1)]\, I_1\{\rho[\tau(\tau + 2)]^{1/2}\} \tag{154}$$

where $I_1(x)$ is the modified Bessel function of the first kind of order one.

The first three moments of the polymer distribution, obtained from the expression for L without any difficulty, are

$$Q_0 = \frac{k_2}{k_5}\frac{M}{\bar{r}^2}\tau; \quad Q_1 = \frac{k_2}{k_5}\frac{M}{\bar{r}}\tau; \quad Q_2 = \frac{k_2}{k_5}M\tau(\tau + 2) \tag{155}$$

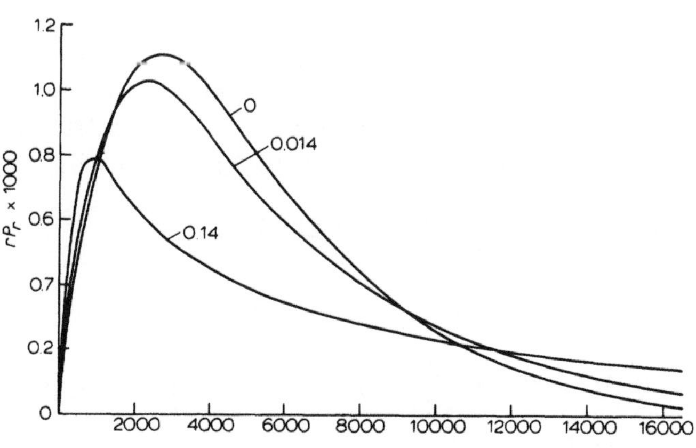

Fig. 2. Weight distributions of branched polymers. Single radicals; termination by disproportionation; $k_1 = 2 \times 10^{-11}$, $k_2 = 10^2$, $k_3 = 2.7 \times 10^{-2}$, $k_4 = 6 \times 10^6$, values of k_5 indicated, all expressed in l mole^{-1} sec^{-1}; $M = 8$ mole l^{-1} constant (from Bamford and Tompa [5]).

so that

$$r_{\mathrm{n}} = \bar{r}; \quad r_{\mathrm{w}} = \bar{r}(\tau + 2) \tag{156}$$

i.e. the number average chain length of the polymer formed is not affected by chain transfer to dead polymer, but the weight average increases with time at a rate proportional to the transfer coefficient k_5.

This kinetic scheme of polymerization is treated further by Bamford and Tompa [5], where it is extended to the case when the initial polymer concentration is not zero, and where the branching distribution of the polymer formed is also investigated.

Figure 2 shows as an illustration three polymer distributions rP_r with values of the rate coefficients indicated; the curves for $k_5 = 0.014$ and 0.14 are taken at a time such that the weight of the polymer formed equals the initial weight of monomer present but assuming the monomer concentration to have been kept constant.

5. The use of the moments of the distribution

5.1 RADICAL POLYMERIZATION, INITIATION BY DIRADICALS, TERMINATION BY COMBINATION

We shall show for the example of the case treated already by Laplace transformation in Sect. 4.1.2 how to obtain the polymer distribution from the moments without going through the generating function or the Laplace transform; this can be useful in cases where the differential equations for the latter cannot be solved analytically or where it is for one reason or another inconvenient to do so. This case will also serve as an example of how the moments can occasionally be obtained by a mathematical tour de force.

We multiply eqn. (131d) by r^n, sum over r and add (131c); we multiply similarly (131b) by r^n, sum over r and add (131a) multiplied by 2^n; and finally we multiply (131e) by r^n and sum over r. We obtain, introducing at once the assumption of the stationary state and using eqn. (118),

$$\frac{\mathrm{d}Y_n}{\mathrm{d}t} = 0 = \gamma k_3 M^2 + k_2 M \sum_{r=2}^{\infty} r^n X_{r-1} + 2k_3 M E_n - (k_2 + k_3 + w) M Y_n + 2k_4 \sum_{r=3}^{\infty} r^n \sum_{s=1}^{r-1} X_s D_{r-s} \tag{157a}$$

$$\frac{\mathrm{d}E_n}{\mathrm{d}t} = 0 = 2^{n-1} k_1 M^2 + 2k_2 M \sum_{r=3}^{\infty} r^n D_{r-1} - 2(k_2 + k_3 + w) M E_n + 2k_4 \sum_{r=4}^{\infty} r^n \sum_{s=2}^{r-2} D_s D_{r-s} \tag{157b}$$

$$\frac{\mathrm{d}Q_n}{\mathrm{d}t} = k_3 M Y_n + \tfrac{1}{2} k_4 \sum_{r=2}^{\infty} r^n \sum_{s=1}^{r-1} X_s X_{r-s} \tag{157c}$$

Sums of the type $\Sigma r^n X_{r-1}$ can be treated as follows

$$\sum_{r=2}^{\infty} r^n X_{r-1} = \sum_{r=1}^{\infty} (r+1)^n X_r = \sum_{r=1}^{\infty} \sum_{i=0}^{n} \binom{n}{i} r^i X_r = \sum_{i=0}^{n} \binom{n}{i} Y_i \quad (158)$$

The term with $i = n$ cancels against a similar term when substituted into (157a); further, Y_i is of the order of magnitude of r_n^i and since in all cases of practical interest r_n is of the order of several hundreds or thousands, all terms can be neglected against the largest, which is here the term with $i = n - 1$. However, care will be needed in cases in which r_n cannot be assumed to be large. The first term on the right-hand side of (157a) is of the order of r_n^0 (unity) and is therefore negligible for all $n \geqslant 1$. Sums of the type $\Sigma r^n \Sigma X_s D_{r-s}$ are treated similarly, viz.

$$\sum_{r=3}^{\infty} r^n \sum_{s=1}^{r-1} X_s D_{r-s} = \sum_{r=3}^{\infty} (r - s + s)^n \sum_{s=1}^{r-1} X_s D_{r-s} =$$

$$\sum_{r=3}^{\infty} \sum_{i=0}^{n} \binom{n}{i} (r-s)^{n-i} s^i \sum_{s=1}^{r-1} X_s D_{r-s} =$$

$$\sum_{i=0}^{n} \binom{n}{i} \sum_{s=1}^{\infty} s^i X_s \sum_{r=s+2}^{\infty} (r-s)^{n-i} D_{r-s} = \sum_{i=0}^{n} \binom{n}{i} Y_i E_{n-i} \quad (159)$$

Equations (157) can then be written

$$(k_3 + w) M Y_n = n k_2 M Y_{n-1} + 2k_3 M E_n + 2k_4 \sum_{i=0}^{n} \binom{n}{i} Y_i E_{n-i} \quad (160a)$$

$$(k_3 + w) M E_n = n k_2 M E_{n-1} + k_4 \sum_{i=0}^{n} \binom{n}{i} E_i E_{n-i} \quad (160b)$$

$$\frac{dQ_n}{dt} = k_3 M Y_n + \tfrac{1}{2} k_4 \sum_{i=0}^{n} \binom{n}{i} Y_i Y_{n-i} \quad (160c)$$

We multiply eqn. (160b) by four and add (160a), write

$$R_n = Y_n + 2E_n \quad (161)$$

and obtain

$$(k_3 + w) M (R_n + 2E_n) = n k_2 M (R_{n-1} + 2E_{n-1}) + 2k_3 M E_n +$$

$$2k_4 \sum_{i=0}^{n} \binom{n}{i} R_i E_{n-i} \quad (162)$$

References p. 576

The term with $i = 0$ in the sum gives $2k_4R_0E_n$, which is $2wME_n$, so that the last two terms of the right hand side cancel against similar terms on the left side leaving

$$(k_3 + w)MR_n = nk_2MR_{n-1} + 2nk_2ME_{n-1} + 2k_4 \sum_{i=1}^{n} \binom{n}{i} R_iE_{n-i} \tag{163}$$

On comparing this with eqn. (160a) written for $n - 1$, viz.

$$(k_3 + w)MY_{n-1} = (n-1)k_2MY_{n-2} + 2k_3ME_{n-1} + 2k_4 \sum_{i=0}^{n-1} \binom{n-1}{i} Y_iE_{n-1-i} \tag{164}$$

one might suspect that R_n should be proportional to Y_{n-1} and one can in fact show that this is the case by transforming the sum in eqn. (163)

$$\sum_{i=1}^{n} \binom{n}{i} R_iE_{n-i} = n \sum_{i=1}^{n} \frac{(n-1)!}{(i-1)!(n-i)!} \frac{R_i}{i} E_{n-1} =$$

$$n \sum_{i=1}^{n} \binom{n-1}{i-1} \frac{R_i}{i} E_{n-i} = n \sum_{i=0}^{n-1} \binom{n-1}{i} \frac{R_{i+1}}{i+1} E_{n-1-i} \tag{165}$$

If we now compare eqns. (163) with (164) we see that

$$R_n = n \frac{k_2}{k_3} Y_{n-1} \tag{166}$$

We add (160a) multiplied by 1/2 to (160c) to give

$$\frac{dQ_n}{dt} = \tfrac{1}{2}k_3MR_n - \tfrac{1}{2}wMY_n + \tfrac{1}{2}nk_2MY_{n-1} + \tfrac{1}{2}k_4 \sum_{i=0}^{n} \binom{n}{i} Y_iR_{n-i} \tag{167}$$

The term with $i = n$ in the sum cancels against the second term, and substituting for R_n from (166) we obtain

$$\frac{dQ_n}{dt} = nk_2MY_{n-1} + \tfrac{1}{2}k_4 \frac{k_2}{k_3} \sum_{i=0}^{n-1} \binom{n}{i} (n-i)Y_iY_{n-1-i}$$

$$= nk_2MY_{n-1} + \tfrac{1}{2}nk_4 \frac{k_2}{k_3} \sum_{i=0}^{n-1} \binom{n-1}{i} Y_iY_{n-1-i}$$

$$= n \frac{k_2}{k_3} \frac{dQ_{n-1}}{dt} \tag{168}$$

This gives

$$\frac{dQ_n}{dt} = n!\left(\frac{k_2}{k_3}\right)^n \frac{dQ_0}{dt} \tag{169}$$

Addition of

$$\frac{dQ_0}{dt} = k_3 M Y_0 + \tfrac{1}{2} k_4 Y_0^2 \tag{170}$$

to the equations corresponding to (117a) and (117b) gives

$$\frac{dQ_0}{dt} = \gamma k_3 M^2 \tag{171}$$

so that

$$Q_n = n!\gamma\left(\frac{k_2}{k_3}\right)^n k_3 M^2 t \tag{172}$$

if M can be regarded as constant.

Having obtained an analytical expression for Q_n, the next step is to obtain the coefficients c_j of the development into a series of Laguerre polynomials. A suitable value for $\bar{r}$ must be chosen and it seems most appropriate to choose $\bar{r} = k_2/k_3$. With this choice all c_j are zero from eqn. (19) except c_0, which equals $Q_0/\bar{r}$; since $L_0(\rho) = 1$, eqn. (16) gives

$$P(\rho) = \frac{k_3}{k_2} Q_0 \exp(-\rho) = \gamma \frac{k_3^2}{k_2} M^2 \exp(-r/\bar{r})t \tag{173}$$

as in eqn. (139).

In this particular case the development into a series of Laguerre polynomials turned out to be particularly simple, since the series reduces to its first term, but clearly once an analytical expression has been obtained for Q_n, the series is not difficult to obtain.

5.2 RADICAL POLYMERIZATION, MONOMER CONCENTRATION NOT CONSTANT

5.2.1 Without chain transfer to dead polymer

We consider the reaction scheme already discussed in Sect. 3.3.1, but do not assume that the monomer concentration is constant. In addition, to keep the treatment as general as possible, we write the rate of initiation as $k_1 M^{0\,2}(M/M^0)^{2\mu}$ where M^0 is the initial monomer concentration, and $\mu = 1$ if initiation is second-order in monomer (thermal initiation), $\mu = 1/2$ if it is first-order (photochemical initiation), and $\mu = 0$ if it is independent of the monomer concentration (photosenzitized initiation); the initiations in these three cases are then written as $k_1 M^2$, $k_1 M^0 M_1$ and $k_1 M^{0\,2}$, respectively. We shall write m for the relative monomer concentration,

$$m = \frac{M}{M^0} \tag{174}$$

and obtain for its dependence on time

$$-M^0 \frac{\mathrm{d}m}{\mathrm{d}t} = k_1 M^{0\,2} m^{2\mu} + k_2 M Y_0 + k_3 M Y_0 =$$

$$\gamma(k_2 m + k_3 m + w m^{\mu}) M^{0\,2} m^{\mu} \tag{175}$$

since Y_0 is now given by

$$Y_0 = \gamma M^0 m^{\mu} \tag{176}$$

If we neglect k_3 and w against k_2, we obtain for m as a function of time

$$m = [1 + \gamma k_2 M^0 \mu t]^{-1/\mu} \tag{177}$$

for $\mu \neq 0$ and

$$m = \exp[-\gamma k_2 M^0 t] \tag{178}$$

for $\mu = 0$. This allows us to calculate the decrease in monomer concentration.

The differential equations for the generating functions Z and R or for the Laplace transforms S and L are not difficult to find, but it does not seem possible to obtain analytical solutions for them. Numerical solutions are in principle always possible, especially using an electronic computer; however, it is no easy matter if not wholly impracticable to try to find the coefficient of z^r in the generating functions by numerical methods and inverse Laplace transformation also requires a great expenditure of computer time, in particular if one considers that the procedure has to be gone through for each value of t for which one requires a solution. It is therefore worth while to see if solutions cannot be obtained by trying to compute the coefficients of an expansion into Laguerre polynomials.

We multiply (80b) by r^n, sum over r and add (80a), with the first term on the right-hand side suitably modified, to obtain

$$(k_2 + k_3 + w m^{\mu}) M Y_n = k_1 M^{0\,2} m^{2\mu} + \gamma k_3 M^0 M + k_2 M \sum_{r=2}^{\infty} r^n X_{r-1} \tag{179}$$

The last term on the right-hand side is transformed as in eqn. (158), the first two can be neglected except for $n = 0$, so that for $n \neq 0$

$$Y_n = n \frac{\alpha}{1-\alpha} Y_{n-1} \tag{180}$$

where we have used a definition of α analogous to that in (83), viz.

$$\alpha = \frac{k_2 m}{k_2 m + k_3 m + w m^{\mu}} \tag{181}$$

We also multiply (80c) by r^n and sum over r, which gives

$$\frac{\mathrm{d}Q_n}{\mathrm{d}t} = k_3 M Y_n + k_4 Y_0 Y_n = (k_3 m + w m^{\mu}) M^0 Y_n \tag{182}$$

If we substitute this equation and the corresponding equation for Q_{n-1} into (180) we obtain

$$\frac{\mathrm{d}Q_n}{\mathrm{d}t} = n \frac{\alpha}{1-\alpha} \frac{\mathrm{d}Q_{n-1}}{\mathrm{d}t} \tag{183}$$

and by application of this equation for n, $n-1$, etc.

$$\frac{\mathrm{d}Q_n}{\mathrm{d}t} = n! \left(\frac{\alpha}{1-\alpha}\right)^n \frac{\mathrm{d}Q_0}{\mathrm{d}t} = n! \gamma (k_3 m + w m^{\mu}) \left(\frac{\alpha}{1-\alpha}\right)^n M^{0\,2} m^{\mu} \tag{184}$$

or, with (175),

$$\frac{\mathrm{d}Q_n}{\mathrm{d}m} = -n! \left(\frac{\alpha}{1-\alpha}\right)^{n-1} M^0 \tag{185}$$

It must be borne in mind that, except for $\mu = 1$, α is a function of m so that Q_n is given by the integral of the right-hand side. Before evaluating these integrals it is interesting to consider that whereas the various average chain lengths of the total polymer produced are given by the appropriate ratios of the Q_n, those of the instantaneous polymer produced are given by the ratios of the $\mathrm{d}Q_n$ and equal therefore those of a polymer produced with the instantaneous value of α.

For $\mu = 1$ and with $\bar{r} = \alpha/(1-\alpha)$ we have

$$Q_n = n! \bar{r}^{n-1} M^0 (1 - m) = n! \bar{r}^{n-1} \Delta M \tag{186}$$

The coefficients of the development of P_r into a series of Laguerre polynomials are all 0 except for c_0, as shown in the previous section, and we obtain

$$P_r = \frac{Q_0}{\bar{r}} \mathrm{e}^{-\rho} = \frac{\Delta M}{\bar{r}^2} \alpha^{r-1} \tag{187}$$

as previously, but of course ΔM is not the same function of t as in Sect. 3.3.1.

For values of μ other than unity the Q_n have to be obtained by evaluation of the appropriate integrals. For $\mu = \frac{1}{2}$ it is convenient to use the substitution

$$u = \frac{w}{k_3 m^{1/2} + w}\,. \tag{188}$$

and this gives

$$Q_n = 2n!\left(\frac{w}{k_3}\right)^2\left(\frac{k_2}{k_3}\right)^{n-1} M^0 \int_{u^0}^{u} \frac{(1-u)^n}{u^3}\,\mathrm{d}u \qquad (189)$$

For $\mu = 0$ the appropriate substitution is

$$v = \frac{w}{k_3 m + w} \qquad (190)$$

and this gives

$$Q_n = n!\,\frac{w}{k_3}\left(\frac{k_2}{k_3}\right)^{n-1} M^0 \int_{v^0}^{v} \frac{(1-v)^{n-1}}{v^2}\,\mathrm{d}v \qquad (191)$$

The evaluation of these integrals for increasing values of n presents no difficulties and the coefficients c_j can then be calculated from (19), up to values of j such that the c_j are below the limit of the precision desired. For $\bar{r}$ it is convenient to take $k_2/(k_3 + w)$, but if convergence is slow, i.e. if the coefficients c_j do not decrease sufficiently rapidly with j, other values, e.g. k_2/k_3, must be tried. The values of the Laguerre polynomials at different values of ρ can be obtained from (17) or from recursion formulae, given in the literature, which is more convenient when using an electronic computer, and P_r is then calculated from (16).

Figure 3, reproduced from Bamford and Tompa [5] shows curves of rP_r for the same values of the rate coefficients as Fig. 1, for such times that $\Delta M/M^0 = 5/9$. Three terms in eqn. (16) were sufficient to give 1 % accuracy in P_r up to $\rho = 8$.

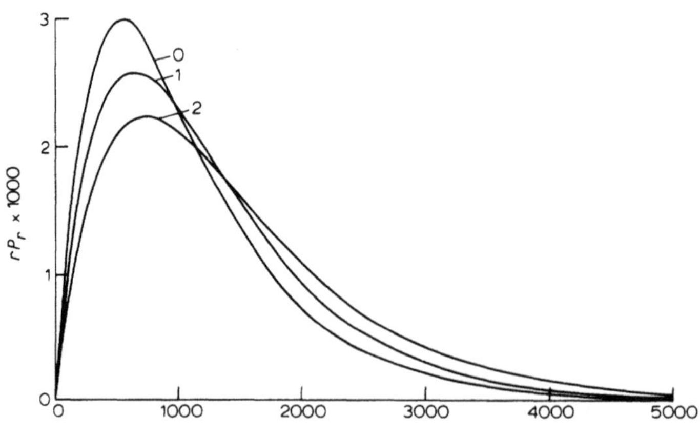

Fig. 3. Weight distributions of polymers produced when change of monomer concentration is taken into account. Single radicals; termination by disproportionation; values of rate coefficients as in Fig. 1; $M^0 = 8$ mole l^{-1}; $\Delta M = 5\,M^0/9$; values of μ indicated (from Bamford and Tompa [5]).

5.2.2 *With chain transfer to dead polymer*

We shall again consider the case of a radical polymerization with chain transfer to dead polymer discussed in Sect. 4.2, but we shall take the change of monomer concentration into account; we shall show on this example how to find the value of Q_n in more complicated cases and how an appropriate choice of $\bar{r}$ can lead to a convergent series even when a more obvious choice gives a divergent series.

The reaction scheme is that of eqn. (140) and the kinetic equations are given by (141), with k_1 replaced by $k_1(M^0)^2m^2$ as in the previous section. Y_0 and $\mathrm{d}m/\mathrm{d}t$ are given by eqns. (176) and (175) of the previous section, and it is easy to verify, by the same series of steps as in Sect. 4.2, that

$$\frac{\mathrm{d}Q_1}{\mathrm{d}t} = -M^0 \frac{\mathrm{d}m}{\mathrm{d}t} \tag{192}$$

as is physically obvious, so that

$$Q_1 = M^0(1-m) \tag{193}$$

always assuming that the initial polymer concentration is zero and incidentally that the stationary-state treatment is justified.

We now proceed to obtain expressions for Y_n and Q_n as in previous sections but shall not describe the steps in detail any more. We obtain

$$[k_3m + wm^\mu + k_5(1-m)]Y_n = nk_2mY_{n-1} + \gamma k_5 m^\mu Q_{n+1} \tag{194a}$$

$$\frac{\mathrm{d}Q_n}{\mathrm{d}t} = [k_3m + wm^\mu + k_5(1-m)]M^0Y_n - \gamma k_5 M^0 m^\mu Q_{n+1} \tag{194b}$$

We write down the second equation replacing n by $n-1$ and eliminate Y_n and Y_{n-1} from the two equation by the use of the first. We also replace $\mathrm{d}t$ by $\mathrm{d}m$ and obtain finally, neglecting α against unity to simplify the equation,

$$[k_3m + wm^\mu + k_5(1-m)]\frac{\mathrm{d}Q_n}{\mathrm{d}m} + nk_5Q_n = nk_2m\frac{\mathrm{d}Q_{n-1}}{\mathrm{d}m} \tag{195}$$

for $n \geqslant 1$ and

$$\frac{\mathrm{d}Q_0}{\mathrm{d}m} = -\frac{k_3m + wm^\mu}{k_2m}M^0 \tag{196}$$

Equation (195) is a recursive differential equation; no analytical solution appears to exist and values of Q_n as functions of m have to be obtained by numerical methods for one value of n after another and in view of the effort involved it is advisable to reduce the total number of moments calculated as far as possible.

TABLE 1

Moments Q_n of the polymer distribution and coefficients c_n of the development into series of Laguerre polynomials for different values of $\bar{r}$. (Rate coefficients k_1, k_2, k_3, k_4 as in Fig. 1, k_5 = 0.14. Single radicals; termination by disproportionation; M^0 = 8 mole 1^{-1}; $\Delta M = 5\,M^0/9$; $\mu = 1/2$.)

n	$\frac{Q_n}{M^0 n! \bar{r}^{n-1}}$	$\bar{r}^2 \cdot n! c_n$			
		$\bar{r} = \bar{r}_0$	$\bar{r} = 1.25\,\bar{r}_0$	$\bar{r} = 1.5\,\bar{r}_0$	$\bar{r} = 1.8\,\bar{r}_0$
0	0.55556	0.556	0.694	0.833	1.000
1	0.55556	0	0.139	0.278	0.444
2	0.79602	0.240	0.220	0.253	0.331
3	1.47300	−0.196	−0.005	0.104	0.205
4	3.18396	0.401	0.152	0.120	0.159
5	7.56376	−0.636	−0.090	0.039	0.107
6	19.1028	1.182	0.183	0.069	0.085
7	50.3402	−2.193	−0.199	0.008	0.060
8	136.855	4.231	0.306	0.047	0.048
9	381.017	−8.304	−0.408	−0.008	0.034
10	1080.886	16.596	0.598	0.038	0.028

The case with $\mu = \frac{1}{2}$ has been treated by Bamford and Tompa [5], with the values of the rate coefficients used in Sect. 5.2.1. The moments of the distribution were computed as described above and from the values at $m = \frac{4}{9}$ the coefficients c_j were obtained with $\bar{r} = \bar{r}_0 = k_2/(k_3 + w)$. For k_5 = 0.014 five terms of the series were required for an accuracy of 1 %, but for k_5 = 0.14 the series diverges (oscillates). The first two columns of Table 1 show the first eleven moments and the coefficients c_j; new coefficients c_j were then computed with $\bar{r}$ =1.25 $\bar{r}_0$, 1.5 $\bar{r}_0$, and 1.8 $\bar{r}_0$ and

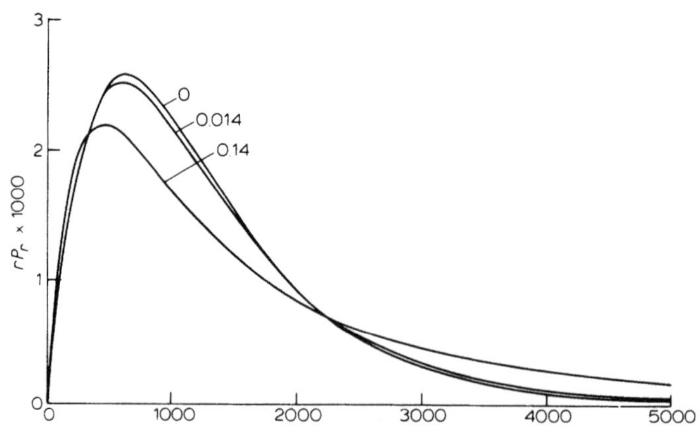

Fig. 4. Weight distributions of branched polymers produced when change of monomer concentration is taken into account. Single radicals; termination by disproportionation. Values of rate coefficients as in Fig. 1; M^0 = 8 mole 1^{-1}; $\Delta M = 5\,M^0/9$; $\mu = \frac{1}{2}$; values of k_5 indicated (from Bamford and Tompa [5]).

these are also shown. Convergence seems to be fastest for $\bar{r} = 1.5\ \bar{r}_0$, and rP_r was computed with these coefficients; it is estimated that these eleven terms give an accuracy of 5 % at $\rho = 4$. The curves for $k_5 = 0.014$ and 0.14 are shown in Fig. 4 (Bamford and Tompa [5]) together with the curve for $k_5 = 0$, taken from Fig. 3.

5.3 RADICAL POLYMERIZATION WITH CHAIN TRANSFER TO DEAD POLYMER, TERMINATION BY COMBINATION: GELATION

If, in a scheme of radical polymerization with chain transfer to polymer, termination is by combination, the last equation of the set (140) of Sect. 4.2 has to be replaced by

$$X_r + X_s \rightarrow P_{r+s};\ k_4 \tag{197}$$

and the corresponding kinetic equation is

$$\frac{\mathrm{d}P_r}{\mathrm{d}t} = k_3 M X_r + \tfrac{1}{2} k_4 \sum_{s=1}^{r-1} X_s X_{r-s} + k_5 Q_1 X_r - k_5 Y_0 r P_r \tag{198}$$

On multiplying this equation by r^n and summing over r it gives

$$\frac{\mathrm{d}Q_n}{\mathrm{d}t} = k_3 M Y_n + \tfrac{1}{2} k_4 \sum_{i=0}^{n} \binom{n}{i} Y_i Y_{n-i} + k_5 Q_1 Y_n - k_5 Y_0 Q_{n+1} \tag{199}$$

or, splitting off the terms $i = 0$ and $i = n$ from the sum in the second term and using for Y_0 and Q_1 eqns. (176) and (193), which are also valid here as is easy to confirm,

$$\frac{\mathrm{d}Q_n}{\mathrm{d}t} = [k_3 m + w m^\mu + k_5(1 - m)] M^0 Y_n + \tfrac{1}{2} k_4 \sum_{i=1}^{n-1} \binom{n}{i} Y_i Y_{n-i} - \gamma k_5 M^0 m^\mu Q_{n+1} \tag{200}$$

also

$$\frac{\mathrm{d}Q_0}{\mathrm{d}t} = \gamma(k_3 m + \tfrac{1}{2} w m^\mu) M^{0\,2} m^\mu \tag{201}$$

We can eliminate again Y_n from eqns. (200) and (194a) to give

$$\frac{\mathrm{d}Q_n}{\mathrm{d}t} = n k_2 M^0 m Y_{n-1} + \tfrac{1}{2} k_4 \sum_{i=1}^{n-1} \binom{n}{i} Y_i Y_{n-i} \tag{202}$$

but we cannot eliminate conveniently all Y_j from this equation and (200) unless $k_5 = 0$. It is of course possible to compute successive values of Y_{n-1} and Q_n starting from the known expressions for Y_0 and Q_1 and then to compute the coefficients of the expansion into Laguerre polynomials. We shall not continue this calculation here since it does not bring anything essentially new but shall only consider the problem of gelation.

Chain transfer to dead polymer molecules produces branched molecules and if termination is by combination the size of the molecules can increase until the bulk of the polymer present is one giant cross-linked molecule; this is gelation. Mathematically, gelation is characterized by the weight average chain length — and a fortiori the higher average chain lengths — going to infinity. We shall therefore calculate Q_2 to obtain an analytical condition for the onset of gelation and since the essential points are brought out by the simplest case we shall neglect the change of monomer concentration and shall assume bimolecular initiation, $\mu = 1$.

Equations (194a) and (202) for Y_1 and Q_2 are

$$\left(k_4 + w + k_5 \frac{Q_1}{M}\right) Y_1 = k_2 Y_0 + \gamma k_5 Q_2 \tag{203}$$

$$\frac{\mathrm{d}Q_2}{\mathrm{d}t} = 2k_2 M Y_1 + k_4 Y_1^2 \tag{204}$$

We introduce the time variable τ used in Sect. 4.2 and defined by eqns. (147) or (148), replace Y_0 by its familiar expression, and obtain

$$Y_1 = \gamma \frac{k_2 M + k_5 Q_2}{(k_3 + w)(\tau + 1)} \tag{205}$$

$$\gamma k_5 \frac{k_2}{k_3 + w} M \frac{\mathrm{d}Q_2}{\mathrm{d}\tau} = 2k_2 M Y_1 + k_4 Y_1^2 \tag{206}$$

Substitution of Y_1 from the first equation into the second gives

$$k_5 \frac{\mathrm{d}Q_2}{\mathrm{d}\tau} = 2\frac{k_2 M + k_5 Q_2}{\tau + 1} + \frac{w}{k_2 M} \frac{(k_2 M + k_5 Q_2)^2}{(k_3 + w)(\tau + 1)^2} \tag{207}$$

This is again Riccati's equation, which one obtains so often when termination is by combination. It is convenient to regard $k_2 M + k_5 Q_2$ as the dependent variable and we solve the equation by the substitution (see Piaggio [21], p. 201)

$$k_2 M + k_5 Q_2 = -\frac{k_3 + w}{w} (\tau + 1)^2 k_2 M \frac{1}{u} \frac{\mathrm{d}u}{\mathrm{d}\tau} \tag{208}$$

which reduces it to

$$\frac{\mathrm{d}^2 u}{\mathrm{d}\tau^2} = 0 \tag{209}$$

The solution is

$$\frac{\mathrm{d}u}{\mathrm{d}\tau} = c_1; \quad u = c_1 \tau + c_2 \tag{210}$$

and, if we write c for c_1/c_2,

$$k_2 M + k_5 Q_2 = -\frac{k_3 + w}{w} (\tau + 1)^2 k_2 M \frac{1}{\tau + c} \tag{211}$$

The constant c is determined from the initial conditions $\tau = 0$, $Q_2^0 = 0$, when

$$k_2 M = -\frac{k_3 + w}{w} k_2 M \frac{1}{c} \tag{212}$$

and substitution into the previous equation gives

$$k_2 M + k_5 Q_2 = \frac{k_3 + w}{k_3 + w - w\tau} k_2 M(\tau + 1)^2 \tag{213}$$

The second moment and the weight average chain length go therefore to infinity when

$$\tau_g = \frac{k_3 + w}{w} \tag{214}$$

which is equivalent to

$$k_5 Q_{1g} = \frac{(k_3 + w)^2}{w} M \tag{215}$$

and this is the condition for the onset of gelation.

6. Application to some more involved cases

6.1 THE COUPLING OF POLYMERS

We consider the end-to-end coupling of polymers having reactive groups A at either end by a coupling agent having reactive groups B at either end, where A and B can condense so as to unite the two molecules. Several aspects of this reaction have been studied by Bamford and Jenkins [22] and we shall follow their treatment fairly closely.

Three types of molecule are involved, those having A groups at either end, those having one A group and one B group at the ends, and those having B groups at either end. In order not to complicate the notation we shall not use the conventions of the preceding chapters; we denote the three types of molecule of chain length r by P_r, Q_r and R_r, respectively, and shall mean by chain length the number of monomer residues of which the initial polymer is composed. The physically most relevant case is that of an initial polymer being coupled by a coupling agent BB, chain length zero, so that there is an initial polymer distribution P_r^0, all Q_r^0 are zero and so are all R_r^0 except R_0^0, and it is with these initial conditions that we shall find the solution. No error or inaccuracy is introduced by regarding the low-molecular coupling agent as having chain length zero.

The reaction scheme is

$$\begin{aligned} &P_r + Q_s \rightarrow P_{r+s}\ ;\ 2k \\ &P_r + R_s \rightarrow Q_{r+s};\ 4k \\ &Q_r + Q_s \rightarrow Q_{r+s};\ 2k \\ &Q_r + R_s \rightarrow R_{r+s};\ 2k \end{aligned} \tag{216}$$

and the kinetic equations are

$$\frac{dP_r}{dt} = -2kQP_r - 4kRP_r + 2k \sum_{s=1}^{r-1} P_sQ_{r-s} \tag{217a}$$

$$\frac{dQ_r}{dt} = -2kPQ_r - 2kQQ_r - 2kRQ_r + 4k \sum_{s=1}^{r} P_sR_{r-s} + k \sum_{s=1}^{r-1} Q_sQ_{r-s} \tag{217b}$$

$$\frac{dR_r}{dt} = -4kPQ_r - 2kQR_r + 2k \sum_{s=1}^{r} Q_sR_{r-s} \tag{217c}$$

We have here used the symbols P, Q, R to denote the moments of order zero, to keep the notation simple. We introduce the concentration A and B of active ends

$$A = 2P + Q; \quad B = 2R + Q \tag{218}$$

and from eqns. (217)

$$\frac{dA}{dt} = \frac{dB}{dt} = -kAB \tag{219}$$

so that the difference between A and B is constant, which is physically evident. We shall denote this difference by $A^0\delta$ where δ can be positive, zero, or negative and shall use the relative concentration $a = A/A^0$ rather than A, so that

$$A = aA^0; \quad B = (a + \delta)A^0 \tag{220}$$

Then

$$\frac{da}{dt} = -ka(a + \delta)A^0 \tag{221}$$

and we replace t by a as independent variable by dividing eqns. (217) by (221), giving

$$A^0 a(a + \delta) \frac{dP_r}{da} = 2QP_r + 4RP_r - 2 \sum_{s=1}^{r-1} P_sQ_{r-s} \tag{222a}$$

$$A^0 a(a + \delta) \frac{dQ_r}{da} = 2PQ_r + 2QQ_r + 2RQ_r - 4 \sum_{s=1}^{r} P_sR_{r-s} - \sum_{s=1}^{r-1} Q_sQ_{r-s} \tag{222b}$$

$$A^0 a(a+\delta)\frac{dR_r}{da} = 4PR_r + 2QR_r - 2\sum_{s=1}^{r} Q_s R_{r-s} \tag{222c}$$

Strictly speaking, separate equations should be written for $r = 1$ and $r = 0$, but if, for example, P_0 and Q_0 are assumed to be zero, eqns. (222) are formally correct for all values of r.

Summing each of these equations over all r gives

$$A^0 a(a+\delta)\frac{dP}{da} = 4PR \tag{223a}$$

$$A^0 a(a+\delta)\frac{dQ}{da} = 2PQ + Q^2 + 2QR - 4PR \tag{223b}$$

$$A^0 a(a+\delta)\frac{dR}{da} = 4PR \tag{223c}$$

With eqns. (218) and (220) eqn. (223a) transforms into

$$A^0 a(a+\delta)\frac{dP}{da} = 2P(2P + A^0\delta) \tag{224}$$

and its solution, with $A^0 = 2P^0$ at $t = 0$, $a = 1$, is

$$P = \tfrac{1}{2}A^0 \frac{a^2(a+\delta)}{(1+\delta)-(1-a)^2} \tag{225a}$$

and one then easily obtains

$$Q = A^0 \frac{a(1-a)(a+\delta)}{(1+\delta)-(1-a)^2} \tag{225b}$$

$$R = \tfrac{1}{2}A^0 \frac{(a+\delta)^2}{(1+\delta)-(1-a)^2} \tag{225c}$$

We substitute these values into eqn. (222), using also eqn. (218), which gives

$$A^0 a(a+\delta)\frac{dPr}{da} = 2A^0(a+\delta)P_r - 2\sum_{s=1}^{r-1} P_s Q_{r-s} \tag{226a}$$

$$A^0 a(a+\delta)\frac{dQ_r}{da} = A^0(2a+\delta)Q_r - 4\sum_{s=1}^{r} P_r R_{r-s} - \sum_{s=1}^{r-1} Q_s Q_{r-s} \tag{226b}$$

$$A^0 a(a+\delta)\frac{dR_r}{da} = 2A^0 a R_r - 2\sum_{s=1}^{r} Q_s R_{r-s} \tag{226c}$$

We now form the equations for the Laplace transforms X, Y, and Z of P_r, Q_r and R_r by multiplying each equation by $\exp(-p\rho)\, d\rho$ and integrating over ρ, leaving $\bar{r}$ undefined for the moment, and obtain

$$A^0 a(a+\delta)\frac{\partial X}{\partial a} = 2A^0(a+\delta)X - 2\bar{r}XY \tag{227a}$$

$$A^0 a(a+\delta)\frac{\partial Y}{\partial a} = A^0(2a+\delta)Y - 4\bar{r}XZ - \bar{r}Y^2 \tag{227b}$$

$$A^0 a(a+\delta)\frac{\partial Z}{\partial a} = 2A^0 aZ - 2\bar{r}YZ \tag{227c}$$

If we divide (227a) by X and (227c) by Z and subtract we obtain

$$A^0 a(a+\delta)\left[\frac{1}{X}\frac{\partial X}{\partial a} - \frac{1}{Z}\frac{\partial Z}{\partial a}\right] = 2A^0\delta \tag{228}$$

the integral of which is

$$\frac{X^0 Z}{XZ^0} = \frac{(a+\delta)^2}{a^2(1+\delta)^2} \tag{229}$$

with the initial conditions $X = X^0$, $Z = Z^0$, $a = 1$. Here X^0 is the Laplace transform of the initial polymer distribution, while Z^0 is, see eqn. (15),

$$Z^0 = \frac{R^0}{\bar{r}} = \frac{1}{2}\frac{A^0}{\bar{r}}(1+\delta) \tag{230}$$

so that

$$Z = \frac{X}{X^0}\frac{A^0}{2\bar{r}}\frac{(a+\delta)^2}{a^2(1+\delta)} \tag{231}$$

To be precise, (230) is the value of Z^0 for $p = 0$; however, since $R_r^0 = 0$ except for $r = 0$, Z^0 by the definition of a Laplace transform does not contain p and so (230) is the value of Z^0.

If this value of Z and the value of Y expressed in terms of X from (227a) are introduced into (227b) and the resulting equation simplified, one obtains

$$\frac{a^4}{2}\frac{1}{X}\frac{\partial^2 X}{\partial a^2} + a^3\frac{1}{X}\frac{\partial X}{\partial a} - \frac{3a^4}{4}\frac{1}{X^2}\left(\frac{\partial X}{\partial a}\right)^2 - \frac{2\bar{r}X^2}{A^0 X^0(1+\delta)} = 0 \tag{232}$$

This differential equation can be solved by the substitution

$$X = \frac{a^2}{x^2} \tag{233}$$

which reduces it to

$$\frac{d^2 x}{da^2} + \frac{2\bar{r}}{x^3 A^0 X^0(1+\delta)} = 0 \tag{234}$$

The text-book method for solving this is to write

$$\frac{d^2x}{da^2} = \frac{d}{da}\left(\frac{dx}{da}\right) = \frac{dx}{da}\frac{d}{dx}\left(\frac{dx}{da}\right) = \frac{1}{2}\frac{d}{dx}\left(\frac{dx}{da}\right)^2 \tag{235}$$

so that

$$\left(\frac{dx}{da}\right)^2 = \frac{2\bar{r}}{x^2A^0X^0(1+\delta)} + C \tag{236}$$

To find the value of C, we need dx/da at $a = 1$; we obtain $\partial X/\partial a$ in terms of dx/da at $a = 1$ from (233) and comparison with the value obtained from (227a) at $a = 1$, with $Y^0 = 0$, shows that $dx/da = 0$ at $a = 1$. This leads to

$$\frac{dx}{da} = \left[\frac{2\bar{r}(1 - x^2X^0)}{x^2A^0X^0(1+\delta)}\right]^{1/2} \tag{237}$$

and the integral of this gives ultimately

$$X = \frac{X^0A^0a^2(1+\delta)}{A^0(1+\delta) - 2\bar{r}X^0(1-a)^2} \tag{238a}$$

The values of Y and Z, obtained easily from (227a) and (231) are

$$Y = \frac{2X^0A^0a(a+\delta)(1-a)}{A^0(1+\delta) - 2\bar{r}X^0(1-a)^2} \tag{238b}$$

$$Z = \frac{1}{2\bar{r}}\,\frac{A^{0\,2}(a+\delta)^2}{A^0(1+\delta) - 2\bar{r}X^0(1-a)^2} \tag{238c}$$

From these values of X, Y, and Z all information on the polymer distribution at any time t — linked to a by (221) — can be obtained if the Laplace transform of the initial polymer distribution is known.

However, a certain amount of information can be derived even without knowing the exact form of X^0. X and Y are in a ratio which is independent of z, so that P_r and Q_r will be in the same ratio, independent of r. The ratio of X to Z is not independent of z since it contains X^0, but on the other hand the polymer distribution R_r contains "polymers of chain length zero", i.e. the original coupling agent, R_0. Its concentration is the constant term in the development of Z into a series of powers of z and since X^0 begins with a term in z (it does not contain polymer of chain length zero) the relevant term in Z is

$$R_0 = \frac{A^0}{2\bar{r}}\,\frac{(a+\delta)^2}{1+\delta} \tag{239}$$

This immediately gives us the rate of decrease of the concentration of the coupling agent.

References p. 576

We introduce the Laplace transform of the polymer without the coupling agent and call it $\overline{Z}$, which is

$$\overline{Z} = Z - R_0 = \frac{X^0 A^0 (1-a)^2 (a+\delta)}{(1+\delta)[A^0(1+\delta) - 2\bar{r}X^0(1-a)^2]} \tag{240}$$

and find that its ratio to X and Y is also independent of z, viz.

$$X : Y : Z = P_r : Q_r : R_r = \\ = a^2(1+\delta)^2 : 2a(1-a)(1+\delta)(a+\delta) : (1-a)^2(a+\delta)^2 \tag{241}$$

Since the concentrations of the r-mers are in a constant ratio, the average chain lengths are the same for the three species; it is not difficult to calculate it and we shall do so for P_r.

From the expression for the moments in term of the Laplace transform we get

$$r_{\mathrm{n}} = -\bar{r}\left(\frac{1}{X}\frac{\partial X}{\partial p}\right)(p = 0) \tag{242}$$

We derive (238a) logarithmically

$$\frac{1}{X}\frac{\partial X}{\partial p} = \frac{1}{X^0}\frac{\partial X^0}{\partial p} + \frac{2\bar{r}(1-a)^2}{A^0(1+\delta) - 2\bar{r}X^0(1-a)^2}\frac{\partial X^0}{\partial p} \tag{243}$$

The value of A^0 ($=2P^0$) equals $2\bar{r}X^0$ for $p = 0$ and thus (243) gives, on putting $p = 0$,

$$r_{\mathrm{n}} = r_{\mathrm{n}}^0 \frac{1+\delta}{(1+\delta) - (1-a)^2} \tag{244}$$

This gives the quantitative relationship between the increase in r_{n} and the decrease in a. When a is zero, $r_{\mathrm{n}} = r_{\mathrm{n}}^0\,(1+\delta)/\delta$ and for $\delta = 0$ this tends to infinity. This is the mathematical expression of the fact that when there are equal amounts of A and B present and the reaction has come to an end, all the polymer present is coupled to yield one giant molecule, the chain length of which is for practical purposes infinite.

Bamford and Jenkins [22] have considered in some detail the case where the initial polymer is the result of a vinyl polymerization, termination by combination, with $k_3 = 0$, and they have only evaluated the expressions for $\delta = 0$. We keep the treatment general; the initial polymer distribution is, see (102),

$$P_r^0 = \frac{\Delta M}{k_3 + w}\frac{1}{\bar{r}^2}\left(k_3 + \tfrac{1}{2}w\frac{r}{\bar{r}}\right)\alpha^{r-1} \tag{245}$$

with $\bar{r} = k_2/(k_3 + w) = 1/(1-\alpha)$. This $\bar{r}$ is not the r_{n} of the polymer, as we have already shown. We have also shown that α^r can be written exp($-\rho$), with an error of the order of $1-\alpha$. The Laplace transform of this distribution is

$$X^0 = \frac{\Delta M}{k_3 + w}\frac{1}{\bar{r}^2}\left[\frac{k_3}{1+p} + \frac{1}{2}\frac{w}{(1+p)^2}\right] \tag{246}$$

Since P^0 is $\bar{r}X^0$ with $p = 0$ and $A^0 = 2P^0$, we have

$$A^0 = \frac{2\Delta M}{\bar{r}}\frac{k_3 + \frac{1}{2}w}{k_3 + w} \tag{247}$$

and substitution of the values of A^0 and X^0 into (238) gives

$$X = \frac{\Delta M}{\bar{r}^2}\frac{k_3 + \frac{1}{2}w}{k_3 + w}a^2(1+\delta) \times$$

$$\frac{k_3(1+p) + \frac{1}{2}w}{(k_3 + \frac{1}{2}w)(1+p)^2(1+\delta) - [k_3(1+p) + \frac{1}{2}w](1-a)^2} \tag{248a}$$

$$Y = \frac{2\Delta M}{\bar{r}^2}\frac{k_3 + \frac{1}{2}w}{k_3 + w}a(a+\delta)(1-a) \times$$

$$\frac{k_3(1+p) + \frac{1}{2}w}{(k_3 + \frac{1}{2}w)(1+p)^2(1+\delta) - [k_3(1+p) + \frac{1}{2}w](1-a)^2} \tag{248b}$$

$$Z = \frac{\Delta M}{\bar{r}^2}\frac{k_3 + \frac{1}{2}w}{k_3 + w}(a+\delta)^2 \times$$

$$\frac{(k_3 + \frac{1}{2}w)(1+p)^2}{(k_3 + \frac{1}{2}w)(1+p)^2(1+\delta) - [k_3(1+p) + \frac{1}{2}w](1-a)^2} \tag{248c}$$

We do not propose to carry the development further; the interested reader is referred to the paper quoted [22].

6.2 POLYMERIZATION ON PREFORMED POLYMER WITH ACTIVE SITES

As a second example of application to a complex case we consider a polymerization in which initiation is on definite active sites of a preformed polymer [23]. We denote by $P_{r,a}$ and $X_{r,a}$ polymer and radical of chain length r and containing a active sites, respectively. We take initiation to be mono-molecular and termination to be by combination. We shall take the monomer concentration to be constant and shall apply the method of the stationary state.

References p. 576

The kinetic scheme is

$$\begin{aligned} P_{r,a} &\to X_{r,a}\,;\ ak_1M \\ X_{r,a} + M &\to X_{r+1,a}\,;\ k_2 \\ X_{r,a} + M &\to P_{r,a} + X_{1,0}\,;\ k_3 \\ X_{r,a} + P_{s,b} &\to P_{r,a} + X_{s,b}\,;\ sk_3 \\ X_{r,a} + X_{s,b} &\to P_{r+s,a+b}\,;\ k_4 \end{aligned} \tag{249}$$

We have assumed here that the rate coefficient for transfer to dead polymer is the same as that to monomer, i.e. that the monomer can be considered as regards transfer reactions as a polymer of chain length unity, that the decrease of the number of active sites by unity due to an initiation step can be neglected, and that polymers and radicals which originate from a transfer to monomer react at the same rates as other polymers and radicals; they are incorporated in the scheme formally by allowing for a the value zero. We have written the rate coefficient for initiation as k_1M for dimensional reasons.

The kinetic equations are

$$\frac{dX_{r,a}}{dt} = k_2MX_{r-1,a}(1-\delta_{r1}) + k_1aMP_{r,a} + k_3rY_{0,0}P_{r,a} + k_3MY_{0,0}\,\delta_{r1}\,\delta_{a0} - k_2MX_{r,a} - k_3MX_{r,a} - k_3Q_{1,0}X_{r,a} - k_4Y_{0,0}X_{r,a} \tag{250a}$$

$$\frac{dP_{r,a}}{dt} = k_3MX_{r,a} + k_3Q_{1,0}X_{r,a} + \tfrac{1}{2}k_4 \sum_{s=1}^{r-1} \sum_{b=0}^{a} X_{s,b}X_{r-s,a-b} - k_1aMP_{r,a} - k_3rY_{0,0}P_{r,a} \tag{250b}$$

$$\frac{dM}{dt} = -k_2MY_{0,0} - k_3MY_{0,0} \tag{250c}$$

We have used here the Kronecker δ's so that we do not have to write the equations separately for $r = 1$ and $a = 0$; these are so defined that δ_{ij} equals unity or zero according as i equals or does not equal j. We have also anticipated the definitions of the moments of the distributions

$$Y_{n,m} = \sum_{r=1}^{\infty} \sum_{a=0}^{\infty} r^n a^m X_{r,a}\,; \quad Q_{n,m} = \sum_{r=1}^{\infty} \sum_{a=0}^{\infty} r^n a^m P_{r,a} \tag{251}$$

It turns out in fact to be convenient to define the moments not only over chain length but also over number of active sites since these latter also appear explicitly in the kinetic equations.

It is of course possible to set up equations for generating functions and Laplace transforms from (250), but these equations turn out to be of such complexity that little progress can be made towards their solution. We shall limit our efforts here to calculate the first few moments and to derive from the expressions obtained a condition for gel formation. Further moments can be calculated by continuing along the path indicated and if a sufficient number of them are calculated the distributions can be obtained by a development into a series of Laguerre polynomials, as discussed in Sect. 5.

We multiply (250a) and (250b) by $r^n a^m$ and sum over all r and all a. This gives

$$\frac{\mathrm{d}Y_{n,m}}{\mathrm{d}t} = 0 = k_1 M Q_{n,m+1} + k_3 Y_{0,0} Q_{n+1,m} + k_3 M Y_{0,0}\delta_{m0} + nk_2 M Y_{n-1,m} - k_3 M Y_{n,m} - k_3 Q_{1,0} Y_{n,m} - k_4 Y_{0,0} Y_{n,m} \tag{252a}$$

$$\frac{\mathrm{d}Q_{n,m}}{\mathrm{d}t} = k_3 M Y_{n,m} + k_3 Q_{1,0} Y_{n,m} + \tfrac{1}{2}k_4 \sum_{i=0}^{n}\sum_{j=0}^{m}\binom{n}{i}\binom{m}{j} Y_{i,j} Y_{n-i,m-j} - k_1 M Q_{n,m+1} - k_3 Y_{0,0} Q_{n+1,m} \tag{252b}$$

Summing these two equations gives

$$\frac{\mathrm{d}Q_{n,m}}{\mathrm{d}t} = k_3 M Y_{0,0}\delta_{m0} + nk_2 M Y_{n-1,m} + \tfrac{1}{2}k_4 \sum_{i=0}^{n}\sum_{j=0}^{m}\binom{n}{i}\binom{m}{j} Y_{i,j} Y_{n-i,m-j} - k_4 Y_{0,0} Y_{n,m} \tag{253}$$

For $n = 0$, $m = 0$ one has

$$\frac{\mathrm{d}Y_{0,0}}{\mathrm{d}t} = 0 = -k_4 Y_{0,0}^2 + k_1 M Q_{0,1} \tag{254a}$$

$$\frac{\mathrm{d}Q_{0,0}}{\mathrm{d}t} = k_3 M Y_{0,0} - \tfrac{1}{2}k_4 Y_{0,0}^2 \tag{254b}$$

so that

$$Y_{0,0} = (\gamma M Q_{0,1})^{1/2} \tag{255}$$

writing as before $\gamma = (k_1/k_4)^{1/2}$.

For $n = 1$, $m = 0$ and $n = 0$, $m = 1$, (253) gives

$$\frac{\mathrm{d}Q_{1,0}}{\mathrm{d}t} = k_2 M Y_{0,0} + k_3 M Y_{0,0}\,; \quad \frac{\mathrm{d}Q_{0,1}}{\mathrm{d}t} = 0 \tag{256}$$

References p. 576

so that, with (250c), both $Q_{1,0} + M$ and $Q_{0,1}$ are constant. This is physically obvious, since the increase in the total polymer $Q_{1,0}$ is compensated by the decrease in monomer and $Q_{0.1}$ is the total amount of active sites. $Y_{0,0}$ is then from (255) also constant.

We get physically new results for $n + m = 2$, viz.

$$\frac{dQ_{2,0}}{dt} = k_3MY_{0,0} + 2k_2MY_{1,0} + k_4Y_{1,0}^2 \tag{257a}$$

$$\frac{dQ_{1,1}}{dt} = k_2MY_{0,1} + k_4Y_{1,0}Y_{0,1} \tag{257b}$$

$$\frac{dQ_{0,2}}{dt} = k_4Y_{0,1}^2 \tag{257c}$$

We need $Y_{1,0}$ and $Y_{0,1}$ from

$$\frac{dY_{1,0}}{dt} = 0 = k_1MQ_{1,1} + k_3Y_{0,0}Q_{2,0} + k_3MY_{0,0} + k_2MY_{0,0} - k_3MY_{1,0} - k_3Q_{1,0}Y_{1,0} - k_4Y_{0,0}Y_{1,0} \tag{258a}$$

$$\frac{dY_{0,1}}{dt} = 0 = k_1MQ_{0,2} + k_3Y_{0,0}Q_{1,1} - k_3MY_{0,1} - k_3Q_{1,0}Y_{0,1} - k_4Y_{0,0}Y_{0,1} \tag{258b}$$

To make the algebraic expressions less cumbersome we simplify the notation drastically: we put $Y_{1,0} = u$, $Y_{0,1} = v$, $Q_{2,0} = x$, $Q_{1,1} = y$, $Q_{0,2} = z$ and write, with obvious meanings for the coefficients a, b, c, g, h, k, which are all constant,

$$u = c + bx + ay \tag{259a}$$

$$v = by + az \tag{259b}$$

$$\frac{dx}{dt} = ku^2 + 2gu + h \tag{260a}$$

$$\frac{dy}{dt} = kuv + gv \tag{260b}$$

$$\frac{dz}{dt} = kv^2 \tag{260c}$$

Thus, by differentiating the two eqns. (259) and substituting from (260)

$$\frac{du}{dt} = ku(bu + av) + g(2bu + av) + bh \tag{261a}$$

$$\frac{dv}{dt} = kv(bu + av) + gbv \tag{261b}$$

We multiply (261a) by b, (261b) by a and add, and write $w = bu + av$ to obtain

$$\frac{dw}{dt} = kw^2 + 2gbw + b^2h \tag{262}$$

Equation (262) reduces to a simple integral

$$t = \int_{w^0}^{w} \frac{dw}{kw^2 + 2gbw + b^2h} \tag{263}$$

and the solution depends on whether the discriminant R

$$R = g^2b^2 - kb^2h = b^2(g^2 - kh) = b^2(k_2^2M^2 - k_3k_4MY_{0,0}) \tag{264}$$

is positive or negative.

In most cases of physical interest k_2 is several orders of magnitude larger than k_3 or k_4 so that R is positive. If we write $g^2 - kh = f^2$, the factors of the denominator of the integral in (263) are $kw + bg \pm bf$ and the solution is

$$\frac{kw + bg - bf}{kw + bg + bf} = \frac{kw^0 + bg - bf}{kw^0 + bg + bf}\, e^{2bft} \tag{265}$$

As t increases the value of the expression on the right-hand side of (265) increases from an initial value below unity and will reach unity for a value of t_g given by

$$t_g = \frac{1}{2bf} \ln \frac{kw^0 + bg + bf}{kw^0 + bg - bf} \tag{266}$$

At this point the left-hand side must also be unity and this is only possible if w becomes infinite. It can also be shown by finding an explicit value for w from (265) that it reaches infinity at a time given by (266).

If w is infinite either u or v or both are infinite and (260) shows that $Q_{2,0}$ and $Q_{1,1}$ certainly and $Q_{0,2}$ possibly increase at an infinite rate and grow beyond all limits. As mentioned before this is the mathematical expression of gelation and t_g is the time which will elapse before gelation sets in.

The path to be followed to compute the distributions numerically is clear now: if w is known, u and v can be calculated and hence x, y and z, though progress will be difficult without the use of a computer. As a next step the four equations for $Q_{n,m}$ with $n + m = 3$ can be solved, using the expressions for $Y_{n,m}$ with $n + m = 2$, and so on. It is impossible to foresee how many moments will be required to obtain reasonably accurate values for the distribution, but the number will increase rapidly as t increases towards t_g.

REFERENCES

1 C. H. Bamford, W. G. Barb, A. D. Jenkins and P. F. Onyon, The Kinetics of Vinyl Polymerization by Radical Mechanisms, Academic Press, London, 1958.
2 L. Küchler, Polymerisationskinetik, Springer, Berlin, 1951.
3 G. M. Burnett, Mechanism of Polymer Reactions, Interscience, London, 1954.
4 C. H. Bamford and H. Tompa, J. Polym. Sci., 10 (1953) 345.
5 C. H. Bamford and H. Tompa, Trans. Faraday Soc., 50 (1954) 1097.
6 J. P. Howe, J. Chem. Phys., 13 (1955) 899.
7 H. Tompa, Int. Symp. Macromol. Chem., Moscow, 1961, Vol. 2, p. 80.
8 J. J. Hermans, Makromol. Chem., 87 (1965) 21.
9 N. R. Amundsen and D. Luss, J. Macromol. Sci., Rev. Macromol. Chem., 2(1) (1968) 145.
10 G. Natta, I. Pasquon and M. Dente, Chim. Ind., 44 (1962) 1, 10.
11 W. W. Graessly, H. M. Mittelhauser and R. Maramba, Makromol. Chem., 86 (1965) 129.
12 W. W. Graessly, and H. M. Mittelhauser, J. Polym. Sci., Part A-2, 5 (1967) 431.
13 W. W. Graessly, R. D. Hartung and W. C. Uy, J. Polym. Sci., Part A-2, 7 (1969) 1919.
14 O. Saito, K. Nagasubramanian and W. W. Graessly, J. Polym. Sci., Part A-2, 7 (1969) 1937.
15 C. D. Hodgeman (Ed.), Handbook of Chemistry and Physics, Chemical Rubber Publishing Co., Ohio, 1961.
16 G. Doetsch, Tabellen zur Laplace Transformation und Anleitung zum Gebrauch, Springer, Berlin, 1947.
17 A. Erdelyi, Tables of Integral Transforms, Vol. 1, McGraw-Hill, London, 1954.
18 H. Margenau and G. M. Murphy, The Mathematics of Physics and Chemistry, D. van Nostrand, Princeton, 1962.
19 A. C. Aitken, Statistical Mathematics, Oliver and Boyd, Edinburgh, 1962.
20 P. J. Flory, J. Am. Chem. Soc., 58 (1936) 1877.
21 H. T. H. Piaggio, An Elementary Treatise on Differential Equations and Their Applications, G. Bell and Sons, London, 1944.
22 C. H. Bamford and A. D. Jenkins, Trans. Faraday Soc., 56 (1960) 907.
23 C. H. Bamford and H. Tompa, to be published.

Index

B

C

D

E

F

I

K

L

M

N

O

P

Q

R

T

W

X

Z

www.ingramcontent.com/pod-product-compliance
Ingram Content Group UK Ltd.
Pitfield, Milton Keynes, MK11 3LW, UK
UKHW020224290926
15470UKWH00001B/372

9 780444 414861